Lecture Notes in Statistics 166

Edited by P. Bickel, P. Diggle, S. Fienberg, K. Krickeberg, I. Olkin, N. Wermuth, and S. Zeger

Springer

New York
Berlin
Heidelberg
Barcelona
Hong Kong
London
Milan
Paris
Singapore
Tokyo

Hira L. Koul

Weighted Empirical Processes in Dynamic Nonlinear Models

Second Edition

Springer

Hira L. Koul
Department of Statistics and Probability
Michigan State University
East Lansing, MI 48824
USA

Library of Congress Cataloging-in-Publication Data
Koul, H.L. (Hira L.)
Weighted empirical processes in dynamic nonlinear models / Hira L. Koul.—2nd ed.
p. cm. — (Lecture notes in statistics ; 166)
Rev. ed. of: Weighted empiricals and linear models. c1992.
ISBN 0-387-95476-7 (softcover : alk. paper)
1. Sampling (Statistics) 2. Linear models (Statistics) 3. Regression analysis. 4. Autoregression (Statistics) I. Koul, H.L. (Hira L.). Weighted empiricals and linear models. II. Title. III. Lecture notes in statistics (Springer-Verlag) ; v. 166.
QA276.6 .K68 2002
519.5—dc21 2002020944

ISBN 0-387-95476-7 Printed on acid-free paper.

Printed in the United States of America.

9 8 7 6 5 4 3 2 1 SPIN 10874053

www.springer-ny.com

Springer-Verlag New York Berlin Heidelberg
A member of BertelsmannSpringer Science+Business Media GmbH

Preface to the Second Edition

The role of the weak convergence technique via weighted empirical processes has proved to be very useful in advancing the development of the asymptotic theory of the so called robust inference procedures corresponding to non-smooth score functions from linear models to nonlinear dynamic models in the 1990's. This monograph is an expanded version of the monograph *Weighted Empiricals and Linear Models*, **IMS Lecture Notes-Monograph, 21** published in 1992, that includes some aspects of this development. The new inclusions are as follows.

Theorems 2.2.4 and 2.2.5 give an extension of the Theorem 2.2.3 (old Theorem 2.2b.1) to the unbounded random weights case. These results are found useful in Chapters 7 and 8 when dealing with homoscedastic and conditionally heteroscedastic autoregressive models, actively researched family of dynamic models in time series analysis in the 1990's. The weak convergence results pertaining to the partial sum process given in Theorems 2.2.6 and 2.2.7 are found useful in fitting a parametric autoregressive model as is expounded in Section 7.7 in some detail. Section 6.6 discusses the related problem of fitting a regression model, using a certain partial sum process. In both sections a certain transform of the underlying process is shown to provide asymptotically distribution free tests.

Other important changes are as follows. Theorem 7.3.1 gives the asymptotic uniform linearity of linear rank statistics in linear autoregressive (LAR) models for any nondecreasing bounded score function φ, compared to its older version Theorem 7.3b.1 that assumed φ to be differentiable with uniformly continuous derivative. The new Section 7.5 is devoted to autoregression quantiles and rank scores. Its contents provide an important extension of the regression quantiles of Koenker-Bassett to LAR models.

The author gratefully acknowledges the help of Kanchan Mukherjee with Section 8.3, Vince Melfi's help with some tex problems and the NSF DMS 0071619 grant support.

East Lansing, Michigan — Hira L. Koul
March 18, 2002

Preface to the First Edition

An empirical process that assigns possibly different non-random (random) weights to different observations is called a *weighted (randomly weighted empirical process*. These processes are as basic to linear regression and autoregression models as the ordinary empirical process is to one sample models. However their usefulness in studying linear regression and autoregression models has not been fully exploited. This monograph addresses this question to a large extent.

There is a vast literature in nonparametric inference that discusses inferential procedures based on empirical processes in k-sample location models. However, their analogs in autoregression and linear regression models are not readily accessible. This monograph makes an attempt to fill this void. The statistical methodologies studied here extend to these models many of the known results in k-sample location models, thereby giving a unified theory.

By viewing linear regression models via certain weighted empirical processes one is naturally led to new and interesting inferential procedures. Examples include minimum distance estimators of regression parameters and goodness - of - fit tests pertaining to the errors in linear models. Similarly, by viewing autoregression models via certain randomly weighted empirical processes one is naturally led to classes of minimum distance estimators of autoregression parameters and goodness - of - fit tests pertaining to the error distribution.

The introductory Chapter 1 gives an overview of the usefulness of weighted and randomly weighted empirical processes in linear models. Chapter 2 gives general sufficient conditions for the weak convergence of suitably standardized versions of these processes to continuous Gaussian processes. This chapter also contains the proof of the asymptotic uniform linearity of weighted empirical processes based on the residuals when errors are heteroscedastic and independent. Chapter 3 discusses the asymptotic uniform linearity of linear rank and signed rank statistics when errors are heteroscedastic and independent. It also includes some results about the weak convergence of weighted empirical processes of ranks and signed ranks. Chapter 4 is devoted to the study of the asymptotic behavior of M- and R-estimators of regression parameters under heteroscedastic and independent errors, via weighted empirical processes. A brief discussion about bootstrap approximations to the distribution of a class of M-estimators appears in Section 4.2.2. This chapter also contains a proof of the consistency of a class of robust estimators for certain scale parameters under heteroscedastic errors.

In carrying out the analysis of variance of linear regression models based on ranks, one often needs an estimator of the functional $\int f d\varphi(F)$, where F is the error distribution function, f its density and φ is a function from $[0, 1]$ to the real line. Some estimators of this functional and the proofs of their consistency in the linear regression setting appear in Section 4.5.

Chapters 5 and 6 deal with minimum distance estimation, via weighed empirical processes, of the regression parameters and tests of goodness - of - fit pertaining to the error distribution. One of the main themes emerging from these two chapters is that the inferential procedures based on weighted empiricals with weights proportional to the design matrix provide the right extensions of k-sample location model procedures to linear regression models.

It is customary to expect that a method that works for linear regression models should have an analogue that will also work in autoregressive models. Indeed many of the inferential procedures based on weighted empirical processes in linear regression that are discussed in Chapters 3 - 6 have precise analogues in autoregression based on certain randomly weighted empirical processes and appear in Chapter 7. In particular, the proof of the asymptotic uniform linearity of the ordinary empirical process of the residuals in autoregression appears here.

All asymptotic uniform linearity results in the monograph are shown to be consequences of the asymptotic continuity of certain basic weighted and randomly weighted empirical processes.

Chapters 2-4 are interdependent. Chapter 5 is mostly self- contained and can be read after reading the Introduction. Chapter 6 uses results from Chapters 2 and 5. Chapter 7 is almost self-contained. The basic result needed for this chapter appears in Section 2.2b.

The first version of this monograph was prepared while I was visiting Department of Statistics, Poona University, India, on sabbatical leave from Michigan State University, during the academic year 1982-83. Several lecture on some parts of this monograph were given at the Indian Statistical institute, New Delhi, and Universities of La Trobe, Australia, and Wisconsin, Madison. I wish to thank Professors S. R. Adke, Richard Johnson, S. K. Mitra, M. S. Prasad and B. L. S. Prakasa Rao for having some discussions pertaining to the monograph. My special thanks go to James Hannan for encouraging me to finish the project and for proof reading parts of the manuscript, to Soumendra Lahiri for helping me with sections on bootstrapping, and to Bob Serfling for taking keen interest in the monograph and for many comments that helped to improve the initial draft.

Ms. Achala Sabne and Ms. Lora Kemler had the pedestrian task of typing the manuscript. Their patient endeavors are gratefully acknowledged. Ms. Kemler's keen eye for details has been an indispensable help.

During the preparation of the monograph the author was partly supported by the University Grants Commission of India and the National Science Foundation, grant numbers NSF 82-01291, DMS-9102041.

East Lansing, Michigan Hira L. Koul
May 28, 1992

Contents

Notation and Conventions

AUL	:=	Asymptotic uniform linearity.
C-S	:=	the Cauchy-Schwarz inequality.
D.C.T.	:=	the Dominated Convergence Theorem.
d.f.('s)	:=	distribution function(s).
Fubini	:=	the Fubini Theorem.
$\|g\|_\infty$	:=	the supremum norm over the domain of a real valued function g.
i.i.d.	:=	independent identically distributed.
L-F CLT	:=	the Lindeberg-Feller Central Limit Theorem.
$\mathcal{N}(\mathbf{0}, \mathbf{C})$	:=	either a r.v. whose distribution is normal with mean vector $\mathbf{0}$ and covariance matrix $\mathbf{C}$ or the corresponding distribution.
$o(1)(o_p(1))$	:=	a sequence of numbers (r.v.'s) converging to zero (in probability).
$O(1)(O_p(1))$	:=	a sequence of numbers (r.v.'s) that is bounded (in probability).
r.v.('s)	:=	random variable(s).
R.W.E.P.('s)	:=	randomly weighted empirical process(es).
τ_a^2	:=	$\sum_{i=1}^n a_{ni}^2$, for an arbitrary real vector $(a_{n1}, \cdots, a_{nn})'$.
$u_p(1)$	:=	a sequence of stochastic processes converging to zero uniformly over the time domain, in probability.
W.E.P.('s)	:=	weighted empirical process(es).
w.r.t.	:=	with respect to.

The p-dimension Euclidean space is denoted by $\mathbb{R}^p$, $p \geq 1$; $\mathbb{R} = \mathbb{R}^1$; $\mathcal{B}^p :=$ the σ- algebra of Borel sets in $\mathbb{R}^p$, $\mathcal{B} = \mathcal{B}^1$; $\lambda :=$ Lebesgue measure on $(\mathbb{R}, \mathcal{B})$. The symbol " $:=$ " stands for "by definition".

For any set $\mathcal{A} \subset \mathbb{R}, \mathcal{D}(\mathcal{A})$ denotes the class of real valued functions on $\mathcal{A}$ that are right continuous and have left limits while $\mathcal{DI}(\mathcal{A})$ denotes the subclass in $\mathcal{D}(\mathcal{A})$ whose members are nondecreasing. $C[0,1]$ $:=$ the class of real valued bounded continuous functions on $[0,1]$.

A vector or a matrix will be designated by a bold letter. A $\mathbf{t} \in \mathbb{R}^p$ is a $p \times 1$ vector, $\mathbf{t}'$ its transpose, $\|\mathbf{t}\|^2 := \sum_{j=1}^p t_j^2$, $|\mathbf{t}| := \max\{|t_j|,\ 1 \leq j \leq p\}$. For any p-square matrix $\mathbf{C}$,

$$\|\mathbf{C}\|_\infty = \sup\{\|\mathbf{t}'\mathbf{C}\|;\ \|\mathbf{t}\| \leq 1\}.$$

For an $n \times p$ matrix $\mathbf{D}$, $\mathbf{d}'_{ni}$ denotes its ith row, $1 \leq i \leq n$, and $\mathbf{D}_c$ the $n \times p$ matrix $\mathbf{D} - \bar{\mathbf{D}}$, whose i^{th} row consists of $(\mathbf{d}_{ni} - \bar{\mathbf{d}}_n)'$, with $\bar{\mathbf{d}}_n := \sum_{i=1}^n \mathbf{d}_{ni}/n, 1 \leq i \leq n$.

Often in a discussion or in a proof the subscript n on the triangular arrays and various other quantities will not be exhibited. The index i in $\sum_i$ and $\max_i$ will vary from 1 to n, unless specified otherwise. All limits, unless specified otherwise, are taken as $n \to \infty$.

For a sequence of r.v.'s $\{X, X_n,\ n \geq 1\}$, $X_n \to_d X$ means that the distribution of X_n converges weakly to that of X. For two r.v.'s X, Y, $X \stackrel{d}{=} Y$ means that the distribution of X is the same as that of Y.

For a sequence of stochastic processes $\{Y, Y_n,\ n \geq 1\}$, $Y_n \Longrightarrow Y$ means that Y_n converges weakly to Y in a given topology. $Y_n \longrightarrow_{f.d.} Y$ means that all finite dimensional distributions of Y_n converge weakly to that of Y.

For convenient reference we list here some of the most often used assumptions in the manuscript. For an arbitrary d.f. F on $\mathbb{R}$, assumptions (**F1**), (**F2**) and (**F3**) are as follows:

(**F1**) *F has uniformly continuous density f* w.r.t. λ.

(**F2**) $f > 0$, a.e. λ.

(**F3**) $\sup_{x\in\mathbb{R}} |xf(x)| \leq k < \infty$,
$\sup_{x\in\mathbb{R}} |xf(x(1+u)) - xf(x)| \to 0$, as $u \to 0$.

These conditions are introduced for the first time just before Corollary 3.2.1 and are used frequently subsequently.

For an $n \times p$ design matrix matrix $\mathbf{X}$, the conditions (**NX**), (**NX1**) and (**NX**$_c$) are as follows:

(**NX**) $(\mathbf{X}'\mathbf{X})^{-1}$ exists, $\forall \ n \geq p$;

$$\max_{1 \leq i \leq n} \mathbf{x}'_{ni}(\mathbf{X}'\mathbf{X})^{-1}\mathbf{x}'_{ni} = o(1).$$

(**NX1**) $(\mathbf{X}'_c\mathbf{X}_c)^{-1}$ exists, $\forall \ n \geq p$;

$$\max_{1 \leq i \leq n} \mathbf{x}'_{ni}(\mathbf{X}'_c\mathbf{X}_c)^{-1}\mathbf{x}_{ni} = o(1).$$

(**NX**$_c$) $(\mathbf{X}'_c\mathbf{X}_c)^{-1}$ exists, $\forall \ n \geq p$;

$$\max_{1 \leq i \leq n} (\mathbf{x}_{ni} - \overline{\mathbf{x}}_n)'(\mathbf{X}'_c\mathbf{X}_c)^{-1}(\mathbf{x}_{ni} - \overline{\mathbf{x}}_n) = o(1).$$

The condition (**NX**) is the most often used from Theorem 2.3.3 onwards. The letter N in these conditions stands for Noether, who was the first person to use (**NX**), in the case $p = 1$, to obtain the asymptotic normality of weighted sums of r.v.'s; see Noether (1949).

1

Introduction

1.1 Weighted Empirical Processes

A weighted empirical process (W.E.P.) corresponding to the random variables (r.v.'s) $X_{n1}, \cdots, X_{nn}$ and the non-random real weights $d_{n1}, ..., d_{nn}$ is defined to be

$$U_d(x) := \sum_{i=1}^{n} d_{ni} I(X_{ni} \le x), \quad x \in \mathbb{R}, n \ge 1.$$

The weights $\{d_{ni}\}$ need not be nonnegative.

The classical example of a W.E.P. is the *ordinary empirical process* that corresponds to $d_{ni} \equiv n^{-1}$. Another example is given by the two sample empirical process obtained as follows: Let m be an integer, $1 \le m \le n, r := n - m; d_{ni} = -r/n, 1 \le i \le m; d_{ni} = m/n, m+1 \le i \le n$. Then the corresponding U_d - process becomes

$$U_d(x) \equiv (mr/n) \left\{ r^{-1} \sum_{i=m+1}^{n} I(X_{ni} \le x) - m^{-1} \sum_{i=1}^{m} I(X_{ni} \le x) \right\},$$

precisely the process that arises in two-sample models.

More generally, W.E.P's arise naturally in linear regression models where, for each $n \ge 1$ and some $\boldsymbol{\beta} \in \mathbb{R}^p$, the data $\{(\mathbf{x}'_{ni}, Y_{ni}), 1 \le i \le n\}$ are related to the error variables $\{e_{ni}, 1 \le i \le n\}$ by the linear relation

(1.1.1) $$Y_{ni} = \mathbf{x}'_{ni}\boldsymbol{\beta} + e_{ni}, \qquad 1 \le i \le n.$$

Here $e_{n1}, \cdots, e_{nn}$ are independent r.v.'s with respective continuous d.f.'s $F_{n1}, \cdots, F_{nn}, \mathbf{x}'_{ni} = (x_{ni1}, \cdots, x_{nip})$ is the ith row of the known $n \times p$ design matrix $\mathbf{X}$ and $\boldsymbol{\beta}$ is the parameter vector of interest.

Consider the vector of W.E.P.'s $\mathbf{V} := (V_1, ..., V_p)'$ where

$$V_j(y, \mathbf{t}) := \sum_{i=1}^{n} x_{nij} I(Y_{ni} \leq y + \mathbf{x}'_{ni}\mathbf{t}), \tag{1.1.2}$$

for $y \in \mathbb{R}$, $\mathbf{t} \in \mathbb{R}^p$, $1 \leq j \leq p$. Clearly, $V_j(\cdot, \mathbf{t})$ is an example of the W.E.P. $U_d(\cdot)$ with $d_{ni} \equiv x_{nij}$ and $X_{ni} \equiv Y_{ni} - \mathbf{x}'_{ni}\mathbf{t}, 1 \leq i \leq n, 1 \leq j \leq p$.

Observe that the data $\{(\mathbf{x}'_{ni}, Y_{ni}), 1 \leq i \leq n\}$ in the model (1.1.1) are readily summarized by the vector of W.E.P.'s $\{\mathbf{V}(y, \mathbf{0}), y \in \mathbb{R}\}$ in the sense that the given data can be recovered from the sample paths of this vector up to a permutation. This in turn suffices for the purpose of inference about $\boldsymbol{\beta}$ in (1.1.1). In this sense the vector of W.E.P's $\{\mathbf{V}(y, \mathbf{0}), y \in \mathbb{R}\}$ is at least as important to linear regression model (1.1.1) as is the ordinary empirical process to one-sample location models. One of the purposes of this monograph is to discuss the role of $\mathbf{V}$ - processes in inference and in proving limit theorems in models (1.1.1) in a unified fashion.

1.2 M-, R- and Scale Estimators

Many inferential procedures involving (1.1.1) can be viewed as functions of $\mathbf{V}$. For example the least squares estimator, or more generally, the class of M- estimators corresponding to the score function ψ, (Huber : 1981), is defined as a solution $\mathbf{t}$ of the equation

$$\int \psi(y)\mathbf{V}(dy, \mathbf{t}) = \text{ a known constant.}$$

Similarly, rank (R) estimators of $\boldsymbol{\beta}$ corresponding to the score function φ are defined to be a solution $\mathbf{t}$ of the equation

$$\int \varphi(H_n(y, \mathbf{t}))\mathbf{V}(dy, \mathbf{t}) = \text{a known constant,} \tag{1.2.1}$$

$$H_n(y, \mathbf{t}) := n^{-1} \sum_{i=1}^{n} I(Y_{ni} \leq y + \mathbf{x}'_{ni}\mathbf{t}), \quad y \in \mathbb{R}, \quad \mathbf{t} \in \mathbb{R}^p.$$

A significant portion of nonparametric inference in models (1.1.1) deals with M- and R- estimators of $\boldsymbol{\beta}$ (Adichie; 1967. Huber; 1973) and linear rank tests of hypotheses about $\boldsymbol{\beta}$ (Hájek - Šidák; 1967). By viewing these procedures as functions of $\{\mathbf{V}(y, \mathbf{t}), y \in \mathbb{R}, \mathbf{t} \in \mathbb{R}^p\}$, it is possible to give a unified treatment of their asymptotic distribution theory, as is done in Chapters 3 and 4 below.

There is a vast literature in nonparametric inference that discusses inferential procedures based on functionals of empirical processes in the k-sample location model such as the books by Puri and Sen (1969), Serfling (1980) and Huber (1981). Yet their appropriate extensions to the linear regression model are not readily accessible. This monograph seeks to fill this void. The methodology and inference procedures studied here extend many known results in the k-sample location model to the model (1.1.1), thereby giving a unified treatment.

An important result needed for study of the asymptotic behavior of R-estimators of $\boldsymbol{\beta}$ is the asymptotic uniform linearity of the linear rank statistics of (1.2.1) in the regression parameter vector. Jurečková (1969, 1971) obtained this result under (1.1.1) with i.i.d. errors. A similar result was proved in Koul (1969, 1971) and van Eeden (1972) for linear signed rank statistics under i.i.d. symmetric errors. Its extension to the case of non-identically distributed errors is not readily available. Theorems 3.2.4 and 3.3.3 prove the asymptotic uniform linearity of linear rank and linear signed rank statistics with bounded scores under the general independent errors model (1.1.1). In the case of i.i.d. errors, the conditions in these theorems on the error d.f. are more general than requiring finite Fisher information for location. The results are proved uniformly over all bounded score functions and are consequences solely of the asymptotic sample continuity of $\mathbf{V}$ - processes and some smoothness of $\{F_{ni}\}$. The uniformity with respect to the score functions is useful when constructing adaptive rank tests that are asymptotically efficient against Pitman alternatives for a large class of error distributions.

Chapter 3 also contains a proof of the asymptotic normality of linear rank and linear signed rank statistics under independent alternatives and for indicator score functions. This proof proceeds via the weak convergence of certain basic W.E.P.'s and complements some of the results in Dupač and Hájek (1969).

Section 4.2.1 discusses the asymptotic distribution of M- estimators under heteroscedastic errors using the asymptotic continuity of **V**- processes. Section 4.2.2 presents some second order results on bootstrap approximations to the distribution of a class of M-estimators.

In order to make M-estimators scale invariant one often needs an appropriate robust scale estimator. One such scale estimator, as recommended by Huber (1981) and others, is

$$s_1 = med\{|Y_{ni} - \mathbf{x}'_{ni}\hat{\boldsymbol{\beta}}|, \ 1 \le i \le n\},$$

where $\hat{\boldsymbol{\beta}}$ is an estimator of $\boldsymbol{\beta}$. The asymptotic distribution of s_1 under heteroscedastic errors is given in Section 4.3. If the errors are i.i.d., this asymptotic distribution does not depend on $\hat{\boldsymbol{\beta}}$ provided the common error distribution is symmetric around 0. This observation naturally leads one to construct a scale estimator based on the symmetrized residuals, thereby giving another scale estimator

$$s_2 := med\{|Y_{ni} - \mathbf{x}'_{ni}\hat{\boldsymbol{\beta}} - Y_{ni} + \mathbf{x}'_{ni}\hat{\boldsymbol{\beta}}|; 1 \le i, j \le n\}.$$

As expected, the asymptotic distribution of s_2 is shown to be free from the estimator $\hat{\boldsymbol{\beta}}$ in the case of i.i.d. errors, not necessarily symmetric. It also appears in Section 4.3.

Section 4.4 discusses the asymptotic distribution of a class of R-estimators under heteroscedastic errors using the asymptotic uniform linearity results of Chapter 3. The R-estimators considered are asymptotically equivalent to Jaeckel's estimators.

The complete rank analysis of the linear regression model (1.1.1) requires an estimate of the scale parameter

$$Q(f) := \int f d\varphi(F)$$

where f is density of the unknown common error d.f. F and φ is a nondecreasing function on $(0, 1)$. This estimate is used to standardize the test statistic and estimate the standard error of the R-estimator corresponding to the score function φ. This parameter also appears in the efficiency comparisons of rank procedures and it is of interest to estimate it, after the fact, in an analysis.

Lehmann (1963), Sen (1966), Koul (1971), among others, provide estimators of $Q(f)$ in the one - and two - sample location models and in the linear regression model. These estimators are given in terms of the lengths of Lebesgue measures of certain confidence intervals or regions. They are usually not easy to compute when the dimension p of $\boldsymbol{\beta}$ is larger than 1.

In Section 4.5, estimators of $Q(f)$, based on kernel type density estimators of f and the empirical d.f. H_n, are defined and their consistency under (1.1.1) with i.i.d. errors is proved. An estimator whose window width is based on the data and is of the order of square root n, is also considered. The consistency proof presented is a sole consequence of the asymptotic continuity of certain W.E.P.'s and some smoothness of the error d.f.'s.

1.3 M.D. Estimators & Goodness-of-fit tests

1.3.1 Minimum distance estimators

The practice of obtaining estimators of parameters by minimizing a certain distance between some functions of observations and parameters has been present in statistics since its beginning. The classical examples of this method are the Least Square and the minimum Chi Square estimators.

The minimum distance (m.d.) estimation method, where one obtains an estimator of a parameter by minimizing some distance between the empirical d.f. and the modeled d.f. was elevated to a general method of estimation by Wolfowitz (1953, 1954, 1957). In these papers he demonstrated that, compared to the maximum likelihood estimation method, the m.d. estimation method yielded consistent estimators rather cheaply in several problems of varied levels of difficulty.

This methodology saw increasing research activity from the mid 1970's when many authors demonstrated various robustness properties of certain m.d. estimators. See, e.g., Beran (1977, 1978), Parr and Schucany (1979), Millar (1981, 1982, 1984), Donoho and Liu (1988 a, b), among others. All of these authors restrict their attention to the one sample setup or to the two sample location model. See Parr (1981) for additional bibliography on m.d. estimation till 1980.

In spite of many advances made in the m.d. estimation methodology in one sample models, little was known till late 1970's as to how to extend this methodology to one of the most applied models, v.i.z., the multiple linear regression model (1.1.1). A *significant* advantage of viewing the model (1.1.1) through $\mathbf{V}$ is that one is naturally led to interesting m.d. estimators of $\boldsymbol{\beta}$ that are natural extensions of their one- and two- sample location model counterparts. To illustrate this, consider the m.d. estimator $\hat{\theta}$ of the one sample location parameter θ, when errors are i.i.d. symmetric around 0, defined by the relation

$$\hat{\theta} := \text{ argmin } \{T_n(t); t \in \mathbb{R}\},$$

with

$$T_n(t) = \int \{n^{-1/2} \sum_{i=1}^{n} I(Y_{ni} \leq y+t) - I(-Y_{ni} < y-t)]\}^2 dG(y), \; t \in \mathbb{R},$$

where $G \in \mathcal{DI}(\mathbb{R})$. Since (1.1.1) is an extension of the one sample location model, it is only natural to seek an extension of $\hat{\theta}$ in this model. Assuming that $\{e_{ni}\}$ are symmetrically distributed around 0, the first thing one is tempted to consider as an extension of $\hat{\theta}$ is $\boldsymbol{\beta}_1^+$ defined by the relation

$$\boldsymbol{\beta}_1^+ := \text{ argmin}\{K_1^+(\mathbf{t}); \mathbf{t} \in \mathbb{R}^p\},$$

with

$$K_1^+(\mathbf{t}) := \int \Big\{n^{-1/2} \sum_{i=1}^{n} \Big[I(Y_{ni} \leq y + \mathbf{x}_{ni}'\mathbf{t}) - I(-Y_{ni} < y - \mathbf{x}_{ni}'\mathbf{t})\Big]\Big\}^2 dG(y), \; \mathbf{t} \in \mathbb{R}^p.$$

However, any extension of $\hat{\theta}$ to the linear regression model should have the property that it reduce to $\hat{\theta}$ when the model is reduced to the one sample location model and, in addition, that it reduce to an appropriate extension of $\hat{\theta}$ to the k-sample location model when the model (1.1.1) is reduced to it. In this sense β_1^+ does not provide the right extension but $\beta_{\mathbf{X}}^+$ does, where

$$(1.3.1) \qquad \beta_{\mathbf{X}}^+ := \operatorname{argmin}\{K_{\mathbf{X}}^+(\mathbf{t});\ \mathbf{t} \in \mathbb{R}^p\},$$

with

$$\begin{aligned} K_{\mathbf{X}}^+(\mathbf{t}) &:= \int \mathbf{V}^{+\prime}(y,\mathbf{t})(\mathbf{X}'\mathbf{X})^{-1}\mathbf{V}^+(y,\mathbf{t})dG(y), \quad \mathbf{t} \in \mathbb{R}^p, \\ \mathbf{V}^{+\prime} &:= (V_1^+, \dots, V_p^+), \\ V_j^+(y,\mathbf{t}) &:= V_j(y,\mathbf{t}) - \sum_{i=1}^n x_{nij} + V_j(-y,\mathbf{t}), \end{aligned}$$

for $1 \le j \le p$, $y \in \mathbb{R}$, $\mathbf{t} \in \mathbb{R}^p$.

In the case errors are not symmetric but i.i.d. according to a known d.f. F, so that $EV_j(y, \boldsymbol{\beta}) \equiv \sum_i x_{nij}F(y)$, a suitable class of m.d. estimators of $\boldsymbol{\beta}$ is defined by the relation

$$\begin{aligned} (1.3.2)\quad \hat{\boldsymbol{\beta}}_{\mathbf{X}} &:= \operatorname{argmin}\ \{K_{\mathbf{X}}(\mathbf{t});\ \mathbf{t} \in \mathbb{R}^p\}, \\ K_{\mathbf{X}}(\mathbf{t}) &:= \int \|\mathbf{W}(y,\mathbf{t})\|^2 dG(y), \\ \mathbf{W}(y,\mathbf{t}) &:= (\mathbf{X}'\mathbf{X})^{-1/2}\{\mathbf{V}(y,\mathbf{t}) - \mathbf{X}'\mathbf{1}F(y)\},\ y \in \mathbb{R},\ \mathbf{t} \in \mathbb{R}^p, \\ \mathbf{1}' &:= (1, \cdots, 1)_{1\times n}. \end{aligned}$$

Chapter 5 discusses the existence, the asymptotic distribution, the robustness and the asymptotic optimality of $\boldsymbol{\beta}_{\mathbf{X}}^+$ and $\hat{\boldsymbol{\beta}}_{\mathbf{X}}$ under (1.1.1) with heteroscedastic errors. For example, if $p = 1$ in (1.1.1) and the design variable is nonnegative then the asymptotic variance of $\boldsymbol{\beta}_{\mathbf{X}}^+$ is smaller than that of β_1^+ for a large class of symmetric error d.f.'s F and integrating measures G. A similar result holds about $\hat{\boldsymbol{\beta}}_{\mathbf{X}}$ and for $p \ge 1$. Chapter 5 also discusses several other m.d. estimators of $\boldsymbol{\beta}$ and their asymptotic theory under (1.1.1) with heteroscedastic errors. These include analogues of $\hat{\boldsymbol{\beta}}_{\mathbf{X}}$ when the common error d.f. is unknown and some m.d. estimators corresponding to certain supremum distances based on $\mathbf{V}$.

1.3.2 Goodness-of-fit testing

Closely related to the problem of m.d. estimation is the problem of testing the goodness-of-fit hypothesis $H_0 : F_{ni} \equiv F_0, F_0$ a known d.f.. One test statistic for this problem is

$$(1.3.3) \qquad \hat{D}_1 := \sup_y |n^{1/2}\{H_n(y, \hat{\boldsymbol{\beta}}) - F_0(y)\}|,$$

where $\hat{\boldsymbol{\beta}}$ is an estimator of $\boldsymbol{\beta}$. This test statistic is suggested by looking at the estimated residuals and mimicking the one sample location model technique. In general, its large sample distribution depends on the design matrix. In addition, it does not reduce to the Kiefer (1959) tests of goodness-of-fit in the k-sample location problem when (1.1.1) is reduced to this model. Test statistics that overcome these deficiencies are

$$\hat{D}_2 := \sup_y |\mathbf{W}^0(y, \hat{\boldsymbol{\beta}})|, \quad \hat{D}_3 := \sup_y \|\mathbf{W}^0(y, \hat{\boldsymbol{\beta}})\|,$$

where $\mathbf{W}^0$ is equal to the $\mathbf{W}$ of (1.3.2) with $F = F_0$. Another natural class of tests is based on $K^0_{\mathbf{X}}(\hat{\boldsymbol{\beta}}_{\mathbf{X}})$, where $K^0_{\mathbf{X}}$ is equals to the $K_{\mathbf{X}}$ of (1.3.2) with $\mathbf{W}$ replaced by $\mathbf{W}^0$ in there.

All of the above and several other goodness-of-fit tests are discussed at some length in Chapter 6. Section 6.2.1 discusses the asymptotic null distributions of the supremum distance statistics $\hat{D}_j, j = 1, 2, 3$. Also discussed in this section are asymptotically distribution free analogues of these tests, in a sense similar to that discussed by Durbin (1973, 1976) and Rao (1972) for the one-sample location model. Section 6.2.2 discusses smooth bootstrap approximations to the null distributions of tests based on W.E.P.'s.

Tests based on L_2 - distances are discussed in Section 6.3. Some modifications of goodness-of-fit tests when F_0 has a scale parameter appear in Section 6.4 while tests of the symmetry of the errors are discussed in Section 6.5.

1.3.3 Regression model fitting

Another problem of interest is that of fitting a regression model. More precisely, let X, Y be r.v.'s with X being p-dimensional. In much of the existing literature the regression function is defined to be the conditional mean function $\mu(x) := E(Y|X = x)$, $x \in \mathbb{R}^p$, assuming it exists, and then one proceeds to fit a parametric model to this function, i.e., one assumes the existence of a parametric family

$$\mathcal{M} = \{m(x, \boldsymbol{\theta}) : x \in \mathbb{R}^p, \boldsymbol{\theta} \in \Theta\}$$

of functions and then proceeds to tests the hypothesis

$$\mu(x) = m(x, \boldsymbol{\theta}_0), \quad \text{for some } \boldsymbol{\theta}_0 \in \Theta \text{ and } \forall\, x \in \mathcal{I},$$

based on n i.i.d. observations $\{(X_i, Y_i);\ 1 \le i \le n\}$ on (X, Y), where $\mathcal{I}$ is a compact subset of $\mathbb{R}^p$.

We consider the problem of fitting the model $\mathcal{M}$ to the ψ- regression function defined as follows. Let ψ be a nondecreasing real valued function such that $E|\psi(Y - r)| < \infty$, for each $r \in \mathbb{R}$. Define the ψ-regression function m_ψ by the requirement that

$$E[\psi(Y - m_\psi(X))|X] = 0, \quad a.s.$$

Observe that, if $\psi(x) \equiv \psi_\alpha(x) := I(x > 0) - (1 - \alpha)$, for an $0 < \alpha < 1$, then $m_\psi(x) \equiv m_\alpha(x)$, the αth quantile of the conditional distribution of Y, given $X = x$, and if $\psi(x) \equiv x$, then $m_\psi = \mu$.

Tests for fitting a model to m_ψ, i.e., for testing the hypothesis

$$H_0 : m_\psi(\mathbf{x}) = m(\mathbf{x}, \boldsymbol{\theta}_0), \quad \text{for some } \boldsymbol{\theta}_0 \in \Theta \text{ and } \forall\, \mathbf{x} \in \mathcal{I},$$

will be based on the weighted empirical process

$$S_{n,\psi}(x, \hat{\boldsymbol{\theta}}) := n^{-1/2} \sum_{i=1}^{n} \psi(Y_i - m(X_i, \hat{\boldsymbol{\theta}}))\, I(\ell(X_i) \le x),$$

where $x \in [-\infty, \infty]$, ℓ is a known function from $\mathbb{R}^p$ to $\mathbb{R}$, and $\hat{\boldsymbol{\theta}}$ is a $n^{1/2}$- consistent estimator of $\boldsymbol{\theta}_0$. This process is an appropriate extension of the usual partial sum process useful in testing for the mean in the one sample problem to the current regression setup.

Section 6.6 discusses the weak convergence of this process and gives a transform of this process whose asymptotic null distribution is known, so that tests based on this transformed process are asymptotically distribution free.

1.4 R.W.E. Processes and Dynamic Models

A *randomly weighted empirical process* (R.W.E.P.) corresponding to the r.v.'s $\zeta_{n1}, \cdots \zeta_{nn}$, the random noise $\delta_{n1}, \cdots, \delta_{nn}$ and the random real weights $h_{n1}, \cdots, h_{nn}$ is defined to be

$$(1.4.1) \qquad V_h(x) := n^{-1} \sum_{i=1}^{n} h_{ni} I(\zeta_{ni} \le x + \delta_{ni}), \quad x \in \mathbb{R},\ n \ge 1.$$

Examples of R.W.E.P.'s are provided by the W.E.P.'s $\{V_j; 1 \le j \le p\}$ of (1.1.2) in the case the design variables are random. More importantly, R.W.E.P.'s arise naturally in autoregressive models. To illustrate this let $\mathbf{Y}_0 = (X_0, \cdots, X_{1-p})\prime$ be an observable random vector. In the pth order linear autoregressive (LAR(p)) model one observes $\{X_i\}$ obeying the relation

$$(1.4.2) \qquad X_i = \rho_1 X_{i-1} + .. + \rho_p X_{i-p} + \varepsilon_i, \quad i \ge 1,\ \boldsymbol{\rho} \in \mathbb{R}^p,$$

for some $\boldsymbol{\rho} = (\rho_1, \cdots, \rho_p)' \in \mathbb{R}^p$, where $\{\varepsilon_i, i \ge 1\}$ are i.i.d. r.v.'s, independent of $\mathbf{Y}_0$.

Processes that play a fundamental role in the robust estimation of $\boldsymbol{\rho}$ in this model are randomly weighted residual empirical processes $\mathbf{T} = (T_1, \cdots, T_p)'$, where, for $x \in \mathbb{R}$, $\mathbf{t} \in \mathbb{R}^p$,

$$(1.4.3) \qquad T_j(x, \mathbf{t}) := n^{-1} \sum_{i=1}^{n} g_j(\mathbf{Y}_{i-1}) I(X_i \le x + \mathbf{t}'\mathbf{Y}_{i-1}),$$

with $\mathbf{Y}'_{i-1} = (X_{i-1}, ..., X_{i-p})$, $i \ge 1$, and where the components g_j, $1 \le j \le p$, of the vector $\boldsymbol{g}' = (g_1, \cdots, g_p)$, are p measurable functions, each from $\mathbb{R}^p$ to $\mathbb{R}$. Clearly, for each $1 \le j \le p$, $T_j(x, \boldsymbol{\rho} + n^{-1/2}\mathbf{t})$ is an example of $V_h(x)$ with $\zeta_{ni} \equiv \epsilon_i, \delta_{ni} \equiv n^{-1/2}\mathbf{t}'\mathbf{Y}_{i-1}$ and $h_{ni} \equiv g_j(\mathbf{Y}_{i-1})$.

It is customary to expect that a method that works for linear regression models should have an analogue that will also work in autoregressive models. Indeed the above inferential procedures based on W.E.P.'s in linear regression have perfect analogues in LAR(p) models in terms of $\mathbf{T}$. The generalized M-estimators of $\boldsymbol{\rho}$ as proposed by Denby and Martin (1979) corresponding to the weight function $\boldsymbol{g}$

and the score function ψ are given as a solution $\mathbf{t}$ of the p equations

$$\int \psi(x)\mathbf{T}(dx, \mathbf{t}) = 0,$$

assuming that $E\psi(\varepsilon) = 0$. Clearly, the classical least square estimator is obtained upon taking $\boldsymbol{g}(\mathbf{y}) \equiv \mathbf{y}$, $\psi(x) \equiv x$ in these equations.

A generalized R-estimator $\hat{\boldsymbol{\rho}}_R$ corresponding to a score function φ is defined by the relation

$$\text{(1.4.4)} \qquad \hat{\boldsymbol{\rho}}_R := \operatorname{argmin} \{\|\mathbf{S}(\mathbf{t})\|; \mathbf{t} \in \mathbb{R}^p\},$$

where

$$\begin{aligned} \mathbf{S}(\mathbf{t}) &:= \int \varphi(F_n(x, \mathbf{t}))\mathbf{T}(dx, \mathbf{t}), \\ F_n(x, \mathbf{t}) &:= n^{-1}\sum_{i=1}^n I(X_i \le x + \mathbf{t}'\mathbf{Y}_{i-1}), \quad x \in \mathbb{R}, \mathbf{t} \in \mathbb{R}^p. \end{aligned}$$

An analogue of an R-estimator of (1.2.1) is obtained by taking $\boldsymbol{g}(\mathbf{x}) \equiv \mathbf{x}$ in (1.4.4).

The m.d. estimators $\boldsymbol{\rho}_{\mathbf{X}}^+$ that are analogues of $\boldsymbol{\beta}_{\mathbf{X}}^+$ of (1.3.1) are defined as minimizers, w.r.t. $\mathbf{t} \in \mathbb{R}^p$, of

$$\begin{aligned} K^+(\mathbf{t}) := \sum_{j=1}^p \int \Big[n^{-1/2}\sum_{i=1}^n X_{i-j}\Big\{&I(X_i \le x + \mathbf{t}'\mathbf{Y}_{i-1}) \\ &- I(-X_i < x - \mathbf{t}'\mathbf{Y}_{i-1})\Big\}\Big]^2 dG(x). \end{aligned}$$

Observe that K^+ involves $\mathbf{T}$ corresponding to $g_j(\mathbf{Y}_{i-1}) \equiv X_{i-j}$.

Chapter 7 discusses these and some other procedures in detail. Section 7.2 contains a result that establishes the asymptotic uniform linearity in $\mathbf{t}$ of the R.W.E.P.'s $\{\mathbf{T}(x, \rho + n^{-1/2}\mathbf{t}), x \in \mathbb{R}, \|t\| \le b\}$ and the residual empirical processes $\{F_n(x, \rho + n^{-1/2}\mathbf{t}), x \in \mathbb{R}, \|\mathbf{t}\| \le b\}$, for every $0 < b < \infty$. These results are used to investigate the asymptotic behavior of G-M- and G-R- estimators in Sections 7.3.1 and 7.3.2 respectively. In order to carry out the rank analysis in LAR(p) models, one needs a consistent estimator of $Q(f)$ where now f is the error density of $\{\varepsilon_i\}$. A class of such estimators is given in Section 7.3.3. A large class of m.d. estimators and their asymptotics appears in Section 7.4. Section 7.5 discusses an extension of an important methodology of the regression quantiles of Koenker-Bassett (1978) and the related rank scores to linear autoregression. Section 7.6 briefly discusses some tests of goodness-of-fit hypotheses pertaining to the error d.f. in linear AR models.

The problem of fitting a given parametric autoregressive model of order 1 to a real valued stationary ergodic Markovian time series X_i, $i = 0, \pm 1, \pm 2, \cdots$ is discussed in Section 7.7. Much of the development here is parallel to that of Section 6.6 above. Define the ψ-autoregressive function m_ψ by the requirement that

$$E[\psi(X_1 - m_\psi(X_0))|X_0] = 0, \quad a.s.$$

Tests of goodness-of-fit for fitting the model $\mathcal{M}$ to m_ψ are to be based on the process

$$M_{n,\psi}(x, \hat{\boldsymbol{\theta}}) := n^{-1/2} \sum_{i=1}^{n} \psi(X_i - m(X_{i-1}, \hat{\boldsymbol{\theta}}))\, I(X_{i-1} \le x),$$

for $x \in [-\infty, \infty]$. Section 7.7 discusses a martingale type transform of this process whose asymptotic null distribution is known.

The decade of 1990's has seen an exponential growth in the applications of nonlinear AR models to economics, finance and other sciences. Tong (1990) illustrates the usefulness of homoscedastic dynamic AR models in a large class of applied examples from physical sciences while Gouriéroux (1997) contains several examples from economics and finance where the ARCH (autoregressive conditional heteroscedastic) model of Engle (1982) and its various generalizations are found useful. Most of the existing literature has focused on developing classical inference procedures in these models. The theoretical development of the analogues of the estimators discussed above that are known to be robust against outliers in the innovations in linear AR models has relatively lagged behind. Chapter 8 discusses the asymptotic distributions of the analogues of M-, R-, and m.d.- estimators and sequential empirical processes of residuals for a class of nonlinear AR and ARCH time series models. The main point of the chapter is to exhibit the significance of the weak convergence approach in providing a unified methodology for establishing these results for non-smooth underlying score functions ψ and φ.

The contents of Chapter 2 are basic to those of Chapters 3, 4, 7, 8 and parts of Chapter 6. Sections 2.2.1, 2.2.2 and 2.2.3, contain, respectively, proofs of the weak convergence of suitably standardized W.E.P.'s, R.W.E.P.'s and the partial sum processes α_n to continuous Gaussian processes. Even though W.E.P.'s are a special case of R.W.E.P.'s, it is beneficial to investigate their weak convergence separately. For example, the weak convergence of U_d is obtained under a fairly general independence setup and minimal conditions on $\{d_{ni}\}$ whereas that of V_h is obtained under some hierarchal dependence structure on $\{\eta_{ni}, h_{ni}, \delta_{ni}\}$ and the boundedness and/or the square integrability of the weights $\{h_{ni}\}$. The process α_n is different from the V_h process because in these processes the weights are the innovations, whereas in the weighted empiricals process of innovations, the jumps occur at the innovations.

In Section 2.3, the asymptotic continuity of certain standardized W.E.P.'s is used to prove the asymptotic uniform linearity of $\mathbf{V}(., \mathbf{t})$ in $\mathbf{t}$, for $\mathbf{t}$ in certain shrinking neighborhoods of $\boldsymbol{\beta}$, under fairly general independent, non-identically distributed errors. This result is found useful in Chapter 4 when discussing M- estimators and in Chapter 6 when discussing supremum distance test statistics for goodness-of-fit hypotheses. The asymptotic continuity is also found useful in Chapter 3 to prove various results about rank and signed rank statistics under heteroscedastic errors. The asymptotic continuity of V_h - processes is found useful in Chapters 7 and 8 when discussing the AR and ARCH models.

Chapter 2 concludes with results on functional and bounded laws of the iterated logarithm pertaining to certain W.E.P.'s. It also includes an inequality due to Marcus and Zinn (1984) that gives an exponential bound on the tail probabilities of W.E.P.'s of independent r.v.'s. This inequality is an extension of the well celebrated Dvoretzky, Kiefer and Wolfowitz (1956) inequality for the ordinary empirical process. A result about the weak convergence of W.E.P.'s when r.v.'s are p-dimensional is also stated. These results are included for completeness, without proofs. They are not used in the subsequent sections. A martingale property of a properly centered U_d process is proved in Section 2.4.

2

Asymptotic Properties of W.E.P.'s

2.1 Introduction

Let, for each $n \geq 1, \eta_{n1}, \cdots, \eta_{nn}$ be independent r.v.'s taking values in $[0,1]$ with respective d.f.'s $G_{n1}, \cdots, G_{nn}$ and $d_{n1}, \cdots, d_{nn}$ be real numbers. Define

$$(2.1.1) \qquad W_d(t) = \sum_{i=1}^{n} d_{ni}\{I(\eta_{ni} \leq t) - G_{ni}(t)\}, \quad 0 \leq t \leq 1.$$

Observe that W_d belongs to $\mathcal{D}[0,1]$ for each n and any triangular array $\{d_{ni}, 1 \leq i \leq n\}$, while V_h of (1.4.1) belongs to $\mathcal{D}(\mathbb{R})$ for each n and any triangular array $\{h_{ni}, 1 \leq i \leq n\}$.

In this chapter we first prove certain weak convergence results about suitably standardized W_d, V_h and α_n processes. This is done in Sections 2.2.1, 2.2.2, and 2.2.3, respectively. Section 2.3.1 uses the asymptotic continuity of a certain W_d - process to obtain the asymptotic uniform linearity result about $\mathbf{V}(\cdot, \mathbf{u})$ of (1.1.2) in $\mathbf{u}$. Analogous result for $\mathbf{T}(\cdot, \mathbf{u})$ of (1.4.3) uses the asymptotic continuity of a certain V_h - process and is proved in Section 7.2. The weak convergence results about α_n is used in Section 7.6 to derive the limiting distribution of some tests for fitting an autoregressive model to the given time series.

A proof of an exponential inequality for a stopped martingale with bounded differences due to Johnson, Schechtman and Zinn (1985) and Levental (1989) is included in Section 2.2.2. This inequality is of a general interest. It is used to carry out a chaining argument pertaining to the weak convergence of V_h with bounded h.

Section 2.4 treats laws of the iterated logarithm pertaining to W_d, the weak convergence of W_d when $\{\eta_i\}$ are in $[0,1]^p$, the weak convergence of W_d w.r.t. some other metrics when $\{\eta_i\}$ are in $[0,1]$, an embedding result for W_d when $\{\eta_i\}$ are i.i.d. uniform $[0,1]$ r.v.'s and a proof of its martingale property. It also includes an exponential inequality for the tail probabilities of W.E.P.'s of independent r.v.'s. This inequality is an extension of the well celebrated Dvoretzky, Kiefer and Wolfowitz (1956) inequality for the ordinary empirical process. These results are stated for the sake of completeness, without proofs. They are not used in the subsequent sections.

2.2 Weak Convergence

2.2.1 W_d - processes

In this section we give two proofs of the weak convergence of suitably standardized $\{W_d\}$ to a limit in $C[0,1]$. Accordingly, let

$$(2.2.1)\quad G_d(t) := \sum_{i=1}^{n} d_{ni}^2 G_{ni}(t),$$

$$C_d(s,t) := \sum_{i=1}^{n} d_{ni}^2 [G_{ni}(s \wedge t) - G_{ni}(s)G_{ni}(t)], \quad 0 \le s,t \le 1.$$

Let ρ denote the *supremum metric.*

Theorem 2.2.1 *Let $\{\eta_{ni}\}, \{d_{ni}\}$ and $\{G_{ni}\}$ be as in Section 2.1. In addition assume that the following hold;*

(**N1**) $\qquad \tau_d^2 := \sum_{i=1}^n d_{ni}^2 = 1, \ \forall\, n \ge 1.$

(**N2**) $\qquad \max_{1\le i\le n} d_{ni}^2 \to 0.$

(**C**) $\qquad \lim_{\delta\to 0} \limsup_n \sup_{0\le t\le 1-\delta} [G_d(t+\delta) - G_d(t)] = 0.$

Then, for every $\epsilon > 0$,

$$(i) \qquad \lim_{\delta \to 0} \limsup_n P\left(\sup_{|t-s|<\delta} | W_d(t) - W_d(s) | > \epsilon \right) = 0.$$

(ii) Moreover, $W_d \Rightarrow$ some W on $(\mathcal{D}[0,1], \rho)$ if and only if for every $0 \le s, t \le 1$, $C_d(s,t)$ converges to some covariance function $C(s,t)$. In this case W is necessarily a continuous Gaussian process with covariance function C and $W(0) = 0 = W(1)$.

Remark 2.2.1 Perhaps a remark about the labeling of the conditions is in order. The letter **N** in (**N1**) and (**N2**) stands for **Noether** who was the first person to use these conditions to obtain the asymptotic normality of certain weighted sums of r.v.'s. See Noether (1949).

The letter **C** in the condition (**C**) stands for the specified *continuity* of the sequence $\{G_d\}$. Observe that the d.f.'s $\{G_i\}$ need not be continuous for each i and n; only $\{G_d\}$ needs to be equicontinuous in the sense of (**C**). Of course if $\{\eta_i\}$ are i.i.d. G then, because of (**N1**), (**C**) is equivalent to the continuity of G. □

The proof of the theorem will be given after the following *two* lemmas.

Lemma 2.2.1 *For any $0 \le s \le t \le u \le 1$ and each $n \ge 1$*

$$\begin{aligned} & E|W_d(t) - W_d(s)|^2 |W_d(u) - W_d(t)|^2 \\ (2.2.2) \qquad & \le 3\,[G_d(u) - G_d(t)][G_d(t) - G_d(s)] \\ (2.2.3) \qquad & \le 3\,[G_d(u) - G_d(s)]^2. \end{aligned}$$

Proof. Fix $0 \le s \le t \le u \le 1$ and let

$$\begin{aligned} p_i &= G_i(t) - G_i(s), & q_i &= G_i(u) - G_i(t), \\ \alpha_i &= I(s < \eta_i \le t) - p_i, & \beta_i &= I(t < \eta_i \le u) - q_i, \quad 1 \le i \le n. \end{aligned}$$

Observe that $E\alpha_i = 0 = E\beta_j$ for all $1 \le i, j \le n$, $\{\alpha_i\}$ are independent as are $\{\beta_i\}$ and that α_i is independent of β_j for $i \ne j$. Moreover,

$$W_d(t) - W_d(s) = \sum_i d_i \alpha_i, \qquad W_d(u) - W_d(t) = \sum_i d_i \beta_i.$$

Now expand and multiply the quadratics and use the above facts to obtain

$$
\begin{aligned}
(2.2.4)\quad & E|W_d(t) - W_d(s)|^2 |W_d(u) - W_d(t)|^2 \\
&= \sum_i d_i^4 E\alpha_i^2\beta_i^2 + \sum\sum_{i\neq j} d_i^2 d_j^2 \; E\alpha_i^2 d_j^2 E\alpha_i^2 E\beta_j^2 \\
&\quad + 2\sum\sum_{i\neq j} d_i^2 d_j^2 E(\alpha_i\beta_i)E(\alpha_j\beta_j).
\end{aligned}
$$

But

$$
\begin{aligned}
E\alpha_i^2 &= p_i(1-p_i), \qquad E\beta_j^2 = q_j(1-q_j), \\
E\alpha_i^2\beta_i^2 &= (1-p_i)^2 p_i q_i^2 + (1-q_i)^2 q_i p_i^2 + p_i q_i(1-q_i-p_i) \\
&\le \{(1-p_i) + (1-q_i) + (1-q_i-p_i)\} p_i q_i \\
&\le 3p_i q_i, \\
E(\alpha_i\beta_i) &= -(1-p_i)p_i q_i - (1-q_i)q_i p_i + p_i q_i(1-q_i-p_i) \\
&= p_i q_i, \qquad 1 \le i \le n, 1 \le j \le n.
\end{aligned}
$$

Therefore,

$$
\begin{aligned}
LHS(2.2.4) &\le 3\left\{\sum_i d_i^4 p_i q_i + \sum\sum_{i\neq j} d_i^2\, d_j^2 p_i q_j\right\} \\
&= 3\left[\sum_i d_i^2 p_i\right]\left[\sum_j d_j^2 q_j\right].
\end{aligned}
$$

This completes the proof of (2.2.2), in view of the definition of $\{p_i, q_j\}$. That of (2.2.3) follows from (2.2.2) and the monotonicity of the G_i, $1 \le i \le n$. □

Lemma 2.2.2 *For every* $\epsilon > 0$ *and* $s \le u$,

$$
\begin{aligned}
(2.2.5)\quad & P[\sup_{s\le t\le u} |W_d(t) - W_d(s)| \ge \epsilon] \\
&\le \kappa\epsilon^{-4}[G_d(u) - G_d(s)]^2 + P[|W_d(u) - W_d(s)| \ge \epsilon/2]
\end{aligned}
$$

where κ *does not depend on* ϵ, n *or on any underlying quantity.*

Proof. Let $\delta = u - s$, $m \geq 1$ be an integer, and for $1 \leq j \leq m$, let

$$\begin{aligned} \xi_j &= W_d((j/m)\delta + s) - W_d(((j-1)/m)\delta + s), \\ S_k &= \sum_{j=1}^{k} \xi_j, \quad M_m = \max_{1 \leq k \leq m} |S_k|. \end{aligned} \tag{2.2.6}$$

The right continuity of W_d implies that for each n and each sample path, $M_m \to \sup\{|W_d(t) - W_d(s)|; s \leq t \leq u\}$ as $m \to \infty$, w.p.1. In view of Lemma 2.2.1, Lemma 9.1.1 in the Appendix is applicable to the above r.v.'s $\{\xi_j\}$ with $\gamma = 2$, $\alpha = 1$ and

$$u_j = 3^{1/2}\{G_d((j/m)\delta + s) - G_d(((j-1)/m)\delta + s)\}, \quad 1 \leq j \leq m.$$

Hence (2.2.5) follows from that lemma and the right continuity of W_d. □

Proof of Theorem 2.2.1. For a $\delta > 0$, let $r = [\delta^{-1}]$, the greatest integer less than or equal to $1/\delta$. Define $t_j = j\delta$, $1 \leq j \leq r$ and $t_0 = 0$. Let $\Gamma_j = W_d(t_j) - W_d(t_{j-1}), 1 \leq j \leq r$. Then

$$\begin{aligned} & P\Big(\sup_{|t-s|<\delta} |W_d(s) - W_d(s)| \geq \epsilon\Big) \\ & \leq \sum_{j=1}^{r} P\Big(\sup_{t_{j-1} \leq s \leq t_j} |W_d(s) - W_d(t_{j-1})| \geq \epsilon/3\Big) \\ & \leq \kappa\epsilon^{-2} \sum_{j=1}^{r} [G_d(t_j) - G_d(t_{j-1})]^2 + \sum_{j=1}^{r} P[|\Gamma_j| \geq \epsilon/6] \\ & \leq \kappa\epsilon^{-2} \sup_{0 \leq t \leq 1-\delta} [G_d(t+\delta) - G_d(t)] + \sum_{j=1}^{r} P[|\Gamma_j| \geq \epsilon/6] \\ & = I_n(\delta) + \prod_n(\delta), \quad \text{(say)}. \end{aligned} \tag{2.2.7}$$

In the above, the first inequality follows from Lemma 9.1.2 of the Appendix, the second inequality follows from Lemma 2.2.2 above and the last inequality follows because, by (**N1**),

$$\sum_{j=1}^{r} [G_d(t_j) - G_d(t_{j-1})] \leq G_d(1) = 1. \tag{2.2.8}$$

Next, observe that

$$\begin{aligned}\sigma_j^2 &:= Var(\Gamma_j)\\ &= \sum_i d_i^2\{G(t_j)-G_i(t_{j-1})\}\{1-G_j(t_j)+G_i(t_{j-1})\},\\ &\leq G_d(t_j)-G_d(t_{j-1}), \quad 1\leq j\leq r,\end{aligned}$$

and, by (2.2.8), that

$$\text{(2.2.9)} \qquad \sum_{j=1}^{r}\sigma_j^4 \leq \sup_{0\leq t\leq 1-\delta}[G_d(t+\delta)-G_d(t)], \quad \text{all } r \text{ and } n.$$

Furthermore, (**N1**) and (**N2**) enable one to apply the Lindeberg - Feller Central limit Theorem (L-F CLT) to conclude that $\sigma_j^{-1}\Gamma_j \to_d Z, Z$ a $\mathcal{N}(0,1)$ r.v. Therefore, for every $\delta > 0$ (or $r < \infty$),

$$\text{(2.2.10)} \qquad \left|\prod_n(\delta) - \sum_{j=1}^{r} P(|Z| \geq (\epsilon/6)\sigma_j^{-1})\right| \to 0 \text{ as } n\to\infty.$$

By the Markov inequality applied to the summands in the second term of (2.2.10) and by (2.2.9),

$$\begin{aligned}\limsup_n \prod_n(\delta) &\leq 3\limsup_n \sum_{j=1}^{r}(6\sigma_j/\epsilon)^4 \quad (EZ^4=3)\\ &\leq \kappa\epsilon^{-4}\limsup_n \sup_{0\leq t\leq 1-\delta}[G_d(t+\delta)-G_d(t)].\end{aligned}$$

The result (i) now follows from this, (2.2.7) and the assumption (**C**).

Proof of (ii). Suppose $\mathcal{C}_d \to \mathcal{C}$. Let m be a positive integer, $0\leq t_1,\cdots,t_m\leq 1$ and $a_1,\cdots,a_m$ be arbitrary but fixed numbers. Consider

$$\text{(2.2.11)} \qquad T_n := \sum_{j=1}^{m} a_j W_d(t_j) = \sum_{i=1}^{n} d_i\nu_i$$

where

$$\nu_i := \sum_{j=1}^{m} a_j\{I(\eta_i\leq t_j)-G_i(t_j)\}, \qquad 1\leq i\leq n.$$

Note that

$$(2.2.12) \qquad |\nu_i| \le \sum_{j=1}^{m} |a_j| < \infty, \qquad ; 1 \le i \le n.$$

Also, $Var(T_n) \to \sigma^2 := \sum_{j=1}^m \sum_{j=1}^m a_j a_r \mathcal{C}(t_j, t_r)$. In view of (**N1**) and (**N2**), the L-F CLT yields that $T_n \to_d \mathcal{N}(0,1)$. Hence all finite dimensional distributions of W_d converge weakly to those of a Gaussian process W with the covariance function $\mathcal{C}$ and $W(0) = 0 = W(1)$. In view of (i), this implies that $W_d \Rightarrow W$ in $(\mathcal{D}[0,1], d)$ with W denoting a continuous Gaussian process tied down at 0 and 1.

Conversely, suppose $W_d \Longrightarrow W$. By (i), W is in $\mathcal{C}[0,1]$. In particular the sequence T_n of (2.2.11) converges in distribution to $T := \sum_{j=1}^m a_j W(t_j)$. Moreover, (2.2.12) and (**N1**) imply that, for all $n \ge 1$,

$$\begin{aligned} ET_n^4 &= E\left(\sum_{i=1}^m d_i \nu_i\right)^4 = \sum_{i=1}^m d_i^4 E\nu_i^4 + 3\sum_{i=1}^m \sum_{j \ne i,1}^m d_i^2 d_j^2 \, E\nu_i^2 \, E\nu_j^2 \\ &\le 3\left(\sum_{j=1}^m |a_j|\right)^4. \end{aligned}$$

Therefore $\{T_n^2, n \ge 1\}$ is uniformly integrable and hence

$$ET_n^2 = \sum_{j=1}^m \sum_{k=1}^m a_j a_k \mathcal{C}_d(t_j, t_k) \to \sum_{j=1}^m \sum_{k=1}^m a_j a_k Cov[W(t_j), W(t_k)]$$

for any set of numbers $0 \le \{t_j\} \le 1$ and any finite real numbers $a_1, \cdots, a_m$. Hence

$$\mathcal{C}_d(s,t) \to Cov[W(s), W(t)] = \mathcal{C}(s,t) \text{ for all } 0 \le s, t \le 1.$$

Now repeat the above argument of the "only if" part to conclude that W must be a tied down Gaussian process $\mathcal{C}[0,1]$. □

Another set of sufficient conditions for the weak convergence of $\{W_d\}$ is given in the following

Theorem 2.2.2 *Under the notation of Theorem 2.2.1, suppose that* **(N1)** *holds. In addition, assume that the following hold:*

(B) $$n \max_{1\le i\le n} d_{ni}^2 = O(1).$$

(D) $n^{-1}\sum_{i=1}^{n} G_{ni}(t) - t$ *is nonincreasing in* t, $0 \le t \le 1, n \ge 1.$

Then also (i) and (ii) of Theorem 2.2.1 hold.

Remark 2.2.2 Clearly **(B)** implies **(N2)**. Moreover

$$\begin{aligned}[G_d(t+\delta) - G_d(t)] &\le n\max_i d_i^2[n^{-1}\sum_i\{G_i(t+\delta) - G_i(t)\}] \\ &= n\max_i d_i^2\Big[n^{-1}\sum_i\{G_i(t+\delta) - (t+\delta)\} \\ &\qquad -n^{-1}\sum_i\{G_i(t) - t\} + \delta\Big] \\ &\le n\max_i d_i^2\,\delta, \quad 0 \le t \le 1-\delta, \text{ by } \textbf{(D)}.\end{aligned}$$

Thus **(B)** and **(D)** together imply **(N2)** and **(C)**. Hence Theorem 2.2.2 follows from Theorem 2.2.1. However, we can also give a different proof of Theorem 2.2.2 which is direct and quite interesting (see (2.2.15) below). This proof will be based on the following *two* lemmas.

Lemma 2.2.3 *Under* **(D)**, *for all* $n \ge 1$, $0 \le s,t \le 1$,

$$E|W_d(t) - W_d(s)|^4 \le k_d^2\{3(t-s)^2 + (t-s)n^{-1}\}, \tag{2.2.13}$$

where $k_d :=:= n\max_{1\le i\le n} d_{ni}^2$.

Proof. Suppose $0 \le s \le t \le 1$. Let α_i and p_i be as in the proof of Lemma 2.2.1. Using the independence of $\{\alpha_i\}$ and the fact that $E\alpha_i = 0$ for all $1 \le i \le n$, one obtains

$$\begin{aligned}E|W_d(t) - W_d(s)|^4 &= E\Big(\sum d_i\alpha_i\Big)^4 \\ &= \sum_i d_i^4 E\alpha_i^4 + 3\sum\sum_{i\ne j} d_i^2 d_j^2 E\alpha_i^2 E\alpha_j^2 \\ &= \sum_i d_i^4\{E\alpha_i^4 - 3E^2(\alpha_i^2)\} + 3\Big(\sum_i d_i^2 E\alpha_i^2\Big)^2\end{aligned}$$

$$
\begin{aligned}
&= \sum_i d_i^4 p_i(1-p_i)(1-6p_i(1-p_i) \\
&\qquad\qquad +3\Big[\sum_i d_i^2 p_i(1-p_i)\Big]^2 \\
&\le k_d^2\Big\{n^{-2}\sum_i p_i + 3\Big(n^{-1}\sum_i p_i\Big)^2\Big\}.
\end{aligned}
$$

But $s \le t$ and (**D**) imply

$$0 \le n^{-1}\sum_i p_i = n^{-1}\sum_i [G_i(t) - G_i(s)] \le (t-s).$$

Hence the

$$L.H.S.(2.2.13) \le k_d^2\{n^{-1}(t-s) + 3(t-s)^2\}, \quad 0 \le s \le t \le 1.$$

The proof is completed by interchanging the role of s and t in the above argument in the case $t \le s$. □

Next, define, for $(i-1)/n \le t \le i/n, \;\; 1 \le i \le n$,

$$
\begin{aligned}
(2.2.14)\; Z_d(t) &= W_d((i-1)/n) \\
&\quad + \{nt-(i-1)\}[W_d(i/n) - W_d((i-1)/n)].
\end{aligned}
$$

Lemma 2.2.4 *The assumption (*__D__*) implies that* $\forall\; 0 \le s,t \le 1$, $n \ge 1$,

$$(2.2.15) \qquad E|Z_d(t) - Z_d(s)|^4 \le 144\, k_d^2\, |t-s|^2.$$

*If, in addition, (*__N1__*) and (*__B__*) hold, then*

$$(2.2.16) \qquad \sup_t |W_d(t) - Z_d(t)| = o_p(1).$$

Proof. Let $n \ge 1$ and $0 \le s,t \le 1$ be arbitrary but fixed. Choose integers $1 \le i,j \le n$ such that

$$(2.2.17) \qquad (i-1)/n \le s \le i/n \quad \text{and} \quad (j-1)/n \le t \le j/n.$$

For the sake of convenience, let

$$
\begin{aligned}
\delta_{k,m} &:= |Z_d(m/n) - Z_d(k/n)| = |W_d(m/n) - W_d(k/n)|, \\
b_{k,m} &= 4k_d^2[(m-k)/n]^2, \quad m,\ k \text{ integers}; \\
\Delta_{u,v} &:= |Z_d(u) - Z_d(v)|,
\end{aligned}
$$

From (2.2.13),

$$E\delta_{k,m}^4 \le k_d^2\{3(m-k)^2/n^2 + n^{-2}|m-k|\} \le 4k_d^2[(m-k)/n]^2 =: b_{k,m}. \tag{2.2.18}$$

The proof of (2.2.15) will be completed by considering the following three cases.

Case 1. $i < j - 1$. Then because of (2.2.14) and (2.2.17),

$$\Delta_{s,t} \le \max\{\delta_{i,j-1}, \delta_{i,j}, \delta_{i-1,j-1}, \delta_{i-1,j}\}$$

which entails that

$$\begin{aligned} E\Delta_{s,t}^4 &\le E\{\delta_{i,j-1}^4 + \delta_{i,j}^4 + \delta_{i-1,j-1}^4 + \delta_{i-1,j}^4\} \\ &\le b_{i,j-1} + b_{i,j} + b_{i-1,j-1} + b_{i-1,j} \\ &\le 4b_{i-1,j} = 16k_d^2[(j-(i-1))/n]^2 \end{aligned} \tag{2.2.19}$$

where the second inequality follows from (2.2.18) and the last from $0 \le j - i - 1 < j - i < j - (i-1)$.

Note that (2.2.17), $i < j - 1$ and i, j integers imply that

$$3(t-s) \ge [j-(i-1)]/n.$$

From this and (2.2.19) one obtains

$$E\Delta_{s,t}^4 \le 144k_d^2(t-s)^2. \tag{2.2.20}$$

Case 2. $i = j$. In this case $(i-1)/n \le s, t \le i/n$. From (2.2.14) one has

$$\Delta_{s,t} = n|t-s|\delta_{i-1,i}$$

so that from (2.2.18)

$$E\Delta_{s,t}^4 < n^4(t-s)^4.4k_d^2.n^{-2} \le 4k_d^2(t-s)^2. \tag{2.2.21}$$

The last inequality follows because $n(t-s) \le 1$.

Case 3. $i = j - 1$. By the triangle inequality

$$\Delta_{s,t} \le 2\max(\Delta_{s,i/n}, \Delta_{i/n,t}).$$

Thus by Case 2, applied once with s and i/n and once with i/n and t, one obtains

$$\begin{aligned} E\Delta_{s,t}^4 &\leq 2^4\{E\Delta_{s,i/n,t}^4\} \\ &\leq 2^6 k_d^2\{(i/n - s)^2 + (t - i/n)^2\} \leq 2^7 k_d^2 (t-s)^2. \end{aligned}$$

In view of this, (2.2.21) and (2.2.20), the proof of (2.2.15) is complete.

To prove (2.2.16), let $d_{i+} = \max(0, d_i), d_{i-} = \max(0, -d_i)$. Then one has $d_i = d_{i+} - d_{i-}$. Decompose W_d and Z_d accordingly. Note that $\max(d_{i+}^2, d_{i-}^2) = d_{i+}^2 + d_{i-}^2 = d_i^2, 1 \leq i \leq n$. This and (**N1**) imply that $\tau_{d+} \leq 1, \tau_{d-} \leq 1$. It also implies that if (**N2**) is satisfied by the $\{d_i\}$ then it is also satisfied by $\{d_{i+}, d_{i-}\}$. By the triangle inequality,

$$||W_d - Z_d||_\infty \leq ||W_{d+} - Z_{d+}||_\infty + ||W_{d-} - Z_{d-}||_\infty.$$

Moreover, $d_{i+} \wedge d_{i-} \geq 0$, for all i. Therefore, it is enough to prove (2.2.16) for $d_i \geq 0$, $1 \leq i \leq n$. Accordingly suppose that is the case. Then

$$(2.2.22) \qquad ||W_d - Z_d||_\infty \leq \mathcal{U}_1 + \mathcal{U}_2,$$

where

$$\begin{aligned} \mathcal{U}_1 &= \max_{1\leq i\leq n} \sup_{(i-1)/n\leq t\leq i/n} |W_d(t) - W_d((i-1)/n)|, \\ \mathcal{U}_2 &= \max_{1\leq i\leq n} \sup_{(i-1)/n\leq t\leq i/n} |W_d(t) - W_d(i/n)| \end{aligned}$$

For $(i-1)/n \leq t \leq i/n$, and $d_i \geq 0, 1 \leq i \leq n$,

$$\begin{aligned} &|W_d(t) - W_d(i/n)| \\ &\leq |\sum_{j=1}^n d_j\ I(t < \eta_j \leq i/n)| + \sum_{j=1}^n d_j\ [G_j(i/n) - G_j(t)] \\ &\leq |W_d(\frac{i}{n}) - W_d(\frac{i-1}{n})| + 2\sum_{j=1}^n d_j\ [G_j(\frac{i}{n}) - G_j(\frac{i-1}{n})] \\ &\leq \delta_{i-1,i} + 2\max_{1\leq j\leq n} d_j, \qquad \text{by } (\mathbf{D}) \end{aligned}$$

This inequality, (2.2.18) and the Markov inequality imply that for every $\epsilon > 0$ and for n sufficiently large such that $2\max_j d_j < \epsilon$, the existence of which is guaranteed by (**B**),

$$
\begin{aligned}
P(\mathcal{U}_2 \geq \epsilon) &\leq P(\max_i \delta_{i-1,i} \geq \epsilon - 2\max_i d_i) \\
&\leq (\epsilon - 2\max_i d_i)^{-4} \sum_{i=1}^{n} E\delta_{i-1,i}^4 \\
&\leq (\epsilon - 2\max_i d_i)^{-4} . 4k_d^2 n^{-1} \to 0.
\end{aligned}
$$

Exactly similar calculations show that $\mathcal{U}_1 = o_p(1)$. □

Proof of Theorem 2.2.2. Observe that $Z_d(0) = 0 = Z_d(1)$ and that $Z_d \in C[0,1]$ for every $n \geq 1$ and each sequence $\{d_i\}$. Hence by (2.2.15) and Theorem 9.1.2 of the Appendix, $\{Z_d\}$ is tight in $C[0,1]$. Thus claim (i) follows from (2.2.16). To prove (ii) just argue as in the proof of (ii) of Theorem 2.2.1 above. □

The following corollary will be useful later on. To state it we need some more notation. Let $F_{n1}, ..., F_{nn}$ denote d.f.'s on $\mathbb{R}$ and X_{ni} be a r.v. with d.f. F_{ni}, $1 \leq i \leq n$. Define, for $x \in \mathbb{R}$, $1 \leq i \leq n$, $0 \leq t \leq 1$,

$$
\begin{aligned}
(2.2.23)\quad H(x) &:= n^{-1}\sum_{i=1}^{n} F_{ni}(x); \quad H^{-1}(t) := \inf\{x; H(x) \geq t\}; \\
L_{ni}(t) &:= F_{ni}(H^{-1}(t)); \quad L_d(t) := \sum_{i=1}^{n} d_{ni}^2 L_{ni}(t), \\
W_d^*(t) &:= \sum_{i=1}^{n} d_{ni}\left\{I(X_{ni} \leq H^{-1}(t)) - L_{ni}(t)\right\}.
\end{aligned}
$$

Corollary 2.2.1 *Assume that*

(2.2.24) $X_{n1}, \cdots, X_{nn}$ *are independent r.v.'s with respective d.f.'s* $F_{n1}, \cdots, F_{nn}$ *on* $\mathbb{R}$.

In addition, suppose that $\{d_{ni}\}, \{F_{ni}\}$ *satisfy* (**N1**), (**N2**) *and*

$$
(\mathbf{C}^*) \qquad \lim_{\delta\to 0}\limsup_n \sup_{0\leq t\leq 1-\delta} [L_d(t+\delta) - L_d(t)] = 0.
$$

Then, for every $\epsilon > 0$,

$$
(2.2.25) \qquad \lim_{\delta\to 0}\limsup_n P(\sup_{|t-s|<\delta} |W_d^*(t) - W_d^*(s)| \geq \epsilon) = 0.
$$

Proof. Follows from Theorem 2.2.1 (i) applied to $\eta_i \equiv H(X_i)$, $G_i \equiv L_i, 1 \leq i \leq n$. □

Remark 2.2.3 Note that if H is continuous then

$$n^{-1} \sum_{i=1}^{n} L_{ni}(t) \equiv t.$$

Therefore,

$$\sup_{0 \leq t \leq 1-\delta} [L_d(t+\delta) - L_d(t)] \leq n \max_i d_{ni}^2 \, \delta.$$

Thus, if we strengthen (**N2**) to require (**B**) then (**C***) is *a priori* satisfied. That is, the conditions of Theorem 2.2.2(i) are satisfied.

If $F_{ni} \equiv F$, where F is a continuous d.f., then $L_{ni}(t) \equiv t$. Therefore, in view of (**N1**), (**C***) is *a priori* satisfied. Moreover $C_d^*(s,t) := Cov(W_d^*(s), W_d^*(t)) = s(1-t), 0 \leq s \leq t \leq 1$. Therefore we obtain

Corollary 2.2.2 *Suppose that $X_{n1}, ..., X_{nn}$ are i.i.d. F, F a continuous d.f.. Suppose that $\{d_{ni}\}$ satisfy (**N1**) and (**N2**). Then $W_d^* \Rightarrow B$ in $(\mathcal{D}[0,1], \rho)$ with B a Brownian bridge in $C[0,1]$.*

Observe that $d_{ni} \equiv n^{-1/2}$ satisfy (**N1**) and (**N2**). In other words the above corollary includes the well celebrated result, viz., the weak convergence of the sequence of the ordinary empirical processes.

Note. A variant of Theorem 2.2.1 was first proved in Koul (1970). The above formulation and proof is based on this work and that of Withers (1975). Theorem 2.2.2 is motivated by the work of Shorack (1973) which deals only with the weak convergence of the W_1-process, the process W_d with $d_{ni} \equiv n^{-1/2}$. The sufficiency of condition (**D**) for (2.2.13) was observed by Eyster (1977). □

2.2.2 V_h - processes

In this subsection we shall investigate the weak convergence of the R.W.E.P.'s $\{V_h(x), x \in \mathbb{R}\}$ of (1.4.1). To state the general result we need some more structure on the underlying r.v.'s.

Accordingly, let $(\Omega, \mathcal{A}, P)$ be a probability space and G be a d.f. on $\mathbb{R}$. For each integer $n \geq 1$, let $(\zeta_{ni}, h_{ni}, \delta_{ni}), 1 \leq i \leq n$, be an array of trivariate r.v.'s defined on $(\Omega, \mathcal{A})$ such that $\{\zeta_{ni}, 1 \leq i \leq n\}$ are i.i.d. G r.v.'s and ζ_{ni} is independent of (h_{ni}, δ_{ni}) for each $1 \leq i \leq n$. Furthermore, let $\{\mathcal{A}_{ni}\}$ be an array of sub σ-fields such that $\mathcal{A}_{ni} \subset \mathcal{A}_{n,i+1}, 1 \leq i \leq n, n \geq 1; (h_{n1}, \delta_{n1})$ is $\mathcal{A}_{n1}$ - measurable; the r.v.'s $\{\zeta_{n1}, \cdots, \zeta_{n,j-1}; (h_{ni}, \delta_{ni}), 1 \leq i \leq j\}$ are $\mathcal{A}_{nj}$ - measurable, $2 \leq j \leq n$; and ζ_{nj} is independent of $\mathcal{A}_{nj}, 1 \leq j \leq n$. Define,

$$(2.2.26)\quad V_h(x) := n^{-1} \sum_{i=1}^{n} h_{ni} I(\zeta_{ni} \leq x + \delta_{ni}),$$

$$J_h(x) := n^{-1} \sum_{i=1}^{n} E\Big[h_{ni} I(\zeta_{ni} \leq x + \delta_{ni})\Big|\mathcal{A}_{ni}\Big]$$

$$= n^{-1} \sum_{i=1}^{n} h_{ni} G(x + \delta_{ni}),$$

$$V_h^*(x) := n^{-1} \sum_{i=1}^{n} h_{ni} I(\zeta_{ni} \leq x), \quad J_h^*(x) := n^{-1} \sum_{i=1}^{n} h_{ni} G(x),$$

$$U_h(x) := n^{1/2}[V_h(x) - J_h(x)],$$

$$U_h^*(x) := n^{1/2}[V_h^*(x) - J_h^*(x)], \quad x \in \mathbb{R}.$$

We shall give three results here. The first one states and proves the weak convergence of these processes for the bounded weights while the second states the same result under weaker assumptions for only square integrable weights. The proofs of these results are similar in spirit yet different. An additional similar theorem along with its proof pertaining to analogues of the V_h process suitable for heteroscedastic autoregressive models is also given in this section. To begin with we state the following

Theorem 2.2.3 *In addition to the above, assume that the following conditions hold:*

$$(2.2.27)\qquad \sup_{n \geq 1} \max_i |h_{ni}| \leq c, \text{ a.s. for some constant } c < \infty.$$

$$(2.2.28)\qquad \max_i |\delta_{ni}| = o_p(1).$$

Then, for every $x \in \mathbb{R}$ at which G is continuous,

$$|U_h(x) - U_h^*(x)| = o_p(1). \tag{2.2.29}$$

In addition, if

$$n^{-1/2}\sum_i |h_{ni}\delta_{ni}| = O_p(1), \tag{2.2.30}$$

(2.2.31) G *has a uniformly continuous a.e. positive Lebesgue density g,*

then

$$\|U_h - U_h^*\|_\infty = o_p(1). \tag{2.2.32}$$

If, in addition,

$$\mathcal{A}_{ni} \subset \mathcal{A}_{n+1,i},\ 1 \le i \le n; n \ge 1. \tag{2.2.33}$$

$$\left(n^{-1}\sum_i h_{ni}^2\right)^{1/2} = \alpha + o_p(1), \quad \alpha \quad \text{a positive r.v.}, \tag{2.2.34}$$

then

$$U_h \Rightarrow \alpha \cdot B(G), \qquad U_h^* \Rightarrow \alpha \cdot B(G) \tag{2.2.35}$$

where B is a Brownian Bridge in $C[0,1]$, independent of α.

The proof of (2.2.29) is straightforward using Chebbychev's inequality, while that of (2.2.32) uses a restricted chaining argument and an exponential inequality for martingales with bounded differences. It will be a consequence of the following *two* lemmas.

Lemma 2.2.5 *Under (2.2.27) - (2.2.31), $\forall \epsilon > 0$ and for $r = 1, 2$,*

$$\lim_n P\left(\sup_{x,y} n^{-1/2}\sum_{i=1}^n |h_{ni}|^r |G(y+\delta_{ni}) - G(x+\delta_{ni})| \le 2c^r\epsilon\right) = 1,$$

where the supremum is taken over the set

$$\{x, y \in \mathbb{R}; n^{1/2}|G(x) - G(y)| \le \epsilon\}.$$

Proof. Let $\epsilon > 0$, $q(u) := g(G^{-1}(u))$, $0 \le u \le 1$; $\gamma_n := \max_i |\delta_i|$,

$$\begin{aligned} \omega_n &:= \sup\{|q(u) - q(v)|; |u - v| \le \epsilon n^{-1/2}\} \\ &= \sup\{|g(x) - g(y)|; |G(x) - G(y)| \le \epsilon n^{-1/2}\}, \\ \Delta_n &:= \sup\{|g(x) - g(y)|; |y - x| \le \gamma_n\}. \end{aligned}$$

By (2.2.31), q is uniformly continuous on $[0, 1]$. Hence, by (2.2.28),

$$\Delta_n = o_p(1), \qquad \omega_n = o(1). \tag{2.2.36}$$

But

$$\begin{aligned} &\sup_{x,y} n^{-1/2} \sum_{i=1}^{n} |h_i|^r |G(y + \delta_i) - G(x + \delta_i)| \\ &\le \sup_{x,y} n^{-1/2} \sum_{i=1}^{n} |h_i|^r |G(y) - G(x)| + n^{-1} \sum_{i=1}^{n} |h_i|^r |\delta_i| [\omega_n + 2\Delta_n] \\ &\le c^r \epsilon + O_p(1).o_p(1), \qquad \text{by (2.2.30) and (2.2.36).} \end{aligned}$$

This completes the proof of the Lemma. □

Lemma 2.2.6 *Let $\{\mathcal{F}_i, i \ge 0\}$ be an increasing sequence of σ - fields, m be a positive integer, $\tau \le m$ be a stopping time relative to $\{\mathcal{F}_i\}$ and $\{\xi_i, 1 \le i \le m\}$ be a sequence of real valued martingale differences w.r.t. $\{\mathcal{F}_i\}$. In addition, suppose that for some constant M, $L < \infty$,*

$$|\xi_i| \le M < \infty, \qquad 1 \le i \le m, \tag{2.2.37}$$

$$\sum_{i=1}^{\tau} E(\xi_i^2 | \mathcal{F}_{i-1}) \le L. \tag{2.2.38}$$

Then, for every $a > 0$,

$$\begin{aligned} P\left(\left|\sum_{i=1}^{\tau} \xi_i\right| > a\right) \\ \le 2\exp\{-(a/2M)\, arcsinh(Ma/2L)\}. \end{aligned} \tag{2.2.39}$$

Proof. Write $\sigma_i^2 = E(\xi_i^2 | \mathcal{F}_{i-1}), i \ge 1$. First, consider the **case** $\tau = m$:

Recall the following elementary facts: For all $x \in \mathbb{R}$,

(a) $\exp(x) - x - 1 \leq 2(\cosh x - 1) \leq x \sinh x$,
(b) $\sinh(x)/x$ is increasing in $|x|$,
(c) $x \leq \exp(x - 1)$.

Because $E(\xi_i|\mathcal{F}_{i-1}) \equiv 0$ and by (2.2.37), for a $\delta > 0$ and for all $1 \leq i \leq m$,

$$\begin{aligned} E\{[\exp(\delta\xi_i) - 1]|\mathcal{F}_{i-1}\} &\leq E\{\delta\xi_1 \sinh(\delta\xi_1)|\mathcal{F}_{i-1}\}, && \text{by } (a), \\ &\leq \sigma_i^2 \delta \sinh(\delta M)/M, && \text{by } (b). \end{aligned} \tag{2.2.40}$$

Use a conditioning argument to obtain

$$\begin{aligned} & E\exp\left\{\delta\sum_{i=1}^{m}\xi_i\right\} \\ &= E\left[\exp\left(\delta\sum_{i=1}^{m-1}\xi_i\right) E\left\{\exp(\delta\xi_m)|\mathcal{F}_{m-1}\right\}\right] \\ &\leq E\left[\exp\left(\delta\sum_{i=1}^{m-1}\xi_i\right) \exp\left(E\left\{\exp(\delta\xi_m)|\mathcal{F}_{m-1}\right\} - 1\right)\right], \\ &\leq E\left[\exp\left(\delta\sum_{i=1}^{m-1}\xi_i\right) \exp(\sigma_m^2.\delta/M.\sinh(\delta M))\right] \\ &\leq E\left[\exp\left(\delta\sum_{i=1}^{m-1}\xi_i\right) \exp\left\{(L - \sum_{i=1}^{m-1}\sigma_i^2).(\delta/M).\sinh(\delta M)\right\}\right], \end{aligned}$$

where the second inequality follows from (2.2.40) while the first follows from (c). Observe that $L - \sum_{i=1}^{j-1}\sigma_i^2$ is $\mathcal{F}_{j-2}$ measurable, for all $j \geq 2$. Hence, iterating the above argument $m - 1$ times yields

$$E\exp\left\{\delta\sum_{i=1}^{m}\xi_i\right\} \leq \exp\left\{L.(\delta/M).\sinh(\delta M)\right\}.$$

Now, by the Markov inequality, $\forall\, a > 0$,

$$\begin{aligned} P\left(\sum_{i=1}^{m}\xi_i \geq a\right) &\leq E\exp\left\{\delta\Big(\sum_{i=1}^{m}\xi_i - a\Big)\right\} \\ &\leq \exp\left\{\delta[L/M.\sinh(\delta M) - a]\right\}. \end{aligned}$$

The choice of $\delta = (1/M)\operatorname{arcsinh}(Ma/2L)$ in this leads to the inequality

$$P\left(\sum_{i=1}^{m} \xi_i \geq a\right) \leq \exp\{(-a/2M)\operatorname{arcsinh}(Ma/2L)\}.$$

An application of this inequality to $\{-\xi_i\}$ will yield the same bound for $P(\sum_{i=1}^{m} \xi_i \leq -a)$, thereby completing the proof of (2.2.39) in the case $\tau = m$. Now consider the general case $\tau \leq m$:

Let $\chi_j = \xi_j I(j \leq \tau)$. Because the event $[j \leq \tau] \in \mathcal{F}_{j-1}$, it follows that $\{\chi_j, \mathcal{F}_j\}$ satisfy the conditions of the previous case. Hence

$$\begin{aligned} P\left(\left|\sum_{i=1}^{\tau} \xi_i\right| \geq a\right) &= P\left(\left|\sum_{i=1}^{m} \chi_i\right| \geq a\right) \\ &\leq \exp\{(-a/2M)\operatorname{arcsinh}(Ma/2L)\}. \qquad \square \end{aligned}$$

Proof of Theorem 2.2.3. For the clarity of the present proof it is important to emphasize the dependence of various underlying processes on n. Accordingly, we shall write V_n, U_n etc. for V_h, U_h etc. in the proof.

On $\mathbb{R}$ define the matric $d(x,y) := |G(x) - G(y)|^{1/2}$. This metric makes $\mathbb{R}$ totally bounded. Thus, to prove the theorem, it suffices to prove

$$\begin{aligned} &(a) \qquad |U_n(y) - U_n^*(y)| = o_p(1), \qquad \forall\ y \in \mathbb{R}, \\ &(b) \quad \forall\ \epsilon > 0\ \exists\ \delta > 0 \ni \\ &\qquad (i) \quad \limsup_n P(\sup_{d(x,y)) \leq \delta} |U_n(y) - U_n(x)| > \epsilon) < \epsilon, \\ &\qquad (ii) \quad \limsup_n P(\sup_{d(x,y) \leq \delta} |U_n^*(y) - U_n^*(x)| > \epsilon) < \epsilon. \end{aligned}$$

Proof of (a). The fact that $U_n - U_n^*$ is a sum of conditionally centered bounded r.v.'s yields that

$$Var(U_n(y) - U_n^*(y)) \leq En^{-1}\sum_{i=1}^{n} h_i^2 |G(y+\delta_i) - G(y)| = o(1),$$

by (2.2.27), (2.2.28), (2.2.31) and the D.C.T.

Proof of (b)(i). The following proof of (b)(i) uses a restricted chaining argument as discussed in Pollard (1984, p. 160-162), and the exponential inequality of Lemma 2.2.6 above.

Fix an $\epsilon > 0$. Let $a_n := [n^{1/2}/\epsilon]$, the greatest integer less than or equal to $n^{1/2}/\epsilon$, and define the grid

$$\mathcal{H}_n := \{y_j; G(y_j) = j\epsilon n^{-1/2}, \quad 1 \le j \le a_n\}, \quad n \ge 1.$$

Also let

$$Z_i(x) := I(\zeta_i \le x + \delta_i) - G(x + \delta_i), \quad x \in \mathbb{R}, \ 1 \le i \le n.$$

Write $h_i \equiv h_{i+} - h_{i-} \equiv \max(0, h_i)$, so that

$$\begin{aligned} U_h(x) &= n^{-1/2} \sum_{i=1}^{n} h_{i+} Z_i(x) - n^{-1/2} \sum_{i=1}^{n} h_{i-} Z_i(x) \\ &= U_h^+(x) - U_h^-(x), \quad \text{say.} \end{aligned}$$

Thus to prove (b)(i), by the triangle inequality, it suffices to prove it for U_n^+ processes. The details of the proof shall be given for the U_n^+ process only; those for the U_n^- being similar.

Next, let, for $k \ge 1$, and $x, \ y \in \mathbb{R}$,

$$D_k(x, y) := \sum_{i=1}^{k} h_{i+}^2 E\left\{ [Z_i(x) - Z_i(y)]^2 \middle| \mathcal{A}_i \right\},$$

and define the sequence of stopping times

$$\tau_n^+ := n \wedge \max\left\{ k \ge 1; \ \max_{x, y \in \mathcal{H}_n} \frac{D_k(x, y)}{d^2(x, y)} < 3\epsilon c^2 n \right\}.$$

Observe that $\tau_n^+ \le n$. To adapt the present situation to that of the Pollard, we first prove that $P(\tau_n^+ < n) \to 0$ (see (2.2.41) below). This allows one to work with $n^{-1/2}(\tau_n^+)^{1/2} U_{\tau_n^+}^+$ instead of U_n^+. By Lemma 2.2.6 and the fact that arcsinh(x) is increasing and concave in x, one obtains that if x, y in $\mathcal{H}_n$ are such that $d^2(x, y) \ge t\,\epsilon\, n^{-1/2}$ then

$$\begin{aligned} &P(n^{-1/2}(\tau_n^+)^{1/2} |U_{\tau_n^+}^+(x) - U_{\tau_n^+}^+(y)| \ge t) \\ &\le 2\exp\left\{ -\frac{t^2}{2cd^2(x, y)} \,\epsilon \operatorname{arcsinh}(1/(6\epsilon^2 c)) \right\}, \qquad \text{for all } \ t > 0. \end{aligned}$$

This enables one to carry out the chaining argument as in Pollard. What remains to be done is to connect between the points in $\mathbb{R}$ and a point in $\mathcal{H}_n$ which will be done in (2.2.42) below. We shall now prove

$$P(\tau_n^+ < n) \to 0. \tag{2.2.41}$$

Proof of (2.2.41). For y_j, y_k in $\mathcal{H}_n$ with $y_j < y_k, d^2(y_j, y_k) \geq (k-j)\,\epsilon\, n^{-1/2}$. Hence, using the fact that $(h_{i+})^2 \leq h_i^2$,

$$\begin{aligned}
&D_n(y_k, y_j)/d^2(y_i, y_k) \\
&\leq \sum_{i=1}^n h_i^2[G(y_k+\delta_i) - G(y_j+\delta_i)]\{(k-j)\epsilon\}^{-1} n^{1/2} \\
&\leq \{(k-j)\epsilon\}^{-1} n^{1/2} \sum_{i=1}^n h_i^2 \sum_{r=j}^{k-1} [G(y_{r+1}+\delta_i) - G(y_r+\delta_i)] \\
&\leq \epsilon^{-1} n^{1/2} \max_{1 \leq r \leq a_n} \sum_{i=1}^n h_i^2 [G(y_{r+1}+\delta_i) - G(y_r+\delta_i)].
\end{aligned}$$

Now apply Lemma 2.2.5 with $r = 2$ to obtain

$$P(\max_{1lek,j \leq a_n} \left\{ \frac{D_k(x,y)}{d^2(x,y)} < 3\epsilon c^2 n \right\} \longrightarrow 1.$$

This completes the proof of (2.2.41).

Next, for each $x \in \mathbb{R}$, let y_{j_x} denote the point in $\mathcal{H}_n$ that is the closest to x in d-metric from the points in $\mathcal{H}_n$ that satisfy $y_{j_x} \leq x$. We shall now prove: $\forall\ \epsilon > 0$,

$$P(\sup_x |U_n^+(x) - U_n^+(y_{j_x})| > 8c\epsilon) \to 0. \tag{2.2.42}$$

Proof of (2.2.42). Now write V_n^+, J_n^+ for V_n, J_n when $\{h_i\}$ in these quantities is replaced by $\{h_{i+}\}$.

The definition of y_{j_x}, G increasing, and the fact that $h_{i+} \leq |h_i|$ for all i, imply that

$$\begin{aligned}
&\sup_x |n^{1/2}[J_n^+(x) - J_n^+(y_{j_x})]| \\
&\leq \max_{1 \leq j \leq a_n} n^{-1/2} \sum_{i=1}^n |h_i| [G(y_{j+1}+\delta_i) - G(y_j+\delta_i)]
\end{aligned}$$

An application of Lemma 2.2.5 with $r = 1$ now yields that

$$(2.2.43) \qquad P(\sup_x |n^{1/2}[J_n^+(x) - J_n^+(y_{j_x})]| > 4c\epsilon) \longrightarrow 0.$$

But $h_{i+} \geq 0, 1 \leq i \leq n$, implies that V_n^+ is nondecreasing in x. Therefore, using the definition of y_{j_x},

$$\begin{aligned} & n^{-1/2}[U_n^+(y_{j_x-1}) - U_n^+(y_{j_x})] + J_n^+(y_{j_x-1}) - J_n^+(y_{j_x}) \\ = \;& V_n^+(y_{j_x-1}) - V_n^+(y_{j_x}) \\ \leq \;& V_n^+(x) - V_n^+(y_{j_x}) \\ \leq \;& V_n^+(y_{j_x+1}) - V_n^+(y_{j_x}) \\ = \;& n^{-1/2}[U_n^+(y_{j_x+1}) - U_n^+(y_{j_x})] + J_n^+(y_{j_x+1}) - J_n^+(y_{j_x}). \end{aligned}$$

Hence,

$$(2.2.44) \qquad \begin{aligned} & \sup_x |n^{1/2}[V_n^+(x) - V_n^+(y_{j_x}]| \\ & \leq 2 \max_{1 \leq j \leq a_n} |U_n^+(y_{j+1}) - U_n^+(y_j)| \\ & \quad +2 \max_{1 \leq j \leq a_n} |n^{1/2}[J_n^+(y_{j+1}) - J_n^+(y_j)]| \end{aligned}$$

Thus, (2.2.42) will follow from (2.2.43), (2.2.44) and

$$(2.2.45) \qquad P(\max_{1 \leq j \leq a_n} |U_n^+(y_{j+1}) - U_n^+(y_j)| > c\epsilon) \to 0.$$

In view of (2.2.41), to prove (2.2.45), it suffices to show that

(2.2.46)
$$P(\max_{1 \leq j \leq a_n} n^{-1/2}(\tau_n^+)^{1/2}|U_{\tau_n^+}^+(y_{j+1}) - U_{\tau_n^+}^+(y_j)| > c\epsilon) \to 0.$$

But the L.H.S. of (2.2.46) is bounded above by

$$\sum_{j=1}^{a_n} P\Big(\Big|\sum_{i=1}^{\tau_n^+} h_{i+}[Z_i(y_{j+1} - Z_i(y_j)]\Big| > c\epsilon\, n^{1/2}\Big).$$

Now apply Lemma 2.2.6 with $\xi_i \equiv h_{i+}[Z_i(y_{j+1}) - Z_i(y_j)]$, $\mathcal{F}_{i-1} \equiv \mathcal{A}_i$, $\tau \equiv \tau_n^+$, $M = c$, $a = c\epsilon\, n^{1/2}$, $m = n$. By the definition of τ_n^+, $L = 3c^2\epsilon^2 n^{1/2}$. Hence, by Lemma 2.2.6,

$$\begin{aligned} & P\Big(\Big|\sum_{i=1}^{\tau_n^+} h_{i+}[Z_i(y_{j+1}) - Z_i(y_j)]\Big| > c\epsilon n^{1/2}\Big) \\ & \leq 2\exp[-\frac{n^{1/2}\epsilon}{2}\text{arcsinh}(1/6\epsilon)]. \end{aligned}$$

Since this bound does not depend on j, it follows that

$$\text{L.H.S. (2.2.46)} \quad \leq 2\epsilon^{-1}n^{1/2}\exp[-\frac{n^{1/2}\epsilon}{2}\text{arcsinh}(1/6\epsilon)] \to 0.$$

This completes the proof of (2.2.42) for U_n^+. As mentioned earlier the proof of (2.2.42) for U_n^- is exactly similar, thereby completing the proof of (b)(i).

Adapt the above proof of (b)(i) with $\delta_i \equiv 0$ to conclude (b)(ii). *Note that* (b)(ii) *holds solely under* (2.2.27) *and the assumption that* G *is continuous and strictly increasing, the other assumptions are not required here.* The proof of (2.2.32) is now complete.

The claim (2.2.35) follows from (b)(ii) above, Lemma 9.1.3 of the Appendix and the Cramér - Wold device. □

As noted in the proof of the above theorem, the weak convergence of U_h^* holds only under (2.2.27), (2.2.34) and the assumption that G is continuous and strictly increasing. For an easy reference later on we state this result as

Corollary 2.2.3 *Let the setup of Theorem 2.2.3 hold. Assume that* G *is continuous and strictly increasing and that (2.2.27), (2.2.34) hold. Then,* $U_h^* \Rightarrow \alpha \cdot B(G)$*, where* B *is a Brownian bridge in* $C[0,1]$*, independent of* α.

Consider the process $\mathcal{U}_h(t) := U_h^*(G^{-1}(t)), 0 \leq t \leq 1$. Now work with the metric $|t-s|^{1/2}$ on $[0,1]$. Upon repeating the arguments in the proof of the above theorem, modified appropriately, one can readily conclude the following

Corollary 2.2.4 *Let the setup of Theorem 2.2.3 hold. Assume that* G *is continuous and that (2.2.27), (2.2.34) hold. Then* $\{\mathcal{U}_h\} \Longrightarrow \alpha.B$*, where* B *is a Brownian Bridge in* $C[0,1]$*, independent of* α.

Remark 2.2.4 Suppose that in Theorem 2.2.2 the r.v.'s $\eta_{n1}, \cdots \eta_{nn}$ are i.i.d. Uniform $[0,1]$. Then, upon choosing $h_{ni} \equiv n^{1/2}d_{ni}$, $\zeta_{ni} \equiv G^{-1}(\eta_{ni})$, one sees that $U_h \equiv W_d(G)$, provided G is continuous. Moreover the condition (**D**) is *a priori* satisfied, (**B**) is equivalent to (2.2.27) and (**N1**) implies (2.2.34) trivially. Consequently, for this special setup, Theorem 2.2.2 is a special case of Corollary 2.2.4. But in general these two results serve different purposes. Theorem 2.2.1 is the most general for the independent setup given there and can not be deduced from Theorem 2.2.3.

The next theorem gives an analog of the above theorem when the weights are not necessarily bounded r.v.'s nor do they need to be square integrable.

Theorem 2.2.4 *Suppose the setup in (2.2.26), and the assumptions (2.2.28), (2.2.30), (2.2.31), and (2.2.34) hold. Then (2.2.32) continues to hold.*

If, in addition, (2.2.33) holds, and

(2.2.47) *For each* $n \geq 1$, $\{h_{ni};\ 1 \leq i \leq n\}$ *are square integrable,*

then (2.2.35) continues to hold.

A proof of this theorem appears in Koul and Ossiander (1994). This proof uses the ideas of bracketing and chaining and is a bit involved, hence not included here. We are rather interested in its numerous applications to time series models as will be seen later in the Chapters 7 and 8 of this monograph.

The next theorem gives an analogue of the above two results useful in the presence of heteroscedasticity. Accordingly, let τ_{ni}, $1 \leq i \leq n$ be another array of r.v.'s independent of ζ_{ni}, for each $1 \leq i \leq n$. Let $\{\mathcal{A}_{ni}\}$ be an array of sub σ-fields such that $\mathcal{A}_{ni} \subset \mathcal{A}_{n,i+1}$, $1 \leq i \leq n$, $n \geq 1$, $(h_{n1}, \delta_{n1}, \tau_{n1})$ is $\mathcal{A}_{n1}$ - measurable; the r.v.'s $\{\zeta_{n1}, \cdots, \zeta_{n,j-1}; (h_{ni}, \delta_{ni}, \tau_{ni}),\ 1 \leq i \leq j\}$ are $\mathcal{A}_{nj}$ - measurable, $2 \leq j \leq n$; and ζ_{nj} is independent of $\mathcal{A}_{nj}$, $1 \leq j \leq n$. Define,

(2.2.48)
$$\tilde{V}_h(x) := n^{-1/2} \sum_{i=1}^{n} h_{ni} I(\zeta_{ni} \leq x + x\tau_{ni} + \delta_{ni}),$$
$$\tilde{J}_h(x) := n^{-1/2} \sum_{i=1}^{n} h_{ni} G(x + x\tau_{ni} + \delta_{ni}),$$
$$\tilde{U}_h(x) := \tilde{V}_h(x) - \tilde{J}_h(x), \qquad x \in \mathbb{R}.$$

Next, we introduce the additional needed assumptions.

(2.2.49) G has Lebesgue density g satisfying the following: $g > 0$ on the set $\{x : 0 < G(x) < 1\}$, $g(G^{-1}(u))$ is uniformly continuous in $0 \le u \le 1$,

(2.2.50) $\sup_{x \in \mathbb{R}} (1 + |x|) g(x) < \infty$,

(2.2.51) $\limsup_{u \to 0} \sup_{x \in \mathbb{R}} |xg(x(1+u)) - xg(x)| = 0.$

(2.2.52) $E\left(n^{-1} \sum_{i=1}^{n} h_{ni}^4\right) = O(1).$

(2.2.53) $\max_{1 \le i \le n} |\tau_{ni}| = o_p(1).$

(2.2.54) $n^{1/2} E\left[n^{-1} \sum_{i=1}^{n} \{h_{ni}^2 (|\delta_{ni}| + |\tau_{ni}|)\}\right]^2 = o(1).$

(2.2.55) $n^{-1/2} \sum_{i=1}^{n} |h_{ni} \tau_{ni}| = O_p(1).$

Remark 2.2.5 Note that (2.2.50) implies that for some constant $0 < C < \infty$,

(2.2.56) $D(s) := \sup_{x \in \mathbb{R}} |G(x + xs) - G(x)| \le C\,|s|, \qquad \forall\ s \in \mathbb{R}.$

Clearly, $D(s) \le 1 \le 2|s|$, for all $|s| \ge 1/2$. Now consider $-1/2 < s < 0$. In this case $-\log(1 + s) \le -2s = 2|s|$, and

$$D(s) = \left| \int_x^{x(1+s)} yg(y) \frac{1}{y} dy \right| \le \sup_{y \in \mathbb{R}} |y| g(y) \{-\log(1+s)\} \le C|s|.$$

A similar and somewhat simpler argument proves the claim in the case $0 < s < 1/2$, using the fact $\log(1 + s) \le s$. □

Theorem 2.2.5 *Suppose the setup in (2.2.26), and the assumptions (2.2.28), (2.2.30) and (2.2.49)-(2.2.55) hold. Then,*

(2.2.57) $$\sup_{x \in \mathbb{R}} |\tilde{U}_h(x) - U_h^*(x)| = o_p(1).$$

Suppose $\tau_{ni} \equiv 0$. Then (2.2.57) is equivalent to the conclusion (2.2.32) of Theorem 2.2.4. But the conditions of Theorem 2.2.5 are clearly much stronger than those needed by Theorem 2.2.4. Theorem 2.2.4 gives the best known result of this type useful in homoscedastic autoregressive (AR) time series models. Theorem 2.2.5 gives on the other hand a similar result useful in autoregressive conditionally heteroscedastic (ARCH) time series models.

As will be seen in the Chapters 7 and 8 of this monograph, these results are used to obtain the asymptotic uniform linearity of the analogues of M and R scores, and of the empirical process of the residuals in standardized underlying parameters in a large class of AR and ARCH models. The latter result about the empirical processes in turn is useful in variety of inference problems in these models, including the investigation of the consistency of the error density estimates. Theorems 2.2.4 and 2.2.5 are also useful in obtaining the asymptotic uniform quadraticity result of a certain class of minimum distance scores in these models.

Proof of Theorem 2.2.5. Assume without loss of generality that all h_{ni} are non-negative. Next, write $\tilde{U}_h(x) = \tilde{U}_h^+(x) + \tilde{U}_h^-(x)$, where $\tilde{U}_h^+(x)$, $\tilde{U}_h^-(x)$ correspond to that part of the sum in $\tilde{U}_h(x)$ which has $\tau_{ni} \geq 0$, $\tau_{ni} < 0$, respectively. Decompose $U_h^*(x)$ similarly. It thus suffices to show that

$$\sup_{x\in\mathbb{R}} |\tilde{U}_h^+(x) - U_h^{*+}(x)| = o_p(1), \tag{2.2.58}$$

$$\sup_{x\in\mathbb{R}} |\tilde{U}_h^-(x) - U_h^{*-}(x)| = o_p(1). \tag{2.2.59}$$

Details will be given only for (2.2.58), they being similar for (2.2.59).

Now, fix a $\delta > 0$ and let $-\infty = x_0 < x_1 < \cdots < x_{r_n-1} < x_{r_n} = \infty$ be a partition of $\mathbb{R}$ where $x_j = G^{-1}(j\delta/n^{1/2})$, $0 \leq j \leq r_n - 1$ and $r_n := [n^{1/2}/\delta] + 1$. Note that

$$n^{1/2}[G(x_j) - G(x_{j-1})] \leq \delta, \ \forall 1 \leq j \leq r_n. \tag{2.2.60}$$

The dependence of x_j's on n is suppressed for the sake of convenience. Also, in the proof below C is a generic constant, possibly different in different context, but never depending on n or the $\{x_j\}$.

Using the monotonicity of the indicator function and the d.f. G, we obtain that for $x_{j-1} < x \leq x_j$,

$$
\begin{aligned}
(2.2.61)\ |\tilde{U}_h^+(x) - U_h^{*+}(x)| \\
\leq\ & |\tilde{U}_h^+(x_j) - U_h^{*+}(x_{j-1})| + |\tilde{U}_h^+(x_{j-1}) - U_h^{*+}(x_j)| \\
& + 2\,|\tilde{J}_h^+(x_j) - \tilde{J}_h^+(x_{j-1})| + 2\,|J_h^{*+}(x_j) - J_h^{*+}(x_{j-1})| \\
=\ & \mathcal{B}_{nj,1} + \mathcal{B}_{nj,2} + 2\mathcal{B}_{nj,3} + 2\mathcal{B}_{nj,4}, \quad \text{say.}
\end{aligned}
$$

Note that by (2.2.52),

$$
\max_{1 \leq j \leq r_n} \mathcal{B}_{nj,4} \leq n^{-1} \sum_{i=1}^{n} |h_{ni}|\, \delta = O_p(\delta).
$$

Next, consider $\mathcal{B}_{nj,1}$. For the sake of brevity, let $t_{ni} = \tau_{ni} + 1$. Then, one can rewrite

$$
\begin{aligned}
\mathcal{B}_{nj,1} \;=\; & n^{-1/2} \sum_{i=1}^{n} h_{ni} \Big\{ I(\zeta_{ni} \leq x_j t_{ni} + \delta_{ni}) - I(\zeta_{ni} \leq x_{j-1}) \\
& \qquad - G(x_j t_{ni} + \delta_{ni}) + G(x_{j-1}) \Big\},
\end{aligned}
$$

which is a sum of martingale differences.

We shall apply the inequality (9.1.4) given in the Appendix below, with $p = 4$, $\mathcal{D}_i = \mathcal{A}_{ni}$, and

$$
\begin{aligned}
D_i \;\equiv\; & n^{-1/2} h_{ni} \Big\{ I(\zeta_{ni} \leq x_j t_{ni} + \delta_{ni}) - I(\zeta_{ni} \leq x_{j-1}) \\
& \qquad - G(x_j t_{ni} + \delta_{ni}) + G(x_{j-1}) \Big\}.
\end{aligned}
$$

Use $|D_i| \leq n^{-1/2} h_{ni}$, $\forall\, i$, and the fact

$$
E(D_i^2 | \mathcal{D}_{i-1}) \;\leq\; n^{-1} h_{ni}^2 |G(x_j t_{ni} + \delta_{ni}) - G(x_{j-1})|,
$$

to obtain, for an $\epsilon > 0$,

$$
\begin{aligned}
& P[|\mathcal{B}_{nj,1}| > \epsilon] \\
& \leq\; C\epsilon^{-4} \Big(n^{-2} \sum_{i=1}^{n} E h_{ni}^4 \\
& \qquad + E\Big[n^{-1} \sum_{i=1}^{n} h_{ni}^2 |G(x_j t_{ni} + \delta_{ni}) - G(x_{j-1})| \Big]^2 \Big).
\end{aligned}
$$

The first term in the above inequality is free from j. To deal with the second term, use the boundedness of g, (2.2.56), and a telescoping argument to obtain that for all $1 \le j \le r_n$, $1 \le i \le n$,

$$\begin{aligned} & G(x_j t_{ni} + \delta_{ni}) - G(x_{j-1}) \\ &= G(x_j) - G(x_{j-1}) + G(x_j t_{ni} + \delta_{ni}) - G(x_j t_{ni}) \\ &\quad + G(x_j t_{ni}) - G(x_j) \\ &\le C\Big[\delta n^{-1/2} + |\delta_{ni}| + |\tau_{ni}|\Big]. \end{aligned}$$

Therefore,

$$\begin{aligned} & \Big[n^{-1}\sum_{i=1}^n h_{ni}^2 |G(x_j t_{ni} + \delta_{ni}) - G(x_{j-1})|\}\Big]^2 \\ & \le C n^{-1}\delta^2\Big(n^{-1}\sum_{i=1}^n h_{ni}^2\Big)^2 + C\Big[n^{-1}\sum_{i=1}^n h_{ni}^2\{|\delta_{ni}| + |\tau_{ni}|\}\Big]^2. \end{aligned}$$

Hence, using $r_n = O(n^{1/2})$,

$$\begin{aligned} & P\Big(\max_{1\le j\le r_n} |\mathcal{B}_{nj,1}| > \epsilon\Big) \\ &\le C\Big(n^{-1/2}(1+\delta^2)\, n^{-1}\sum_{i=1}^n E h_{ni}^4 \\ &\quad + n^{1/2} E\Big[n^{-1}\sum_{i=1}^n h_{ni}^2(|\delta_{ni}| + |\tau_{ni}|)\Big]^2\Big), \end{aligned}$$

which in turn, together with (2.2.52) and (2.2.54), implies that $\max_{1\le j\le r_n} |\mathcal{B}_{nj,1}| = o_p(1)$. A similar statement holds for $\mathcal{B}_{nj,2}$.

Next, consider

$$\begin{aligned} \mathcal{B}_{nj,3} &= n^{-1/2}\sum_{i=1}^n h_{ni}\Big[G(x_j t_{ni} + \delta_{ni}) - G(x_{j-1} t_{ni} + \delta_{ni})\Big] \\ &\le n^{-1}\sum_{i=1}^n h_{ni}\delta + n^{-1/2}\sum_{i=1}^n h_{ni}\tau_{ni}[x_j g(x_j) - x_{j-1} g(x_{j-1})] \\ &\quad + n^{-1/2}\sum_{i=1}^n h_{ni}[G(x_j t_{ni} + \delta_{ni}) - G(x_j t_{ni}) - \delta_{ni} g(x_j t_{ni})] \\ &\quad + n^{-1/2}\sum_{i=1}^n h_{ni}[G(x_j t_{ni}) - G(x_j) - \tau_{ni} x_j g(x_j)] \end{aligned}$$

$$- n^{-1/2} \sum_{i=1}^{n} h_{ni}[G(x_{j-1}t_{ni}) - G(x_{j-1}) \\ -\tau_{ni}x_{j-1}g(x_{j-1})]$$

$$-n^{-1/2} \sum_{i=1}^{n} h_{ni}[G(x_{j-1}t_{ni} + \delta_{ni}) - G(x_{j-1}t_{ni}) \\ -\delta_{ni}g(x_{j-1}t_{ni})]$$

$$+n^{-1/2} \sum_{i=1}^{n} h_{ni}\delta_{ni}[g(x_j t_{ni}) - g(x_{j-1}t_{ni})].$$

Now, let $m_n := \max_{1 \le i \le n} |\delta_{ni}|$, $\mu_n := \max_{1 \le i \le n} |\tau_{ni}|$. Note that, uniformly in $1 \le j \le r_n$, the sum of the absolute values of the third and sixth term in the above bound is bounded above by

$$C n^{-1/2} \sum_{i=1}^{n} |h_{ni}\delta_{ni}| \sup_{|x-y| \le m_n} |g(x) - g(y)| = o_p(1),$$

by (2.2.30) and the uniform continuity of g, implied by the the boundedness of g and the uniform continuity of $g(G^{-1})$ on $[0, 1]$.

Next, we bound the fourth term; the fifth term can be handled similarly. Clearly, the absolute value of the fourth term is bounded above by

$$n^{-1/2} \sum_{i=1}^{n} |h_{ni}\tau_{ni}x_j| \int_0^1 |g(x_j + tx_j\tau_{ni}) - g(x_j)| dt$$

$$\le n^{-1/2} \sum_{i=1}^{n} |h_{ni}\tau_{ni}| \int_0^1 \sup_{x \in \mathbb{R}} \{|x||g(x + tx\delta) - g(x)|\} \, dt = o_p(1),$$

by (2.2.51) and (2.2.55).

Finally, consider the seventh term; the second term can be dealt with similarly. To begin with observe that by (2.2.56),

$$\max_{1 \le i \le n, 1 \le j \le r_n} |G(x_j t_{ni}) - G(x_j)| \le C \max_{1 \le i \le n} |\tau_{ni}|.$$

Hence by the uniform continuity of $g(G^{-1})$ assured by (2.2.49), and by (2.2.53),

$$\max_{1 \le i \le n, 1 \le j \le r_n} |g(x_j t_{ni}) - g(x_j)| \\ = \max_{1 \le i \le n, 1 \le j \le r_n} |g(G^{-1}(G(x_j t_{ni}))) - g(G^{-1}(G(x_j)))| = o_p(1).$$

Upon combining all these bounds and using $E(n^{-1}\sum_{i=1}^n h_{ni}) = O(1)$, we obtains

$$\max_{1\le j\le m} |\mathcal{B}_{nj,3}| \le \delta\, O_p(1) + o_p(1).$$

All of the above facts together with the arbitrariness of δ thus imply (2.2.57), thereby completing the proof of Theorem 2.2.5. □

Note: The inequality (2.2.39) and its proof has its origins in the papers of Johnson, Schechtman and Zinn (1985) and Levental (1989). The proof of Theorem 2.2.3 has its roots in Levental and Koul (1989) and Koul (1991). Theorem 2.2.5 was first proved in Koul and Mukherjee (2001). □

2.2.3 α_n- processes

In this sub-section we shall prove the weak convergence of a sequence of partial sum processes useful in autoregressive model fitting. Accordingly, let X_i, $i = 0, \pm 1, \cdots$ be a strictly stationary ergodic Markov process with stationary d.f. G, and let $\mathcal{F}_i = \sigma(X_i, X_{i-1}, \ldots)$ be the σ-field generated by the observations obtained up to time i. Furthermore, let, for each $n \ge 1$, $\{Z_{ni}, 1 \le i \le n\}$ be an array of r.v.'s adapted to $\{\mathcal{F}_i\}$ such that (Z_{ni}, X_i) is strictly stationary in i, for each $n \ge 1$, and satisfying

$$E\{Z_{ni}|\mathcal{F}_{i-1}\} = 0, \qquad 1 \le i \le n. \tag{2.2.62}$$

Our goal here is to establish the weak convergence of the process

$$\alpha_n(x) = n^{-1/2}\sum_{i=1}^n Z_{ni} I(X_{i-1} \le x), \qquad x \in \mathbb{R}. \tag{2.2.63}$$

The process α_n is also a weighted empirical process. It is also called a marked empirical process of the X_{i-1}'s, the marks being given by the martingale difference array $\{Z_{ni}\}$. An example of this process is the partial sum process of the innovations where $Z_{ni} \equiv Z_i = (X_i - E(X_i|X_{i-1}))$, assuming the expectation exists. Further examples appear in Section 7.7 below.

The process α_n takes its value in the Skorokhod space $D(-\infty, \infty)$. Extend it continuously to $\pm\infty$ by putting

$$\alpha_n(-\infty) = 0 \text{ and } \alpha_n(+\infty) = n^{-1/2}\sum_{i=1}^n Z_{ni}.$$

Then α_n becomes a process in $\mathcal{D}[-\infty, \infty]$.

We now formulate the assumptions that guarantee the weak convergence of α_n to a continuous limit in $\mathcal{D}[-\infty, \infty]$. To this end, let

$$L_y(x) := P(X_1 - \varphi(X_0) \le x | X_0 = y), \quad x, y \in \mathbb{R},$$

where φ is a real valued measurable function. Because of the Markovian assumption, L_{X_0} is the d.f. of $X_1 - \varphi(X_0)$, given $\mathcal{F}_0$. For example, if $\{X_i\}$ is integrable, we may take $\varphi(x) \equiv \mathbb{E}[X_{i+1} \mid X_i = x]$, so that $\varphi_{i+1} := X_{i+1} - \varphi(X_i)$ are just the innovations generating the process $\{X_i\}$. In the context of time series analysis the innovations are often i.i.d. in which case L_y does not depend on y. However, for our general result of this section, we may let L_y depend on y.

This section contains two Theorems. Theorem 2.2.6 deals with the general weights Z_{ni} satisfying some moment conditions and Theorem 2.2.7 with the bounded Z_{ni}. The following assumptions will be needed in the first theorem:

For some $\eta > 0$, $\delta > 0$, $K < \infty$ and for all n sufficiently large:

$$EZ_{n1}^4 \le K, \tag{2.2.64}$$

$$EZ_{n1}^4 |X_0|^{1+\eta} \le K, \tag{2.2.65}$$

$$E\Big\{Z_{n2}^2 Z_{n1}^2 |X_1|\Big\}^{1+\delta} \le K. \tag{2.2.66}$$

There exists a function φ from $\mathbb{R}$ to $\mathbb{R}$ such that the corresponding family of functions $\{L_y, \, y \in \mathbb{R}\}$ admit Lebesgue densities l_y which are uniformly bounded:

$$\sup_{x,y} l_y(x) \le K < \infty. \tag{2.2.67}$$

There exists a continuous nondecreasing function τ^2 on $\mathbb{R}$ to $[0, \infty)$ such that $\forall \, x \in [-\infty, \infty]$,

$$n^{-1} \sum_{i=1}^{n} E[Z_{ni}^2 | \mathcal{F}_{i-1}] \, I(X_{i-1} \le x) = \tau^2(x) + o_p(1). \tag{2.2.68}$$

The assumptions (2.2.64) - (2.2.67) are needed to guarantee the continuity of the weak limit and the tightness of the process α_n, while (2.2.68) is needed to identify the weak limit. Here, B denotes the Brownian motion on $[0, \infty)$.

Theorem 2.2.6 *Under (2.2.64) - (2.2.68),*

$$\alpha_n \Longrightarrow B \circ \tau^2 \text{ in the space } \mathcal{D}[-\infty, \infty], \tag{2.2.69}$$

where $B \circ \tau^2$ is a continuous Brownian motion on $\mathbb{R}$ with respect to time τ^2.

Proof. For convenience, we shall now not exhibit the dependence of Z_{ni} on n. Apply the CLT for martingales given in Lemma 9.1.3 of the Appendix, to show that all finite dimensional distributions converge weakly to the right limit, under (2.2.64) and (2.2.68).

As to tightness, fix $-\infty \le t_1 < t_2 < t_3 \le \infty$ and assume, without loss of generality, that the moment bounds of (2.2.64)- (2.2.66) hold for all $n \ge 1$ with $K \ge 1$. Then we have

$$\begin{aligned}
&[\alpha_n(t_3) - \alpha_n(t_2)]^2[\alpha_n(t_2) - \alpha_n(t_1)]^2 \\
&= n^{-2}\Big[\sum_{i=1}^{n} Z_i I(t_2 < X_{i-1} \le t_3)\Big]^2\Big[\sum_{i=1}^{n} Z_i I(t_1 < X_{i-1} \le t_2)\Big]^2 \\
&= n^{-2}\sum_{i,j,k,l} \xi_i\xi_j\zeta_k\zeta_l,
\end{aligned}$$

with $\xi_i = Z_i I(t_2 < X_{i-1} \le t_3)$, $\zeta_i = Z_i I(t_1 < X_{i-1} \le t_2)$. Now, if the largest index among i, j, k, l is not matched by any other, then $\mathbb{E}\{\xi_i\xi_j\zeta_k\zeta_l\} = 0$. Also, since the two intervals $(t_2, t_3]$ and $(t_1, t_2]$ are disjoint, $\xi_i\zeta_i \equiv 0$. We thus obtain

$$\begin{aligned}
&E\Big\{n^{-2}\sum_{i,j,k,l} \xi_i\xi_j\zeta_k\zeta_l\Big\} \\
&\quad = n^{-2}\sum_{i,j<k} E\{\zeta_i\zeta_j\xi_k^2\} + n^{-2}\sum_{i,j<k} E\{\xi_i\xi_j\zeta_k^2\}.
\end{aligned} \tag{2.2.70}$$

The moment assumption (2.2.64) guarantees that the above expectations exist. In this proof, the constant K is a generic constant, which may vary from expression to expression but never depends on n or the chosen t's.

We shall only bound the first sum in (2.2.70), the second being dealt with similarly. To this end, fix a $2 \le k \le n$ for the time being and write

$$(2.2.71)\quad \sum_{i,j<k} E\{\zeta_i\zeta_j\xi_k^2\}$$

$$= E\Big\{\Big(\sum_{i=1}^{k-1}\zeta_i\Big)^2\xi_k^2\Big\} = E\Big\{\Big(\sum_{i=1}^{k-1}\zeta_i\Big)^2\mathbb{E}(\xi_k^2|\mathcal{F}_{k-1})\Big\}$$

$$\leq 2E\Big\{\Big(\sum_{i=1}^{k-2}\zeta_i\Big)^2 E(\xi_k^2|\mathcal{F}_{k-1})\Big\} + 2E\Big\{\zeta_{k-1}^2 E(\xi_k^2|\mathcal{F}_{k-1})\Big\}.$$

The first expectation equals

$$(2.2.72)\qquad E\Big\{\Big(\sum_{i=1}^{k-2}\zeta_i\Big)^2 I(t_2 < X_{k-1} \leq t_3)\mathbb{E}(Z_k^2|\mathcal{F}_{k-1})\Big\}.$$

Write

$$E(Z_k^2|\mathcal{F}_{k-1}) = r(X_{k-1}, X_{k-2}, \ldots)$$

for an appropriate function r. Note that due to stationarity r is the same for each k. Condition on $\mathcal{F}_{k-2}$ and use the Markov property and Fubini's Theorem to show that (2.2.72) is the same as

$$E\Big\{\Big(\sum_{i=1}^{k-2} V_i\Big)^2 \int_{t_2}^{t_3} r(x, X_{k-2}, \ldots)\; l_{X_{k-2}}(x - \varphi(X_{k-2}))dx\Big\}$$

$$= \int_{t_2}^{t_3} \mathbb{E}\Big\{\Big(\sum_{i=1}^{k-2} V_i\Big)^2 r(x, X_{k-2}, \ldots)\; l_{X_{k-2}}(x - \varphi(X_{k-2}))\Big\}dx$$

$$\leq \int_{t_2}^{t_3} \Big\{E\Big(\sum_{i=1}^{k-2}\zeta_i\Big)^4\Big\}^{\frac{1}{2}}$$

$$\times\Big\{E\Big(r(x, X_{k-2}, \cdots)l_{X_{k-2}}(x - \varphi(X_{k-2}))\Big)^2\Big\}^{\frac{1}{2}}dx,$$

where the last inequality follows from the Cauchy-Schwarz inequality. Since the ζ_i's form a centered martingale difference array, Burkholder's inequality (Chow and Teicher: 1978, p. 384) and the moment inequality yield

$$E\Big(\sum_{i=1}^{k-2}\zeta_i\Big)^4 \leq K\, E\Big(\sum_{i=1}^{k-2}\zeta_i^2\Big)^2 \leq K\,(k-2)^2\, E\zeta_1^4.$$

We also have

$$EV_1^4 = E\big(Z_1^4\, I(t_1 < X_0 \le t_2)\big) = [L_1(t_2) - L_1(t_1)],$$

where

$$L_1(t) = EZ_1^4 I(X_0 \le t), \qquad -\infty \le t \le \infty.$$

Let, for $-\infty \le t \le \infty$,

$$L_2(t) = \int_{-\infty}^{t} \Big\{E\Big(r(x, X_{k-2}, \cdots)\; l_{X_{k-2}}(x - \varphi(X_{k-2}))\Big)^2\Big\}^{\frac{1}{2}} dx.$$

Note that due to stationarity, L_2 is the same for each k. It thus follows that (2.2.72) is bounded from the above by

$$(2.2.73) \qquad K\ (k-2)\ [L_1(t_2) - L_1(t_1)]^{\frac{1}{2}}\ [L_2(t_3) - L_2(t_2)].$$

The assumption (2.2.64) implies that L_1 is a continuous nondecreasing bounded function on the real line. Clearly, L_2 is also nondecreasing and continuous. We shall now show that $L_2(\infty)$ is finite. For this, let h be a strictly positive continuous Lebesgue density on the real line such that $h(x) \sim |x|^{-1-\eta}$ as $x \to \pm\infty$, where η is as in (2.2.65). By Hölder's inequality,

$$L_2(\infty) \le \Big[\int_{-\infty}^{\infty} E\Big(r(x, X_{k-2}, \ldots)\; l_{X_{k-2}}(x - \varphi(X_{k-2}))\Big)^2 h^{-1}(x) dx\,\Big]^{\frac{1}{2}}.$$

Use the assumption (2.2.67) to bound one power of $l_{X_{k-2}}$ from the above so that the last integral in turn is less than or equal to

$$K\, E\Big\{r^2(X_{k-1}, X_{k-2}, \cdots)\; h^{-1}(X_{k-1})\Big\} \le K\, E\Big\{Z_1^4\; |h^{-1}(X_0)|\Big\}.$$

The finiteness of the last expectation follows, however, from assumption (2.2.65).

We now bound the second expectation in (2.2.70). Since Z_{k-1} is measurable w.r.t. $\mathcal{F}_{k-1}$, there exists some function s such that

$$Z_{k-1}^2 = s(X_{k-1}, X_{k-2}, \cdots).$$

Put $u = rs$ with r as before. Then we have, with the δ as in (2.2.66),

$$\begin{aligned}
&E\Big\{\zeta_{k-1}^2 E(\xi_k^2|\mathcal{F}_{k-1})\Big\} \\
&= E\Big\{I(t_2 < X_{k-1} \le t_3) I(t_1 < X_{k-2} \le t_2)\ u(X_{k-1}, X_{k-2}, \cdots)\Big\} \\
&= \int_{t_2}^{t_3} E\Big\{I(t_1 < X_{k-2} \le t_2)\ u(x, X_{k-2}, \cdots) \\
&\qquad\qquad \times l_{X_{k-2}}(x - \varphi(X_{k-2}))\Big\} dx \\
&\le [G(t_2) - G(t_1)]^{\frac{\delta}{1+\delta}} \\
&\qquad \times \int_{t_2}^{t_3} E^{\frac{1}{1+\delta}}\Big\{u^{1+\delta}(x, X_{k-2}, \cdots) l_{X_{k-2}}^{1+\delta}(x - \varphi(X_{k-2}))\Big\} dx.
\end{aligned}$$

Put, for $-\infty \le t \le \infty$,

$$L_3(t) = \int_{-\infty}^{t} E^{\frac{1}{1+\delta}}\Big\{u^{1+\delta}(x, X_{k-2}, \ldots) l_{X_{k-2}}^{1+\delta}(x - \varphi(X_{k-2}))\Big\} dx.$$

Arguing as above, now let q be a positive continuous Lebesgue density on the real line such that $q(x) \sim |x|^{-1-\frac{1}{\delta}}$ as $x \to \pm\infty$. By Hölder's inequality,

$$\begin{aligned}
&L_3(\infty) \\
&\le \Big[\int_{-\infty}^{\infty} E\Big\{u^{1+\delta}(x, X_{k-2}, \ldots)\ l_{X_{k-2}}^{1+\delta}(x - \varphi(X_{k-2}))\Big\} q^{-\delta}(x) dx\Big]^{\frac{1}{1+\delta}} \\
&\le K\ \Big\{E\Big(u^{1+\delta}(X_{k-1}, \ldots)\ q^{-\delta}(X_{k-1})\Big)\Big\}^{\frac{1}{1+\delta}} \\
&\le K\ \Big\{E\Big(Z_2^{2(1+\delta)} Z_1^{2(1+\delta)} q^{-\delta}(X_1)\Big)\Big\}^{\frac{1}{1+\delta}},
\end{aligned}$$

where the last inequality follows from Hölder's inequality applied to conditional expectations. The last expectation is, however, finite by assumption (2.2.66). Thus L_3 is also a nondecreasing continuous bounded function on the real line and we obtain

$$E\Big\{\zeta_{k-1}^2 E\{\xi_k^2|\mathcal{F}_{k-1}\}\Big\} \le K\ [G(t_2) - G(t_1)]^{\frac{\delta}{1+\delta}} [L_3(t_3) - L_3(t_2)].$$

Upon combining this with (2.2.70) and (2.2.73) and summing over $k = 2$ to $k = n$, we obtain

$$
\begin{aligned}
& n^{-2} \sum_{i,j<k} E\{\zeta_i \zeta_j \xi_k^2\} \\
& \leq K \Big\{ [L_1(t_2) - L_1(t_1)]^{\frac{1}{2}} \, [L_2(t_3) - L_2(t_2)] \\
& \qquad + [G(t_2) - G(t_1)]^{\frac{\delta}{1+\delta}} \, [L_3(t_3) - L_3(t_2)] \Big\}.
\end{aligned}
$$

One has a similar bound for the second sum in (2.2.70). Thus summarizing we see that the sums in (2.2.70) satisfies Chentsov's criterion for tightness. For relevant details see Billingsley (1968: Theorem 15.6). This completes the proof of Theorem 2.2.6. □

The next theorem covers the case of uniformly bounded $\{Z_{ni}\}$. In this case we can avoid the moment conditions (2.2.65) and (2.2.66) and replace the condition (2.2.67) by a weaker condition.

Theorem 2.2.7 *Suppose the r.v.'s $\{Z_{ni}\}$ are uniformly bounded and (2.2.68) holds. In addition, suppose there exists a measurable function φ from $\mathbb{R}$ to $\mathbb{R}$ such that the corresponding family of functions $\{L_y,\, y \in \mathbb{R}\}$ has Lebesgue densities $\{l_y, y \in \mathbb{R}\}$ satisfying*

$$
\int \Big[E\, l_{X_0}^{1+\delta}(x - \varphi(X_0)) \Big]^{\frac{1}{1+\delta}} \, dx < \infty, \tag{2.2.74}
$$

for some $\delta > 0$. Then also the conclusion (2.2.69) holds.

Proof. Proceed as in the proof of the previous theorem up to (2.2.71). Now use the boundedness of $\{Z_k\}$ and argue as for (2.2.73) to conclude that (2.2.71) is bounded from the above by

$$
\begin{aligned}
& K \int_{t_2}^{t_3} E\Big\{ \Big(\sum_{i=1}^{k-2} \zeta_i \Big)^2 \, l_{X_{k-2}}\Big(x - \varphi(X_{k-2}) \Big) \Big\} \, dx \\
& \leq K \int_{t_2}^{t_3} \Big(E\Big| \sum_{i=1}^{k-2} \zeta_i \Big|^{2\frac{1+\delta}{\delta}} \Big)^{\frac{\delta}{1+\delta}} \Big\{ E l_{X_0}^{1+\delta}(x - \varphi(X_0)) \Big\}^{\frac{1}{1+\delta}} \, dx \\
& \leq K\, (k-2)\, [G(t_2) - G(t_1)]^{\frac{\delta}{1+\delta}} \, [\Gamma(t_3) - \Gamma(t_2)],
\end{aligned}
$$

with δ as in (2.2.74). Here

$$\Gamma(t) = \int_{-\infty}^{t} \Big\{E\, l_{X_0}^{1+\delta}(x - \varphi(X_0))\Big\}^{\frac{1}{1+\delta}} dx, \qquad -\infty \le t \le \infty,$$

Note that (2.2.74) implies that Γ is strictly increasing, continuous and bounded on $\mathbb{R}$. Similarly

$$\begin{aligned}
&E\Big\{\zeta_{k-1}^2\, E(\xi_k^2|\mathcal{F}_{k-1})\Big\} \\
&\le\; K\, E\Big\{\zeta_{k-1}^2\, I(t_2 < X_{k-1} \le t_3)\Big\} \\
&=\; E\Big\{I(t_1 < X_{k-2} \le t_2)\, E\Big[Z_{k-1}^2\, I(t_2 < X_{k-1} \le t_3)\Big|\mathcal{F}_{k-2}\Big]\Big\} \\
&\le\; K\, E\Big\{I(t_1 < X_{k-2} \le t_2)\Big[\int_{t_2}^{t_3} l_{X_{k-2}}(x - \varphi(X_{k-2}))\, dx\Big]\Big\} \\
&=\; K \int_{t_2}^{t_3} E\Big\{I(t_1 < X_{k-2} \le t_2)\, l_{X_{k-2}}(x - \varphi(X_{k-2}))\Big\}\, dx \\
&\le\; K\, [G(t_2) - G(t_1)]^{\frac{\delta}{1+\delta}}\, [\Gamma(t_3) - \Gamma(t_2)].
\end{aligned}$$

Upon combining the above bounds we obtain that (2.2.70) is bounded from the above by

$$K(k-1)[G(t_2) - G(t_1)]^{\frac{\delta}{1+\delta}}\, [\Gamma(t_3) - \Gamma(t_2)].$$

Summation from $k = 2$ to $k = n$ thus yields

$$n^{-2} \sum_{i,j<k} \mathbb{E}\{\zeta_i \zeta_j \xi_k^2\} \le K\, [G(t_2) - G(t_1)]^{\frac{\delta}{1+\delta}}\, [\Gamma(t_3) - \Gamma(t_2)].$$

The rest of the details are as in the proof of Theorem 2.2.4. □

Note: Theorems 2.2.6 and 2.2.7 first appeared in Koul and Stute (1999).

2.3 AUL of Residual W.E.P.'s

In this section we shall obtain the AUL (asymptotic uniform linearity) of residual W.E.P.'s. It will be observed that the asymptotic continuity property of the type specified in Theorem 2.2.1(i) is the basic tool to obtain this result. Accordingly let $\{X_{ni}\}, \{F_{ni}\}, \{H\}$ and $\{L_{ni}\}$ be as in (2.2.23) and define

$$
\begin{aligned}
(2.3.1)\quad S_d(t, \mathbf{u}) &:= \sum_{i=1}^{n} d_{ni} I(X_{ni} \le H^{-1}(t) + \mathbf{c}'_{ni}\mathbf{u}), \\
\mu_d(t, \mathbf{u}) &:= \sum_{i=1}^{n} d_{ni} F_{ni}(H^{-1}(t) + \mathbf{c}'_{ni}\mathbf{u}), \\
Y_d(t, \mathbf{u}) &:= S_d(t, \mathbf{u}) - \mu_d(t, \mathbf{u}), \quad 0 \le t \le 1, \ \mathbf{u} \in \mathbb{R}^p,
\end{aligned}
$$

where $\{\mathbf{c}_{ni}, 1 \le i \le n\}$ are $p \times 1$ vectors of real numbers. We also need

$$
\begin{aligned}
(2.3.2)\quad S_d^0(x, \mathbf{u}) &:= \sum_{i=1}^{n} d_{ni} I(X_{ni} \le x + \mathbf{c}'_{ni}\mathbf{u}), \\
\mu_d^0(x, \mathbf{u}) &:= \sum_{i=1}^{n} d_{ni} F_{ni}(x + \mathbf{c}'_{ni}\mathbf{u}), \\
Y_d^0(x, \mathbf{u}) &:= S_d^0(x, \mathbf{u}) - \mu_d^0(x, \mathbf{u}), \quad -\infty \le x \le \infty, \ \mathbf{u} \in \mathbb{R}^p.
\end{aligned}
$$

Clearly, if H is strictly increasing then $S_d^0(x, \mathbf{u}) \equiv S_d(H(x), \mathbf{u})$. Similar remark applies to other functions.

Throughout the text, any W.E.P. with weights $d_{ni} \equiv n^{-1/2}$ will be indicated by the subscript 1. Thus, e.g., $\forall -\infty \le x \le \infty$, $\mathbf{u} \in \mathbb{R}^p$,

$$
\begin{aligned}
(2.3.3)\, S_1^0(x, \mathbf{u}) &= n^{-1/2} \sum_{i=1}^{n} I(X_{ni} \le x + \mathbf{c}'_{ni}\mathbf{u}), \\
Y_1^0(x, \mathbf{u}) &= n^{-1/2} \sum_{i=1}^{n} \{I(X_{ni} \le x + \mathbf{c}'_{ni}\mathbf{u}) - F_{ni}(x + \mathbf{c}'_{ni}\mathbf{u})\}.
\end{aligned}
$$

Theorem 2.3.1 *In addition to (2.2.24), (**N1**), (**N2**), and (**C***) assume that d.f.'s $\{F_{ni}, 1 \le i \le n\}$ have densities $\{f_{ni}, 1 \le i \le n\}$ w.r.t. λ such that the following hold:*

$$(2.3.4)\qquad \lim_{\delta \to 0} \limsup_{n} \max_{1 \le i \le n} \sup_{|x-y| \le \delta} |f_{ni}(x) - f_{ni}(y)| = 0,$$

$$(2.3.5)\qquad \max_{i,n} \|f_{ni}\|_\infty \le k < \infty.$$

In addition, assume that

$$\max_{1 \le i \le n} \|\mathbf{c}_{ni}\| = o(1), \tag{2.3.6}$$

$$\sum_{i=1}^{n} \|d_{ni}\mathbf{c}_{ni}\| = O(1). \tag{2.3.7}$$

Then, for every $0 < b < \infty$,

$$\sup |S_d(t, \mathbf{u}) - S_d(t, \mathbf{0}) - \mathbf{u}' \sum_{i=1}^{n} d_{ni}\mathbf{c}_{ni}q_{ni}(t)| = O_p(1), \tag{2.3.8}$$

where $q_{ni} := f_{ni}H^{-1}$, $1 \le i \le n$, *and the supremum is taken over* $0 \le t \le 1$, $\|u\| \le b$.

Consequently, if H *is strictly increasing on* $\mathbb{R}$, *then*

$$\sup |S_d^0(x, \mathbf{u}) - S_d^0(x, 0) - \mathbf{u}' \sum_{i=1}^{n} d_{ni}\mathbf{c}_{ni}f_{ni}(x)| = o_p(1). \tag{2.3.9}$$

where the supremum is taken over $-\infty \le x \le \infty$, $\|u\| \le b$.

Theorem 2.3.1 is a consequence of the following *four* lemmas. In these lemmas the setup is as in the theorem.

In what follows, $\sup_{t,\mathbf{u}}$ stands for the supremum over $0 \le t \le 1$ and $\|\mathbf{u}\| \le b$, unless mentioned otherwise. Let, for $0 \le t \le 1$, $\mathbf{u} \in \mathbb{R}^p$,

$$\boldsymbol{\nu}_d(t) := \sum_{i=1}^{n} d_{ni}\mathbf{c}_{ni}q_{ni}(t) \tag{2.3.10}$$

$$R_d(t, \mathbf{u}) := S_d(t, \mathbf{u}) - S_d(t, 0) - \mathbf{u}'\boldsymbol{\nu}_d(t).$$

Lemma 2.3.1 *Under (2.3.4) - (2.3.7),*

$$\sup_{t,\mathbf{u}} |\mu_d(t, \mathbf{u}) - \mu_d(t, \mathbf{0}) - \mathbf{u}'\boldsymbol{\nu}_d(t)| = o(1). \tag{2.3.11}$$

Proof. Let $\delta_n = b \max_i \|\mathbf{c}_i\|$. By (2.3.4), $\{F_i\}$ are uniformly differentiable for sufficiently large n, uniformly in $1 \le i \le n$. Hence, the L.H.S. of (2.3.11) is bounded from the above by

$$\sum_{i=1}^{n} \|d_i\mathbf{c}_i\| \max_i \sup_{|x-y| \le \delta_n} |f_i(x) - f_i(y)| = o(1),$$

by (2.3.4), (2.3.6) and (2.3.7). □

Lemma 2.3.2 *Under* (**N1**), (**N2**), (**C***), (2.3.4) - (2.3.7), $\forall \|\mathbf{u}\| \leq b$,

$$\sup_{0\le t\le 1} |Y_d(t, \mathbf{u}) - Y_d(t, \mathbf{0})| = o_p(1). \tag{2.3.12}$$

Proof. Fix a $\|\mathbf{u}\| \leq b$. The lemma will follow if we show

(i) $Y_d(t, \mathbf{u}) - Y_d(t, \mathbf{0}) = o_p(1)$ for each $0 \le t \le 1$,

(ii) $\forall \epsilon > 0$, and for $\mathbf{a} = \mathbf{u}$ or $\mathbf{a} = \mathbf{0}$,

$$\lim_{\delta\to 0} \limsup_n P\left(\sup_{|t-s|<\delta} \left|Y_d(t, \mathbf{a}) - Y_d(s, \mathbf{a})\right| \ge \epsilon\right) = 0.$$

Since $Y_d(\cdot, \mathbf{0}) = W_d^*(\cdot)$ of (2.2.23), for $\mathbf{a} = \mathbf{0}$, (ii) follows from (2.2.25) of Corollary 2.2.1. To verify (ii) for $\mathbf{a} = \mathbf{u}$, take $\eta_i = H(X_i - \mathbf{c}_i'\mathbf{u})$, $1 \le i \le n$, in (2.2.1). Then $Y_d(\cdot, \mathbf{u}) \equiv W_d(\cdot)$ of (2.2.1) and $G_i(\cdot) = F_i(H^{-1}(\cdot) + \mathbf{c}_i'\mathbf{u})$, $1 \le i \le n$. Moreover, by (2.3.4)-(2.3.6),

$$\begin{aligned} &\sup_{0\le t\le 1-\delta} \sum_{i=1}^n d_i^2[F_i(H^{-1}(t+\delta) + \mathbf{c}_i'\mathbf{u}) - F_i(H^{-1} + \mathbf{c}_i'\mathbf{u})] \\ &\le\; 2bk \max_i \|\mathbf{c}_i\| + \sup_{0\le t\le 1-\delta} [L_d(t+\delta) - L_d(t)], \\ &=\; o(1), \end{aligned} \tag{2.3.13}$$

as $n \to \infty$, and then $\delta \to 0$, by (**C***). Hence, (**C**) is satisfied by the above $\{G_i\}$. The other conditions being (**N1**) and (**N2**) which are also assumed here, it follows that the above $\{\eta_i\}$ and $\{W_d\}$ satisfy the conditions of Theorem 2.2.1(i). Thus (ii) for $\mathbf{a} = \mathbf{u}$ follows from this theorem.

To **obtain (i)**, note that again by (2.3.4)-(2.3.6),

$$\begin{aligned} Var[Y_d(t, \mathbf{u}) - Y_d(t, \mathbf{0})] \;&\le\; \sum_{i=1}^n d_i^2 |F_i(H^{-1}(t) + \mathbf{c}_i'\mathbf{u}) - F_i(H^{-1}(t))| \\ &\le\; b\,k \max_i \|\mathbf{c}_i\| = o(1). \end{aligned}$$

This and the Chebychev inequality imply (i) and hence (2.3.12). □

To state and prove the next lemma we need some more notation. Let $\kappa_{ni} = \|\mathbf{c}_{ni}\|$, $1 \le i \le n$, and define for $0 \le t \le 1$, $\mathbf{u} \in \mathbb{R}^p$, $a \in \mathbb{R}$,

$$
\begin{aligned}
(2.3.14)\quad S_d^*(t,\mathbf{u},a) &= \sum_{i=1}^n d_{ni} I(X_{ni} \le H^{-1}(t) + \mathbf{c}_{ni}'\mathbf{u} + a\kappa_{ni}),\\
\mu_d^*(t,\mathbf{u},a) &= ES_d^*(t,\mathbf{u},a),\\
Y_d^*(t,\mathbf{u},a) &= S_d^*(t,\mathbf{u},a) - \mu_d^*(t,\mathbf{u},a).
\end{aligned}
$$

Lemma 2.3.3 *Under* (**N1**), (**N2**), (**C***), *(2.3.4) - (2.3.7),* $\forall\ \epsilon > 0$, $0 < b < \infty$, $|a| < \infty$ *and* $\|\mathbf{u}\| \le b$,

$$
(2.3.15)\quad \lim_{\delta\to 0}\limsup_n P\left(\sup_{|t-s|<\delta} |Y_d^*(t,\mathbf{u},a) - Y_d^*(s,\mathbf{u},a)| \ge \epsilon\right) = 0.
$$

Proof. In Theorem 2.2.1(i), take $\eta_i = H(X_i - c_i'\mathbf{u} - a\kappa_i), 1 \le i \le n$. Then $W_d(\cdot) = Y_d^*(\cdot,\mathbf{u},a)$ and $G_i(\cdot) = F_i(H^{-1}(\cdot) + \mathbf{c}_i,\mathbf{u} + a\kappa_i), 1 \le i \le n$. Again similar to (2.3.13),

$$
\begin{aligned}
\sup_{0\le t\le 1-\delta}[G_d(t+\delta) - G_d(t)] &\le 2k(b+a)\max_i \|\mathbf{c}_i\| \\
&\quad + \sup_{0\le t\le 1-\delta}[L_d(t+\delta) - L_d(t)] \\
&= o(1), \qquad \text{by (2.3.6) and } (\mathbf{C}^*).
\end{aligned}
$$

Hence (2.3.15) follows from Theorem 2.2.1(i). □

Lemma 2.3.4 *Under* (**N1**), (**N2**), (**C***), *(2.3.4)-(2.3.7),* $\forall \epsilon > 0$ *there is a* $\delta > 0$ *such that for every* $0 < b < \infty$, $\|\mathbf{v}\| \le b$,

$$
(2.3.16)\qquad \limsup_n P\left(\sup_{t,\|u-v\|\le\delta} |R_d(t,\mathbf{u}) - R_d(t,\mathbf{v})| \ge \epsilon\right) = 0,
$$

where R_d *is defined at (2.3.10).*

Proof. Assume, without loss of generality, that $d_i \ge 0, 1 \le i \le n$. For, otherwise write $d_i = d_{i+} - d_{i-}$, $1 \le i \le n$, where $\{d_{i+}, d_{i-}\}$ are as in the proof of Lemma 2.2.4. Then $S_d = \tau_{d+}S_{d+} - \tau_{d-}S_{d-}$, $R_d = \tau_{d+}R_{d+} - \tau_{d-}R_{d-}$, where $\tau_{d+}^2 = \Sigma_i(d_{i+})^2$, $\tau_{d-}^2 = \Sigma_i(d_{i-})^2$. In view of (**N1**), $\tau_{d+} \le 1, \tau_{d-} \le 1$. Moreover, if $\{d_i\}$ satisfy (**N2**) and (2.3.7) above, so do $\{d_{i+}, d_{i-}\}$ because $d_{i+}^2 \vee d_{i-}^2 = d_{i+}^2 + d_{i-}^2 = d_i^2, 1 \le i \le n$. Hence the triangle inequality will yield (2.3.16), if proved for R_{d+} and R_{d-}. But note that $d_{i+} \wedge d_{i-} \ge 0$ for all i.

Now $\|\mathbf{u} - \mathbf{v}\| \le \delta$ implies

$$(2.3.17) \quad -\delta\kappa_i + \mathbf{c}_i'\mathbf{v} \le \mathbf{c}_i'\mathbf{u} \le \delta\kappa_i + \mathbf{c}_i'\mathbf{v}, \quad \kappa_i = \|\mathbf{c}_i\|, \quad 1 \le i \le n.$$

Therefore, because $d_i \ge 0$ for all i,

$$S_d^*(t, \mathbf{v}, -\delta) \le S_d(t, \mathbf{u}) \le S_d^*(t, \mathbf{v}, \delta) \quad \text{for all } t,$$

yielding

$$\begin{aligned} (2.3.18) \qquad & L_1(t, \mathbf{u}, \mathbf{v}) \\ &:= S_d^*(t, \mathbf{v}, -\delta) - S_d(t, \mathbf{v} - (\mathbf{u} - \mathbf{v})'\boldsymbol{\nu}_d(t) \\ &\le R_d(t, \mathbf{u}) - R_d(t, \mathbf{v}) \\ &\le S_d^*(t, \mathbf{v}, \delta) - S_d(t, \mathbf{v}) - (\mathbf{u} - \mathbf{v})'\boldsymbol{\nu}_d(t) \\ &=: L_2(t, \mathbf{u}, \mathbf{v}). \end{aligned}$$

We shall show that there is a $\delta > 0$ such that for every $\|\mathbf{v}\| \le b$,

$$(2.3.19) \qquad P\left(\sup_{t, \|u - v\| \le \delta} |L_j(t, \mathbf{u}, \mathbf{v})| \ge \epsilon\right) = o(1), \quad j = 1, 2.$$

We shall prove (2.3.19) for L_2 only as it is similar for L_1. Observe that

$$\begin{aligned} (2.3.20) \quad & L_2(t, \mathbf{u}, \mathbf{v}) \\ & \le |Y_d^*(t, \mathbf{v}, \delta) - Y_d^*(t, \mathbf{v}, 0)| \\ & \quad + |\mu_d^*(t, \mathbf{v}, \delta) - \mu_d^*(t, \mathbf{v}, 0)| + |(u - \mathbf{v})'\boldsymbol{\nu}_d(t)| \end{aligned}$$

The Mean Value Theorem, (2.3.4), (2.3.5) and $\|\mathbf{u} - \mathbf{v}\| \le \delta$ imply

$$\begin{aligned} (2.3.21) \quad \sup |\mu_d^*(t, \mathbf{v}, \delta) - \mu_d^*(t, \mathbf{v}, 0)| &\le \delta k \sum_{i=1}^n \|d_i \mathbf{c}_i\|, \\ \sup_t |(\mathbf{u} - \mathbf{v})'\boldsymbol{\nu}_d(t)| &\le k\delta \sum_{i=1}^n \|d_i \mathbf{c}_i\|. \end{aligned}$$

Let $M(t)$ denote the first term on the R.H.S. of (2.3.20). I.e.,

$$M(t) = Y_d^*(t, \mathbf{v}, \delta) - Y_d^*(t, \mathbf{v}, 0), \quad 0 \le t \le 1.$$

We shall first prove that

$$\sup_t |M(t)| = o_p(1). \tag{2.3.22}$$

To begin with,

$$\begin{aligned} Var(M(t)) &\le \sum_{i=1}^n d_i^2 [F_i(H^{-1}(t) + \mathbf{c}_i'\mathbf{v} + \delta k_i) - F_i(H^{-1}(t) + \mathbf{c}_i'\mathbf{v})] \\ &\le \delta k \max_i \kappa_i, \qquad \text{by (2.3.4), (2.3.5),} \\ &= o(1), \qquad \text{by (2.3.7).} \end{aligned}$$

Hence

$$M(t) = o_p(1), \qquad \forall\, 0 \le t \le 1. \tag{2.3.23}$$

Next, note that, for a $\gamma > 0$,

$$\begin{aligned} \sup_{|t-s|<\gamma} |M(t) - M(s)| \le & \sup_{|t-s|<\gamma} |Y_d^*(t, \mathbf{v}, \delta) - Y_d^*(s, \mathbf{v}, \delta)| \\ & + \sup_{|t-s|<\gamma} |Y_d^*(t, \mathbf{v}, 0) - Y_d^*(s, \mathbf{v}, 0)|. \end{aligned}$$

Apply Lemma 2.3.3 twice, once with $a = \delta$ and once with $a = 0$, to obtain that $\forall\, \epsilon > 0$,

$$\lim_{\gamma \to 0} \limsup_n P\left(\sup_{|t-s|<\gamma} |M(t) - M(s)| \ge \epsilon \right) = 0.$$

This and (2.3.23) prove (2.3.22).

Now choose $\delta > 0$ so that

$$\limsup_n \delta k \sum_{i=1}^n \|d_i \mathbf{c}_i\| \le \epsilon/3. \qquad \text{(Use (2.3.7) here).} \tag{2.3.24}$$

From this, (2.3.20), (2.3.21) and (2.3.22) one readily obtains

$$\begin{aligned} \limsup_n P & \left(\sup_{t, \|u-v\| \le \delta} |L_2(t, \mathbf{u}, \mathbf{v})| \ge \epsilon \right) \\ & \le \limsup_n P\left(\sup_t |M(t)| \ge \epsilon/3 \right) = 0. \end{aligned}$$

This prove (2.3.19) for L_2. A similar argument proves (2.3.19) for L_1 with the same δ as in (2.3.24), thereby completing the proof of the Lemma. □

Proof of Theorem 2.3.1. Fix an $\epsilon > 0$ and choose a $\delta > 0$ satisfying (2.3.24). By the compactness of the ball $N(b) := \{\mathbf{u} \in \mathbb{R}^p; \|\mathbf{u}\| \le b\}$ there exist points $\mathbf{v}_1, \cdots \mathbf{v}_r$ in $N(b)$ such that for any $\mathbf{u} \in N(b)$, $\|\mathbf{u} - \mathbf{v}_j\| \le \delta$ for some $j = 1, 2, \cdots, r$. Thus

$$\begin{aligned} &\limsup_n P(\sup_{t,\mathbf{u}} |R_d(t,\mathbf{u})| \ge \epsilon) \\ &\le \sum_{j=1}^{r} \limsup_n P(\sup_{t, \|\mathbf{u}-\mathbf{v}_j\| \le \delta} |R_d(t, \mathbf{u} - R_d(t, \mathbf{v}_j))| \ge \epsilon/2) \\ &\quad + \sum_{j=1}^{r} \limsup_n P(\sup_t |R_d(t, \mathbf{v}_j)| \ge \epsilon/2) = 0 \end{aligned}$$

by Lemmas 2.3.2 and 2.3.4. □

Remark 2.3.1 A reexamination of the above proof reveals that Theorem 2.3.1 is a sole consequence of the continuity of certain W.E.P.'s and the smoothness of $\{F_{ni}\}$. It does not use the full force of the weak convergence of these W.E.P.'s. □

Remark 2.3.2 By the relationship

$$R_d(t, \mathbf{u}) = Y_d(t, \mathbf{u}) - Y_d(t, \mathbf{0}) + \mu_d(t, \mathbf{0}) - \mathbf{u}'\boldsymbol{\nu}_d(t)$$

and by Lemma 2.3.1, (2.3.8) of Theorem 2.3.1 is equivalent to

$$\text{(2.3.25)} \qquad \sup_{0 \le t \le 1, \|\mathbf{u}\| \le b} |Y_d(t, \mathbf{u}) - Y_d(t, \mathbf{0})| = o_p(1).$$

This will be useful when dealing with W.E.P.'s based on ranks in Chapter 3. □

The above theorem needs to be extended and reformulated when dealing with a linear regression model with an unknown scale parameter or with M - estimators in the presence of a preliminary scale estimator. To that end, define, for $x,\ s \in \mathbb{R},\ 0 \le t \le 1,\ \mathbf{u} \in \mathbb{R}^p$,

$$S_d(s, t, \mathbf{u}) \ := \ \sum_{i=1}^{n} d_{ni} I(X_{ni} \le (1 + sn^{-1/2}) H^{-1}(t) + \mathbf{c}_{ni}'\mathbf{u}),$$

$$S_d^0(s,x,\mathbf{u}) \;:=\; \sum_{i=1}^n d_{ni} I(X_{ni} \le (1+sn^{-1/2})x + \mathbf{c}'_{ni}\mathbf{u}),$$

and define $Y_d(s,t,\mathbf{u})$, $\mu_d(s,t,\mathbf{u})$ similarly. We are now ready to prove

Theorem 2.3.2 *In addition to the assumptions of Theorem 2.3.1, assume that*

$$\text{(2.3.26)} \qquad \max_{i,n}\sup_x |xf_{ni}(x)| \le k < \infty.$$

Then

$$\text{(2.3.27)} \quad \sup\Big| S_d(s,t,\mathbf{u}) - S_d(0,t,\mathbf{0}) - \sum_{i=1}^n d_{ni}\{sn^{-1/2}H^{-1}(t) + \mathbf{c}'_{ni}\mathbf{u}\}q_{ni}(t)\Big| = o_p(1).$$

where the supremum is taken over $|s| \le b$, $\|\mathbf{u}\| \le b$, $0 \le t \le 1$.

Consequently, if H *is strictly increasing for all* $n \ge 1$, *then*

$$\text{(2.3.28)} \quad \sup\Big| S_d^0(s,x,\mathbf{u}) - S_d^0(0,x,\mathbf{0}) - \sum_{i=1}^n d_{ni}\{sn^{-1/2}x + \mathbf{c}'_{ni}\mathbf{u}\}f_{ni}(x)\Big| = o_p(1).$$

where the supremum is taken over $|s| \le b$, $\|\mathbf{u}\| \le b$ *and* $x \in \mathbb{R}$.

Sketch of Proof. The argument is quite similar to that of Theorem 2.3.1. We briefly indicate the modifications of the previous proof.

An analogue of Lemma 2.3.1 will now assert

$$\sup\Big|\mu_d(s,t,\mathbf{u}) - \mu_d(1,t,\mathbf{0}) - \{(n^{-1/2}\Sigma d_i q_i(t)H^{-1}(t))s + \mathbf{u}'\nu_d(t)\}\Big| = o(1).$$

This uses (2.3.4), (2.3.5), (2.3.6), (2.3.7), (2.3.26) and (**N1**).

An analogue of Lemma 2.3.2 is obtained by applying Theorem 2.2.1 (i) to $\eta := H(X_i - \mathbf{c}'_i\mathbf{u})\sigma_n^{-1})$, $1 \le i \le n$, $\sigma_n := (1 + sn^{-1/2})$. This states that for every $|s| \le b$ and every $\|\mathbf{u}\| \le b$,

$$\sup_{0\le t\le 1} |Y_d(s,t,\mathbf{u}) - Y_d(s,t,\mathbf{0})| = o_p(1).$$

In verifying (**C**) for these $\{\eta_i\}$, one has an analogue of (2.3.13):

$$\sup_{0\le t\le 1-\delta} [G_d(t+\delta) - G_d(t)]$$
$$\le 2k\{B \max_i \|\mathbf{c}_i\| + bn^{-1/2}\} + \sup_{0\le t\le 1-\delta} [L_d(t+\delta) - L_d(t)].$$

Note that here $G_d(t) \equiv \sum_{i=1}^n d_i^2 F_i(\sigma_n H^{-1}(t) + \mathbf{c}_i'\mathbf{u})$.

One similarly has an analogue of Lemma 2.3.3. Consequently, from Theorem 2.3.1 one can conclude that for each fixed $s \in [-b, b]$,

$$\sup_{0\le t\le 1, \|\mathbf{u}\|\le b} |R_d(s,t,\mathbf{u})| = o_p(1), \tag{2.3.29}$$

where $R_d(s,t,\mathbf{u})$ equals the L.H.S. of (2.3.27) without the supremum. To complete the proof, once again exploit the compactness of $[-b, b]$ and the monotonic structure that is present in S_d and μ_d. Details are left for interested readers. □

Consider now the specialization of Theorems 2.3.1 and 2.3.2 to the case when $F_{ni} \equiv F$, F a d.f. Note that in this case (**N1**) implies that $L_d(t) \equiv t$ so that (**C***) is *a priori* satisfied. To state these specializations we need the following assumptions:

(**F1**) *F has uniformly continuous density f* w.r.t. λ.

(**F2**) $f > 0$, a.e. λ.

(**F3**) $\sup_{x\in\mathbb{R}} |xf(x)| \le k < \infty$.

Note that (**F1**) implies that f is bounded and that (**F2**) implies that F is strictly increasing.

Corollary 2.3.1 *Let $X_{n1}, \cdots, X_{nn}$ be i.i.d. F. In addition, suppose that (**N1**), (**N2**), (2.3.6), (2.3.7) and (**F1**) hold. Then (2.3.8) holds with $q_{ni} = f(F^{-1})$.*

*If, in addition, (**F2**) holds, then (2.3.9) holds with $f_{ni} \equiv f$.* □

Corollary 2.3.2 *Let $X_{n1}, \cdots, X_{nn}$ be i.i.d. F. In addition, suppose that (**N1**), (**N2**), (2.3.6), (2.3.7), (**F1**) and (**F3**) hold. Then (2.3.27) holds with $H \equiv F$ and $q_{ni} \equiv f(F^{-1})$.*

*If, in addition, (**F2**) holds, then (2.3.28) holds with $f_{ni} \equiv f$.* □

We shall now apply the above results to the model (1.1.1) and the $\{V_j\}$ - processes of (1.1.2). The results thus obtained are useful in studying the asymptotic distributions of certain goodness-of-fit tests and a class of M-estimators of $\boldsymbol{\beta}$ of (1.1.1) when there is an unknown scale parameter also. We need the following assumption about the design matrix $\mathbf{X}$.

(**NX**) $(\mathbf{X}'\mathbf{X})^{-1}$ *exists,* $\forall\ n \geq p$; $\max_i \mathbf{x}_{ni}'(\mathbf{X}'\mathbf{X})^{-1}\mathbf{x}_{ni} = o(1)$.

This is Noether's condition for the design matrix $\mathbf{X}$. Now, let

$$
\begin{aligned}
(2.3.30)\quad \mathbf{A} &:= (\mathbf{X}'\mathbf{X})^{-1/2}, \qquad \mathbf{D} := \mathbf{X}\mathbf{A},\\
\mathbf{q}'(t) &:= (q_{n1}(t), \cdots, q_{nn}(t)), \qquad \boldsymbol{\Lambda}(t) := \mathrm{diag}(\mathbf{q}(t)),\\
\boldsymbol{\Gamma}_1(t) &:= \mathbf{A}\mathbf{X}'\boldsymbol{\Lambda}(t)\mathbf{X}\mathbf{A},\\
\boldsymbol{\Gamma}_2(t) &:= n^{-1/2}H^{-1}(t)\mathbf{D}'\mathbf{q}(t), \qquad 0 \leq t \leq 1.
\end{aligned}
$$

Write $\mathbf{D} = ((d_{ij})), 1 \leq i \leq n,\ 1 \leq j \leq p$, and let $\mathbf{d}_{(j)}$ denote the j-th column of $\mathbf{D}$. Note that $\mathbf{D}'\mathbf{D} = \mathbf{I}_{p\times p}$. This in turn implies that

$$
(2.3.31)\qquad (\mathbf{N1}) \text{ is satisfied by } \mathbf{d}_{(j)} \text{ for all } 1 \leq j \leq p.
$$

Moreover, with $\mathbf{a}_{(j)}$ denoting the j-th column of $\mathbf{A}$,

$$
\begin{aligned}
(2.3.32)\quad & \max_i d_{ij}^2\\
&= \max_i (\mathbf{x}'\mathbf{a}_{(j)})^2 \leq \max_i \sum_{j=1}^{p} (\mathbf{x}_i'\mathbf{a}_{(j)})^2\\
&= \max_i \mathbf{x}_i' \left(\sum_{j=1}^{p} \mathbf{a}_{(j)}\mathbf{a}_{(j)}' \right) \mathbf{x}_i\\
&= \max_i \mathbf{x}_i'(\mathbf{X}'\mathbf{X})^{-1}\mathbf{x}_i = o(1), \qquad \text{by } (\mathbf{NX}).
\end{aligned}
$$

Let

$$
(2.3.33)\qquad L_j(t) := \sum_{i=1}^{n} d_{ij}^2 F_i(H^{-1}(t)), \quad 0 \leq t \leq 1,\ 1 \leq j \leq p.
$$

We are now ready to state

Theorem 2.3.3 *Let $\{(\mathbf{x}'_{ni}, Y_{ni}), 1 \le i \le n\}, \boldsymbol{\beta}, \{F_{ni}, 1 \le i \le n\}$ be as in the model (1.1.1). In addition, assume that $\{F_{ni}\}$ satisfy (2.3.4), (2.3.5) and that* (**C***) *is satisfied by each L_j of (2.3.33), $1 \le j \le p$.*

Then, for every $0 < b < \infty$,

$$(2.3.34)\quad \sup \|\mathbf{A}\{\mathbf{V}(H^{-1}(t), \boldsymbol{\beta} + \mathbf{A}\mathbf{u}) - \mathbf{V}(H^{-1}(t), \boldsymbol{\beta})\} - \Gamma_1(t)\mathbf{u}\| = o_p(1).$$

where the supremum is over $0 \le t \le 1$, $\|\mathbf{u}\| \le b$.

If, in addition, H is strictly increasing for all $n \ge 1$, then, for every $0 < b < \infty$,

$$(2.3.35)\quad \sup \|\mathbf{A}\{\mathbf{V}(x, \boldsymbol{\beta} + \mathbf{A}\mathbf{u}) - \mathbf{V}(x, \boldsymbol{\beta})\} - \Gamma_1(H(x))\mathbf{u}\| = o_p(1).$$

where the supremum is over $-\infty \le x \le \infty$, $\|\mathbf{u}\| \le b$.

Theorem 2.3.4 *Suppose that $\{(\mathbf{x}'_{ni}, Y_{ni}), 1 \le i \le n\}$ and $\boldsymbol{\beta} \in \mathbb{R}^p$ obey the model*

$$(2.3.36)\qquad Y_{ni} = \mathbf{x}'_{ni}\boldsymbol{\beta} + \gamma\epsilon_{ni}, \qquad 1 \le i \le n, \gamma > 0,$$

with $\{\epsilon_{ni}\}$ independent r.v.'s having d.f.'s $\{F_{ni}\}$. Assume that (**NX**) *holds. In addition, assume that (2.3.4), (2.3.5), (2.3.26) are satisfied by $\{F_{ni}\}$ and that* (**C***) *is satisfied by each L_j of (2.3.33), $1 \le j \le p$. Then for every $0 < b < \infty$,*

$$(2.3.37)\qquad \begin{aligned} &\mathbf{A}\left[\mathbf{V}(\alpha H^{-1}(t), \boldsymbol{\beta} + \mathbf{A}\mathbf{u}\gamma) - \mathbf{V}(\gamma H^{-1}(t), \boldsymbol{\beta})\right] \\ &\quad = \boldsymbol{\Gamma}_1(t)\,\mathbf{u} - \boldsymbol{\Gamma}_2(t)\;v + u_p(1), \end{aligned}$$

where $v := n^{1/2}(\alpha - \gamma)\gamma^{-1}, \alpha > 0$.

If, in addition, H is strictly increasing for every $n \ge 1$, then

$$(2.3.38)\qquad \begin{aligned} &\mathbf{A}\left[\mathbf{V}(\alpha x, \boldsymbol{\beta} + \mathbf{A}\mathbf{u}\gamma) - \mathbf{V}(\gamma x, \boldsymbol{\beta})\right] \\ &\quad = \boldsymbol{\Gamma}_1(H(x))\,\mathbf{u} - \boldsymbol{\Gamma}_2(H(x))\,v + u_p(1). \end{aligned}$$

In (2.3.37) and (2.3.38), $u_p(1)$ are a sequences of stochastic processes in $(t, x, \mathbf{u}, v)$ tending to zero uniformly in probability over $0 \le t \le 1$, $-\infty \le x \le \infty$, $\|\mathbf{u}\| \le b$ and $|v| \le b$.

Proof of Theorem 2.3.3. Apply Theorem 2.3.1 to $X_i = Y_i - \mathbf{x}_i'\boldsymbol{\beta}$, $\mathbf{c}_i = \mathbf{x}_i'A$, $1 \le i \le n$. Then F_i is the d.f. of X_i and the j-th components of $\mathbf{AV}(H^{-1}(t), \boldsymbol{\beta} + \mathbf{Au})$ and $\mathbf{AV}(H^{-1}(t), \boldsymbol{\beta})$ are $S_d(t, \mathbf{u}), S_d(t, \mathbf{0})$ of (2.3.1), respectively, with $d_i = d_{ij}$ of (2.3.30), $1 \le i \le n$, $1 \le j \le p$. Therefore (2.3.34) will follows by p applications of (2.3.8), one for each $\mathbf{d}_{(j)}$, provided the assumptions of Theorem 2.3.1 are satisfied. But in view of (2.3.31) and (2.3.32), the assumption (**NX**) implies (**N1**), (**N2**) for $d_{(j)}, 1 \le j \le p$. Also, (2.3.6) for the specified $\{c_i\}$ is equivalent to (**NX**). Finally, the C-S inequality and (2.3.31) verifies (2.3.7) in the present case. This makes Theorem 2.3.1 applicable and hence (2.3.34) follows. □

Proof of Theorem 2.3.4. Follows from Theorem 2.3.2 when applied to $X_i = (Y_i - \mathbf{x}_i'\boldsymbol{\beta})\gamma^{-1}$, $\mathbf{c}_i' = \mathbf{x}_i'A$, $1 \le i \le n$, in a fashion similar to the proof of Theorem 2.3.3 above. □

The following corollaries follow from Corollaries 2.3.1 and 2.3.2 in the same way as the above Theorems 2.3.3 and 2.3.4 follow from Theorems 2.3.1 and 2.3.2. These are stated for an easy reference later on. Let $C(b) := \{\mathbf{s} \in \mathbb{R}^p, \|\mathbf{A}^{-1}(\mathbf{s} - \boldsymbol{\beta})\| \le b\}$.

Corollary 2.3.3 *Suppose that the model (1.1.1) with $F_{ni} \equiv F$ holds. Assume that the design matrix* $\mathbf{X}$ *and the d.f. F satisfying* (**NX**) *and* (**F1**). *Then,* $\forall\ 0 < b < \infty$,

$$\text{(2.3.39)} \qquad \begin{aligned} &\mathbf{A}\left[\mathbf{V}(F^{-1}(t), \mathbf{s}) - \mathbf{V}(F^{-1}(t), \boldsymbol{\beta})\right] \\ &= f(F^{-1}(t))\mathbf{A}^{-1}(\mathbf{s} - \boldsymbol{\beta}) + u_p(1), \end{aligned}$$

where $u_p(1)$ is a sequence of stochastic processes in $(t, \mathbf{s})$ tending to zero uniformly in probability over $0 \le t \le 1$; $\mathbf{s} \in C(b)$.

If, in addition, F satisfies (**F2**), *then*

$$\text{(2.3.40)} \quad \|\mathbf{A}\{\mathbf{V}(x, \mathbf{s}) - \mathbf{V}(x, \boldsymbol{\beta})\} - f(x)\mathbf{A}^{-1}(\mathbf{s} - \boldsymbol{\beta})\| = u_p(1),$$

where $u_p(1)$ is a sequence of stochastic processes in $(x, \mathbf{s})$ tending to zero uniformly in probability over $-\infty \le x \le \infty$; $\mathbf{s} \in C(b)$. □

Corollary 2.3.4 *Suppose that the model (2.3.36) with $F_{ni} \equiv F$ holds and that the design matrix* $\mathbf{X}$ *and the d.f. F satisfy* (**NX**),

*(**F1**) and (**F3**). Then (2.3.37) holds with*

$$(2.3.41)\quad \begin{aligned} \Gamma_1(t) &= f(F^{-1}(t))\mathbf{I}_{p\times p}, \\ \Gamma_2(t) &= F^{-1}(t)f(F^{-1}(t))\mathbf{A}\mathbf{X}'\mathbf{1}, \qquad 0 \le t \le 1. \end{aligned}$$

*If, in addition, F satisfies (**F2**), then (2.3.38) holds with $\mathbf{\Gamma}_j(H) \equiv \mathbf{\Gamma}_j(F), j = 1, 2,$ i.e.,*

$$\mathbf{A}\left[\mathbf{V}(\alpha x, \boldsymbol{\beta} + \mathbf{A}\mathbf{u}\gamma) - \mathbf{V}(\gamma x, \boldsymbol{\beta})\right] = f(x)\mathbf{u} - xf(x)v + u_p(1),$$

where v is as in (2.3.37) and $u_p(1)$ is as in (2.3.38). □

We end this section by stating an AUL result about the ordinary residual empirical processes H_n of (1.2.1) for an easy reference later on.

Corollary 2.3.5 *Suppose that the model (1.1.1) with $F_{ni} \equiv F$ holds. Assume that the design matrix $\mathbf{X}$ and the d.f. F satisfying (**NX**) and (**F1**). Then, $\forall\ 0 < b < \infty$,*

$$(2.3.42)\quad \begin{aligned} &n^{1/2}\left[H_n(F^{-1}(t), s) - H_n(F^{-1}(t), \boldsymbol{\beta})\right] \\ &= f(F^{-1}(t)) \cdot n^{-1/2}\sum_{i=1}^{n} \mathbf{x}'_{ni}\mathbf{A} \cdot \mathbf{A}^{-1}(\mathbf{s} - \boldsymbol{\beta})| + u_p(1), \end{aligned}$$

where $u_p(1)$ is as in Corollary 2.3.3.

*If, in addition, F satisfies (**F2**), then, $\forall\ 0 < b < \infty$,*

$$(2.3.43)\quad \begin{aligned} &n^{1/2}\left[H_n(x, \mathbf{s} - H_n(x, \boldsymbol{\beta})\right] \\ &= f(x) \cdot n^{-1/2}\sum_{i=1}^{n} \mathbf{x}'_{ni}\mathbf{A} \cdot \mathbf{A}^{-1}(\mathbf{s} - \boldsymbol{\beta}) + u_p(1), \end{aligned}$$

where $u_p(1)$ is as in (2.3.40).

Proof. The proof follows from Theorem 2.3.1 by specializing it to the case where $d_{ni} \equiv n^{-1/2}$ and the rest of the entities as in the proof of Theorem 2.3.3. □

Note: Ghosh and Sen (1971) and Koul and Zhu (1991) prove an almost sure version of (2.3.40) in the case $p = 1$ and $p > 1$, respectively. □

2.4 Some Additional Results for W.E.P.'S.

For the sake of general interest, here we state some further results about W.E.P.'s. To begin with, we have

2.4.1 Laws of the iterated logarithm

In this subsection, we assume that

$$(2.4.1) \qquad d_{ni} \equiv d_i, \quad \eta_{ni} \equiv \eta_i, \quad G_{ni} \equiv G_i, \quad 1 \le i \le n.$$

Define

$$(2.4.2)\ \mathcal{U}_n(t) \ := \ \sum_{i=1}^n d_i\{I(\eta_i \le t) - G_i(t)\}, \quad \sigma_n^2 := \sum_{i=1}^n d_i^2,$$

$$\xi_n(t) \ := \ \mathcal{U}_n(t)/\{2\sigma_n^2 \ell n \ell n \sigma_n^2\}^{1/2}, \quad n \ge 1,\ 0 \le t \le 1.$$

Let $r(s,t) := s \wedge t - st$, $0 \le s,\ t \le 1$, and $H(r)$ be the reproducing kernel Hilbert space generated by the kernel r with $\|\cdot\|_r$ denoting the associated norm on $H(r)$. Let

$$K = \{f \in H(r); \|f\|_r \le 1\}.$$

Let ρ denote the uniform metric.

Theorem 2.4.1 *If $\eta_1, \eta_2, \cdots$ are i.i.d. uniform on $[0,1]$ r.v.'s and $d_1, d_2, \cdots$ are any real numbers satisfying*

$$\lim_n \sigma_n^2 = \infty, \qquad \lim_n (\max_{1\le i\le n} d_i^2) \frac{\ell n \ell n \sigma_n^2}{\sigma_n^2} = 0,$$

then

$$P(\rho(\xi_n, K) \to 0 \text{ and the set of limit points of } \{\xi_n\} \text{ is } K) = 1. \square$$

This theorem was proved by Vanderzanden (1980, 1984) using some of the results of Kuelbs (1976) and certain martingale properties of ξ_n.

Theorem 2.4.2 *Let $\eta_1, \eta_2, \cdots$ be independent nonnegative random variables. Let $\{d_i\}$ be any real numbers. Then*

$$\limsup_n \sup_{t \ge 0} \sigma_n^{-1} |\mathcal{U}_n(t_-)| < \infty \ \text{ a.s.}. \qquad \square$$

A proof of this appears in Marcus and Zinn (1984). Actually they prove some other interesting results about w.e.p.'s with weights which are r.v.'s and functions of t. Most of their results, however, are concerned with the bounded law of the iterated logarithm. They also proved the following inequality that is similar to, yet a generalization of, the classical Dvoretzky - Kiefer - Wolfowitz exponential inequality for the ordinary empirical process. Their proof is valid for triangular arrays and real r.v.'s.

Exponential inequality. *Let $X_{n1}, X_{n2}, \cdots, X_{nn}$ be independent r.v.'s with respective d.f.'s $F_{n1}, \cdots, F_{nn}$ and $\{d_{ni}\}$ be any real numbers satisfying* (**N1**). *Then,* $\forall\ \lambda > 0,\ \ \forall\ n \geq 1,$

$$
\begin{aligned}
P&\left(\sup_{x\in\mathbb{R}}\left|\sum_{i=1}^{n} d_{ni}\{I(X_{ni} \leq x) - F_{ni}(x)\}\right| \geq \lambda\right) \\
&\leq\ [1 + (8\pi)^{1/2}\lambda]\exp(-\lambda^2/8). \qquad \square
\end{aligned}
$$

The above two theorems immediately suggest some interesting probabilistic questions. For example, is Vanderzanden's result valid for nonidentical r.v.'s $\{\eta_i\}$? Or can one remove the assumption of nonnegative $\{\eta_i\}$ in Theorem 2.4.1?

2.4.2 Weak convergence of W.E.P.'s in $\mathcal{D}[0,1]^p$, in ρ_q - metric and an embedding result.

Next, we state a weak convergence result for multivariate r.v.'s. For this we revert back to triangular arrays. Now suppose that $\eta_{ni} \in [0,1]^p, 1 \leq i \leq n$, are independent r.v.'s of dimension p. Define

$$
(2.4.3) \qquad W_d(\mathbf{t}) := \sum_{i=1}^{n} d_{ni}\{I(\eta_{ni} \leq \mathbf{t}) - G_{ni}(\mathbf{t})\}, \quad \mathbf{t} \in [0,1]^p.
$$

Let G_{nij} be the j-th marginal of $G_{ni}, 1 \leq i \leq n, 1 \leq j \leq p$.

Theorem 2.4.3 *Let $\{\eta_{ni}, 1 \leq i \leq n\}$ be independent p- variate r.v.'s and $\{d_{ni}\}$ satisfy* (**N1**) *and* (**N2**). *Moreover suppose that for each $1 \leq j \leq p$,*

$$
\lim_{\delta\to 0}\limsup_{n}\sup_{0\leq t\leq 1-\delta}\sum_{i=1}^{n} d_{ni}^2\{G_{nij}(t+\delta) - G_{nij}(t)\} = 0.
$$

Then, for every $\epsilon > 0$

1. $\lim_{\delta\to 0}\limsup_n P(\sup_{|s-t|<\delta}|W_d(\mathbf{t}) - W_d(\mathbf{s})| > \epsilon) = 0.$

2. *Moreover,* $W_d \Longrightarrow$ *some* W *on* $(\mathcal{D}[0,1]^p, \rho)$ *if, and only if, for each* $\mathbf{s}, \mathbf{t} \in [0,1]^p$,

$$Cov(W_d(\mathbf{s}), W_d(\mathbf{t})) \to Cov(W(\mathbf{s}), W(\mathbf{t})) =: \mathcal{C}(\mathbf{s}, \mathbf{t}).$$

In this case W *is necessarily a Gaussian process,* $P(W \in \mathcal{C}[0,1]^p) = 1$, $W(\mathbf{0}) = 0 = W(\mathbf{1})$. □

Theorem 2.4.3 is essentially proved in Vanderzanden (1980), using results of Bickel and Wichura (1971).

Mehra and Rao (1975), Withers (1975), and Koul (1977), among others, obtain the weak convergence results for $\{W_d\}$ - processes when $\{\eta_{ni}\}$ are weakly dependent. See Dehling and Taqqu (1989) and Koul and Mukherjee (1992) for similar results when $\{\eta_{ni}\}$ are long range dependent.

Shorack (1971) proved the weak convergence of W_d/q - process in the ρ-metric, where $q \in Q$, with

$$\begin{aligned}\mathcal{Q} \;:=\; \Big\{&q, q \text{ a continuous function on } [0,1], q \geq 0, q(t) = q(1-t),\\ &q(t)\uparrow,\ t^{-1/2}q(t)\downarrow \text{ for } 0 \leq t \leq 1/2, \int_0^1 q^{-2}(t)dt < \infty\Big\}.\end{aligned}$$

Theorem 2.4.4 *Suppose that* $\eta_{n1}, \cdots, \eta_{nn}$ *are independent random variables in* $[0,1]$ *with respective d.f.'s* $G_{n1}, \cdots G_{nn}$ *such that*

$$n^{-1}\sum_{i=1}^{n} G_{ni}(t) = t, \quad 0 \leq t \leq 1.$$

In addition, suppose that $\{d_{ni}\}$ *satisfy* (**N1**) *and* (**B**) of Theorem 2.2.2. *Then,*

(i) $\forall\ \epsilon > 0,\ \forall\ q \in \mathcal{Q}$,

$$\lim_{\delta\to 0}\limsup_n P\left(\sup_{|t-s|<\delta}\left|\frac{W_d(t)}{q(t)} - \frac{W_d(s)}{q(s)}\right| > \epsilon\right) = 0.$$

(ii) $q^{-1}W_d \Longrightarrow q^{-1}W$, W *a continuous Gaussian process with covariance function* $\mathcal{C}$ *if, and only if* $\mathcal{C}_d \to \mathcal{C}$. □

Shorack (1991) and Einmahl and Mason (1991) proved the following embedding result.

Theorem 2.4.5. *Suppose that $\eta_{n1}, \cdots \eta_{nn}$ are i.i.d. Uniform $[0,1]$ r.v.'s. In addition, suppose that $\{d_{ni}\}$ satisfy* (**N1**) *and that*

$$\sum_{i=1}^{n} d_{ni} = 0, \qquad n\sum_{i=1}^{n} d_{ni}^4 = O(1).$$

Then on a rich enough probability space there exist a sequence of versions $\mathcal{W}_d$ of the processes W_d and a fixed Brownian bridge B on $[0,1]$ such that

$$\sup_{1/n \le t \le 1-1/n} n^{\nu} \frac{|\mathcal{W}_d(t) - B(t)|}{\{t(1-t)\}^{1/2-\nu}} = O_p(1), \qquad for \ all \ \ 0 \le \nu < 1.$$

The closed interval $1/n \le t \le 1 - 1/n$ may be replaced by the open interval $\min\{\eta_{nj}; 1 \le j \le n\} < t < \max\{\eta_{nj}; 1 \le j \le n\}$. □

2.4.3 A martingale property

In this subsection we shall prove a martingale property of W.E.P.'s. Let $X_{n1}, X_{n2}, \cdots, X_{nn}$ be independent real r.v.'s with respective d.f.'s $F_{n1}, \cdots, F_{nn}$; $d_{n1}, \cdots, d_{nn}$ be real numbers. Let $a \le b$ be fixed real numbers. Define

$$M_n(t) := \sum_{i=1}^{n} d_{ni}\{I(X_{ni} \in (a,t]\}\{1 - p_{ni}(a,t]\}^{-1},$$

$$R_n(t) := \sum_{i=1}^{n} d_{ni}\{I(X_{ni} \in (t,b] - p_{ni}(t,b]\}\{1 - p_{ni}(t,b]\}^{-1}, \quad t \in \mathbb{R},$$

where

$$p_{ni}(s,t] := F_{ni}(t) - F_{ni}(s), \quad 0 \le s \le t \le 1, \quad 1 \le i \le n.$$

Let $T_1 \subset [a,\infty), T_2 \subset (-\infty, b]$ be such that $M_n(t)[R_n(t)]$ is well-defined for $t \in T_1[t \in T_2]$. Let

$$\mathcal{F}_{1n}(t) := \sigma - \text{field}\{I(X_{ni} \in (a,s]),\ a \le s \le t,\ i = 1, \cdots, n\}, \quad t \in T_1,$$

$$\mathcal{F}_{2n}(t) := \sigma - \text{field}\{I(X_{ni} \in (s,b]),\ t \le s \le b,\ \ i = 1, \cdots, n\},\ t \in T_2.$$

Martingale Lemma. *Under the above setup,* $\forall n \geq 1$, $\{M_n(t), \mathcal{F}_{1n}(t), t \in T_1\}$ *is a martingale and* $\{R_n(t), \mathcal{F}_{2n}(t), t \in T_2\}$ *is a reverse martingale.*

Proof. Write $q_i(a,s] = 1 - p_i(a,s]$. Because $\{X_i\}$ are independent, for $a \leq s \leq t$,

$$
\begin{aligned}
&E\Big\{M_n(t)\Big|\mathcal{F}_{1n}(s)\Big\} \\
&= \sum_{i=1}^{n} d_i\, I\Big(X_i \in (a,s]\Big) \\
&\qquad \times \left[\frac{E\Big\{\Big[I\Big(X_i \in\in (a,t]\Big) - p_i(a,t]\Big]\Big|X_i \in (a,s]\Big\}}{q_i(a,t]}\right. \\
&\qquad \left. + I\Big(X_i \notin (a,s]\Big) E\Big\{\Big[I(X_i \in (a,t) - p_i(a,t]\Big]\Big|X_i \notin (a,s]\Big]\Big\}\right] \\
&= \sum_{i=1}^{n} d_i \Bigg\{ I\Big(X_i \in (a,s]\Big) \\
&\qquad + \frac{I\Big(X_i \notin (a,s]\Big)\left\{\frac{p_i(s,t]}{q_i(a,s]} - p_i(a,t]\right\}}{q_i(a,t]}\Bigg\} \\
&= \sum_{i=1}^{n} d_i \frac{I\Big(X_i \in (a,s]\Big) - q_i(a,s]}{q_i(a,s)} = M_n(s).
\end{aligned}
$$

A similar argument yields the result about R_n. □

Note: The above martingale lemma is well known when the r.v.'s $\{X_{ni}\}$ are i.i.d. and $d_{ni} \equiv n^{-1/2}$. In the case $\{X_{ni}\}$ are i.i.d. and $\{d_{ni}\}$ are arbitrary, the observation about $\{M_n\}$ being a martingale first appeared in Sinha and Sen (1979). The above martingale Lemma appears in Vanderzanden (1980, 1984).

Theorem 2.4.1 above generalizes a result of Finkelstein (1971) for the ordinary empirical process to W.E.P.'s of i.i.d. r.v.'s.. In fact, the set K is the same as the set $\mathbf{K}$ of Finkelstein. □□

3

Linear Rank and Signed Rank Statistics

3.1 Introduction

Let $\{X_{ni}, F_{ni}\}$ be as in (2.2.23) and $\{\mathbf{c}_{ni}\}$ be $p \times 1$ real vectors. The rank and the absolute rank of the ith residual for $1 \le i \le n$, $\mathbf{u} \in \mathbb{R}^p$, are defined, respectively, as

$$
\begin{aligned}
(3.1.1) \qquad R_{i\mathbf{u}} &= \sum_{j=1}^{n} I(X_{nj} - \mathbf{u}'\mathbf{c}_{nj} \le X_{ni} - \mathbf{u}'\mathbf{c}_{ni}), \\
R_{i\mathbf{u}}^{+} &= \sum_{j=1}^{n} I(|X_{nj} - \mathbf{u}'\mathbf{c}_{nj}| \le |X_{ni} - \mathbf{u}'\mathbf{c}_{ni}|),
\end{aligned}
$$

Let φ be a nondecreasing real valued function on $[0, 1]$ and define

$$
\begin{aligned}
(3.1.2) \qquad T_d(\varphi, \mathbf{u}) &= \sum_{i=1}^{n} d_{ni}\varphi\left(\frac{R_{i\mathbf{u}}}{n+1}\right), \\
T_d^{+}(\varphi, \mathbf{u}) &= \sum_{i=1}^{n} d_{ni}\varphi^{+}\left(\frac{R_{i\mathbf{u}}^{+}}{n+1}\right) s(X_{ni} - \mathbf{u}'\mathbf{c}_{ni}),
\end{aligned}
$$

for $\mathbf{u} \in \mathbb{R}^p$, where

$$\varphi^{+}(s) = \varphi((s+1)/2), \; 0 \le s \le 1; \quad s(x) = I(x > 0) - I(x < 0).$$

The processes $\{T_d(\varphi, \mathbf{u}), \mathbf{u} \in \mathbb{R}^p\}$ and $\{T_d^+(\varphi, \mathbf{u}), \mathbf{u} \in \mathbb{R}^p\}$ are used to define rank (R) estimators of $\boldsymbol{\beta}$ in the linear regression model (1.1.1). See, e.g., Adichie (1967), Koul (1971), Jurečková (1971) and Jaeckel (1972). One key property used in studying these R-estimators is the asymptotic uniform linearity (AUL) of $T_d(\varphi, \mathbf{u})$ and $T_d^+(\varphi, \mathbf{u})$ in $\mathbf{u}$ over bounded sets. Such results have been proved by Jurečková (1969) for $T_d(\varphi, \mathbf{u})$ for general but fixed functions φ, by Koul (1969) for $T_d^+(I, \mathbf{u})$ (where I is the identity function) and by van Eeden (1971) for $T_d^+(\varphi, \mathbf{u})$ for general but fixed φ functions. In all of these papers $\{X_{ni}\}$ are assumed to be i.i.d.

In Sections 3.2 and 3.3 below we prove the AUL of $T_d(\varphi, \cdot)$, $T_d^+(\varphi, \cdot)$, uniformly in those φ which have $\|\varphi\|_{tv} < \infty$, and under fairly general independent setting. These proofs reveal that this AUL property is also a consequence of the asymptotic continuity of certain W.E.P.'s and the smoothness of $\{F_{ni}\}$.

Besides being useful in studying the asymptotic distributions of R-estimators of $\boldsymbol{\beta}$ these results are also useful in studying some rank based minimum distance estimators, some goodness-of-fit tests for the error distributions of (1.1.1) and the robustness of R-estimators against certain heteroscedastic errors.

3.2 AUL of Linear Rank Statistics

At the outset we shall assume

$$(3.2.1)\quad \varphi \in \mathcal{C} \;:=\; \Big\{\varphi : [0,1] \to \mathbb{R},\ \varphi \in \mathcal{DI}[0,1], \quad \text{with } \|\varphi\|_{tv} := \varphi(1) - \varphi(0) = 1\Big\}.$$

Define the W.E.P. based on ranks, with weights $\{d_{ni}\}$,

$$(3.2.2)\quad Z_d(t, \mathbf{u}) \;:=\; \sum_{i=1}^{n} d_{ni} I(R_{i\mathbf{u}} \le nt), \quad 0 \le t \le 1, \mathbf{u} \in \mathbb{R}^p.$$

Note that with $n\bar{d}_n = \sum_{i=1}^n d_{ni}$,

$$(3.2.3)\quad T_d(\varphi, \mathbf{u}) \;=\; \int \varphi(nt/(n+1)) Z_d(dt, \mathbf{u}) \;=\; -\int Z_d((n+1)t/n, \mathbf{u}) d\varphi(t) + n\bar{d}_n \varphi(1).$$

This representation shows that in order to prove the AUL of $T_d(\varphi, \cdot)$, it suffices to prove it for $Z_d(t, \cdot)$, uniformly in $0 \le t \le 1$. Thus, we shall first prove the AUL property for the Z_d-process. Define, for $x \in \mathbb{R}$, $0 \le t \le 1$, $\mathbf{u} \in \mathbb{R}^p$,

$$
\begin{aligned}
(3.2.4) \qquad H_{n\mathbf{u}}(x) &:= n^{-1}\sum_{i=1}^{n} I(X_{ni} - \mathbf{c}_{ni}'\mathbf{u} \le x), \\
H_{\mathbf{u}}(x) &:= n^{-1}\sum_{i=1}^{n} F_{ni}(x + \mathbf{c}_{ni}'\mathbf{u}), \\
H_{n\mathbf{u}}^{-1}(t) &= \inf\{x; H_{n\mathbf{u}}(x) \ge t\}, \\
H_{\mathbf{u}}^{-1}(t) &= \inf\{x; H_{\mathbf{u}}(x) \ge t\}.
\end{aligned}
$$

Note that $H_{\mathbf{0}}$ is the H of (2.2.23). We shall write H_n for $H_{n\mathbf{0}}$. Recall that for any d.f. G,

$$G(G^{-1}(t)) \ge t,\ 0 \le t \le 1 \quad \text{and} \quad G^{-1}(G(x)) \le x, \quad x \in \mathbb{R}.$$

This fact and the relation $nH_{nu}(X_i - \mathbf{c}_i'\mathbf{u}) \equiv R_{i\mathbf{u}}$ yield that $\forall\ 0 \le t \le 1,\ 1 \le i \le n$,

$$(3.2.5) \quad \left[X_i - \mathbf{c}_i'\mathbf{u} \ge H_{n\mathbf{u}}^{-1}(t)\right] \Rightarrow [R_{i\mathbf{u}} \ge nt] \Rightarrow \left[X_i - \mathbf{c}_i'\mathbf{u} \ge H_{n\mathbf{u}}^{-1}(t)\right].$$

For technical convenience, it is desirable to center the weights of linear rank statistics appropriately. Accordingly, let

$$(3.2.6) \qquad w_{ni} := (d_{ni} - \bar{d}_n), \quad 1 \le i \le n.$$

Then, with Z_w denoting the Z_d when weights are $\{w_{ni}\}$,

$$Z_d(t, \mathbf{u}) = Z_w(t, \mathbf{u}) + \bar{d}_n \cdot [nt], \quad 0 \le t \le 1,\ \mathbf{u} \in \mathbb{R}^p.$$

Hence, for all $0 \le t \le 1$, $\mathbf{u} \in \mathbb{R}^p$,

$$(3.2.7) \qquad Z_d(t, \mathbf{u}) - Z_d(t, \mathbf{0}) = Z_w(t, \mathbf{u}) - Z_w(t, \mathbf{0}).$$

Next define, for arbitrary real weights $\{d_{ni}\}$, and for $0 \le t \le 1$, $\mathbf{u} \in \mathbb{R}^p$,

$$(3.2.8) \qquad \mathcal{V}_d(t, \mathbf{u}) := \sum_{i=1}^{n} d_{ni} I\left(X_{ni} - \mathbf{c}_{ni}'\mathbf{u} \le H_{n\mathbf{u}}^{-1}(t)\right).$$

By (3.2.5) and direct algebra, for any weights $\{d_{ni}\}$,

$$\sup_{t,\mathbf{u}} |\mathcal{Z}_d(t,\mathbf{u}) - \mathcal{V}_d(t,\mathbf{u})| \le 2 \max_i |d_{ni}|. \tag{3.2.9}$$

Consider the condition

$$\textbf{(N3)} \qquad \tau_w^2 = 1, \qquad \max_i w_{ni}^2 \to 0.$$

In view of (3.2.7) and (3.2.9), (**N3**) implies that the problem of proving the AUL for the $\mathcal{Z}_d$-process is reduced to proving it for the $\mathcal{V}_w$ -process.

Next, recall the definition (2.3.1) and define

$$\tilde{\mathcal{T}}_d(t,\mathbf{u}) := \mathcal{V}_d(t,\mathbf{u}) - \mu_d(t,\mathbf{u}), \quad 0 \le t \le 1, \ \mathbf{u} \in \mathbb{R}^p. \tag{3.2.10}$$

Note the basic decomposition: for any real numbers $\{d_{ni}\}$ and for all $0 \le t \le 1, \mathbf{u} \in \mathbb{R}^p$,

$$\tilde{\mathcal{T}}_d(t,\mathbf{u}) = Y_d(HH_{n\mathbf{u}}^{-1}(t),\mathbf{u}) + \mu_d(HH_{n\mathbf{u}}^{-1}(t),\mathbf{u}) - \mu_d(t,\mathbf{u}), \tag{3.2.11}$$

provided H is strictly increasing for all $n \ge 1$. This decomposition is basic to the following proof of the AUL property of $\mathcal{Z}_d$.

Theorem 3.2.1 *Suppose that* $\{X_{ni}, F_{ni}\}$ *satisfy* (2.2.24), (**N3**) *holds, and* $\{\mathbf{c}_{ni}\}$ *satisfy* (2.3.6) *and* (2.3.7) *with* $d_{ni} \equiv w_{ni}$. *In addition, assume that* (**C***) *holds with* $d_{ni} \equiv w_{ni}$, H *is strictly increasing, the densities* $\{f_{ni}\}$ *of* $\{F_{ni}\}$ *satisfy* (2.3.5) *and that*

$$\lim_{\delta\to 0} \limsup_n \max_i \sup_{|H(x)-H(y)|<\delta} |f_{ni}(x) - f_{ni}(y)| = 0. \tag{3.2.12}$$

Then, for every $0 < b < \infty$,

$$\begin{aligned} \sup |\tilde{\mathcal{T}}_w(t,\mathbf{u}) - Y_w(t,\mathbf{0}) - \mu_w(HH_{n\mathbf{u}}^{-1}(t),\mathbf{0}) + \mu_w(t,\mathbf{0})| \\ = o_p(1), \end{aligned} \tag{3.2.13}$$

where the supremum is being taken over $0 \le t \le 1$, $\|\mathbf{u}\| \le b$.

Before proceeding to prove the theorem, we prove the following lemma which is of independent interest. In this result, no assumptions other than independence of $\{X_{ni}\}$ are being used.

Lemma 3.2.1 *Let $H, H_n, H_{\mathbf{u}}$ and $H_{n\mathbf{u}}$ be as in* (3.2.4) *above. Assuming only* (2.2.24), *we have*

$$\|H_n - H\|_\infty \to 0, \quad \text{a.s..} \tag{3.2.14}$$

If, in addition, (2.3.6) *holds and if, for any* $0 < b < \infty$,

$$\sup_{|x-y|\leq 2m_n b} |H(x) - H(y)| \to 0, \quad (m_n = \max_i \|\mathbf{c}_i\|), \tag{3.2.15}$$

then,

$$\sup_{|x|<\infty, \|u\|\leq b} |H_{n\mathbf{u}}(x) - H_{\mathbf{u}}(x)| \to 0, \quad \text{a.s..} \tag{3.2.16}$$

Proof. Note that $H_n(x) - H(x)$ is a sum of centered independent Bernoulli r.v.'s. Thus $E[H_n(x) - H(x)]^4 = O(n^{-2})$. Apply the Markov inequality with the 4th moment and the Borel - Cantelli lemma to obtain

$$|H_n(x) - H(x)| \to 0, \quad \text{a.s., for every } \ x \in \mathbb{R}.$$

Now proceed as in the proof of the Glivenko - Cantelli Lemma (Loéve (1963), p. 21) to conclude (3.2.14).

To prove (3.2.16), note that $\|\mathbf{u}\| \leq b$ implies that

$$-m_n b \leq \mathbf{c}_i'\mathbf{u} \leq m_n b, \quad 1 \leq i \leq n.$$

The monotonicity of $H_{n\mathbf{u}}$ and $H_{\mathbf{u}}$ yields that for $\|\mathbf{u}\| \leq b, \ x \in \mathbb{R}$,

$$\begin{aligned}
H_n(x - bm_n) &- H(x - bm_n) + H(x - bm_n) - H(x + bm_n) \\
&\leq H_{n\mathbf{u}}(x) - H_{\mathbf{u}}(x) \\
&\qquad \leq H_n(x + bm_n) - H(x + bm_n) \\
&\qquad\qquad\qquad + H(x + bm_n) - H(x - bm_n).
\end{aligned}$$

Hence (3.2.16) follows from (3.2.15) and the following inequality:

$$\text{L.H.S. (3.2.16)} \leq 2 \sup_{|x|<\infty} |H_n(x) - H(x)| + \sup_{|x-y|\leq 2m_n b} |H(x) - H(y)|.$$

□

Proof of Theorem 3.2.1. From (3.2.11), for all $0 \le t \le 1$, $\mathbf{u} \in \mathbb{R}^p$,

$$\begin{aligned}\tilde{T}_w(t,\mathbf{u}) &= [Y_w(HH_{n\mathbf{u}}^{-1}(t),\mathbf{u}) - Y_w(HH_{n\mathbf{u}}^{-1}(t),\mathbf{0}] \\ &\quad +[Y_w(HH_{n\mathbf{u}}^{-1}(t),\mathbf{0} - Y_w(t,\mathbf{0})] \\ &\quad +Y_w(t,\mathbf{0}) - [\mu_w(t,\mathbf{u}) - \mu_w(t,\mathbf{0}) - \mathbf{u}'\boldsymbol{\nu}_w(t)] \\ &\quad +[\mu_w(HH_{n\mathbf{u}}^{-1}(t),\mathbf{u}) - \mu_w(HH_{n\mathbf{u}}^{-1}(t),\mathbf{0}) \\ &\qquad\qquad -\mathbf{u}'\boldsymbol{\nu}_w(HH_{n\mathbf{u}}^{-1}(t))] \\ &\quad +\mu_w(HH_{n\mathbf{u}}^{-1}(t),\mathbf{0}) - \mu_w(t,\mathbf{0}) + \mathbf{u}'\boldsymbol{\nu}_w(t).\end{aligned}$$

Therefore

$$\begin{aligned}\text{L.H.S. (3.2.13)} &\le \sup|Y_w(t,\mathbf{u}) - Y_w(t,\mathbf{0}) - Y(t,\mathbf{0})| \\ &\quad + \sup|Y_w(HH_{n\mathbf{u}}^{-1}(t),\mathbf{0}) - Y(t,\mathbf{0})| \\ &\quad + 2\sup|\mu_w(t,\mathbf{u}) - \mu_w(t,\mathbf{0}) - \mathbf{u}'\boldsymbol{\nu}_w(t)| \\ &\quad + \sup|\mathbf{u}'[\boldsymbol{\nu}_w(HH_{nu}^{-1}(t)) - \boldsymbol{\nu}_w(t)]| \\ &= A_1 + A_2 + A_3 + A_4, \quad \text{say,}\end{aligned} \tag{3.2.17}$$

where, as usual, the supremum is being taken over $0 \le t \le 1$, $\|\mathbf{u}\| \le b$. In what follows, the range of x and y over which the supremum is being taken is $\mathbb{R}$, unless specified otherwise.

Now, (2.3.5) implies that $|H(x) - H(y)| \le |x - y|k$. This and (2.3.6) together imply (3.2.15). It also implies that

$$\sup_{|x-y|<\delta} |f_{ni}(y) - f_{ni}(x)| \le \sup_{|H(x)-H(y)|<k\delta} |f_{ni}(y) - f_{ni}(x)|.$$

for all $1 \le i \le n$ and all $\delta > 0$. Hence, by (3.2.12), it follows that $\{f_{ni}\}$ satisfy (2.3.4). Now apply Lemma 2.3.1 and (2.3.25), with $d_{ni} = w_{ni}$, $1 \le i \le n$, to conclude that

$$A_j = o_p(1), \qquad j = 1, 3. \tag{3.2.18}$$

Next, observe that

$$\begin{aligned}\sup|HH_{n\mathbf{u}}^{-1}(t) - t| &\le \sup_{x,\mathbf{u}}|H_{n\mathbf{u}}(x) - H_{\mathbf{u}}(x)| + n^{-1} \\ &\quad + \sup_{x,\mathbf{u}}|H_{\mathbf{u}}(x) - H(x)|,\end{aligned}$$

$$\sup_{x,\mathbf{u}}|H_u(x) - H(x)| \le \sup_x|H(x + m_n b) - H(x - m_n b)|.$$

Hence, in view of Lemma 3.2.1, we obtain

$$(3.2.19) \qquad \sup_{t,\mathbf{u}} |HH_{n\mathbf{u}}^{-1}(t) - t| \to 0, \quad \text{a.s.}$$

(We need to use the convergence in probability only).

Now, fix a $\delta > 0$ and let $B_n^\delta = [\sup_{t,\mathbf{u}} |HH_{n\mathbf{u}}^{-1}(t) - t| < \delta]$. By (3.2.19),

$$(3.2.20) \qquad \limsup_n P\left((B_n^\delta)^c\right) = 0.$$

Next, observe that $Y_d(\cdot, \mathbf{0}) = W_d^*(\cdot)$ of (2.2.23). Hence, with A_2 as in (3.2.17), for every $\eta > 0$,

$$\begin{aligned} &\limsup_n P(|A_2| \geq \eta) \\ &\leq \ \limsup_n P\left(\sup_{|t-s|<\delta} |W_w^*(t) - W_w^*(s)| \geq \eta, \ B_n^\delta \right). \end{aligned}$$

Upon letting $\delta \to 0$ in this inequality, (2.2.25) implies

$$(3.2.21) \qquad A_2 = o_p(1).$$

Next, we have

$$\begin{aligned} (3.2.22) \quad & \lim_{\delta\to 0} \limsup_n \sup_{|t-s|<\delta} \|\boldsymbol{\nu}_w(t) - \boldsymbol{\nu}_w(s)\| \\ &\leq \ \lim_{\delta\to 0} \limsup_n \max_i \sup_{|H(x)-H(y)|<\delta} |f_{ni}(y) - f_{ni}(x)| \\ &\qquad\qquad \times \sum_{i=1}^n \|w_i \mathbf{c}_i\| \\ &= 0 \ , \qquad \text{by (3.2.12) and (2.3.7).} \end{aligned}$$

From (3.2.22) and (3.2.20) one obtains, similar to (3.2.21), that $A_4 = o_p(1)$. This completes the proof of the theorem. □

From a practical point of view, it is worthwhile to state the AUL result in the i.i.d. case separately. Accordingly, we have

Theorem 3.2.2 *Suppose that $X_{n1}, \cdots, X_{nn}$ are i.i.d. F. In addition, assume that* (**F1**), (**F2**), (**N3**), (2.3.6) *and* (2.3.7) *with $d_{ni} \equiv w_{ni}$ hold. Then,* $\forall \ 0 < b < \infty$,

$$(3.2.23) \quad \sup_{0\leq t\leq 1, \|\mathbf{u}\|\leq b} \left| Z_d(t, \mathbf{u}) - Z_d(t, \mathbf{0}) - \mathbf{u}' \sum_{i=1}^n w_{ni} \mathbf{c}_{ni} q(t) \right| = o_p(1),$$

$$\text{(3.2.24)} \sup_{\varphi \in \mathcal{C}, \|\mathbf{u}\| \le b} \left| T_d(\varphi, \mathbf{u}) - T_d(\varphi, \mathbf{0}) + \mathbf{u}' \sum_{i=1}^n w_{ni} \mathbf{c}_{ni} \int q \, d\varphi \right| = o_p(1).$$

where $q = f(F^{-1})$.

Proof. Take $F_{ni} \equiv F$ in Theorem 3.2.1. Then **(F1)** and **(F2)** imply that q is uniformly continuous on $[0, 1]$ and ensure the satisfaction of all assumptions pertaining to F in Theorem 3.2.1. In addition, $\mu_w(t, \mathbf{0}) = 0$, $0 \le t \le 1$. Thus, Theorem 3.2.1 is applicable and one obtains

$$\sup_{t, \mathbf{u}} |\tilde{T}_w(t, \mathbf{u}) - Y_w(t, \mathbf{0}| = o_p(1)$$

which in turn yields

$$\text{(3.2.25)} \qquad \sup_{t, \mathbf{u}} |\tilde{T}_w(t, \mathbf{u}) - \tilde{T}_w(t, \mathbf{0})| = o_p(1).$$

Next, let $\boldsymbol{\rho} = \sum_{i=1}^n w_{ni} \mathbf{c}_{ni}$. From (3.2.7),

$$\text{L.H.S. (3.2.23)} = \sup_{t, \mathbf{u}} |Z_w(t, \mathbf{u}) - Z_w(t, \mathbf{0}) - \mathbf{u}' \boldsymbol{\rho} q(t)|.$$

Hence,

$$\begin{aligned}
&\text{L.H.S.(3.2.23)} \\
&\le \sup_{t, \mathbf{u}} \Big\{ |Z_w(t, \mathbf{u}) - \mathcal{V}_w(t, \mathbf{u})| + |Z_w(t, \mathbf{0}) - \mathcal{V}_w(t, \mathbf{0})| \\
&\qquad + |\tilde{T}_w(t, \mathbf{u}) - \tilde{T}_w(t, \mathbf{0})| + |\mu_w(t, \mathbf{u}) - \mathbf{u}' \boldsymbol{\rho} \, q(t)| \Big\} \\
&= o_p(1),
\end{aligned}$$

by (3.2.9), (3.2.10), **(N3)**, (3.2.25) and Lemma 2.3.1 applied to $F_{ni} \equiv F$, $d_{ni} \equiv w_{ni}$.

To conclude (3.2.24), observe that by (3.2.23), the uniform continuity of q and (2.3.7) with $d_{ni} \equiv w_{ni}$,

$$\begin{aligned}
\text{L.H.S.(3.2.24)} &\le \sup_{t, \mathbf{u}} \Big\{ |Z_d(t, \mathbf{u}) - Z_d(t, \mathbf{0}) - \mathbf{u}' \boldsymbol{\rho} q(t)| \\
&\qquad + |\mathbf{u}' \boldsymbol{\rho}| \; |q((n+1)t/n) - q(t)| \Big\} \\
&= o_p(1). \qquad \square
\end{aligned}$$

Remark 3.2.1 Theorem 3.2.2 continues to hold if F depends on n, provided now that the $\{q\}$ are uniformly equicontinuous on $[0, 1]$. $\square$

Remark 3.2.2 An analogue of Theorem 3.2.2 was first proved in Koul (1970) under somewhat stronger conditions on various underlying entities. In Jurečková (1969) one finds yet another variant of (3.2.24) for a fixed but a fairly general function φ and with p in $\mathbf{c}_{ni}$ equal to 1. Because of the importance of the AUL property of $T_d(\varphi, \cdot)$, it is worthwhile to compare Theorem 3.2.2 above with that of Jurečková Theorem 3.1 (1969). For the sake of completeness we state it as

Theorem 3.2.3 (Theorem 3.1, Jurečková (1969)). *Let $X_{n1}, \cdots, X_{nn}$ be i.i.d. F. In addition, assume the following:*

(a) *F has an absolutely continuous density f whose a.e. derivative $\dot{f}$ satisfies*

$$0 < I(f) < \infty, \quad I(f) := \int (\dot{f}/f)^2 dF.$$

(b) *$\{w_{ni}\}$ satisfy* (**N3**).

(c) 1. $\sum_{i=1}^n (c_{ni} - \bar{c}_n)^2 \le M < \infty$, *(recall here c_{ni} is 1×1),*
2. $\max_i (c_{ni} - \bar{c}_n)^2 = o(1)$, $\bar{c}_n = n^{-1} \sum_{i=1}^n c_{ni}$.

(d) *φ is a nondecreasing square integrable function on $(0,1)$ with*

$$\int_0^1 (\varphi(t) - \overline{\varphi})^2 dt > 0, \quad \overline{\varphi} := \int_0^1 \varphi(u) du.$$

(e) *Either* $(d_{ni} - d_{nj})(c_{ni} - c_{nj}) \ge 0, \ \forall\, 1 \le i, j \le n$,
or $(d_{ni} - d_{nj})(c_{ni} - c_{nj}) \le 0, \ \forall\, 1 \le i, j \le n$.

Then, $\forall\ 0 < b < \infty$,

$$\sup_{\|\mathbf{u}\| \le b} \left| T_d(\varphi, u) - T_d(\varphi, 0) + u \sum_{i=1}^n w_{ni} c_{ni} b(\varphi, f) \right| = o_p(1),$$

where $b(\varphi, f) := -\int_{-\infty}^{\infty} \varphi(F(x)) \dot{f}(x) dx$. □

The strongest point of Theorem 3.2.3 is that it allows for unbounded score functions, such as the "Normal scores" that corresponds to $\varphi = \Phi^{-1}$, Φ being the d.f. of a $\mathcal{N}(0,1)$ r.v. However, this is balanced by requiring (a), (c1) and (e). Note that (b) and (c1) together imply (2.3.7) with $d_{ni} = w_{ni}, 1 \le i \le n$. Moreover, Theorem 3.2.2 does not require anything like (e).

Claim 3.2.1 (a) *implies that* f *is* $Lip(1/2)$.

First, from Hájek - Šidák (1967), pp. 19-20, we recall that (a) implies that $f(x) \to 0$ as $|x| \to \infty$. Now, absolute continuity and nonnegativity of f implies that

$$|f(x) - f(y)| \le \int_x^y |(\dot{f}/f)| dF, \quad x < y.$$

Therefore, by the Cauchy-Schwarz inequality, for $x < y$,

$$(3.2.26) \quad |f(x) - f(y)| \le \left\{ \int_x^y (\dot{f}/f)^2 dF \cdot [F(y) - F(x)] \right\}^{1/2}$$

$$(3.2.27) \quad \le I^{1/2}(f).$$

Letting $y \to \infty$ in (3.2.27) yields

$$(3.2.28) \quad \|f\|_\infty \le I^{1/2}(f).$$

Now (3.2.26) and (3.2.28) together imply

$$|f(x) - f(y)| \le I^{1/2}(f) \{ \int_x^y f(t) dt \}^{1/2} \le I^{3/4}(f)(y - x)^{1/2}.$$

A similar inequality holds for $x > y$, thereby giving

$$|f(x) - f(y)| \le I^{3/4}(f)|y - x|^{1/2}, \quad \forall\, x, y \in \mathbb{R},$$

and proving the claim. Consequently, (a) *implies* (**F1**).

Note that f can be uniformly continuous, bounded, positive a.e., yet need not satisfy $I(f) < \infty$. For example, consider

$$\begin{aligned} f(x) &:= (1 - x)/2, \quad 0 \le x \le 1, \\ &:= (x - 2j + 1)/2^{j+2}, \quad 2j - 1 \le x \le 2j \\ &:= (2j + 1 - x)/2^{j+2}, \quad 2j \le x \le 2j + 1, \quad j \ge 1; \\ f(x) &:= f(-x), \quad x \le 0. \end{aligned}$$

The above discussion shows that both Theorems 3.2.2 and 3.2.3 are needed. Neither displaces the other. If one is interested in the AUL property of, say, Normal scores type rank statistics, then Theorem 3.2.3 gives an answer. On the other hand if one is interested in the AUL property of, say, the Wilcoxon type rank statistics, then Theorem 3.2.2 provides a better result.

The proof of Theorem 3.2.3 uses contiguity and projection technique *a la* Hájek (1962) to approximate $T_d(\varphi, u)$ for each fixed u. Then condition (e) implies the monotonicity of $T_d(\varphi, \cdot)$ which yields the uniformity with respect to u. Such a proof is harder to extend to the case where u and c_{ni} are $p \times 1$ vectors; this has been done by Jurečková (1971).

The proof of Theorem 3.2.2 exploits the monotonicity inherent in the W.E.P.'s Y_d and certain smoothness properties of F. It would be desirable to *extend this proof to include unbounded* φ. □

We now return to Theorem 3.2.1 with general $\{F_{ni}\}$. We wish to state an AUL theorem for $\{Z_d\}$ and $\{T_d(\varphi, \cdot)\}$ under general $\{F_{ni}\}$. Theorem 3.2.1 still does not quite do it because there is $\mathbf{u}$ in μ_w - expressions. We need to carry out an expansion of these terms in order to recover a term that is linear in $\mathbf{u}$. To that effect we have

Lemma 3.2.2 *In addition to the assumptions of Theorem* 3.2.1, *suppose that*

$$(3.2.29) \qquad n^{-1/2} \sum_{i=1}^{n} \|\mathbf{c}_{ni}\| = O(1).$$

Then, $\forall\ 0 < b < \infty$,

$$(3.2.30) \qquad \sup_{0 \le t \le 1, \|\mathbf{u}\| \le b} \left| n^{1/2}(HH_{n\mathbf{u}}^{-1}(t) - t) + Y_1(t, \mathbf{0}) + \mathbf{u}'\boldsymbol{\nu}_1(t) \right| = o_p(1),$$

where $Y_1, \boldsymbol{\nu}_1$ etc. *are* Y_d, $\boldsymbol{\nu}_d$ *of* (2.3.1), (2.3.10) *with* $d_{ni} \equiv n^{-1/2}$.

Consequently,

$$(3.2.31) \qquad \sup_{0 \le t \le 1, \|\mathbf{u}\| \le b} |n^{1/2}(HH_{n\mathbf{u}}^{-1}(t) - t)| = O_p(1).$$

Proof. Write $Y_1(\cdot), \mu_1(\cdot)$ for $Y_1(\cdot, \mathbf{0}), \mu_1(\cdot, \mathbf{0})$, respectively. Let I

denote the identity function and set $\Delta_{nu} := n^{1/2}(H_{nu}H_{nu}^{-1} - I)$. Then

$$\begin{aligned}
& n^{1/2}(HH_{nu}^{-1} - I) \\
&= n^{1/2}(HH_{nu}^{-1} - H_{\mathbf{u}}H_{nu}^{-1} - H_{nu}H_{nu}^{-1}) + \Delta_{nu} \\
&= -[\mu_1(HH_{nu}^{-1}, \mathbf{u}) - \mu_1(HH_{nu}^{-1}) - \mathbf{u}'\boldsymbol{\nu}_1(HH_{nu}^{-1})] + \Delta_{nu} \\
& \quad -\mathbf{u}'[\boldsymbol{\nu}_1(HH_{nu}^{-1}) - \boldsymbol{\nu}_1 - Y_1 \\
& \quad -[Y_1(HH_{nu}^{-1}, \mathbf{u}) - Y_1(HH_{nu}^{-1})] - [Y_1(HH_{nu}^{-1}) - Y_1]. \qquad (3.2.32)
\end{aligned}$$

Now, note that $\sup_{t,\mathbf{u}} |\Delta_{nu}(t)| \le n^{-1/2}$. Hence

$$\begin{aligned}
& \sup_{t,\mathbf{u}} |n^{1/2}(HH_{nu}^{-1}(t) - t) + Y_1(t) + \mathbf{u}'\boldsymbol{\nu}_1(t)| \\
& \le \sup_{t,\mathbf{u}} |\mu_1(t, \mathbf{u}) - \mu_1(t) - \mathbf{u}'\boldsymbol{\nu}_1(t)| \\
& \quad + B \sup_{t,\mathbf{u}} \|\boldsymbol{\nu}_1(HH_{nu}^{-1}(t)) - \boldsymbol{\nu}_1(t)\| + \sup_{t,\mathbf{u}} |Y_1(t, \mathbf{u}) - Y_1(t)| \\
& \quad + \sup_{t,\mathbf{u}} |W_1^*(HH_{nu}^{-1}(t)) - W_1^*(t)|, \qquad (3.2.33)
\end{aligned}$$

where we have used the fact that $Y_1(t) = W_1^*(t)$ of (2.2.23). The first term on the r.h.s. of (3.2.33) tends to zero by Lemma 2.3.1 when applied with $d_{ni} \equiv n^{-1/2}$. The third term tends to zero in probability by (2.3.25) applied with $d_{ni} \equiv n^{-1/2}$. To show that the other two terms go to zero in probability, use Lemma 3.2.1, (2.2.25) and an analogue of (3.2.22) for $\boldsymbol{\nu}_1$ and an argument similar to the one that yielded (3.2.21) and (3.2.23) above. Thus we have (3.2.30). Since $\sup_{t,\mathbf{u}} |Y_1(t, \mathbf{0}) + \mathbf{u}'\boldsymbol{\nu}_1(t)| = O_p(1)$, (3.2.31) follows. □

Lemma 3.2.3 *In addition to the assumptions of Theorem 3.2.1 and (3.2.29), suppose that for every $0 < k < \infty$,*

$$\max_i \sup_{|t-s| \le kn^{-1/2}} n^{1/2}|L_{ni}(t) - L_{ni}(s) - (t-s)\ell_{ni}(s)| = o_p(1) \qquad (3.2.34)$$

where $L_{ni} := F_{ni}H^{-1}$, $\ell_{ni} := f_{ni}(H^{-1})/h(H^{-1})$, $1 \le i \le n$, with $h := n^{-1}\sum_{i=1}^n f_{ni}$. Moreover, suppose that, with

$$\tilde{w}(t) := n^{-1}\sum_{i=1}^n w_{ni}\ell_{ni}(t), \quad 0 \le t \le 1,$$

(3.2.35) $$\sup_{0\le t\le 1} n^{1/2}|\tilde{w}(t)| = O(1).$$

Then, $\forall\ 0 < b < \infty$,

(3.2.36) $$\sup\Big|\mu_w(HH_{n\mathbf{u}}^{-1}(t)) - \mu_w(t) + \{Y_1(t) + \mathbf{u}'\boldsymbol{\nu}_1(t)\}n^{1/2}\tilde{w}(t)\Big| = o_p(1),$$

where $\mu_w(t), Y_1(t)$ *stand for* $\mu_w(t, \mathbf{0}), Y_1(t, \mathbf{0})$, *respectively, and where the supremum is being taken over* $0 \le t \le 1, \|\mathbf{u}\| \le b$.

Proof. Let $M_{\mathbf{u}} := \mu_w(HH_{n\mathbf{u}}^{-1}) - \mu_w$. From (3.2.31) it follows that $\forall\ \epsilon > 0,\ \exists K_\epsilon$ and $N_{1\epsilon}$ such that

(3.2.37) $$P(A_n^\epsilon) \ge 1 - \epsilon, \quad n \ge N_1\epsilon,$$

where

$$A_n^\epsilon = \sup_{t,\mathbf{u}} |HH_{n\mathbf{u}}^{-1}(t) - t| \le K_\epsilon n^{-1/2}].$$

By assumption (3.2.34), there exists $N_{2\epsilon}$ such that $n \ge N_{2\epsilon}$ implies

(3.2.38) $$\max_i \sup_{|t-s|\le K_\epsilon n^{-1/2}} n^{1/2}|L_i(t) - L_i(s) - (t-s)\ell_i(s)| < \epsilon.$$

Define

$$Z_{\mathbf{u}i}^\epsilon := \{L_i(HH_{n\mathbf{u}}^{-1}) - L_i - [HH_{n\mathbf{u}}^{-1} - I]\ell_i\}I(A_n^\epsilon), \quad 1 \le i \le n.$$

In view of (3.2.38) and (3.2.37),

(3.2.39) $$P\left(\max_i \sup_{t,\mathbf{u}} n^{1/2}|Z_{\mathbf{u}i}^\epsilon(t)| > \epsilon\right) < \epsilon,$$

for all $n \ge N_\epsilon := N_{1\epsilon} \vee N_{2\epsilon}$. Moreover,

$$\begin{aligned} M_{\mathbf{u}} &= M_{\mathbf{u}}I(A_n^\epsilon) + M_{\mathbf{u}}I((A_n^\epsilon)^c) \\ &= \sum_{i=1}^n w_i Z_{\mathbf{u}i}^\epsilon + Z_{\mathbf{u}0} + n^{1/2}[HH_{n\mathbf{u}}^{-1} - I]n^{1/2}\tilde{w}, \end{aligned}$$

where

$$Z_{\mathbf{u}0}^\epsilon := \{M_{\mathbf{u}} - n^{1/2}[HH_{n\mathbf{u}}^{-1} - I] \cdot n^{1/2}\tilde{w}\}I((A_n^\epsilon)^c).$$

Note that

$$P(\sup_{t,\mathbf{u}} |Z_{\mathbf{u}0}^{\epsilon}| \neq 0) \leq P((A_n^{\epsilon})^c) < \epsilon, \quad \forall\, n > N_{\epsilon}.$$

By the C-S inequality, (**N3**) and (3.2.39),

$$P\left(\sup_{t,\mathbf{u}} \left|\sum_{i=1}^{n} w_i Z_{\mathbf{u}i}^{\epsilon}(t)\right| > \epsilon\right) \leq \epsilon, \quad \forall \ n > N_{\epsilon}.$$

These derivations together with Lemma 3.2.2 and (3.2.35) complete the proof of (3.2.36). □

We combine Theorem 3.2.1, Lemmas 3.2.2 and 3.2.3 to obtain the following

Theorem 3.2.4 *Under the notation and assumptions of Theorem* 3.2.1, *Lemmas* 3.2.2 *and* 3.2.3, $\forall\ 0 < b < \infty$,

$$(3.2.40) \quad \sup \left|Z_d(t,\mathbf{u}) - Z_d(t,\mathbf{0}) - \mathbf{u}' \sum_{i=1}^{n} (d_{ni} - \tilde{d}_n(t)) \mathbf{c}_{ni} q_{ni}(t)\right| = o_p(1),$$

$$(3.2.41) \quad \sup \left|T_d(\varphi,\mathbf{u}) - T_d(\varphi,\mathbf{0}) + \mathbf{u}' \int \sum_{i=1}^{n} (d_{ni} - \tilde{d}_n(t)) \mathbf{c}_{ni} q_{ni}(t) d\varphi(t)\right| = o_p(1),$$

where the supremum in (3.2.40) *is over* $0 \leq t \leq 1, \|\mathbf{u}\| \leq b$, *in* (3.2.41) *over* $\varphi \in \mathcal{C}, \|\mathbf{u}\| \leq b$, *and where* $\tilde{d}_n(t) := n^{-1} \sum_{i=1}^{n} d_{ni} \ell_{ni}(t)$, $q_{ni} := f_{ni}(H^{-1}(t)),\ 0 \leq t \leq 1,\ 1 \leq i \leq n.$

Proof. Let $\boldsymbol{\rho}(t) := \sum_{i=1}^{n} (d_i - \tilde{d}(t)) \mathbf{c}_i q_i(t)$. Note that the fact that $n^{-1} \sum_{i=1}^{n} \ell_i(t) \equiv 1$ implies that $\boldsymbol{\rho}(t) = \sum_{i=1}^{n} (w_i - \tilde{w}(t)) \mathbf{c}_i q_i(t)$, where $\{w_i\}$ are as in (3.2.6). From (3.2.7)-(3.2.9),

$$\begin{aligned}(3.2.42)\ \text{L.H.S.}(3.2.40) &= \sup_{0 \leq t \leq 1, \|\mathbf{u}\| \leq b} |Z_w(t,\mathbf{u}) - Z_w(t,\mathbf{0}) - \mathbf{u}'\boldsymbol{\rho}(t)| \\ &\leq 4 \max_i |w_i| + \sup_{0 \leq t \leq 1, \|\mathbf{u}\| \leq b} |\mathcal{V}_w(t,\mathbf{u}) - \mathcal{V}_w(t,\mathbf{0}) - \mathbf{u}'\boldsymbol{\rho}(t)|.\end{aligned}$$

Now, from Theorem 3.2.1 and Lemma 3.2.3, uniformly in $0 \le t \le 1, \|\mathbf{u}\| \le b$,

$$\text{(3.2.43)} \qquad \sup |\tilde{T}_w(t, \mathbf{u}) - Y_w(t) + \{Y_1(t) + \boldsymbol{\nu}_1(t)\mathbf{u}\} n^{1/2} \tilde{w}(t)| = o_p(1),$$

where $Y_d(t)$ stands for $Y_d(t, \mathbf{0})$ for arbitrary weights $\{d_{ni}\}$. Therefore,

$$\begin{aligned}
&\sup |\mathcal{V}_w(t, \mathbf{0}) - \mathcal{V}_w(t, \mathbf{0}) - \mathbf{u}'\boldsymbol{\rho}(t)| \\
&\quad = \sup |\tilde{T}_w(t, \mathbf{u}) - \tilde{T}_w(t, \mathbf{0}) + \mu_w(t, \mathbf{u}) - \mu_w(t, \mathbf{0}) - \mathbf{u}'\boldsymbol{\rho}(t)| \\
&\quad \le \sup |\tilde{T}_w(t, \mathbf{u}) - \tilde{T}_w(t, \mathbf{0}) + \mathbf{u}'\boldsymbol{\nu}_1(t) n^{1/2} \tilde{w}(t)| \\
&\qquad + \sup |\mu_w(t, \mathbf{u}) - \mu_w(t, \mathbf{0}) - \mathbf{u}'\boldsymbol{\nu}_w(t)| = o_p(1),
\end{aligned}$$

by (3.2.43) and Lemma 2.3.1 and the fact that $\boldsymbol{\rho}(t) = \boldsymbol{\nu}_w(t) - \boldsymbol{\nu}_1(t) n^{1/2} \tilde{w}(t)$. This completes the proof (3.2.40). The proof of (3.2.41) follows from (3.2.40) in the same fashion as does that of (3.2.24) from (3.2.23). □

Remark 3.2.3 As in Remark 2.2.3, suppose we strengthen **(N3)** to require

$$\textbf{(B1)} \qquad n \max_i w_{ni}^2 = O(1), \qquad \tau_w^2 = 1.$$

Then **(C*)** and (3.2.35) are *a priori* satisfied by L_w. □

Remark 3.2.4 If one is interested in the i.i.d. case only, then Theorem 3.2.2 gives a better result than Theorem 3.2.4. □

3.3 AUL of Linear Signed Rank Statistics

In this section our aim is to prove analogs of Theorems 3.2.2 and 3.2.4 for the signed rank processes $\{T_d^*(\varphi, \mathbf{u}), \mathbf{u} \in \mathbb{R}^p\}$, using as many results from the previous sections as possible. Many details are quite similar. Define, for $\mathbf{u} \in \mathbb{R}^p$, $0 \le t \le 1$, $x \ge 0$,

$$\text{(3.3.1)} \qquad \begin{aligned}
Z_d^+(t, \mathbf{u}) &:= \sum_{i=1}^n d_{ni} I(R_{i\mathbf{u}}^+ \le nt)\, s(X_{ni} - \mathbf{c}_{ni}'\mathbf{u}), \\
J_{n\mathbf{u}}(x) &:= n^{-1} \sum_{i=1}^n I(|X_{ni} - \mathbf{c}_{ni}'\mathbf{u}| \le x) \\
&= H_{n\mathbf{u}}(x) - H_{n\mathbf{u}}(-x),
\end{aligned}$$

$$\begin{aligned}
J_{\mathbf{u}}(x) &:= n^{-1}\sum_{i=1}^{n}[F_{ni}(x+\mathbf{c}_{ni}'\mathbf{u}) - F_{ni}(-x+\mathbf{c}_{ni}'\mathbf{u})] \\
&= H_{\mathbf{u}}(x) - H_{\mathbf{u}}(-x), \\
\mathcal{V}_d^+(t,\mathbf{u}) &:= \sum_{i=1}^{n} d_{ni} I(|X_{ni}-\mathbf{c}_{ni}'\mathbf{u}| \le J_{n\mathbf{u}}^{-1}(t))\, s(X_{ni}-\mathbf{c}_{ni}'\mathbf{u}), \\
S_d^+(t,\mathbf{u}) &:= \sum_{i=1}^{n} d_{ni} I(|X_{ni}-\mathbf{c}_{ni}'\mathbf{u}| \le J^{-1}(t))\, s(X_{ni}-\mathbf{c}_{ni}'\mathbf{u}), \\
\mu_d^+(t,\mathbf{u}) &:= \sum_{i=1}^{n} d_{ni}\mu_{ni}^+(t,\mathbf{u}) = ES_d^+(t,\mathbf{u}), \\
\mu_{ni}^+(t,\mathbf{u}) &:= F_{ni}(J^{-1}(t)+\mathbf{c}_{ni}\mathbf{u}) + F_{ni}(-J^{-1}(t)+\mathbf{c}_{ni}'\mathbf{u}) \\
&\qquad\qquad -2F_{ni}(\mathbf{c}_{ni}'\mathbf{u}), \qquad 1 \le i \le n.
\end{aligned}$$

In the above and sequel, J and J_n stand for $J_{\mathbf{0}}$ and $J_{n\mathbf{0}}$, respectively. We also need

$$\begin{aligned}
(3.3.2)\quad Y_d^+(t,\mathbf{u}) &:= S_d^+(t,\mathbf{u}) - \mu_d^+(t,\mathbf{u}), \\
\tilde{T}_d^+(t,\mathbf{u}) &:= \mathcal{V}_d^+(t,\mathbf{u}) - \mu_d^+(t,\mathbf{u}), \quad 0 \le t \le 1, \ \mathbf{u} \in \mathbb{R}^p.
\end{aligned}$$

Analogous to (3.2.11), we have the basic decomposition: For $0 \le t \le 1, \ \mathbf{u} \in \mathbb{R}^p$,

$$(3.3.3)\quad \tilde{T}_d^+(t,\mathbf{u}) = Y_d^+(JJ_{n\mathbf{u}}^{-1}(t),\mathbf{u}) + \mu_d^+(JJ_{n\mathbf{u}}^{-1}(t),\mathbf{u}) - \mu_d^+(t,\mathbf{u}),$$

Now, note that, w.p. 1, for all $0 \le t \le 1, \mathbf{u} \in \mathbb{R}^p$,

$$\begin{aligned}
(3.3.4)\quad Y_d^+(t,\mathbf{u}) &= Y_d(HJ^{-1}(t),\mathbf{u}) \\
&\quad + Y_d(H(-J^{-1}(t)),\mathbf{u}) - 2Y_d(H(0),\mathbf{u}),
\end{aligned}$$

where Y_d is as in (2.3.1). Therefore, by Theorem 2.3.1 (see (2.3.25)), under the assumptions of that theorem and strictly increasing nature of J and H,

$$(3.3.5)\qquad \sup_{t,\mathbf{u}} |Y_d^+(t,\mathbf{u}) - Y_d^+(t,\mathbf{0})| = o_p(1).$$

One also has, in view of the continuity of $\{F_{ni}\}$, a relation like (3.3.4) between μ_d^+ and μ_d. Thus by Lemma 2.3.1, under the assumptions there,

$$(3.3.6)\qquad \sup_{t,\mathbf{u}} |\mu_d^+(t,\mathbf{u}) - \mu_d^+(t,\mathbf{0}) - \mathbf{u}'\boldsymbol{\nu}^+(t)| = o_p(1).$$

where, for $0 \le t \le 1$,

$$\nu_d^+(t) \;:=\; \sum_{i=1}^n d_{ni}\mathbf{c}_{ni}[f_{ni}(J^{-1}(t)) + f_{ni}(-J^{-1}(t)) - 2f_{ni}(0)].$$

We also have an analogue of Lemma 3.2.1:

Lemma 3.3.1 *Without any assumption except* (2.2.24),

$$\sup_{0\le x\le\infty} |J_n(x) - J(x)| \to 0 \;\text{ a.s.}$$

If, in addition, (2.3.6) *and* (3.2.15) *hold, then*

$$\sup_{0\le x\le\infty, \|\mathbf{u}\|\le b} |J_{n\mathbf{u}}(x) - J_{\mathbf{u}}(x)| \to 0 \;\text{ a.s..}$$

Using this lemma, arguments like those in Theorem 3.2.1 and the above discussion, one obtains

Theorem 3.3.1 *Suppose that* $\{X_{ni}, F_{ni}\}$ *satisfy* (2.2.24), (2.3.5) *and that* $\{d_{ni},\ \mathbf{c}_{ni}\}$ *satisfy* (**N1**), (**N2**), (2.3.6) *and* (2.3.7). *In addition, assume that*

$$\text{(3.3.7)} \qquad \lim_{\delta\to 0}\limsup_n \max_i \sup_{|J(x)-J(y)|<\delta} |f_{ni}(x) - f_{ni}(y)| = 0$$

and that H *is strictly increasing for every* n. *Then, for every* $0 < b < \infty$,

$$\text{(3.3.8)} \quad \sup_{0\le t\le 1, \|\mathbf{u}\|\le b} |\tilde{T}_d^+(t,\mathbf{u}) - Y_d^+(t,\mathbf{0}) - \mu_d^+(JJ_{n\mathbf{u}}^{-1}(t),\mathbf{0}) + \mu_d^+(t,\mathbf{0})| = o_p(1). \qquad \square$$

We remark here that (3.3.7) implies (3.2.12).

Next, note that if $\{F_i\}$ are symmetric about 0, then

$$\text{(3.3.9)} \qquad \mu_d^+(t,\mathbf{0}) = 0, \quad 0 \le t \le 1, \quad n \ge 1.$$

Upon combining (3.3.9), (3.3.8) with (3.3.6) one obtains

Theorem 3.3.2 *In addition to the assumptions of Theorem* 3.2.1, *suppose that* $\{F_{ni}, 1 \le i \le n\}$ *are symmetric about* 0.

Then, for every $0 < b < \infty$,

$$\text{(3.3.10)} \qquad \sup_{0\le t\le 1, \|\mathbf{u}\|\le b} \Big| Z_d^+(t,\mathbf{u}) - Z_d^+(t,\mathbf{0}) - \mathbf{u}'\sum_{i=1}^n d_{ni}\mathbf{c}_{ni}\nu_{ni}^+(t)\Big| = o_p(1),$$

$$(3.3.11)\quad \sup\Big|T_d^+(\varphi,\mathbf{u}) - G_d^+(\varphi,\mathbf{0}) + \mathbf{u}'\sum_{i=1}^n d_{ni}\mathbf{c}_{ni}\int_0^1 \nu_{ni}^+(t)d\varphi^+(t)\Big| = o_p(1),$$

where

$$\nu_{ni}^+(t) := 2[f_{ni}(J^{-1}(t)) - f_{ni}(0)], \quad 1 \le i \le n, \quad 0 \le t \le 1,$$

and where the supremum in (3.3.11) is taken over $\varphi \in \mathcal{C}, \|\mathbf{u}\| \le b$.

Proof. Using a relation like (3.2.5) between $R_{i\mathbf{u}}^+$ and $J_{n\mathbf{u}}$, by (**N2**), one obtains, as in (3.2.9),

$$\sup_{t,\mathbf{u}} |Z_d^+(t,\mathbf{u}) - \mathcal{V}_d^+(t,\mathbf{u})| \le 2\max_i |d_i| = o(1),$$

Thus (3.3.9) follows from this, (3.3.8), (3.3.7) and (3.3.6). The conclusion (3.3.11) follows from (3.3.9) in the same way as (3.2.24) follows from (3.2.23). □

Because of the importance of the i.i.d. symmetric case, we specialize the above theorem to yield

Corollary 3.3.1 *Let* F *be a d.f. symmetric around zero, satisfying* (**F1**), (**F2**) *and let* $X_{n1}, \cdots, X_{nn}$ *be i.i.d.* F. *In addition, assume that* $\{d_{ni}, \mathbf{c}_{ni}\}$ *satisfy* (**N1**), (**N2**), (2.3.6) *and* (2.3.7). *Then, for every* $0 < b < \infty$,

$$(3.3.12)\quad \sup_{0\le t\le 1, \|\mathbf{u}\|\le b} \Big|Z_d^+(t,\mathbf{u}) - Z_d^+(t,\mathbf{0}) - \mathbf{u}'\Sigma_i d_{ni}\mathbf{c}_{ni} q^+(t)\Big| = o_p(1),$$

$$(3.3.13)\quad \sup_{\varphi\in\mathcal{C}, \|\mathbf{u}\|\le b} \Big|T_d^+(\varphi,\mathbf{u}) - T_d^+(\varphi,\mathbf{0}) + \Sigma_i d_{ni}\mathbf{c}_{ni}'\mathbf{u}\int_0^1 q^+(t)d\varphi^+(t)\Big| = o_p(1),$$

where $q^+(t) := 2[f(F^{-1}((t+1)/2)) - f(0)]$, $0 \le t \le 1$. □

Remark 3.3.1 van Eeden (1972) proved an analogue of (3.3.13) without the supremum over φ, but for square integrable φ's. She also needs conditions like those in Theorem 3.2.3 above. Thus Remark 3.2.1 is equally applicable here when comparing Corollary 3.2.1 with van Eeden's results. □

Now, we return to Theorem 3.3.1 and expand the μ_d^+-terms further so as to recover an extra linearity term. Define, for $0 \leq t \leq 1, \mathbf{u} \in \mathbb{R}^p$,

$$Y_d^*(t, \mathbf{u}) := \sum_{i=1}^{n} d_{ni}[I(|X_{ni} - \mathbf{c}_{ni}\mathbf{u}| \leq J^{-1}(t)) - F_{i\mathbf{u}}^+(J^{-1}(t))],$$

$$\boldsymbol{\nu}_d^*(t) := \sum_{i=1}^{n} d_{ni}\mathbf{c}_{ni}[f_{ni}(J^{-1}(t)) - f_{ni}(-J^{-1}(t))],$$

where

$$F_{i\mathbf{u}}^+(x) := F_{ni}(x + \mathbf{c}_i'\mathbf{u}) - F_{ni}(-x + \mathbf{c}_i'\mathbf{u}), \quad x \geq 0.$$

Note the relation: For arbitrary $\{d_{ni}\}$,

$$Y_d^*(t, \mathbf{u}) \equiv Y_d(HJ^{-1}(t), \mathbf{u}) - Y_d(H(-J^{-1}(t)), \mathbf{u}). \tag{3.3.14}$$

From (3.3.14) and (2.3.25) applied with $d_{ni} = n^{-1/2}$, we obtain

$$\sup_{t,\mathbf{u}} |Y_1^*(t, \mathbf{u}) - Y_1^*(t, \mathbf{0})| = o_p(1). \tag{3.3.15}$$

Note that in the case $d_{ni} \equiv n^{-1/2}$, (2.3.7) reduces to (3.2.29).

Next, under (3.3.7) and (2.3.7), just as (3.2.22),

$$\lim_{\delta \to 0} \limsup_{n} \sup_{|t-s|<\delta} \|\boldsymbol{\nu}_d^*(t) - \boldsymbol{\nu}_d^*(s)\| = 0, \tag{3.3.16}$$

for the given $\{d_{ni}\}$ and for $d_{ni} \equiv n^{-1/2}$.

Using (3.3.15), (3.3.16) and calculations similar to those done in the proof of Lemma 3.2.2, we obtain

Lemma 3.3.2 *Under the conditions of Theorem* (3.2.) *and* (3.2.29)

$$\sup_{t,\mathbf{u}} |n^{1/2}(JJ_{n\mathbf{u}}^{-1}(t) - t) + Y_1^*(t, \mathbf{0}) + \mathbf{u}'\boldsymbol{\nu}_1^*(t)| = o_p(1). \tag{3.3.17}$$

Consequently,

$$\sup_{t,\mathbf{u}} \left| n^{1/2} \left[JJ_{n\mathbf{u}}^{-1}(t) - t \right] \right| = O_p(1). \tag{3.3.18}$$

□

Similarly arguing as in Lemma 3.2.3, we obtain the following Lemma 3.3.3. In it $\mu_d^+(t)$, $\mu_1^+(t)$ etc. stand for $\mu_d^+(t, \mathbf{0}), \mu_1^+(t, \mathbf{0})$ etc. of (3.3.1).

Lemma 3.3.3 *In addition to the assumptions of Theorem* 3.2.1, (3.2.29), *assume that for every* $0 < k < \infty$,

$$\text{(3.3.19)} \quad \max_i \sup_{|t-s| \le kn^{-1/2}} n^{1/2} |\mu_{ni}^+(t) - \mu_{ni}^+(s) - (t-s)\ell_{ni}^+(s)| = o(1)$$

where $\{\mu_{ni}^+\}$ *are as in* (3.3.1),

$$\ell_{ni}^+(s) := [f_{ni}(J^{-1}(s)) - f_{ni}(-J^{-1}(s))]/h^+(J^{-1}(s)), \quad 0 \le s \le 1,$$

$$h^+(x) := n^{-1} \sum_{i=1}^n [f_{ni}(x) - f_{ni}(-x)], \quad x \ge 0.$$

Moreover, with $\tilde{d}_n^+(t) := n^{-1} \sum_{i=1}^n d_{ni} \ell_{ni}^+(t), 0 \le t \le 1$, *assume that*

$$\text{(3.3.20)} \qquad \sup_{0 \le t \le 1} |n^{1/2} \tilde{d}_n^+(t)| = O(1).$$

Then, for every $0 < b < \infty$,

$$\text{(3.3.21)} \ \sup |\mu_d^+(JJ_{n\mathbf{u}}^{-1}(t)) - \mu_d^+(t) + \{Y_1^*(t) + \mathbf{u}'\boldsymbol{\nu}_1^*(t)\} n^{1/2} \tilde{d}_n^+(t)| = o_p(1),$$

where the supremum is taken over the set $0 \le t \le 1, \|\mathbf{u}\| \le b$. □

Finally, an analogue of Theorem 3.2.3 is

Theorem 3.3.3 *Under the assumptions of Theorem* 3.3.1, (3.2.29), (3.3.19) *and* (3.3.20), *for every* $0 < B < \infty$,

$$\text{(3.3.22)} \quad \sup_{0 \le t \le 1, \|\mathbf{u}\| \le b} \left| Z_d^+(t, \mathbf{u}) - Z_d^+(t, \mathbf{0}) - \mathbf{u}'[\boldsymbol{\nu}_d^+(t) - \boldsymbol{\nu}_1^*(t) n^{1/2} \tilde{d}_n^+(t)] \right| = o_p(1),$$

$$\text{(3.3.23)} \quad \sup_{\varphi \in \mathcal{C}, \|\mathbf{u}\| \le b} \left| T_d^+(\varphi, \mathbf{u}) - T_d^+(\varphi, \mathbf{0}) + \mathbf{u}' \int_0^1 [\boldsymbol{\nu}_d^+(t) - \boldsymbol{\nu}_1^*(t) n^{1/2} \tilde{d}_n^+(t)] d\varphi^+(t) \right| = o_p(1).$$

□

Remark 3.3.2 Unlike the case in Theorem 3.2.3, there does not appear to be a nice simplification of the term $\boldsymbol{\nu}_d^+ - \boldsymbol{\nu}_1^* n^{1/2} \tilde{d}_n^+$. However, it can be rewritten as follows:

$$\begin{aligned}
&\boldsymbol{\nu}_d^+(t) - \boldsymbol{\nu}_1^*(t)n^{1/2}\tilde{d}_n^+(t) \\
&= \sum_{i=1}^n d_i\mathbf{c}_i[f_i(J^{-1}(t)) + f_i(-J^{-1}(t)) - 2f_i(0)] \\
&\quad + \Sigma_i(d_i - \tilde{d}_n^+(t))\mathbf{c}_i[f_i(J^{-1}(t)) - f_i(-J^{-1}(t))].
\end{aligned}$$

This representation is somewhat revealing in the following sense. The first term is due to the shift $\mathbf{u}'\mathbf{c}_i$ in the r.v. X_i and the second term is due to the nonidentical and asymmetric nature of the distribution of $X_i, 1 \le i \le n$. □

Remark 3.3.3 If one is interested in the symmetric case or in the i.i.d. symmetric case then Theorem 3.3.2 and Corollary 3.3.1, respectively, give better results than Theorem 3.3.3. □

3.4 Weak Convergence of Rank and Signed Rank W.E.P.'s.

Throughout this section we shall use the notation of Sections 3.2 - 3.3 with $\mathbf{u} = \mathbf{0}$. Thus, e.g., $Z_d(t)$, $Z_d^+(t)$, etc. will represent $Z_d(t, \mathbf{0})$, $Z_d^+(t, \mathbf{0})$, etc. of (3.2.2) and (3.3.1), i.e., for $0 \le t \le 1$,

$$(3.4.1) \quad Z_d(t) = \sum_{i=1}^n d_{ni} I(R_{ni} \le nt),$$

$$Z_d^+(t) = \sum_{i=1}^n d_{ni} I(R_{ni}^+ \le nt) s(X_{ni}),$$

$$\mathcal{V}_d(t) = \sum_{i=1}^n d_{ni} I(X_{ni} \le H_n^{-1}(t)), \quad \mu_d(t) = \sum_{i=1}^n d_{ni} L_{ni}(t),$$

where R_{ni} (R_{ni}^+) is the rank of X_{ni} ($|X_{ni}|$) among $X_{n1}, \cdots, X_{nn}$ ($|X_{n1}|, \cdots, |X_{nn}|$).

We shall first prove the asymptotic normality of Z_d and Z_d^+ for a fixed t, say $t = v$, $0 < v < 1$. Note that this corresponds to proving the asymptotic normality of simple linear rank and signed rank statistics corresponding to the score function φ that is degenerate at v.

To begin with consider $Z_d(v)$. In the following theorem v is a fixed number in $(0,1)$.

Theorem 3.4.1 *Suppose that* $\{X_{ni}\}, \{F_{ni}\}, \{L_{ni}\}, L_d$ *are as in* (2.2.23) *and* (2.2.24). *Assume that* $\{d_{ni}\}$ *satisfy* (**N1**), (**N2**) *and that* H *is strictly increasing for each* n. *Also assume that*

$$\lim_{\delta\to 0}\limsup_n [L_d(v+\delta) - L_d(v-\delta)] = 0, \tag{3.4.2}$$

and that there are nonnegative numbers $\ell_{ni}(v), 1 \le i \le n$, *such that for every* $0 < k < \infty$,

$$\max_i \sup_{|t-v|\le kn^{-1/2}} n^{1/2}|L_{ni}(t) - L_{ni}(v) - (t-v)\ell_{ni}(v)| \\ = o(1). \tag{3.4.3}$$

Denoting

$$\begin{aligned} \tilde{d}_n(v) &:= n^{-1}\sum_{i=1}^n d_{ni}\ell_{ni}(v), \\ \sigma_d^2(v) &:= \sum_{i=1}^n (d_{ni} - \tilde{d}_n(v))^2 L_{ni}(v)(1 - L_{ni}(v)), \end{aligned}$$

assume that

$$n^{1/2}|\tilde{d}_n(v)| = O(1). \tag{3.4.4}$$

$$\liminf_n \sigma_d^2(v) > 0. \tag{3.4.5}$$

Then,

$$\{\sigma_d(v)\}^{-1}\{Z_d(v) - \mu_d(v)\} \longrightarrow_d \mathcal{N}(0,1).$$

The proof of Theorem 3.4.1 is a consequence of the following *three* lemmas. In these lemmas the setup is the same as in Theorem 3.4.1.

Lemma 3.4.1 *Under the sole assumption of* (2.2.24),

$$\sup_{0\le t\le 1} |HH_n^{-1}(t) - t| = o_p(1).$$

Proof. Upon taking $\mathbf{u} = \mathbf{0}$ in (3.2.19), one obtains

$$\sup_{0\le t\le 1} |HH_n^{-1}(t) - t| \le \sup_{-\infty\le x\le+\infty} |H_n(x) - H(x)| + n^{-1} = o_p(1),$$

by (3.2.14) of Lemma 3.2.1. □

Lemma 3.4.2 *Let* $Y_d(t)$ *denote the* $Y_d(t, \mathbf{0})$ *of* (2.3.1). *Then, under* (3.4.3), *for every* $\epsilon > 0$,

$$\lim_{\delta \to 0} \limsup_n P\Big(\sup_{|t-v|<\delta} |Y_d(t) - Y_d(v)| > \epsilon \Big) = 0.$$

Proof. Apply Lemma 2.2.2 to $\eta_{ni} = H(X_{ni})$, $G_{ni} = L_{ni}$, to obtain that $Y_d \equiv W_d$ of that lemma and that

$$\begin{aligned} & P\Big(\sup_{|t-v|<\delta} |Y_d(t) - Y_d(v)| > \epsilon \Big) \\ \leq \ & \kappa \epsilon^{-2} \Big[L_d(v+\delta) - L_d(v-\delta) \Big]^2 + P\Big(|Y_d(v-\delta) - Y_d(v)| > \epsilon/2 \Big) \\ & \quad + P\Big(|Y_d(v+\delta) - Y_d(v-\delta)| > \epsilon/4 \Big) \\ \leq \ & (\kappa + 20)\epsilon^{-2} \Big[L_d(v+\delta) - L_d(v-\delta) \Big], \quad \text{(by Chebyshev).} \end{aligned}$$

The Lemma now follows from the assumption (3.4.3).

Lemma 3.4.3 *Under* (3.4.3), *for every* $\epsilon > 0$,

$$\limsup_n P(|Y_d(HH_n^{-1}(v)) - Y_d(v)| > \epsilon) = 0.$$

Proof. Follows from Lemmas 3.4.1 and 3.4.2. □

Remark 3.4.1 Lemmas 3.4.2 could be deduced from Corollary 3.3.1 which gives the tightness of the process Y_d under stronger condition (**C***). But here we are interested in the behavior of Y_d only in the neighborhood of one point v and the above lemma proves the continuity of Y_d at the point v at which (3.4.3) holds. Similarly, many of the approximations that follow could of course be deduced from proofs of Theorem 3.2.1 and 3.2.2. But these theorems obtain results uniformly in $0 \leq t \leq 1$ under rather stronger conditions than would be needed in the present case. Of course various decompositions used in their proofs will be useful here also. □

Proof of Theorem 3.4.1. In view of (3.2.9) and (**N2**), it suffices to prove that $\{\sigma_d(v)\}^{-1}\tilde{T}_d(v) \longrightarrow_d N(0,1)$, where

$$\tilde{T}_d(v) = \mathcal{V}_d(v) - \mu_d(v).$$

But, from (3.2.11) applied with $\mathbf{u} = \mathbf{0}$,

$$\begin{aligned} \tilde{T}_d(v) &= Y_d(HH_n^{-1}(v)) + \mu_d(HH_n^{-1}(v)) - \mu_d(v), \quad \text{w.p.1} \\ &= Y_d(v) + o_p(1) + \mu_d(HH_n^{-1}(v)) - \mu_d(v), \quad \text{by (3.4.5).} \end{aligned}$$

Apply the identity (3.2.32) with $\mathbf{u} = \mathbf{0}$ and Lemma 3.4.3 with $d_i \equiv n^{-1/2}$ to obtain,

$$\begin{aligned} (3.4.6) \qquad n^{1/2}[HH_n^{-1}(v) - v] &= -Y_1(HH_n^{-1}(v)) + o_p(1) \\ &= -Y_1(v) + o_p(1). \end{aligned}$$

Since $Y_1(v) \rightarrow_d \mathcal{N}(0, v(1-v))$, $|Y_1(v)| = O_p(1)$. Again, argue as for (3.2.36) with $\mathbf{u} \equiv \mathbf{0},\ t \equiv v$ (i.e., without the supremum on the L.H.S. and with $\mathbf{u} \equiv \mathbf{0},\ t \equiv v$), to conclude that

$$\mu_d(HH_n^{-1}(v)) - \mu_d(v) = -Y_1(v) n^{1/2} \tilde{d}(v) + o_p(1).$$

Combine this with (3.4.6) to obtain

$$\begin{aligned} (3.4.7) \quad \tilde{T}_d(v) & \\ &= Y_d(v) - n^{1/2}\tilde{d}(v) Y_1(v) + o_p(1) \\ &= \sum_{i=1}^{n} (d_{ni} - \tilde{d}(v)) \{ I(X_{ni} \le H^{-1}(v)) - L_{ni}(v) \} + o_p(1). \end{aligned}$$

The theorem now follows from (3.4.5) and the fact that $\{\sigma_d(v)\}^{-1} \times$ { leading term in the RHS of (3.4.7) } $\longrightarrow_d \mathcal{N}(0,1)$ by the L-F CLT, in view of (**N1**) and (**N2**). □

Remark 3.4.2 If $\{F_{ni}\}$ have densities $\{f_{ni}\}$ then $\ell_{ni}(v)$ can be taken to be $f_{ni}(H^{-1}(v))/h(H^{-1}(v))$, just as in (3.2.34). However, if one is interested in the asymptotic normality of linear rank statistic corresponding to the jump score function, with jump at v, then we need $\{L_{ni}\}$ to be smooth only at that jump point.

The above Theorem 3.4.1 bears strong resemblance to Theorem 1 of Dupač-Hájek (1969). The assumptions (**N1**), (**N2**), (3.4.3), and (3.4.5) correspond to (2.2), (2.13) and (2.22) of Dupač - Hájek. Condition (3.4.2) above is not quite comparable to condition (2.12) of Dupač-Hájek but it appears to be less restrictive. In any case, (2.12) and (2.13) together imply the boundedness of $\{\ell_i(v)\}$ and hence the condition (3.4.4) above. Taken together, then, the assumptions of the above theorem are somewhat weaker than those of Dupač - Hájek. On the other hand, the conclusions of the Dupač - Hájek Theorem 1 are stronger than those of the above theorem in that it asserts not only $\{Z_d(v) - \mu_d(v)\}\sigma_d^{-1}(v) \to_d \mathcal{N}(0,1)$ but also that $E[\sigma_d^{-1}(v)(Z_d(v) - \mu_d(v))]^r \to 0$, for $r = 1, 2$, as $n \to \infty$. However, if one is only interested in the asymptotic normality of $\{Z_d(v)\}$ then the above theorem appears to be more desirable. Moreover, in view of the decomposition (3.2.11), the proof presented below makes the role played by conditions (3.4.2) and (3.4.3) transparent.

The assumption about H being strictly increasing is not really an assumption because, without loss of generality, one may assume that $\{F_i\}$ are not flat on a common interval. For, if all $\{F_i\}$ were flat on a common interval, then deletion of this interval would not change the distribution of $R_{n1}, \cdots, R_{nn}$ and hence of $\{Z_d\}$. □

Next, we turn to the asymptotic normality of $Z_d^+(v)$. Again, put $\mathbf{u} = \mathbf{0}$ in the definition (3.3.1) to obtain,

$$
\text{(3.4.8)} \quad
\begin{aligned}
\mathcal{V}_d^+(t) &= \sum_{i=1}^n d_{ni} I(|X_{ni}| \le J_n^{-1}(t)) s(X_{ni}), \\
S_d^+(t) &= \sum_{i=1}^n d_{ni} I(|X_{ni}| \le J^{-1}(t)) s(X_{ni}), \\
\mu_{ni}^+(t) &= F_{ni}(J^{-1}(t)) + F_{ni}(-J^{-1}(t)) - 2F_{ni}(0), \\
\mu_d^+(t) &= \sum_{i=1}^n d_{ni} \mu_{ni}^+(t), \qquad 0 \le t \le 1, \\
Y_d^+ &= S_d^+ - \mu_d^+,
\end{aligned}
$$

Like (3.2.9), we have

$$
\text{(3.4.9)} \qquad \sup_{0 \le t \le 1} |Z_d^+(t) - \mathcal{V}_d^+(t)| \le 2 \max_i |d_i|.
$$

Because of (**N2**), it suffices to consider $\mathcal{V}_d^+$ only. Observe that

$$
Y_d^+(t) = Y_d(HJ^{-1}(t)) + Y_d(H(-J^{-1}(t))) - 2Y_d(H(0)),
$$

where Y_d is as in (2.3.1). Rewrite

$$
\begin{aligned}
(3.4.10) \qquad Y_d^+(t) &= \{Y_d(HJ^{-1}(t)) - Y_d(H(0))\} \\
&\quad -\{Y_d(H(0)) - Y_d(-J^{-1}(t))\} \\
&= Y_{d1}^*(t) - Y_{d2}^*(t), \text{ say.}
\end{aligned}
$$

This representation motivates the following notation as it is required in the subsequent lemma. Let $p_i := F_i(0), q_i := 1 - p_i$ and define for $0 \le t \le 1,\ 1 \le i \le n$,

$$
\begin{aligned}
(3.4.11) \qquad L_{i1}^+(t) &:= \{F_i(J^{-1}(t)) - p_i\}/q_i, & q_i > 0, \\
&= 0, & q_i = 0; \\
L_{i2}^+(t) &:= \{p_i - F_i(-J^{-1}(t))\}/p_i, & p_i > 0, \\
&= 0, & p_i = 0;
\end{aligned}
$$

Observe that $\mu_i^+(v) = q_i L_{i1}^+(v) - p_i L_{i2}^+(v), 1 \le i \le n$. Also define,

$$
\begin{aligned}
L_i^+(t) &:= q_i L_{i1}^+(t) + p_i L_{i2}^+(t) = P(|X_i| \le J^{-1}(t)), \quad 1 \le i \le n, \\
L_{d1}^+(t) &:= \sum_{i=1}^n d_{i1}^2 q_i L_{i1}^+(t), \quad L_{i1}^+(t) := \sum_{i=1}^n d_i^2 p_i L_{i2}^+(t), \quad 0 \le t \le 1.
\end{aligned}
$$

Argue as for the proof of Lemma 2.2.2 and use the triangle and the Chebbychev inequalities to conclude

Lemma 3.4.4 *For every $\epsilon > 0$ and $0 < v < 1$ fixed,*

$$
\begin{aligned}
&P\Big(\sup_{|t-v|<\delta} |Y_{dj}^*(t) - Y_{dj}^*(v)| > \epsilon\Big) \\
(3.4.12) \qquad &\le (\kappa + 20)\epsilon^{-2}[L_{dj}^+(v+\delta) - L_{dj}^+(v-\delta)], \quad j = 1, 2,
\end{aligned}
$$

where κ does not depend on ϵ, δ or any other underlying entities. □

Theorem 3.4.2 *Let $X_{n1}, \cdots, X_{nn}$ be independent r.v.'s with respective continuous d.f.'s $F_{n1}, \cdots, F_{nn}$ and $d_{n1}, \cdots, d_{nn}$ be real numbers. Assume that $\{d_{ni}\}$ satisfy* **(N1)**, **(N2)**. *In addition, assume the following.*

For v fixed in $(0,1)$,

(3.4.13) $\lim_{\delta\to 0}\limsup_n |L_{dj}^+(v+\delta) - L_{dj}^+(v-\delta)| = 0, \quad j = 1,2.$

(3.4.14) *There exist numbers $\{\ell_{ij}^+(v),\ 1 \le i \le n;\ j = 1,2\}$ such such that $\forall\ 0 < k < \infty,\ j = 1,2,$*

$$\max_i \sup_{|t-v|\le kn^{-1/2}} n^{1/2}|L_{ij}^+(t) - L_{ij}^+(v) - (t-v)\ell_{ij}^+(v)| = o(1).$$

With

$$\begin{aligned}
\tilde{d}_n^+(v) &:= n^{-1}\sum_{i=1}^n d_{ni}\{q_i\ell_{i1}^+(v) - p_i\ell_{i2}^+(v)\}, \\
\tau^2(v) &:= \sum_{i=1}^n \Big\{d_{ni}^2\Big[L_{ni}^+(v) - \{\mu_{ni}^+(v)\}^2\Big] \\
&\qquad + (\tilde{d}_n^+(v))^2 L_{ni}^+(v)(1-L_{ni}^+(v)) \\
&\qquad\quad - 2d_{ni}\ \tilde{d}_n^+(v)\ \mu_{ni}^+(v)(1-L_{ni}^+(v))\Big\},
\end{aligned}$$

(3.4.15) (a) $\liminf_n \tau^2(v) > 0.$

(b) $\limsup_n n^{1/2}|\tilde{d}_n^+(v)| < \infty.$

Then,

$$\{\tau(v)\}^{-1}[Z_d^+(v) - \mu_d^+(v)] \longrightarrow_d \mathcal{N}(0,1) \tag{3.4.16}$$

where μ_d^+ is as in (3.4.8).

Proof. The proof of this theorem is similar to that of Theorem 3.4.1 so we shall be brief. To begin with, by (3.4.9) and (**N2**) it suffices to prove that $\{\tau(v)\}^{-1}\tilde{T}_d^+(v) \to_d \mathcal{N}(0,1)$, where $\tilde{T}_d^+(v) := \mathcal{V}_d^+(v) - \mu_d^+(v)$.

Apply Lemma 3.4.1 above to the r.v.'s $|X_{n1}|, \cdots, |X_{nn}|$, to conclude that

$$\sup_{0\le t\le 1} |J(J_n^{-1}(t)) - t| = o_p(1).$$

From this, (3.4.10), (3.4.12) and (3.4.13),

$$\begin{aligned}
\tilde{T}_d^+(v) &= Y_d^+(JJ_n^{-1}(v)) + \mu_d^+(JJ_n^{-1}(v)) - \mu_d^+(v). \\
&= Y_d^+(v) + [\mu_d^+(JJ_n^{-1}(v)) - \mu_d^+(v)] + o_p(1).
\end{aligned}$$

Again, apply arguments like those that yielded (3.4.6) to $\{|X_{ni}|\}$ to obtain

$$n^{1/2}[JJ_n^{-1}(v) = v] = -Y_1^*(v) + o_p(1),$$

where $Y_1^*(v)$ is as in (3.3.19) with $t = v$ and $\mathbf{u} = \mathbf{0}$. Consequently,

$$\tilde{T}_d^+(v) = Y_d^+(v) - n^{1/2}\tilde{d}_n^+(v)Y_1^*(v) + o_p(1) = K_d^+(v) + o_p(1)$$

where

$$\begin{aligned} K_d^+(v) &= Y_d^+(v) - n^{1/2}\tilde{d}_n^+(v)Y_1^*(v) \\ &= \sum_{i=1}^n \Big\{ d_{ni}[I(J(|X_{ni}|) \le v)s(X_{ni}) - \mu_{ni}^+(v)] \\ &\qquad\qquad -\tilde{d}_n^+(v)[I(J(|X_{ni}|) \le v) - L_{ni}^+(v)]\Big\}. \end{aligned}$$

Note that $Var(K_d^+(v)) = \tau^2(v)$. The proof of the theorem is now completed by using the L-F CLT which is justified, in view of (**N1**), (**N2**), and (3.4.15a). □

Remark 3.4.3 Observe that if $\{F_i\}$ are symmetric about 0 then $\mu_i^+ \equiv 0 \equiv \tilde{d}_n^+$ and $\tau^2(v) = \Sigma_i d_{ni}^2 L_{ni}^+(v)$. □

Remark 3.4.4 An alternative proof of (3.4.16), using the techniques of Dupač and Hájek (op.cit.), appears in Koul and Staudte, Jr. (1972a). Thus comments like those in Remark 3.4.1 are appropriate here also. □

Next, we turn to the *weak convergence* of $\{Z_d\}$ and $\{Z_d^+\}$. These results will be stated without proofs as their proofs are consequences of the results of the previous sections in this chapter.

Theorem 3.4.3 (*Weak convergence of Z_d*). *Let $X_{n1}, \cdots, X_{nn}$ be independent r.v.'s with respective continuous d.f.'s $F_{n1}, \cdots, F_{nn}$. With notation as in* (2.2.23), *assume that* (**N1**), (**N2**), (**C***) *hold. In addition, assume the following: There are measurable functions $\{\ell_{ni}, 1 \le i \le n\}$ on $[0,1]$, such that $\forall\, 0 < k < \infty$,*

$$\text{(3.4.17)} \quad \max_i \sup_{|t-s| \le kn^{-1/2}} n^{1/2}|L_{ni}(t) - L_{ni}(s) - (t-s)\ell_{ni}(s)| = 0.$$

Moreover, assume that

$$\text{(3.4.18)} \qquad \limsup_n \sup_{0 \le t \le 1} n^{1/2}|\tilde{d}_n(t)| < \infty,$$

(3.4.19) $$\lim_{\delta\to 0}\limsup_{n}\sup_{|t-s|<\delta} n^{1/2}|\tilde{d}_n(t)-\tilde{d}_n(s)| = 0,$$

(3.4.20) $$\liminf_{n}\sigma^2(t) > 0, \qquad 0<t<1.$$

Finally, with $K_d(t) := \sum_{i=1}^{n}(d_{ni}-\tilde{d}_n(t))\{I(X_{ni}\le H^{-1}(t))-L_{ni}(t)\}$, *assume that*

(3.4.21) $$\begin{aligned} C(t,s) &= \lim_{n} Cov(K_d(t),K_d(s)) \\ &= \lim_{n}\sum_{i=1}^{n}(d_{ni}-\tilde{d}_n(t))(d_{ni}-\tilde{d}_n(s))L_{ni}(s)(1-L_{ni}(t)), \end{aligned}$$

exists for all $0\le s\le t\le 1$.

Then, $Z_d-\mu_d$ *converges weakly to a mean zero, covariance* C *continuous Gaussian process on* $[0,1]$, *tied down at* 0 *and* 1. □

Remark 3.4.5 In (3.4.17), without loss of generality it may be assumed that $n^{-1}\Sigma_i\ell_{ni}(s) = 1, 0\le s\le 1$. For, if (3.4.17) holds for some $\{\ell_{ni}, 1\le i\le n\}$, then it also holds for $\ell_{ni}(s)\equiv\ell^*_{ni}(s) := n^{1/2}[L_{ni}(s+n^{-1/2})-L_{ni}(s)]$, $1\le i\le n, 0\le s\le 1$. Because $n^{-1}\sum_{i=1}^{n}L_{ni}(s)\equiv s$, $n^{-1}\sum_{i=1}^{n}\ell^*_{ni}(s)\equiv 1$. □

Remark 3.4.6 Conditions (**C***), (**N1**) and (3.4.18) may be replaced by the condition (**B**), because, in view of the previous remark,

$$n^{1/2}|\tilde{d}_n(t)| = |n^{-1/2}\sum_{i=1}^{n}d_{ni}\ell_{ni}(t)| \le n^{1/2}\max_i|d_{ni}|, \quad 0\le t\le 1. \quad \square$$

Remark 3.4.7 In the case F_{ni} have density f_{ni}, one can choose

$$\ell_{ni} = f_{ni}(H^{-1})/n^{-1}\sum_{j=1}^{n}f_{nj}(H^{-1}), \quad 1\le i\le n.$$

Remark 3.4.8 In the case $F_{ni}\equiv F$, F a continuous and strictly increasing d.f., $L_{ni}(t)\equiv t$, $\ell_{ni}(t)\equiv 1$, so that (**C***) and (3.4.17) - (3.4.20) are trivially satisfied. Moreover, $C(s,t) = s(1-t)$, $0\le s\le t\le 1$, so that (3.4.21) is satisfied. Thus Theorem 3.4.3 includes Theorem V.3.5.1 of Hájek and Sidák (1967). □

Theorem 3.4.4 (*Weak convergence of* Z_d^+). *Let* $X_{n1},\cdots,X_{nn}$ *be independent r.v.'s with respective d.f.'s* $F_{n1},\cdots,F_{nn}$ *and let* d_{n1},

$\cdots$, d_{nn} be real numbers. Assume that (**N1**) *and* (**N2**) *hold and that the following hold.*

(3.4.22) $\lim_{\delta\to 0}\limsup_n \sup_{0\le t\le 1-\delta} [L_{dj}^+(t+\delta) - L_{dj}^+(t)] = 0, \quad j = 1, 2.$

(3.4.23) *There are measurable functions* ℓ_{ij}^+, $1 \le i \le n$, $j = 1, 2$, *on* $[0, 1]$ *such that for any* $0 < k < \infty$,

$$\max_i \sup_{|t-s|\le kn^{-1/2}} n^{1/2}|L_{ij}^+(t) - L_{ij}^+(s) - (t-s)\ell_{ij}^+(s)| = o(1).$$

(3.4.24) $\limsup_n \sup_{0\le t\le 1} n^{1/2}|\tilde{d}_n^+(t)| < \infty,$

(3.4.25) $\lim_{\delta\to 0}\limsup_n \sup_{|t-s|<\delta} n^{1/2}|\tilde{d}_n^+(t) - \tilde{d}_n^+(s)| = 0.$

(3.4.26) $\liminf_n \tau^2(t) > 0, \quad 0 < t < 1.$

(3.4.27) $\lim_n Cov(K_d^+(s), K_d^+(t)) = C^+(s,t)$ *exists*, $0 \le s \le t \le 1.$

Then, $Z_d^+ - \mu_d^+$ *converges weakly to a continuous mean zero covariance* C^+ *Gaussian process, tied down at* 0. □

Remark 3.4.9 Remarks 3.4.5 through 3.4.7 are applicable here also, with appropriate modifications. □

Remark 3.4.10 Suppose that $F_{ni} \equiv F$, F continuous, and $d_{ni} \equiv n^{-1/2}$. Then

$$\begin{aligned}\sup_{0\le t\le 1} &|Z_d^+(t) - \mu_d^+(t)| \\ &= \sup_{0<x<\infty} n^{1/2}|\{H_n(x) - H_n(0)\} - \{H_n(0) - H_n(-x)\} \\ &\qquad -\{F(x) - F(0)\} - \{F(0) - F(-x)\}|,\end{aligned}$$

precisely the statistic τ_n^* proposed by Smirnov (1947) to test the hypothesis of symmetry about F. Smirnov considered only the null distribution. Theorem 3.4.4 allows one to study its asymptotic distribution under fairly general independent alternatives.

For arbitrary $\{d_{ni}\}$, subject to (**N1**) and (**N2**), the statistic $\sup\{|Z_d^+(t) - \mu_d^+(t)|; 0 \le t \le 1\}$ may be considered a generalized Smirnov statistic for testing symmetry.

4

M, R and Some Scale Estimators

4.1 Introduction

In the last four decades statistics has seen the emergence and consolidation of many competitors of the Least Square estimator of $\boldsymbol{\beta}$ of (1.1.1). The most prominent are the so-called M- and R- estimators. The class of M-estimators was introduced by Huber (1973) and its computational aspects and some robustness properties are available in Huber (1981). The class of R-estimators is based on the ideas of Hodges and Lehmann (1963) and has been developed by Adichie (1967), Jurečková (1971) and Jaeckel (1972).

One of the attractive features of these estimators is that they are robust against certain outliers in errors. All of these estimators are translation invariant, whereas only R-estimators are scale invariant.

Our purpose here is to illustrate the usefulness of the results of Chapter 2 in deriving the asymptotic distributions of these estimators under a fairly general class of heteroscedastic errors. The Section 4.2.1 gives the asymptotic distributions of M-estimators while those of R-estimators are given in Section 4.4. Among other things, the results obtained enable one to study their qualitative robustness against an array of non-identical error d.f.'s converging to a fixed error d.f. The sufficient conditions given here are fairly general for the underlying score functions and the design variables.

Efron (1979) introduced a general resampling procedure, called the bootstrap, for estimating the distribution of a pivotal statistic. Singh (1981) showed that the bootstrap estimate B_n is *second order accurate*, i.e. provides more accurate approximation to the sampling distribution G_n of the standardized sample mean than the usual normal approximation in the sense that $\sup\{|G_n(x) - B_n(x)|; x \in \mathbb{R}\}$ tends to zero at a faster rate than that of the square-root of n. This kind of result holds more generally as noted by Babu and Singh (1983, 1984).

Section 4.2.2 discusses similar results pertaining to a class of M-estimators of $\boldsymbol{\beta}$ when the errors in (1.1.1) are i.i.d. It is noted that the the distribution of the bootstrap estimators obtained by resampling the residuals according to a W.E.P. is second order accurate. A similar result is shown to hold for Shorack's (1982) modified bootstrap estimators.

In an attempt to make M-estimators scale invariant one often needs a preliminary robust scale estimator. Two such estimators are the MAD (median of absolute deviations of residuals) and the MASD (median of absolute symmetrized deviations of residuals). The asymptotic distributions of these estimators under heteroscedastic errors appear in Section 4.3.

In carrying out the analysis of variance of an experimental design or a linear model based on ranks one needs an estimator of the asymptotic variance of certain rank statistics, see, e.g., Hettmansperger (1984). These variances involve the functional $Q(f) = \int f d\varphi(F)$ where φ is a known function, F a common error d.f. having a density f. Some estimators of $Q(f)$ under (1.1.1) are presented in Section 4.5. Again, the results of Chapter 2 are found useful in proving their consistency.

4.2 M-Estimators

4.2.1 First order approximations: Asymptotic normality

This subsection contains the asymptotic distributions of M - estimators of $\boldsymbol{\beta}$ when the errors in (1.1.1) are heteroscedastic. The following subsection 4.2.2 gives some results on the bootstrap approximations to these distributions.

Let the model (1.1.1) hold. Let ψ be a nondecreasing function from $\mathbb{R}$ to $\mathbb{R}$. The corresponding M- estimator $\hat{\boldsymbol{\Delta}}$ of $\boldsymbol{\beta}$ is defined to be a zero of the M - score $\int \psi(y)\mathbf{V}(dy, \mathbf{t})$, where $\mathbf{V}$ is as in (1.1.2). Our objective is to investigate the asymptotic behavior of $\mathbf{A}^{-1}(\hat{\boldsymbol{\Delta}} - \boldsymbol{\beta})$ when the errors in (1.1.1) are heteroscedastic. Our method is still the usual one, v.i.z., to obtain the expansion of the M - score uniformly in $\mathbf{t} \in \{\mathbf{t}; \|\mathbf{A}^{-1}(\mathbf{t} - \boldsymbol{\beta})\| \le b\}, 0 < b < \infty$, to observe that there is a zero of the M - score, $\hat{\boldsymbol{\Delta}}$, in this set and then to apply this expansion to obtain the approximation for $\mathbf{A}^{-1}(\hat{\boldsymbol{\Delta}} - \boldsymbol{\beta})$ in terms of the given M - score at the true $\boldsymbol{\beta}$. To make all this precise, we need to standardize the M - score. For that reason we need some more notation. Let

$$
\begin{aligned}
(4.2.1) \qquad \boldsymbol{\Lambda}^*(y) &:= diag(f_{n1}(y), \cdots, f_{nn}(y)), \quad y \in \mathbb{R}, \\
\mathbf{C} &:= \mathbf{A}\mathbf{X}' \int \boldsymbol{\Lambda}^*(y) d\psi(y) \mathbf{X}\mathbf{A}, \\
\mathbf{T}(\psi, \mathbf{t}) &:= -\mathbf{C}^{-1}\mathbf{A} \int \psi(y)\mathbf{V}(dy, \mathbf{t}), \\
\overline{\mathbf{T}}(\psi, \mathbf{t}) &:= \mathbf{A}^{-1}(\mathbf{t} - \boldsymbol{\beta}) - \mathbf{T}(\psi, \boldsymbol{\beta}), \quad \mathbf{t} \in \mathbb{R}^p.
\end{aligned}
$$

An approximation to $\hat{\boldsymbol{\Delta}}$ is given by the zero $\overline{\boldsymbol{\Delta}}$ of $\overline{\mathbf{T}}(\psi, \mathbf{t})$, v.i.z.,

$$
(4.2.2) \qquad \mathbf{A}^{-1}(\overline{\boldsymbol{\Delta}} - \boldsymbol{\beta}) = \mathbf{T}(\psi, \boldsymbol{\beta}).
$$

A basic result needed to make this precise is the AUL of $\mathbf{T}(\psi, \mathbf{t})$ in $\mathbf{A}^{-1}(\mathbf{t} - \boldsymbol{\beta})$, given in Theorem 4.2.1 below. Let

$$
(4.2.3) \qquad \Psi := \{\psi : \mathbb{R} \mapsto \mathbb{R}, \ \psi \in \mathcal{DI}(\mathbb{R}), \ |\psi|_{tv} \ bounded\}.
$$

Theorem 4.2.1 *Let $\{(\mathbf{x}_{ni}', Y_{ni}), 1 \le i \le n\}, \boldsymbol{\beta}, \{F_{ni}, 1 \le i \le n\}$ be as in the model (1.1.1) satisfying all conditions of Theorem 2.3.3. In* *addition, assume the following:*

$$\text{(4.2.4)} \qquad \limsup_n \|\mathbf{C}^{-1}\|_\infty < \infty.$$

Then, $\forall\ 0 < b < \infty$,

$$\text{(4.2.5)} \quad \sup_{\psi,\mathbf{u}} \|\mathbf{T}(\psi, \boldsymbol{\beta} + \mathbf{A}\mathbf{u}) - \overline{\mathbf{T}}(\psi, \boldsymbol{\beta} + \mathbf{A}\mathbf{u})\| = o_p(1),$$

where the supremum is taken over all $\psi \in \Psi$ *and* $\|\mathbf{u}\| \le b$.

Proof. Rewrite, after integration by parts,

$$\begin{aligned}\mathbf{T}(\psi, \mathbf{t}) - \overline{\mathbf{T}}(\psi, \mathbf{t}) &= \int \mathbf{C}^{-1}\mathbf{A}\,[\,\mathbf{V}(y, \mathbf{t}) \\ &\qquad -V(y, \boldsymbol{\beta}) - \Gamma_1(H(y))\mathbf{A}^{-1}(\mathbf{t} - \boldsymbol{\beta})]d\psi(y).\end{aligned}$$

Now (4.2.5) readily follows from this and (2.3.35). □

In order to use this theorem, we must be able to argue that $\|\mathbf{A}^{-1}(\hat{\boldsymbol{\Delta}} - \boldsymbol{\beta})\| = O_p(1)$. To that effect, define

$$\begin{aligned}\mu_i &:= E\psi(e_i), \quad \tau_i^2 = Var\{\psi(e_i)\}, \quad 1 \le i \le n, \\ \mathbf{b}_n &:= E\mathbf{T}(\psi, \boldsymbol{\beta}) = -\mathbf{C}^{-1}\mathbf{A}\sum_{i=1}^n \mathbf{x}_i\mu_i,\end{aligned}$$

and observe that

$$E\|\mathbf{A}^{-1}(\overline{\boldsymbol{\Delta}} - \boldsymbol{\beta}) - \mathbf{b}_n\|^2 = \mathbf{C}^{-1}\sum_{i=1}^n \mathbf{x}_i'(\mathbf{X}'\mathbf{X})^{-1}\mathbf{x}_i\tau_i^2\mathbf{C}^{-1} = O(1),$$

by (4.2.3), (4.2.4) and the fact that $\sum_{i=1}^n \mathbf{x}_i'(\mathbf{X}'\mathbf{X})^{-1}\mathbf{x}_i \equiv p < \infty$. Hence by the Markov inequality, $\forall\ \epsilon > 0,\ \exists\ 0 < K_\epsilon < \infty,\ \ni$,

$$P(\|\mathbf{A}^{-1}(\overline{\boldsymbol{\Delta}} - \boldsymbol{\beta}) - \mathbf{b}_n\| \le K_\epsilon) \ge 1 - \epsilon,\ \forall\ n \ge 1.$$

Thus, assuming that

$$\text{(4.2.6)} \qquad \sum_{i=1}^n \mathbf{x}_i\mu_i = \mathbf{0},$$

and arguing via Brouwer's fixed point theorem as in Huber (1981, p. 169), one concludes, in view (4.2.5), that $\|\mathbf{A}^{-1}(\hat{\boldsymbol{\Delta}} - \boldsymbol{\beta})\| = O_p(1)$.

A routine application of (4.2.5) enables one to conclude that

$$\mathbf{A}^{-1}(\hat{\boldsymbol{\Delta}} - \boldsymbol{\beta}) = \mathbf{T}(\psi, \boldsymbol{\beta}) + o_p(1). \tag{4.2.7}$$

Note that, under (4.2.6), with $\mathbf{T}_0(\psi, \boldsymbol{\beta}) = \mathbf{C}\mathbf{T}(\psi, \boldsymbol{\beta})$,

$$E\mathbf{T}_0(\psi, \boldsymbol{\beta})\mathbf{T}_0'(\psi, \boldsymbol{\beta}) = \mathbf{A}\sum_{i=1}^{n} \mathbf{x}_i'\mathbf{x}_i\tau_i^2\mathbf{A} = \mathbf{A}\mathbf{X}'\mathcal{T}\mathbf{X}\mathbf{A}$$

where $\mathcal{T} = diag(\tau_1^2, \cdots, \tau_n^2)$. Moreover, for any $\boldsymbol{\lambda} \in \mathbb{R}^p$,

$$\boldsymbol{\lambda}'\mathbf{T}_0(\psi, \boldsymbol{\beta}) = \sum_{i=1}^{n}\sum_{j=1}^{p} \lambda_j d_{ij}\psi(e_i) = \sum_{i=1}^{n} \boldsymbol{\lambda}'\mathbf{A}\mathbf{x}_i'\psi(e_i)$$

where $\{d_{ij}\}$ are as in (2.3.30). In view of (2.3.31) and (2.3.32), (**NX**) and (4.2.6) imply that $\boldsymbol{\lambda}'\mathbf{T}_0(\psi, \boldsymbol{\beta})$ is asymptotically normally distributed with mean 0 and the asymptotic variance $\boldsymbol{\lambda}'\mathbf{A}\mathbf{X}'\mathcal{T}\mathbf{X}\mathbf{A}\lambda$. Thus by the Cramér - Wold device [Theorem 7.7, p. 49, Billingsley (1968)], (4.2.4) and (4.2.7),

$$\boldsymbol{\Sigma}^{-1/2}\mathbf{A}^{-1}(\hat{\boldsymbol{\Delta}} - \boldsymbol{\beta}) \longrightarrow_d N(\mathbf{0}, \mathbf{I}_{p\times p}), \tag{4.2.8}$$

with $\boldsymbol{\Sigma} := \mathbf{C}^{-1}\mathbf{A}\mathbf{X}'\mathcal{T}\mathbf{X}\mathbf{A}\mathbf{C}^{-1}$. We summarize the above discussion as a

Proposition 4.2.1 *Suppose that the d.f.'s $\{F_{ni}\}$ of the errors and the design matrix* $\mathbf{X}$ *of (1.1.1) satisfy (4.2.4), (4.2.6) and the assumptions of Theorem 2.3.3 including that H is strictly increasing for each $n \geq 1$. Then (4.2.8) holds.*

Now, consider the case of the *i.i.d. errors* in (1.1.1) with $F_{ni} \equiv F$. Then,

$$\begin{aligned} \tau_i^2 &= \int \psi^2 dF - \Big(\int \psi dF\Big)^2 = \tau^2, \text{ (say)}, \quad 1 \leq i \leq n, \\ \mathbf{C} &= \Big(\int f d\psi\Big)\mathbf{I}_{p\times p}, \quad \boldsymbol{\Sigma} = \left(\int f d\psi\right)^{-2} \tau^2\mathbf{I}_{p\times p}. \end{aligned} \tag{4.2.9}$$

Consequently (4.2.4) is equivalent to requiring $\int f d\psi > 0$. Next, observe that (4.2.6) becomes

$$\sum_{i=1}^{n} \mathbf{x}_i \int \psi dF = \mathbf{0}. \tag{4.2.10}$$

Obviously, this is satisfied if either $\sum_{i=1}^{n} \mathbf{x}_i = \mathbf{0}$, i.e. if $\mathbf{X}$ is a centered design matrix or if $\int \psi dF = 0$, the often assumed condition. The former excludes the possibility of the presence of the location parameter in (1.1.1). Thus to summarize, we have

Proposition 4.2.2 *Suppose that in (1.1.1), $F_{ni} \equiv F$. In addition, assume that $\mathbf{X}$ and F satisfy (NX), (F1), (4.2.10) and that $\int f d\psi > 0$. Then,*

$$\mathbf{A}^{-1}(\hat{\boldsymbol{\Delta}} - \boldsymbol{\beta}) \to_d \mathcal{N}\Big(\mathbf{0}, \frac{\tau^2}{\left(\int f d\psi\right)^2} \mathbf{I}_{p\times p}\Big). \qquad \square$$

Condition (4.2.10) suggests another way of defining M- estimators of $\boldsymbol{\beta}$ in (1.1.1) in the case of the *i.i.d. errors.* Let

$$\text{(4.2.11)} \qquad \begin{aligned} \overline{\mathbf{x}}_{nj} &:= n^{-1}\sum_{i=1}^{n} x_{nij}, \quad 1 \le j \le p; \\ \overline{\mathbf{x}}_n' &:= (\overline{x}_{n1}, \cdots, \overline{x}_{np}), \\ \overline{\mathbf{X}}' &:= [\overline{\mathbf{x}}_n, \cdots, \overline{\mathbf{x}}_n]_{n\times p}, \quad \mathbf{X}_c := \mathbf{X} - \overline{\mathbf{X}}. \end{aligned}$$

Assume

(NX1) $\qquad (\mathbf{X}_c'\mathbf{X}_c)^{-1}$ exists for all $n \ge p$,
$\qquad \max_i \mathbf{x}_{ni}'(\mathbf{X}_c'\mathbf{X}_c)^{-1}\mathbf{x}_{ni} = o(1)$.

Let

$$\text{(4.2.12)} \quad \begin{aligned} \mathbf{T}^*(\psi, \mathbf{t}) &:= \mathbf{A}_1 \sum_{i=1}^{n} (\mathbf{x}_{ni} - \overline{\mathbf{x}}_n)\psi(Y_i - \mathbf{x}_{ni}'\mathbf{t}), \quad \mathbf{t} \in \mathbb{R}^p, \\ \mathbf{A}_1 &:= (\mathbf{X}_c'\mathbf{X}_c)^{-1/2}. \end{aligned}$$

Define an M-estimator $\boldsymbol{\Delta}^*$ as a solution $\mathbf{t}$ of

$$\text{(4.2.13)} \qquad \mathbf{T}^*(\psi, \mathbf{t}) = \mathbf{0}.$$

Apply Corollary 2.3.1 p times, j^{th} time with $d_{ni} = i^{th}$ element of the jth column of $\mathbf{X}_c\mathbf{A}_1, 1 \le i \le n, 1 \le j \le p$, to conclude an analogue of (4.2.5) above, v.i.z.,

$$\sup_{\psi, \|\mathbf{A}_1^{-1}(\mathbf{t}-\boldsymbol{\beta})\| \le b} \|\mathbf{T}^*(\psi, \mathbf{t}) - \overline{\mathbf{T}}^*(\psi, \mathbf{t})\| = o_p(1)$$

where

$$\overline{\mathbf{T}}^*(\psi, \mathbf{t}) := \mathbf{A}_1^{-1}(\mathbf{t} - \boldsymbol{\beta}) - \Big(\int f d\psi\Big)^{-1} \mathbf{T}^*(\psi, \boldsymbol{\beta}).$$

The proof of this claim is exactly similar to that of (4.2.5) with appropriate modifications of replacing $\mathbf{X}$ by $\mathbf{X}_c$ and $\mathbf{A}$ by $\mathbf{A}_1$ and using $F_{ni} \equiv F$ in the discussion there.

Now, clearly, $F_{ni} \equiv F$ implies that $E\mathbf{T}^*(\psi, \boldsymbol{\beta}) = \mathbf{0}$,

$$E\|\mathbf{T}^*(\psi, \boldsymbol{\beta})\|^2 = \sum_{i=1}^{n} (\mathbf{x}_i - \overline{\mathbf{x}})' \mathbf{A}_1' \mathbf{A}_1 (\mathbf{x}_i - \overline{\mathbf{x}}) \tau^2 = O(1).$$

Hence, $\|\mathbf{T}^*(\psi, \boldsymbol{\beta})\| = O_p(1)$. If $\overline{\boldsymbol{\Delta}}^*$ is the zero of $\overline{\mathbf{T}}^*(\psi, \cdot)$, then

$$\mathbf{A}_1^{-1}(\overline{\boldsymbol{\Delta}}^* - \boldsymbol{\beta}) = \Big(\int f d\psi\Big)^{-1} \mathbf{T}^*(\psi, \boldsymbol{\beta}).$$

Argue, as for (4.2.7) and (4.2.8) to conclude the following

Proposition 4.2.3 *Suppose that in (1.1.1), $F_{ni} \equiv F$. In addition, assume that* $\mathbf{X}$ *and F satisfy* (**NX1**), (**F1**) *and* (**F2**). *Then,*

$$\mathbf{A}_1^{-1}(\boldsymbol{\Delta}^* - \boldsymbol{\beta}) \longrightarrow_d \mathcal{N}\Big(\mathbf{0}, \tau^2\Big(\int f d\psi\Big)^{-2} \mathbf{I}_{p\times p}\Big).$$

Remark 4.2.1 Note that the Proposition 4.2.2 does not require the condition $\int \psi dF = 0$. An advantage of this is that $\boldsymbol{\Delta}^*$ can be used as a preliminary estimator when constructing adaptive estimators of $\boldsymbol{\beta}$. An *adaptive* estimator is one that achieves the Hájek - Le Cam (Hájek 1972, Le Cam 1972) lower bound over a large class of error distributions. Often a minimal condition required to construct an adaptive estimator of $\boldsymbol{\beta}$ is that F have finite Fisher information, i.e., that F satisfy (3.2.a) of Theorem 3.2.3. See, e.g., Bickel (1982), Fabian and Hannan (1982) and Koul and Susarla (1983). Recall, from Remark 3.2.2, that this implies (**F1**).

On the other hand, the condition (**NX1**) does not allow for any location term in the linear regression model. □

So far we have been dealing with the linear regression model with known scale. Now consider the model (2.3.36) where γ is an unknown scale parameter. Let s be an $n^{1/2}$ - consistent estimator of γ, i.e.

$$(4.2.14) \qquad |n^{1/2}(s-\gamma)\gamma^{-1}| = O_p(1).$$

Define an M-estimator $\hat{\mathbf{\Delta}}_1$ of $\boldsymbol{\beta}$ as a solution $\mathbf{t}$ of

$$(4.2.15) \quad \sum_{i=1}^{n} \mathbf{x}_i \psi((Y_i - \mathbf{x}_i'\mathbf{t})s^{-1}) = \mathbf{0} \quad \text{or} \quad \int \psi(y)\mathbf{V}(sdy, \mathbf{t}) = \mathbf{0}.$$

To keep exposition simple, now we shall not exhibit ψ in some of the functions defined below. With $\mathbf{C}$ is as in (4.2.1) above, define, for an $\alpha > 0, \mathbf{t} \in \mathbb{R}^p$,

$$
\begin{aligned}
(4.2.16) \qquad \mathbf{f}'(y) &:= (f_1(y), \cdots, f_n(y)), \\
\mathbf{C}_1 &:= n^{-1/2}\mathbf{A}\mathbf{X}' \int y\mathbf{f}(y)d\psi(y), \\
\mathbf{S}(\alpha, \mathbf{t}) &:= \mathbf{A} \int \psi(y)\mathbf{V}(\alpha dy, \mathbf{t}), \\
\overline{\mathbf{S}}(\alpha, \mathbf{t}) &:= \mathbf{A}^{-1}(\mathbf{t} - \boldsymbol{\beta})\gamma^{-1} + \mathbf{C}^{-1}\mathbf{C}_1 n^{1/2}(\alpha - \gamma)\gamma^{-1} \\
&\quad -\mathbf{C}^{-1}\mathbf{S}(\gamma, \boldsymbol{\beta}),
\end{aligned}
$$

Note that by (**NX**), (**F1**), (**F3**), and (4.2.3),

$$(4.2.17) \qquad \|\mathbf{C}_1\| = O(1).$$

The following theorem is a direct consequence of Theorem 2.3.4. In it $N_1 := \{(\alpha, \mathbf{t}) : \alpha > 0, \mathbf{t} \in \mathbb{R}^p, \|\mathbf{A}^{-1}(\mathbf{t} - \boldsymbol{\beta})\| \le b\gamma, |n^{1/2}(\alpha - \gamma)| \le b\gamma\}, 0 < b < \infty$.

Theorem 4.2.2 *Let $\{(\mathbf{x}_{ni}', Y_{ni}), 1 \le i \le n\}, \boldsymbol{\beta}, \gamma, \{F_{ni}, 1 \le i \le n\}$ be as in (2.3.36) satisfying all the conditions of Theorem 2.3.4. Moreover, assume (4.2.3) and (4.2.4) hold. Then, for every $0 < b < \infty$,*

$$(4.2.18) \qquad \sup \|\mathbf{S}(\alpha, \mathbf{t}) - \overline{\mathbf{S}}(\alpha, \mathbf{t})\| = o_p(1).$$

where the supremum is taken over all $\psi \in \Psi$, and $(\alpha, \mathbf{t}')' \in N_1$. □

Now argue as in the proof of the Proposition 4.2.1 to conclude

Proposition 4.2.4 *Suppose that the design matrix* $\mathbf{X}$ *and d.f.'s* $\{F_{ni}\}$ *of* $\{\epsilon_{ni}\}$ *in (2.3.36) satisfy (4.2.5), (4.2.6) and the assumptions of Theorem 2.3.4 including that* H *is strictly increasing for each* $n \geq 1$. *In addition assume that there exists an estimate* s *of* s *satisfying (4.2.14). Then*

$$(4.2.19) \quad \mathbf{A}^{-1}(\hat{\Delta}_1 - \boldsymbol{\beta})\gamma^{-1} = \mathbf{C}^{-1}\mathbf{S}(\gamma, \boldsymbol{\beta}) - \mathbf{C}^{-1}\mathbf{C}_1 n^{1/2}(s-\gamma)\gamma^{-1} + o_p(1),$$

where $\hat{\boldsymbol{\Delta}}_1$ *now is a solution of (15).* □

Remark 4.2.2 In (4.2.6), F_i is now the d.f. of ϵ_i, and not of $\gamma\epsilon_i, 1 \leq i \leq n$. □

Remark 4.2.3 Effect of symmetry on $\hat{\boldsymbol{\Delta}}_1$. As is clear from (4.2.19), in general the asymptotic distribution of $\hat{\mathbf{D}}_1$ depends on s. However, suppose that for all $1 \leq i \leq n,\ y \in \mathbb{R}$,

$$d\psi(y) \;=\; -d\psi(-y), \quad f_i(y) \equiv f_i(-y).$$

Then $\int y f_i(y) d\psi(y) = 0,\ 1 \leq i \leq n$, and, from (4.2.16), $\mathbf{C}_1 = \mathbf{0}$. Consequently, in this case,

$$\mathbf{A}^{-1}(\hat{\boldsymbol{\Delta}}_1 - \boldsymbol{\beta}) = \gamma \mathbf{C}^{-1}\mathbf{S}(\gamma, \boldsymbol{\beta}) + o_p(1).$$

Hence, with $\boldsymbol{\Sigma}$ as in (4.2.8), we obtain

$$\boldsymbol{\Sigma}^{1/2}\mathbf{A}^{-1}(\hat{\boldsymbol{\Delta}}_1 - \boldsymbol{\beta})\gamma^{-1} \longrightarrow_d \mathcal{N}(\mathbf{0}, \mathbf{I}_{p\times p}).$$

Note that this asymptotic distribution differs from that of (4.2.8) only by the presence of γ^{-1}. In other words, in the case of symmetric errors $\{\epsilon_i\}$ and the skew symmetric score functions $\{\psi\}$, the asymptotic distribution of M-estimator of $\boldsymbol{\beta}$ of (2.3.38) with a preliminary $n^{1/2}$ - consistent estimator of the scale parameter is the same as that of $\gamma^{-1}\times$ M - estimator of $\boldsymbol{\beta}$ of (1.1.1). □

4.2.2 Bootstrap approximations

Before discussing the specific bootstrap approximations we shall describe the concept of Efron's bootstrap a bit more generally in the one sample setup.

Let $\xi_1, \xi_2, \cdots, \xi_n$ be n i.i.d. G r.v.'s, G_n be their empirical d.f. and $T_n = T_n(\boldsymbol{\xi}_n, G)$ be a function of $\boldsymbol{\xi}' := (\xi_1, \xi_2, \cdots, \xi_n)$ and G such that $T_n(\boldsymbol{\xi}, G)$ is a r.v. for every G. Let $\zeta_1, \zeta_2, \cdots, \zeta_n$ denote i.i.d. G_n r.v.'s and $\boldsymbol{\zeta}'_n := (\zeta_1, \zeta_2, \cdots, \zeta_n)$. The *bootstrap* d.f. B_n of $T_n(\boldsymbol{\xi}, G)$ is the d.f. of $T_n(\boldsymbol{\zeta}_n, G_n)$ under G_n. Efron (1979) showed, via numerical studies, that in several examples B_n provides better approximation to the d.f. Γ_n of $T_n(\boldsymbol{\xi}_n, G)$ under G than the normal approximation. Singh (1981) substantiated this observation by proving that in the case of the standardized sample mean the bootstrap estimate B_n is *second order accurate*, i.e.

$$\text{(4.2.20)} \qquad \sup\{|\Gamma_n(x) - B_n(x)|;\ x \in \mathbb{R}\} = o(n^{-1/2}), \text{ a.s..}$$

Recall that the Edgeworth expansion or the Berry-Esseen bound gives that

$$\sup\{\Gamma_n(x) - \Phi(x)|; x \in \mathbb{R}\} = O(n^{-1/2}),$$

where Φ is the d.f. of a $\mathcal{N}(0,1)$ r.v. See, e.g., Feller (1966), Ch. XVI). Babu and Singh (1983, 1984), among others, pointed out that this phenomenon is shared by a large class of statistics. For further reading on bootstrapping we refer the reader to Efron (1982), Hall (1992).

We now turn to the problem of bootstrapping M-estimators in a linear regression model. For the sake of clarity we shall restrict our attention to a *simple* linear regression model only. Our main purpose is to show how a certain weighted empirical sampling distribution naturally helps to overcome some inherent difficulties in defining bootstrap M-estimators. What follows is based on the work of Lahiri (1989). No proofs will be given as they involve intricate technicalities of the Edgeworth expansion for independent non-identically distributed r.v.'s.

Accordingly, assume that $\{e_i, i \geq 1\}$ are i.i.d. F r.v.'s, $\{x_{ni}, 1 \leq i \leq n\}$ are the known design points, $\{Y_{ni}, 1 \leq i \leq n\}$ are observable r.v.'s such that for a $\beta \in \mathbb{R}$,

$$Y_{ni} = x_{ni}\beta + e_i, \quad i \geq 1. \tag{4.2.21}$$

The score function ψ is assumed to satisfy

$$\int \psi dF = 0. \tag{4.2.22}$$

Let $\hat{\Delta}_n$ be an M-estimator obtained as a solution t of

$$\sum_{i=1}^{n} x_{ni}\psi(Y_{ni} - x_{ni}t) = 0, \tag{4.2.23}$$

and F_n be an estimator of F based on the residuals $\hat{e}_{ni} := Y_{ni} - x_{ni}\hat{\Delta}_n, 1 \leq i \leq n$. Let $\{e^*_{ni}, 1 \leq i \leq n\}$ be i.i.d. F_n r.v.'s and define

$$Y^*_{ni} = x_{ni}\hat{\Delta}_n + e^*_{ni}, \qquad 1 \leq i \leq n. \tag{4.2.24}$$

The bootstrap M-estimator Δ^*_n is defined to be a solution t of

$$\sum_{i=1}^{n} x_{ni}\psi(Y^*_{ni} - x_{ni}t) = 0. \tag{4.2.25}$$

Recall, from the previous section, that in general (4.2.22) ensures the absence of the asymptotic bias in $\hat{\Delta}_n$. Analogously, to ensure the absence of the asymptotic bias in Δ^*_n, we need to have F_n such that

$$\int \psi dF_n = E_n\psi(e^*_{n1}) = 0, \tag{4.2.26}$$

where E_n is the expectation under F_n. In general, the choice of F_n that will satisfy (4.2.26) and at the same time be a reasonable estimator of F depends heavily on the forms of ψ and F. When bootstrapping the least square estimator of β, i.e., when $\psi(x) \equiv x$, Freedman (1981) ensure (4.2.26) by choosing F_n to be the empirical d.f. of the centered residuals $\{\hat{e}_{ni} - \hat{e}_{n.}, 1 \leq i \leq n\}$, where $\hat{e}_{n.} := n^{-1}\sum_{j=1}^{n} \hat{e}_{nj}$. In fact, he shows that if one does not center the residuals, the bootstrap distribution of the least squares estimator does not approximate the corresponding original distribution.

Clearly, the ordinary empirical d.f. $\hat{H}_n$ of the residuals $\{\hat{e}_{ni}; 1 \leq i \leq n\}$ does not ensure the validity of (4.2.26) for general designs and a general ψ. We are thus forced to look at appropriate modifications of the usual bootstrap. Here we describe two modifications. One chooses the resampling distribution appropriately and the other modifies the defining equation (4.2.6) *a la* Shorack (1982). Both provide the second order correct approximations to the distribution of standardized $\hat{\Delta}_n$.

Weighted Empirical Bootstrap

Assume that the design points $\{x_{ni}\}$ are either all non-negative or all non-positive. Let $\omega_x = \sum_{i=1}^n |x_{ni}|$ be positive and define

$$(4.2.27) \qquad F_{1n}(y) := \omega_x^{-1} \sum_{i=1}^n |x_{ni}| I(\hat{e}_{ni} \le y), \quad y \in \mathbb{R}.$$

Take the resampling distribution F_n to be F_{1n}. Then, clearly,

$$\begin{aligned} E_{1n}\psi(e_{n1}^*) &= \omega_x^{-1} \sum_{i=1}^n |x_{ni}| \psi(\hat{e}_{ni}) \\ &= \mathrm{sign}(x_1)\omega_x^{-1} \sum_{i=1}^n x_{ni}\psi(Y_{ni} - x_{ni}\hat{\Delta}) = 0, \end{aligned}$$

by the definition of $\hat{\Delta}_n$. That is, F_{1n} satisfies (4.2.26) for any ψ.

Modified Scores Bootstrap

Let F_n be any resampling distribution based on the residuals. Define the bootstrap estimator Δ_{ns} to be a solution t of the equation

$$(4.2.28) \qquad \sum_{i=1}^n x_{ni}[\psi(Y_{ni}^* - x_{ni}t) - E_n\psi(e_{ni}^*)] = 0.$$

In other words the score function is now a *priori* centered under F_n and hence (4.2.26) holds for any F_n and any ψ.

We now describe the second order correctness of these procedures. To that effect we need some more notation and assumptions. To begin with let $\tau_x^2 := \sum_{i=1}^n x_{ni}^2$ and define

$$m_x := \max_{1 \le i \le n} |x_{ni}|; \quad b_{1x} := \sum_{i=1}^n x_{ni}^3/\tau_x^3, \quad b_x := \sum_{i=1}^n |x_{ni}^3|/\tau_x^3.$$

For a d.f. F and any sampling d.f. F_n, define, for an $x \in \mathbb{R}$,

$$\begin{aligned}
\gamma(x) &:= E\psi(e_1 - x), \quad \omega(x) = \sigma^2(x) := E\{\psi(e_1 - x) - \gamma(x)\}^2, \\
\omega_1(x) &:= E\{\psi(e_1 - x) - \gamma(x)\}^3, \quad \gamma_n(x) := E_n\psi(e^*_{n1} - x), \\
\omega_n(x) &:= \sigma^2_n(x) = E_n\{\psi(e^*_{n1} - x) - \gamma_n(x)\}^2, \\
\omega_{1n}(x) &:= E_n\{\psi(e^*_{n1} - x) - \gamma_n(x)\}^3, \\
A_n(c) &:= \{i : 1 \le i \le n, |x_{ni}| > c\tau_x b_x\}, \quad \kappa_n(c) := \#A_n(c), \ c > 0.
\end{aligned}$$

For any real valued function g on $\mathbb{R}$, let $\dot{g}, \ddot{g}$ denote its first and second derivatives at 0 whenever they exist, respectively. Also, write γ_n, ω_n etc. for $\gamma_{\mathcal{N}}(0), \omega_{\mathcal{N}}(0)$, etc. Finally, let $\alpha := -\dot{\gamma}/\sigma$, $\alpha_n := -\dot{\gamma}_n/\sigma_n$ and, define for $x \in \mathbb{R}$, $H_2(x) := x^2 - 1$, and

$$\mathcal{P}_n(x) := \Phi(x) - b_{1x}\left[\left\{\frac{\ddot{\gamma}_n}{\sigma_n} - \frac{\dot{\gamma}_n\dot{\omega}_n}{\sigma^3_n}\right\}\frac{x^2}{2\alpha^2_n} + \frac{\omega_{1n}}{6\sigma^3_n}H_2(x)\right]\varphi(x).$$

In the following theorems, a.s. means for almost all sequences $\{e_i; i \ge 1\}$ of i.i.d. F r.v.'s.

Theorem 4.2.3 *Let the model (4.2.21) hold. In addition, assume that ψ has uniformly continuous bounded second derivative and that the following hold:*

(a) $\tau^2_x \to \infty$. *(b)* $\alpha > 0$.

(c) There exists a constant $0 < c < 1$, such that $ln\tau_x = o(\kappa_n(c))$.

(d) $m_x ln\tau_x = o(\tau_x)$.

(e) There exist constants $\theta > 0$, $\delta > 0$, and $q < 1$, such that

$$\sup[|E\exp\{it\psi(e_1 - x)\}| : |x| < \delta, |t| > \theta] < q.$$

(f) $\sum_{n=1}^{\infty} \exp(-\lambda\omega^2_x/\tau^2_x) < \infty, \ \forall\, \lambda > 0.$

Then, with Δ^*_n defined as a solution of (4.2.25) with $F_n = F_{1n}$,

$$\begin{aligned}
&\sup_y |P_{1n}(\alpha_n\tau_x(\Delta^*_n - \hat{\Delta}_n) \le y) - \mathcal{P}_n(y)| = o(m_x/\tau_x), \\
&\sup_y |P_{1n}(\alpha\tau_x(\hat{\Delta}_n - \beta) \le y) - P_{1n}(\tau_x(\Delta^*_n - \hat{\Delta}_n) \le y)| \\
&= o(m_x/\tau_x), \quad a.s.,
\end{aligned}$$

where P_{1n} denotes the bootstrap probability under F_{1n}, and where the supremum is over $y \in \mathbb{R}$. □

Next we state the analogous result for Δ_{ns}.

Theorem 4.2.4 *Suppose that all of the hypotheses of Theorem 4.2.3 except (f) hold and that Δ_{ns} is defined as a solution of (4.2.28) with $F_n = \hat{H}_n$, the ordinary empirical of the residuals. Then,*

$$\sup_y |\hat{P}_n(\alpha_n \tau_x(\Delta_{ns} - \hat{\Delta}_n) \leq y) - \mathcal{P}_n(y)| = o(m_x/\tau_x),$$

$$\sup_y |\hat{P}_n(\alpha \tau_x(\hat{\Delta}_n \beta) \leq y) - \hat{P}_n(\tau_x(\Delta_{ns} - \hat{\Delta}_n) \leq y)|$$
$$= o(m_x/\tau_x), \quad a.s.,$$

where $\hat{P}_n$ denotes the bootstrap probability under $\hat{H}_n$. □

The proofs of these theorems appear in Lahiri (1989) where he also discusses analogous results for a non-smooth ψ. In this case he chooses the sampling distribution to be a smooth estimator obtained from the kernel type density estimator. Lahiri (1992) gives extensions of the above theorems to multiple linear regression models.

Here we briefly comment about the assumptions (a) - (f). As is seen from the previous section, (a) and (b) are minimally required for the asymptotic normality of M-estimators. Assumptions (c), (e) and (f) are required to carry out the Edgeworth expansions while (d) is slightly stronger than Noether's condition (**NX**) applied to (4.2.21). In particular, $x_i \equiv 1$ and $x_i \equiv i$ satisfy (a), (c), (d) and (f).

A sufficient condition for (e) to hold is that F have a positive density and ψ have a continuous positive derivative on an open interval in $\mathbb{R}$.

4.3 Distributions of Some Scale Estimators

Here we shall now discuss some robust scale estimators.

Definitions. An estimator $\hat{\beta}(\mathbf{X}, \mathbf{Y})$ based on the design matrix $\mathbf{X}$ and the observation vector $\mathbf{Y}$ of β is said to be *location invariant* if

$$\text{(4.3.1)} \qquad \hat{\beta}(\mathbf{X}, \mathbf{Y} + \mathbf{X}\mathbf{b}) = \hat{\beta}(\mathbf{X}, \mathbf{Y}) + \mathbf{b}, \quad \forall\, \mathbf{b} \in \mathbb{R}^p.$$

It is said to be *scale invariant* if

$$\text{(4.3.2)} \qquad \hat{\beta}(\mathbf{X}, a\mathbf{Y}) = a\hat{\beta}(\mathbf{X}, \mathbf{Y}), \quad \forall\, a \in \mathbb{R},\; a \neq 0.$$

A scale estimator $s(\mathbf{X}, \mathbf{Y})$ of a scale parameter γ is said to be *location invariant* if

$$s(\mathbf{X}, \mathbf{Y} + \mathbf{X}\mathbf{b}) = s(\mathbf{X}, \mathbf{Y}), \quad \forall \ \mathbf{b} \in \mathbb{R}^p. \tag{4.3.3}$$

It is said to be *scale invariant* if

$$s(\mathbf{X}, a\mathbf{Y}) = a\, s(\mathbf{X}, \mathbf{Y}), \quad \forall \ a > 0. \tag{4.3.4}$$

Now observe that M-estimators $\hat{\mathbf{\Delta}}$ and $\mathbf{\Delta}^*$ of $\boldsymbol{\beta}$ of Section 4.2.1 are location invariant but not scale invariant. The estimators $\hat{\mathbf{\Delta}}_1$ defined at (4.2.15), are location and scale invariant whenever s satisfies (4.3.3) and (4.3.4). Note that if s does not satisfy (4.3.3) then $\hat{\mathbf{\Delta}}_1$ need not be location invariant. Some of the candidates for s are

$$\begin{aligned} s &:= \Big\{(n-p)^{-1}\sum_{i=1}^{n}(Y_i - \mathbf{x}_i'\hat{\boldsymbol{\beta}})^2\Big\}^{1/2}, \\ s_1 &:= med\Big\{|Y_i - \mathbf{x}_i'\hat{\boldsymbol{\beta}}|; 1 \le i \le n\Big\}, \\ s_2 &:= med\Big\{\Big|Y_i - Y_j - (\mathbf{x}_i - \mathbf{x}_j)'\hat{\boldsymbol{\beta}}\Big|; 1 \le i < j \le n\Big\}, \end{aligned} \tag{4.3.5}$$

where $\hat{\boldsymbol{\beta}}$ is a preliminary estimator satisfying (4.3.1) and (4.3.2).

Estimator s^2, with $\hat{\boldsymbol{\beta}}$ as the least square estimator, is the usual estimator of the error variance, assuming it exists. It is known to be non-robust against outliers in the errors. In robustness studies one needs scale estimators that are not sensitive to outliers in the errors. Estimator s_1 has been mentioned by Huber (1981, p. 175) as one such candidate. The asymptotic properties of s_1, s_2 will be discussed shortly. Here we just mention that each of these estimators estimates a different scale parameter but that is not a point of concern if our goal is only to have location and scale invariant M-estimators of $\boldsymbol{\beta}$.

An alternative way of having location and scale invariant M-estimators of $\boldsymbol{\beta}$ is to use simultaneous M-estimation method for estimating $\boldsymbol{\beta}$ and γ of (2.3.36) as discussed in Huber (1981). We mention here, without giving details, that it is possible to study the asymptotic joint distribution of these estimators under heteroscedastic errors by using the results of Chapter 2.

We shall now study the asymptotic distributions of s_1 and s_2 under the model (1.1.1). With F_i denoting the d.f. of e_i, $H = n^{-1}\sum_{i=1}^n F_i$, let

$$(4.3.6) \quad p_1(y) := H(y) - H(-y),$$

$$(4.3.7) \quad p_2(y) := \int [H(y+x) - H(-y+x)]dH(x), \quad y \geq 0.$$

Define γ_1 and γ_2 by the relations

$$(4.3.8) \qquad p_1(\gamma_1) = 1/2, \qquad p_2(\gamma_2) = 1/2.$$

Note that in the case $F_i \equiv F$, γ_1 is median of the distribution of $|e_1|$ and γ_2 is median of the distribution of $|e_1 - e_2|$. In general, γ_j, p_j, etc. depend on n, but we suppress this for the sake of convenience.

The asymptotic distribution of s_j is obtained by the usual method of connecting the event $\{s_j \leq a\}$ with certain events based on certain empirical processes, as is done when studying the asymptotic distribution of the sample median, $j = 1, 2$. Accordingly, let, for $y \geq 0$,

$$(4.3.9) \quad S(y) := \sum_{i=1}^n I\left(|Y_i - \mathbf{x}_i'\hat{\beta}| \leq y\right),$$

$$T(y) := \sum \sum_{1 \leq i \leq j \leq n} I\left(|Y_i - Y_j - (\mathbf{x}_i - \mathbf{x}_j)'\hat{\beta}| \leq y\right).$$

Then, for an $a > 0$,

$$(4.3.10) \quad \{s_1 \leq a\} = \{S(a) \geq (n+1)2^{-1}\}, \quad n \text{ odd},$$

$$\{S(a) \geq n2^{-1}\} \subseteq \{s_1 \leq a\} \subseteq \{S(a) \geq n2^{-1} - 1\}, \quad n \text{ even}.$$

Similarly, with $N := n(n-1)/2$, for an $a > 0$,

$$(4.3.11) \quad \{s_2 \leq a\} = \{T(a) \geq (N+1)2^{-1}\}, \quad N \text{ odd}$$

$$\{T(a) \geq N2^{-1}\} \subseteq \{s_2 \leq a\} \subseteq \{T(a) \geq N2^{-1} - 1\}, \quad N \text{ even}$$

Thus, to study the asymptotic distributions of $s_j, j = 1, 2$, it suffices to study those of $S(y)$ and $T(y)$, $y \geq 0$.

In what follows we shall be using the notation of Chapter 2 with the following modifications. As before, we shall write S_1^0, μ_1^0 etc. for

S_d^0, μ_d^0 etc. of (2.3.2) whenever $d_{ni} \equiv n^{-1/2}$. Moreover, in (2.3.2), we shall take

$$(4.3.12) \qquad X_i = Y_i - \mathbf{x}_i'\boldsymbol{\beta} = e_i, \;\; \mathbf{c}_i = \mathbf{A}\mathbf{x}_i, \;\; 1 \le i \le n.$$

With these modifications, for all $n \ge 1$, $y \ge 0$,

$$\begin{aligned} S(y) &= S_1^0(y, \mathbf{v}) - S_1^0(-y, \mathbf{v}) = n^{-1/2}\sum_{i=1}^n I(|e_i - \mathbf{c}_i'\mathbf{v}| \le y), \\ 2n^{-1}T(y) &= \int [S_1^0(y + x, \mathbf{v}) - S_1^0(-y + x, \mathbf{v})] S_1^0(dx, \mathbf{v}) - 1, \end{aligned}$$

with probability 1, where $\mathbf{v} = \mathbf{A}^{-1}(\hat{\boldsymbol{\beta}} - \boldsymbol{\beta})$. Let

$$\begin{aligned} (4.3.13) \;\; \mu_1^0(y, \mathbf{u}) &= \mu_1(H(y), \mathbf{u}), \\ Y_1^0(y, \mathbf{u}) &= Y_1(H(y), \mathbf{u}), \quad y \in \mathbb{R}; \\ K(y, \mathbf{u}) &= \int [Y_1^0(y + x, \mathbf{u}) - Y_1^0(-y + x, \mathbf{u})] dH(x), \\ W(y, \mathbf{u}) &= Y_1^0(y, \mathbf{u}) - Y_1^0(-y, \mathbf{u}), \;\; y \ge 0, \;\; \mathbf{u} \in \mathbb{R}^p, \\ g_i(x) &:= \{f_i(\gamma_2 + x) - f_i(-\gamma_2 + x)\}, \\ h_i(x) &:= \{f_i(\gamma_2 + x) + f_i(-\gamma_2 + x)\}, \end{aligned}$$

for $1 \le i \le n$, $x \in \mathbb{R}$. We shall write $W(y), \mathbf{K}(y)$ etc. for $W(y, \mathbf{0})$, $\mathbf{K}(y, \mathbf{0})$ etc.

Theorem 4.3.1 *Assume that (1.1.1) holds with* $\mathbf{X}$ *and* $\{F_{ni}\}$ *satisfying (***NX***), (2.3.4) and (2.3.5). Moreover, assume that* H *is strictly increasing for each* n *and that*

$$\lim_{\delta \to 0} \limsup_{n} \sup_{0 \le s \le 1-\delta} [H(H^{-1}(s + \delta) \pm \gamma_2) - H(H^{-1}(s) \pm \gamma_2)] = 0. \qquad (4.3.14)$$

About $\{\hat{\boldsymbol{\beta}}\}$ *assume that*

$$(4.3.15) \qquad \|\mathbf{A}^{-1}(\hat{\boldsymbol{\beta}} - \boldsymbol{\beta})\| = O_p(1).$$

Then, $\forall \ a \in \mathbb{R}$,

$$(4.3.16) \quad P(n^{1/2}(s_1 - \gamma_1) \le a\gamma_1)$$
$$= P\Big(W(\gamma_1) + n^{-1/2}\sum_{i=1}^{n} \mathbf{x}_i'\mathbf{A}\{f_i(\gamma_1) - f_i(-\gamma_1)\} \cdot \mathbf{v}$$
$$\ge -a \cdot \gamma_1 n^{-1}\sum_{i=1}^{n}[f_i(\gamma_1) + f_i(-\gamma_1)]\Big) + o(1),$$

$$(4.3.17) \quad P(n^{1/2}(s_2 - \gamma_2) \le a\gamma_2)$$
$$= \quad P\Big(2K(\gamma_2) + n^{-3/2}\sum_i\sum_j \mathbf{c}_{ij}\int g_i(x)dF_j(x) \cdot \mathbf{v}$$
$$\ge -\gamma_2 a n^{-1}\sum_{i=1}^{n}\int h_i(x)dH(x)\Big) + o(1).$$

where $\mathbf{c}_{ij} = (\mathbf{x}_i - \mathbf{x}_j)'\mathbf{A}, 1 \le i, j \le n.$

Proof. We shall give the proof of (4.3.17) only; that of (4.3.16) being similar and less involved. Fix an $a \in \mathbb{R}$ and let $Q_n(a)$ denote the left hand side of (4.3.17). Assume that n is large enough so that $a_n := (an^{-1/2} + 1)\gamma_2 > 0$. Then, by (4.3.11),

$$Q_n(a) \quad = \quad P\big(T(a_n) \ge (N+1)/2\big), \qquad N \text{ odd}$$

$$P(T(a_n) \ge N/2) \quad \le \quad Q_n(a) \le P(T(a_n) \ge N2^{-1} - 1), \ N \text{ even}.$$

It thus suffices to study $P(T(a_n) \ge N2^{-1} + b)$, $b \in \mathbb{R}$. Now, let

$$\begin{aligned} T_1(y) &:= n^{-1/2}[2n^{-1}T(y) + 1] - n^{1/2}p_2(y), \quad y \ge 0, \\ k_n &:= (N + 2b)n^{-3/2} + n^{-1/2} - n^{1/2}p_2(a_n). \end{aligned}$$

Then, direct calculations show that

$$(4.3.18) \qquad P(T(a_n) \ge N2^{-1} + b) = P(T_1(a_n) \ge k_n).$$

We now *analyze* k_n: By (4.3.8),

$$k_n = -n^{1/2}[p_2(a_n) - p_2(\gamma_2)] + O(n^{-1/2}).$$

But

$$
\begin{aligned}
& n^{1/2}[p_2(a_n) - p_2(\gamma_2)] \\
&= n^{1/2} \int [\{H(a_n + x) - H(\gamma_2 + x)\} \\
&\qquad\qquad - \{H(-a_n + x) - H(-\gamma_2 + x)\}] dH(x).
\end{aligned}
$$

By (2.3.4) and (2.3.5), the sequence of distributions $\{p_2\}$ is tight on $(\mathbb{R}, \mathcal{B})$, implying that $\gamma_2 = O(1), n^{-1/2}\gamma_2 = o(1)$. Consequently,

$$
\begin{aligned}
& n^{1/2} \int \{H(\pm a_n + x) - H(\pm\gamma_2 + x)\} dH(x) \\
&= a\gamma_2 n^{-1} \sum_{i=1}^{n} \int f_i(\pm\gamma_2 + x) dH(x) + o(1),
\end{aligned}
$$

$$
\text{(4.3.19)}\; k_n = -a\gamma_2 n^{-1} \sum_{i=1}^{n} \int [f_i(\gamma_2 + x) + f_i(-\gamma_2 + x)] dH(x) + o(1).
$$

Next, we approximate $T_1(a_n)$ by a sum of independent r.v.'s. The proof is similar to the one used in approximating linear rank statistics of Section 3.4. From the definition of T_1,

$$
\begin{aligned}
& T_1(y) \\
&= n^{-1/2} \int [S_1^0(y + x, \mathbf{v}) - S_1^0(-y + x, \mathbf{v})] S_1^0(dx, \mathbf{v}) \\
&= n^{-1/2} \int [Y_1^0(y + x, \mathbf{v}) - Y_1^0(-y + x, \mathbf{v})] S_1^0(dx, \mathbf{v}) \\
&\quad + n^{-1/2} \int [\mu_1^0(y + x, \mathbf{v}) - \mu_1^0(-y + x, \mathbf{v})] Y_1^0(dx, \mathbf{v}) \\
&\quad + n^{-1/2} \int [\mu_1^0(y + x, \mathbf{v}) - \mu_1^0(-y + x, \mathbf{v})] \mu_1^0(dx, \mathbf{v}) \\
&\qquad\qquad - n^{1/2} p_2(y) \\
&= E_1(y) + E_2(y) + E_3(y), \quad \text{say}.
\end{aligned}
$$

But

$$
\begin{aligned}
& E_3(y) \\
&:= n^{-1/2} \int [\mu_1^0(y + x, \mathbf{v}) - \mu_1^0(-y + x, \mathbf{v})] \mu_1^0(dx, \mathbf{v}) - n^{1/2} p_2(y)
\end{aligned}
$$

$$= n^{-3/2} \sum_{i=1}^{n} \sum_{j} \int \Big\{ F_1(y + x + c'_{ij}\mathbf{v}) - F_i(-y + x + \mathbf{c}'_{ij}\mathbf{v})$$

$$- F_i(y+x) + F_1(-y+x) \Big\} dF_j(x)$$

$$= n^{-3/2} \sum_{i=1}^{n} \sum_{j} \mathbf{c}'_{ij}\mathbf{v} \int [f_i(y+x) - f_i(-y+x)] dF_j(x) + u_p(1),$$

by (2.3.3), (**NX**) and (4.3.15). In this proof, $u_p(1)$ means $o_p(1)$ uniformly in $|y| \le k$, for every $0 < k < \infty$.

Integration by parts, (4.3.15), (2.3.25), H increasing and the fact that $\int n^{-1/2} \mu_1^0(dx, \mathbf{v}) = 1$ yield that

$$\begin{aligned} E_2(y) & \\ := \; & n^{-1/2} \int \{\mu_1^0(y + x, \mathbf{v}) - \mu_1^0(-y + x, \mathbf{v})\} Y_1^0(dx, \mathbf{v}) \\ = \; & n^{-1/2} \int \{Y_1^0(y + x, \mathbf{v}) - Y_1^0(-y + x, \mathbf{v})\} \mu_1^0(dx, \mathbf{v}) \\ = \; & \int \{Y_1^0(y + x) - Y_1^0(-y + x)\} dH(x) + u_p(1). \end{aligned}$$

Similarly,

$$\text{(4.3.20)} \quad E_1(y) = n^{-1/2} \int \{Y_1^0(y + x) - Y_1^0(-y + x)\} S_1^0(dx) + u_p(1).$$

Now observe that $n^{-1/2} S_1^0 = H_n$, the ordinary empirical d.f. of the errors $\{e_i\}$. Let

$$\begin{aligned} E_{11}(y) & \\ := \; & \int \{Y_1^0(y + x) - Y_1^0(-y + x)\} d(H_n(x) - H(x)) \\ = \; & \mathcal{Z}(y) - \mathcal{Z}(-y), \end{aligned}$$

where

$$\mathcal{Z}(\pm y) := \int Y_1^0(\pm y + x) d[H_n(x) - H(x)], \quad y \ge 0.$$

We shall show that

$$\text{(4.3.21)} \qquad \mathcal{Z}(\pm a_n) = o_p(1).$$

But

$$(4.3.22)\quad \begin{aligned} &|\mathcal{Z}(\pm a_n) - \mathcal{Z}(\pm\gamma_2)| \\ &= \Big| \int [Y_1(H(\pm a_n + x)) - Y_1(H(\pm\gamma_2 + x))] \\ &\qquad\qquad \times d(H_n(x) - H(x)) \Big| \\ &\le 2 \sup_{|y-s|\le|a|n^{-1/2}\gamma} |Y_1(H(y)) - Y_1(H(z))| = o_p(1), \end{aligned}$$

because of (2.3.4)-(2.3.5) and Corollary 2.3.1 applied with $d_{ni} \equiv n^{-1/2}$. Thus, to prove (4.3.21), it suffices to show that

$$(4.3.23)\qquad \mathcal{Z}(\pm\gamma_2) = o_p(1).$$

But

$$\begin{aligned} &|\mathcal{Z}(\pm\gamma_2)| \\ &= \Big| \int_0^1 [Y_1(H(\pm\gamma_2 + H_n^{-1}(t))) - Y_1(H(\pm\gamma_2 + H^{-1}(t)))]dt \Big| \\ &\le \sup_{0\le t\le 1} \Big| Y_1(H(\pm\gamma_2 + H^{-1}(HH_n^{-1}(t))) - Y_1(H(\pm\gamma_2 + H^{-1}(t))) \Big| \\ &= o_p(1). \end{aligned}$$

by the assumption (4.3.14), Lemma 3.4.1 and Corollary 2.3.1 applied with $d_{ni} \equiv n^{-1/2}$. This proves (4.3.21). Consequently, from (4.3.20) and an argument like (4.3.22), it follows that

$$\begin{aligned} E_1(a_n) &= \int \{Y_1^0(a_n + x) - Y_1^0(-a_n + x)\}dH(x) + o_p(1) \\ &= \int \{Y_1^0(\gamma_2 + x) - Y_1^0(-\gamma_2 + x)\}dH(x) + o_p(1). \end{aligned}$$

From these derivations and the definition (4.3.13), we obtain

$$(4.3.24)\quad \begin{aligned} &T_1(a_n) \\ &= 2K(\gamma_2) \\ &\quad + n^{-3/2} \sum_{i,j} \mathbf{c}_{ij}' \mathbf{A} \int \{f_i(\gamma_2 + x) - f_i(-\gamma_2 + x)\}dF_j(x) \cdot \mathbf{v} \\ &\qquad + o_p(1). \end{aligned}$$

Now, from the definition of k_n and (4.3.19), it follows that the $\lim k_n$ does not depend on b. Thus the limit of the l.h.s. of (4.3.18) is the same for $b = -1, 0, 1/2$, and, in view of (4.3.18), (4.3.19) and (4.3.24), it is given by the first term on the r.h.s. of (4.3.17). □

Remark 4.3.1 Observe that, in view of (4.3.8),

$$\begin{aligned} W(\gamma_1) &= n^{-1/2}\sum_{i=1}^{n}\{I(|e_i| \le g_1) - 1/2\}, \\ K(\gamma_2) &= \int \{H(\gamma_2 + x) - H(-\gamma_2 + x)\}dY_1^0(x) \\ &= n^{-1/2}\sum_{i=1}^{n}\{H(\gamma_2 + e_i) - H(-\gamma_2 + e_i) - 1/2\}. \end{aligned}$$

Thus, $W(\gamma_1)$ and $K(\gamma_2)$ are the sums of bounded independent centered r.v.'s and by the L-F CLT one obtains

$$(4.3.25) \quad \sigma_1^{-1}W(\gamma_1 \to_d \mathcal{N}(0,1) \quad \text{and} \quad \sigma_2^{-1}K(\gamma_2) \to_d \mathcal{N}(0,1),$$

where

$$\begin{aligned} \sigma_1^2 &:= Var\,(W(\gamma_1)) \\ &= n^{-1}\sum_{i=1}^{n}\{F_i(\gamma_1) - F_i(-\gamma_1)\}\{1 - F_i(\gamma_1) + F_1(-\gamma_1)\}, \\ \sigma_2^2 &:= Var\,(K(\gamma_2)) \\ &= n^{-1}\sum_{i=1}^{n}\int [H(\gamma_2 + x) - H(-\gamma_2 + x)]^2 dF_1(x) - (1/4). \ \square \end{aligned}$$

Remark 4.3.2 If $\{F_i\}$ are all symmetric about zero, then from (4.3.25), (4.3.16) and (4.3.17), it follows that the asymptotic distribution of s_1 and s_2 does not depend on the initial estimator $\hat{\boldsymbol{\beta}}$ of $\boldsymbol{\beta}$. In fact, in this case we can deduce that

$$(4.3.26) \qquad \tau_1^{-1}n^{1/2}(s_1 - \gamma_1)\gamma_1^{-1} \to_d \mathcal{N}(0,1),$$

$$\tau_2^{-1}n^{1/2}(s_2 - \gamma_2)\gamma_2^{-1} \to_d \mathcal{N}(0,1),$$

where

$$\tau_1^2 := \sigma_1^2\{2\gamma_1 h(\gamma_1)\}^{-2}, \quad \tau_2^2 := \sigma_2^2\{\gamma_2 \int h(\gamma_2 + x)dH(x)\}^{-2},$$

$$h(x) := n^{-1}\sum_{i=1}^n f_i(x). \qquad \square$$

Remark 4.3.3 *i.i.d. case.* In the case $F_i \equiv F$, the asymptotic distribution of s_1 depends on $\hat{\boldsymbol{\beta}}$ unless F is symmetric around zero. However, the asymptotic distribution of s_2 does not depend on $\hat{\boldsymbol{\beta}}$. This is so because in this case the coefficient of $\mathbf{v}$ in (4.3.17) is

$$n^{-3/2}\sum_i\sum_j(\mathbf{x}_i - \mathbf{x}_j)'\mathbf{A}\int[f(\gamma_2 + x) - f(-\gamma_2 + x)]dF(x) = \mathbf{0}. \square$$

That the asymptotic distribution of s_2 is independent of $\hat{\boldsymbol{\beta}}$ is not surprising because s_2 is essentially a symmetrized variant of s_1. We summarize this property of s_2 as

Corollary 4.3.1 *If in the model (1.1.1), $F_{ni} \equiv F$, F satisfies (**F1**), (**F2**) and $\mathbf{X}$ satisfies (**NX**), then $\tau_2^{-1}n^{1/2}(s_2 - \gamma_2) \to_d \mathcal{N}(0,1)$, where now*

$$\tau_2^2 = \frac{\int[F(g_2 + x) - F(-\gamma_2 + x)]^2 dF(x) - 1/4}{\left(\int f(\gamma_2 + x)dF(x)\right)^2}.$$

Note that γ_2 is now the median of the distribution of $|e_1 - e_2|$. Also, observe that the condition (4.3.14) now is equivalent to

$$\sup_{0 \le s \le 1-\delta}\Big[P(F(e_1 - y) \le s + \delta) - P(F(e_1 - y) \le s)\Big] \to 0,$$
$$\text{as } \delta \to 0, \ \forall\, y \in \mathbb{R},$$

which is implied by the assumptions on F. $\square$

4.4 R-Estimators

Consider the model (1.1.1) and the vector of linear rank statistics

$$\text{(4.4.1)} \qquad \mathbf{T}(\mathbf{t}) := \mathbf{A}\sum_{i=1}^n(\mathbf{x}_{ni} - \overline{\mathbf{x}}_n)\varphi(R_{i\mathbf{t}}/(n+1)), \quad \mathbf{t} \in \mathbb{R}^p,$$

where $\mathbf{A}_1$ is as in (4.2.12) and R_{it} is the rank of $Y_{ni} - \mathbf{x}'_{ni}\mathbf{t}$ among $\{Y_{nj} - \mathbf{x}_{nj}\mathbf{t}, 1 \leq j \leq n\}$.

One of the classes of R-estimators of $\boldsymbol{\beta}$ is defined by the relation

$$\text{(4.4.2)} \qquad \inf_{\mathbf{t}} |\mathbf{T}(\mathbf{t})|_1 = |\mathbf{T}(\hat{\boldsymbol{\beta}}_1)|_1 = \sum_{j=1}^{p} |T_j(\mathbf{t})| = 0,$$

T_j being the jth component of $\mathbf{T}$ of (4.4.1). The estimators $\hat{\boldsymbol{\beta}}_1$ were initially studied by Adichie (1967) for the case $p = 1$ and by Jurečková (1971) for $p \geq 1$.

Another class of R-estimators can be defined by the relation

$$\text{(4.4.3)} \qquad \inf_{\mathbf{t}} \|\mathbf{T}(\mathbf{t})\| = \|\mathbf{T}(\hat{\boldsymbol{\beta}})\|.$$

Yet another class of estimators, introduced by Jaeckel (1972), is defined by the relation

$$\text{(4.4.4)} \qquad \inf_{\mathbf{t}} \mathcal{J}(\mathbf{t}) = \mathcal{J}(\hat{\boldsymbol{\beta}}_3),$$

$$\mathcal{J}(\mathbf{t}) := \sum_{i=1}^{n} (Y_{ni} - \mathbf{x}'_{ni}\mathbf{t})\varphi(R_{it}/(n+1)), \quad \mathbf{t} \in \mathbb{R}^p.$$

Jaeckel (op.cit.) showed that for every observation vector $(Y_1, \cdots, Y_n)$ and for every $n \geq p$, $\sum_{i=1}^{n} \varphi(i/n+1) \equiv 0$ implies that $\mathcal{J}(\mathbf{t})$ is nonnegative, continuous and convex function of $\mathbf{t}$. If, in addition, $\mathbf{X}_c$ has the full rank p then the set $\{\mathbf{t}; \mathcal{J}(\mathbf{t}) \leq b\}$ is bounded for every $0 \leq b < \infty$, where $\mathbf{X}_c$ is defined at (4.2.11). Consequently, $\hat{\boldsymbol{\beta}}_3$ exists.

Moreover, the almost everywhere derivative of $\mathcal{J}(\mathbf{t})$ is $-\mathbf{A}_1^{-1}$ $\mathbf{T}(\mathbf{t})$, so that at $\hat{\boldsymbol{\beta}}_3$, $\mathbf{T}$ is nearly equal to zero and hence $\hat{\boldsymbol{\beta}}_1$, $\hat{\boldsymbol{\beta}}_2$, and $\hat{\boldsymbol{\beta}}_3$ are essentially the same estimators. Jaeckel showed, using the AUL property of $T(\mathbf{t})$ due to Jurečková (1971), that indeed $\|\mathbf{A}_1(\hat{\boldsymbol{\beta}}_1 - \hat{\boldsymbol{\beta}}_3)\| = o_p(1)$.

Here we shall discuss the asymptotic distribution of $\{\hat{\boldsymbol{\beta}}_2\}$ under general heteroscedastic errors. The main tool is the AUL Theorem 3.2.4. We shall also conclude that $\|\mathbf{A}_1(\hat{\boldsymbol{\beta}}_2 - \hat{\boldsymbol{\beta}}_3)\| = o_p(1)$ under (1.1.1) with general independent errors.

To begin with note that $\mathbf{T}$ of (4.4.1) is a p-vector $(T_1, \cdots, T_p)'$

where $T_j(\mathbf{t})$ is a $T_d(\varphi, \mathbf{u})$ - statistic of (3.1.2) with

$$\begin{aligned}(4.4.5)\qquad X_{ni} &= Y_{ni} - \mathbf{x}'_{ni}\boldsymbol{\beta}, \quad \mathbf{u} = \mathbf{A}_1^{-1}(\mathbf{t} - \boldsymbol{\beta}),\\ d_{ni} &= \mathbf{a}'_{(j)}(\mathbf{x}_{ni} - \overline{\mathbf{x}}_n), \quad \mathbf{c}_{ni} = A_1(\mathbf{x}_{ni} - \overline{\mathbf{x}}_n), \quad 1 \le i \le n;\\ \mathbf{a}_{(j)} &= j^{th} \text{ column of } \mathbf{A}_1, \ 1 \le j \le p.\end{aligned}$$

Thus specializing Theorem 3.2.4 to this case readily yields

Lemma 4.4.1 *Suppose that (1.1.1) holds with F_{ni} as a d.f. of e_{ni}, $1 \le i \le n$. In addition, assume that*

$$(\mathbf{NX}_c)\qquad (\mathbf{X}_c'\mathbf{X}_c)^{-1} \text{ exists for all } n \ge p,$$
$$\max_{1 \le i \le n}(\mathbf{x}_{ni} - \overline{\mathbf{x}}_n)'\mathbf{X}_c'\mathbf{X}_c)^{-1}(\mathbf{x}_{ni} - \overline{\mathbf{x}}_n) = o(1).$$

About $\{F_{ni}\}$ assume that H is strictly increasing for each n and that (2.3.5), (3.2.12), (3.2.34), (3.2.35) hold and that for all $j = 1, \cdots, p$,

$$(4.4.6)\qquad \lim_{\delta \to 0} \limsup_n \sup_{0 \le s \le 1-\delta} [L_j(s+\delta) - L_j(s)] = 0,$$

where for $0 \le s \le 1, \ 1 \le j \le p$,

$$L_j(s) := \sum_{i=1}^n \left(\mathbf{a}'_{(j)}(\mathbf{x}_{ni} - \overline{\mathbf{x}}_n)\right)^2 F_{ni}(H^{-1}(s)).$$

Then, for every $0 < b < \infty$,

$$(4.4.7)\qquad \sup_{\varphi \in \mathcal{C}, \|\mathbf{A}_1^{-1}(\mathbf{t}-\boldsymbol{\beta})\| \le b} \|\mathbf{T}(\mathbf{t}) - \mathbf{T}(\boldsymbol{\beta}) + \mathbf{K}_n\mathbf{A}_1^{-1}(\mathbf{t} - \boldsymbol{\beta})\| = o_p(1)$$

where

$$\mathbf{K}_n := \mathbf{A}_1 \int_0^1 \sum_{i=1}^n (\mathbf{x}_{ni} - \tilde{\mathbf{x}}_n(s))(\mathbf{x}_{ni} - \overline{\mathbf{x}}_n)' q_{ni}(s) d\varphi(s) \mathbf{A}_1$$

$$\tilde{\mathbf{x}}_n(s) := n^{-1} \sum_{i=1}^n \mathbf{x}_{ni}\ell_{ni}(s),$$

ℓ_{ni} as in (3.4.3) and $q_{ni} := f_{ni}(H^{-1}), 1 \le i \le n.$ □

In order to prove the asymptotic normality of $\hat{\boldsymbol{\beta}}_2$, we need to show that $\|\mathbf{A}_1^{-1}(\hat{\boldsymbol{\beta}}_2 - \boldsymbol{\beta})\| = O_p(1)$. To this effect let

$$\begin{aligned}\boldsymbol{\mu} &:= \mathbf{A}_1 \sum_{i=1}^{n} (\mathbf{x}_{ni} - \bar{\mathbf{x}}_n) \int F_{ni}(H^{-1}) d\varphi, \\ \mathbf{S} &:= \mathbf{T}(\boldsymbol{\beta}) - \boldsymbol{\mu}.\end{aligned}$$

Observe that the distribution of $(\hat{\boldsymbol{\beta}}_2 - \boldsymbol{\beta})$ does not depend on $\boldsymbol{\beta}$, even when $\{e_{ni}\}$ are not identically distributed.

Lemma 4.4.2 *In addition to the assumptions of Lemma 4.4.1 suppose that*

(4.4.8) $$\|\mathbf{S} + \boldsymbol{\mu}\| = O_p(1),$$

(4.4.9) $$\liminf_n \inf_{\|\mathbf{e}\|=1} |\mathbf{e}'\mathbf{K}_n \mathbf{e}| \ge \alpha \quad \textit{for an} \quad \alpha > 0,$$

(4.4.10) $$\mathbf{K}_n^{-1} \ \textit{exists for all} \ n \ge p, \quad \|\mathbf{K}_n^{-1}\| = O(1).$$

Then, for every $\epsilon > 0, 0 < z < \infty$, *there exist a* $0 < b < \infty$ *it and* N_ϵ *such that*

(4.4.11) $$P(\inf_{\|\mathbf{u}\|>b} \|\mathbf{T}(\mathbf{A}_1\mathbf{u} + \boldsymbol{\beta})\| \ge z) \ge 1 - \epsilon, \quad n \ge N_\epsilon.$$

Proof. Fix an $\epsilon > 0, 0 < z < \infty$. Without loss of generality assume $\boldsymbol{\beta} = \mathbf{0}$. Observe that by the C-S inequality

$$\inf_{\|\mathbf{u}\|>b} \|\mathbf{T}(\mathbf{A}_1\mathbf{u})\|^2 \ge \inf_{\|\mathbf{e}\|=1, |r|>b} (\mathbf{e}'\mathbf{T}(r\mathbf{A}_1\mathbf{e}))^2.$$

thus it suffices to prove that there exist a $0 < b < \infty$ and N_ϵ such that

(4.4.12) $$P\left(\inf_{\|\mathbf{e}\|=1, |r|>b} (\mathbf{e}'\mathbf{T}(r\mathbf{A}_1\mathbf{e}))^2 \ge z\right) \ge 1 - \epsilon, \ n > N_\epsilon.$$

Let, for $\mathbf{t} \in \mathbb{R}^p$, $\hat{\mathbf{T}}(\mathbf{t}) := \mathbf{T}(\mathbf{0}) - \mathbf{K}_n \mathbf{A}_1^{-1}\mathbf{t}$, so that, by (4.4.7) for every $0 < b < \infty$,

(4.4.13) $$\sup_{\|\mathbf{e}\|=1, |r|\le b} |\mathbf{e}'\mathbf{T}(r\mathbf{A}_1\mathbf{e}) - \mathbf{e}'\hat{\mathbf{T}}(r\mathbf{A}_1\mathbf{e})| = o_p(1).$$

But

$$\mathbf{e}'\hat{\mathbf{T}}(r\mathbf{A}_1\mathbf{e}) = \mathbf{e}'(\mathbf{S}+\boldsymbol{\mu}) - \mathbf{e}'\mathbf{K}_n\mathbf{e}r.$$

By (4.4.8), there exist a K_ϵ and an $N_{1\epsilon}$ such that

$$P(|\mathbf{S}+\boldsymbol{\mu}| \le \mathbf{K}_\epsilon) \ge 1-\epsilon/2, \quad n \ge N_{1\epsilon}.$$

Choose b to satisfy

$$b \ge (K_\epsilon + z^{1/2})\alpha^{-1}, \quad \alpha \text{ as in (4.4.9)}. \tag{4.4.14}$$

Then

$$\begin{aligned} P\left(\inf_{\|\mathbf{e}\|=1,|r|>b} (\mathbf{e}'\hat{\mathbf{T}}(r\mathbf{A}_1\mathbf{e}))^2 \ge z\right) \tag{4.4.15} \\ \ge \; P\left(\|\mathbf{S}+\boldsymbol{\mu}\| \le -z^{1/2} + b \inf_{\|\mathbf{e}\|=1} |\mathbf{e}'\mathbf{K}_n\mathbf{e}|\right) \\ \ge \; P\left(\|\mathbf{S}+\boldsymbol{\mu}\| \le K_\epsilon\right) \ge 1-\epsilon/2, \quad \forall n \ge N_{1\epsilon}. \end{aligned}$$

Therefore by (4.4.13) ad (4.4.15) there exist N_ϵ and b as in (4.4.14) such that

$$P\left(\inf_{\|\mathbf{e}\|=1,|r|>b} \left(\mathbf{e}'\mathbf{T}(r\mathbf{A}_1\mathbf{e})\right)^2 \ge z\right) \ge 1-\epsilon, \quad n \ge N_\epsilon. \tag{4.4.16}$$

But

$$\mathbf{e}'\mathbf{T}(r\mathbf{A}_1\mathbf{e}) = \mathbf{e}'\mathbf{A}_1 \sum_{i=1}^n (\mathbf{x}_i - \overline{\mathbf{x}})\varphi\left(\frac{R_{ir}^*}{n+1}\right) = \sum_{i=1}^n d_i\varphi\left(\frac{R_{ir}^*}{n+1}\right),$$

where $d_i = \mathbf{e}'\mathbf{A}_1(\mathbf{x}_i - \overline{\mathbf{x}})$, R_{ir}^* is the rank of $Y_i - r(\mathbf{x}_i - \overline{\mathbf{x}})\mathbf{A}_1\mathbf{e}$. But such a linear rank statistic is nondecreasing in r, for every $\mathbf{e}$. See, e.g., Hájek (1969; Theorem 7E, Chapter II). This together with (4.4.16) enables one to conclude (4.4.12) and hence (4.4.11). □

Theorem 4.4.1 *Suppose that (1.1.1) holds and that the design matrix* $\mathbf{X}$ *and the error d.f.'s* $\{F_{ni}\}$ *satisfy the assumptions of Lemmas 4.4.1 and 4.4.2 above. Then*

$$\mathbf{A}_1^{-1}(\hat{\boldsymbol{\beta}}_2 - \boldsymbol{\beta}) - \mathbf{K}_n^{-1}\boldsymbol{\mu} = \mathbf{K}_n^{-1}\mathbf{S} + o_p(1). \tag{4.4.17}$$

Proof. Follows from Lemmas 4.4.1 and 4.4.2. □

Remark 4.4.1 Arguing as in Jaeckel combined with an argument of Lemma 4.4.2, one can show that $\|\mathbf{A}_1^{-1}(\hat{\boldsymbol{\beta}}_2 - \hat{\boldsymbol{\beta}}_3)\| = O_p(1)$. Consequently, under the conditions of Lemmas 4.4.1 and 4.4.2, $\hat{\boldsymbol{\beta}}_2$ and the Jaeckel estimator $\hat{\boldsymbol{\beta}}_3$ also satisfy (4.4.17). □

Remark 4.4.2 Consider the case when $F_{ni} \equiv F, F$ a d.f. satisfying (**F1**), (**F2**). Then $\boldsymbol{\mu} = \mathbf{0}$ and $\mathbf{S} = \mathbf{T}(\boldsymbol{\beta})$. Moreover, under ($\mathbf{NX}_c$) all other assumptions of Lemmas 4.4.1 and 4.4.2 are a priori satisfied. Note that here

$$\ell_{ni} \equiv 1, \quad \tilde{\mathbf{x}}_n(s) \equiv \bar{\mathbf{x}}_n \quad \text{and} \quad \mathbf{K}_n \equiv \int f d\varphi(F) \cdot \mathbf{I}_{p\times p}.$$

Moreover, from Theorem 3.4.3 above, it follows that $\mathbf{S}$ converges in distribution to a $\mathcal{N}(\mathbf{0},\, \sigma_\varphi^2 \mathbf{I}_{p\times p})$, r.v., where

$$\sigma_\varphi^2 = \int_0^1 \varphi^2(u)du - \Big(\int_0^1 \varphi(u)du\Big)^2.$$

We summarize the above discussion as

Corollary 4.4.1 *Suppose that (1.1.1) with $F_{ni} \equiv F$ holds. Suppose that F and $\mathbf{X}$ satisfy (**F1**), (**F2**), and ($\mathbf{NX}_c$). In addition, suppose that φ is nondecreasing bounded on $[0,1]$ and $\int f d\varphi(F) > 0$. Then*

$$\mathbf{A}_1^{-1}(\hat{\boldsymbol{\beta}}_2 - \boldsymbol{\beta}) = \Big(\int f d\varphi(F)\Big)^{-1} \mathbf{T}(\boldsymbol{\beta}) + o_p(1).$$

Hence,

$$\mathbf{A}_1^{-1}(\hat{\boldsymbol{\beta}}_2 - \boldsymbol{\beta}) \rightarrow_d \mathcal{N}(\mathbf{0}, \tau^2 \mathbf{I}_{p\times p}), \; \tau^2 = \sigma_\varphi^2 \Big(\int f d\varphi(F)\Big)^{-2}. \; □$$

This result is quite general as far as the conditions on the design matrix $\mathbf{X}$ and F are concerned but not that general as far as the score function φ is concerned. □

Remark 4.4.3 Robustness against heteroscedastic gross errors. First we give a working definition of qualitative robustness.

Consider the model (1.1.1). Suppose that we have modeled the errors $\{e_{ni}, 1 \le i \le n\}$ to be i.i.d. F whereas their actual d.f.'s are $\{F_{ni}, 1 \le i \le n\}$. Let $P^n := \prod_{i=1}^n F$, $Q^n := \prod_{i=1}^n F_{ni}$ denote the corresponding product probability measure.

Definition 4.4.1. A sequence of estimators $\hat{\boldsymbol{\beta}}$ is said to be *qualitatively robust* for $\boldsymbol{\beta}$ at F against Q^n if it is consistent for $\boldsymbol{\beta}$ under P^n and under those Q^n that satisfy $\mathcal{D}_n := \max_i \sup_y |F_{ni}(y) - F(y)| \to 0$.

The above definition is a variant of that of Hampel (1971). One could use the notions of weak convergence on product probability spaces to give a bit more general definition. For example we could insist that the Prohorov distance between Q^n and P^n should tend to zero instead of requiring $\mathcal{D}_n \to 0$. We do not pursue this any further here.

The result (4.4.17) can be used to study the qualitative robustness of $\hat{\boldsymbol{\beta}}_2$ against certain heteroscedastic errors. Consider, for example, the gross errors model where, for some $0 \le \delta_{ni} \le 1$, with $\max_i \delta_{ni} \to 0$,

$$F_{ni} = (1 - \delta_{ni})F + \delta_{ni}G, \quad 1 \le i \le n,$$

and, where G is d.f. having a uniformly continuous a.e. positive density. If, in addition, $\{\delta_{ni}\}$ satisfy

$$\|\mathbf{A}_1 \sum_{i=1}^n (\mathbf{x}_{ni} - \bar{\mathbf{x}}_n)\delta_{ni}\| = O(1), \tag{4.4.18}$$

then one can readily see that $\|\mathbf{K}_n^{-1}\| = O(1)$ and $\|\boldsymbol{\mu}\| = O(1)$. It follows from (4.4.17) that $\hat{\boldsymbol{\beta}}_2$ is qualitatively robust against the above heteroscedastic gross errors at every F that has uniformly continuous a.e. positive density. Examples of δ_{ni} satisfying (4.4.18) would be

$$\delta_{ni} \equiv n^{-1/2} \quad \text{or } \delta_{ni} = p^{-1/2}\|\mathbf{A}_1(\mathbf{x}_{ni} - \bar{\mathbf{x}}_n)\|, \ 1 \le i \le n.$$

It may be argued that the latter choice of contaminating proportions $\{\delta_{ni}\}$ is more natural to linear regression than the former. A similar remark is applicable to $\hat{\boldsymbol{\beta}}_1$ and $\hat{\boldsymbol{\beta}}_3$. □

4.5 Estimation of $Q(f)$

Consider the model (1.1.1) with $F_{ni} \equiv F$, where F is a d.f. with density f on $\mathbb{R}$. Define

$$Q(f) = \int f\, d\varphi(F), \tag{4.5.1}$$

where $\varphi \in \mathcal{C}$ of (3.2.1).

As is seen from Corollary 4.4.1, the parameter Q appears in the asymptotic variance of R - estimators. The complete rank analysis of the model (1.1.1) requires an estimate of Q. This estimate is used to standardize rank test statistics when carrying out the ANOVA of linear models using Jaeckel's dispersion $\mathcal{J}$ of (4.4.4). See, for example, Hettmansperger (1984) and references therein for the rank based ANOVA.

Lehmann (1963) and Sen (1966) give estimators of Q in the one and two sample location models. These estimators are given in terms of lengths of confidence intervals based on linear rank statistics.

Cheng and Serfling (1981) discuss several estimators of Q when observations are i.i.d. F, i.e., when there are no nuisance parameters. Some of these estimators are obtained by replacing f by a kernel type density estimator and F by an empirical d.f. in Q. Scheweder (1975) discusses similar estimates of Q in the one sample location model.

In this section we discuss two types of estimators of Q. Both use a kernel type density estimator of f based on the residuals and the ordinary residual empirical d.f. to estimate F. The difference is in the way the window width and the kernel are chosen. In one the window width is partially based on the data and is of the order of square root of n and the kernel is the histogram type whereas in the other the kernel and the window width are arbitrary. It will be observed that the AUL result about the residual empirical process of Corollary 2.3.5 is the basic tool needed to prove the consistency of these estimators.

We begin with the class of estimators where the window width is **partly based on the data**. Define

$$p(y) := \int [F(y+x) - F(-y+x)]d\varphi(F(x)), \quad y \geq 0. \tag{4.5.2}$$

Since φ is a d.f., $p(y) \equiv P(|e - e^*| \leq y)$, where e, e^* are independent r.v.'s with respective d.f.'s F and $\varphi(F)$. Consequently, under (**F1**), the density of p at 0 is $2Q$. This suggests that an estimate of Q can be obtained by estimating the slope of p at 0.

Recall the definition of the residual empirical process $H_n(y, \mathbf{t})$ from (1.2.1). Let $\hat{\boldsymbol{\beta}}$ be an estimator of $\boldsymbol{\beta}$ and define

$$(4.5.3) \qquad \hat{H}_n(y) := H_n(y, \hat{\boldsymbol{\beta}}), \quad y \in \mathbb{R}.$$

A natural estimator of p is obtained by substituting $\hat{H}_n$ for F in p, v.i.z.,

$$\hat{p}_n(y) \quad := \quad \int [\hat{H}_n(y + x) - \hat{H}_n(-y + x)] d\varphi(\hat{H}_n(x)), \quad y \geq 0.$$

Let $-\infty = \hat{e}_{(0)} < \hat{e}_{(1)} \leq \hat{e}_{(2)} \leq \cdots \leq \hat{e}_{(n)} < \hat{e}_{(n+1)} = \infty$ denote the ordered residuals $\{\hat{e}_i := Y_i - \mathbf{x}_i'\hat{\boldsymbol{\beta}}, \quad 1 \leq i \leq n\}$. Since $\varphi(\hat{H}_n)$ assigns mass $\{\varphi(j/n) - \varphi(j-1)/n))\}$ to each $\hat{e}_{(j)}$ and zero mass to each of the intervals $(\hat{e}_{(j-1)}, \hat{e}_{(j)})$, $1 \leq j \leq n+1$; it follows that $\forall\, y \in \mathbb{R}$,

$$\begin{aligned}
(4.5.4) \quad & \hat{p}_n(y) \\
= \; & \sum_{i=1}^{n} \{\varphi(j/n) - \varphi(j-1)/n)\}[\hat{H}_n(y + \hat{e}_{(j)}) - \hat{H}_n(-y + \hat{e}_{(j)})] \\
= \; & n^{-1} \sum_{i=1}^{n} \left\{\varphi\left(\frac{j}{n}\right) - \varphi\left(\frac{j-1}{n}\right)\right\} \sum_{i=1}^{n} I(|\hat{e}_{(i)} - \hat{e}_{(j)}| \leq y).
\end{aligned}$$

From this one sees that $\hat{p}_n(y)$ has the following interpretation. For each j, one first computes the proportion of $\{\hat{e}_{(i)}\}$ falling in the interval $[-y + \hat{e}_{(j)}, y + \hat{e}_{(j)}]$ and then $\hat{p}_n(y)$ gives the weighted average of such proportions. Formula (4.5.4) is clearly suitable for computations.

Now, if $\{h_n\}$ is a sequence of positive numbers tending to zero, an estimator of Q is given by

$$\mathcal{Q}_n = \hat{p}_n(h_n)/2h_n.$$

This estimator can be viewed from the density estimation point of view also. Consider a kernel-type density estimator f_n of f based on the residuals $\{\hat{e}_i\}$:

$$f_n(x) := (2nh_n)^{-1}\sum_{i=1}^{n} I(|x-\hat{e}_i| \le h_n),$$

which uses the window $w_n(x) = (1/2)\cdot I(|x| \le h_n)$. Then a natural estimator of Q is

$$\int f_n d\varphi(\hat{H}_n) = \sum_{i=1}^{n}\left\{\varphi\left(\frac{j}{n}\right) - \varphi\left(\frac{j-1}{n}\right)\right\} f_n(\hat{e}_{(j)}) = \mathcal{Q}_n.$$

Scheweder (1975) studied the asymptotic properties of this estimator in the one sample location model. Observe that in this case the estimator of Q does not depend on the estimator of the location parameter which makes it relatively easier to derive its asymptotic properties.

In $\mathcal{Q}_n$, there is an arbitrariness due to the choice of the window width h_n. Here we recommend that h_n be determined from the spread of the data as follows. Let $0 < \alpha < 1$, t_n^α be α-th quantile of $\hat{p}_n$ and define the estimator Q_n^α of Q as

$$Q_n^\alpha := \hat{p}_n(n^{-1/2}t_n^\alpha)/(2n^{-1/2}t_n^\alpha). \tag{4.5.5}$$

The quantile t_n^α is an estimator of the α-th quantile t^α of p. Note that if $\varphi(s) \equiv s$, then t^α is the $\alpha-$th quantile of the distribution of $|e_1 - e_2|$ and t_n^α is the α-th quantile of the empirical d.f. $\hat{p}_n$ of the r.v.'s $\{|\hat{e}_i - \hat{e}_j|,\ 1 \le i, j \le n\}$. Thus, e.g., $t_n^{.5} = s_2$ of (4.3.5). Similarly, if $\varphi(s) = I(s \ge 0.5)$ then t^α (t_n^α) is α-th quantile of the d.f. of $|e_1|$ (empirical d.f. of $|\hat{e}_i|$, $1 \le i \le n$). Again, here $t_n^{.5}$ would correspond to s_1 of (4.3.5). In any case, in general, t^α is a scale parameter in the sense of Bickel and Lehmann (1975).

Note that the estimator Q_n^α is location and scale invariant in the sense of (4.3.3) and (4.3.4), as long as $\hat{\boldsymbol{\beta}}$ is location and scale invariant in the sense of (4.3.1) and (4.3.2). This follows from the fact that under (4.3.1) and (4.3.2), $\hat{p}_n(y, a\mathbf{Y} + \mathbf{X}\mathbf{b}) = \hat{p}_n(y/a, \mathbf{Y})$, $t_n^\alpha(a\mathbf{Y} + \mathbf{X}\mathbf{b}) = a\, t_n^\alpha(\mathbf{Y})$, for all $y \in \mathbb{R}$, $a > 0$, $\mathbf{b} \in \mathbb{R}^p$.

The consistency of $\mathcal{Q}_n^\alpha$ is asserted in the following

Theorem 4.5.1 *Let (1.1.1) hold with $F_{ni} \equiv F$. In addition to* (**NX**)*,* (**F1**) *and* (**F2**)*, assume that $\hat{\boldsymbol{\beta}}$ is an estimator of $\boldsymbol{\beta}$ satisfying (4.3.15). Then,*

$$\sup_{\varphi \in \mathcal{C}} |\mathcal{Q}_n^{\alpha} - Q(f)| = o_p(1). \tag{4.5.6}$$

The proof of (4.5.6) will be a consequence of the following *three* lemmas.

Lemma 4.5.1 *Under the assumptions of Theorem 4.5.1, $\forall\ 0 \le a < \infty$*

$$\sup_{\varphi \in \mathcal{C}, 0 \le z \le a} |n^{1/2}\{\hat{p}_n(n^{-1/2}z) - p(n^{-1/2}z)\}| = o_p(1). \tag{4.5.7}$$

Consequently, $\forall\ 0 \le a < \infty$,

$$\sup_{\varphi \in \mathcal{C}, 0 \le z \le a} |n^{1/2}\hat{p}_n(n^{-1/2}z) - 2zQ(f)| = o_p(1). \tag{4.5.8}$$

Proof. We shall apply Corollary 2.3.5. Let

$$\mathbf{v} = \mathbf{A}^{-1}(\hat{\boldsymbol{\beta}} - \boldsymbol{\beta}), \quad \mathbf{b}_n' = n^{-1/2}\sum_{i=1}^{n} \mathbf{x}_{ni}'\mathbf{A}.$$

Then, from (2.3.43), (4.5.3) and (4.3.15), we obtain

$$\sup_{-\infty \le y \le \infty} |n^{1/2}\{\hat{H}_n(y) - H_n(y)\} - \mathbf{b}_n'\mathbf{v}f(y)| = o_p(1). \tag{4.5.9}$$

where

$$H_n(y) \equiv H_n(y, \boldsymbol{\beta}) \equiv n^{-1}\sum_{i=1}^{n} I(e_{ni} \le y), \quad y \in \mathbb{R}.$$

Also, we will use the notation (2.3.1) with $\mathbf{u} = \mathbf{0}$ and

$$d_{ni} \equiv n^{-1/2}, \quad X_{ni} = Y_{ni} - \mathbf{x}_{ni}'\boldsymbol{\beta}, \quad F_{ni} \equiv F. \tag{4.5.10}$$

Then $Y_1(t, \mathbf{0}) \equiv n^{1/2}[H_n(F^{-1}(t)) - t]$, $0 \le t \le 1$. Write $Y_1(\cdot)$ for $Y_1(\cdot, \mathbf{0})$. Now, (4.5.9) and φ bounded imply that,

$$
\begin{aligned}
& n^{1/2}\{\hat{p}_n(y) - p(y)\} \\
& \quad = n^{1/2} \int \{H_n(y+x) - H_n(-y+x)\} d\varphi(\hat{H}_n(x)) \\
& \qquad + \mathbf{b}_n' \mathbf{v} \int [f(y+x) - f(-y+x)] d\varphi(\hat{H}_n(x)) \\
& \qquad - n^{1/2} p(y) + u_p(1) \\
(4.5.11) & \quad = R_{n1}(y) + R_{n2}(y) + R_{n3}(y) + u_p(1),
\end{aligned}
$$

where $u_p(1)$ stands for a sequence of random processes that converge to zero, uniformly in $y \in \mathbb{R}$, $\varphi \in \mathcal{C}$, in probability, and where

$$
\begin{aligned}
R_{n1}(y) &= \int \{Y_1(F(y+x)) - Y_1(F(-y+x))\} d\varphi(\hat{H}_n(x)) \\
R_{n2}(y) &= \mathbf{b}_n' \mathbf{v} \int [f(y+x) - f(-y+x)] d\varphi(\hat{H}_n(x)) \\
R_{n3}(y) &= n^{1/2} \Big\{ \int [F(y+x) - F(-y+x)] d\varphi(\hat{H}_n(x)) \\
&\qquad - \int [F(y+x) - F(-y+x)] d\varphi(F(x)) \Big\},
\end{aligned}
$$

for $y \in \mathbb{R}$. From (**F1**), (**F2**), the boundedness of φ, and the asymptotic continuity of Y_1, which follows from Corollary 2.2.1, applied to the quantities given in (4.5.10), we obtain, with $k = 2a\|f\|_\infty$,

$$
\begin{aligned}
(4.5.12) \quad \sup_{0 \le z \le a, \varphi \in \mathcal{C}} |R_{n1}(n^{-1/2} z)| &\le \sup_{|t-s| \le kn^{-1/2}} |Y_1(t) - Y_1(s)| \\
&= o_p(1).
\end{aligned}
$$

Again, (**F1**) and the boundedness of φ imply, in a routine fashion, that

$$
(4.5.13) \qquad \sup_{0 \le z \le a, \varphi \in \mathcal{C}} |R_{n1}(n^{-1/2} z)| = o_p(1).
$$

Now consider R_{n3}. By the MVT, (**F1**) and the boundedness of φ, the first term of $R_{n3}(n^{-1/2} z)$ can be written as

$$
2z \int f(\xi_{xzn}) d\varphi(\hat{H}_n(x)) = 2z \int f(x) d\varphi(\hat{H}_n(x)) + u_p(1),
$$

where $\{\xi_{xzn}\}$ are real numbers such that $|\xi_{xzn} - x| \le an^{-1/2}$. Do the same with the second integral and put two together to obtain

$$\begin{aligned} R_{n3}(n^{-1/2}z) &= 2z\left\{\int f d\varphi(\hat{H}_n) - \int f d\varphi(F)\right\} + u_p(1) \\ &= 2z\left\{\int_0^1 [q(F\hat{H}_n^{-1}(t)) - q(t)] d\varphi(t)\right\} + u_p(1). \end{aligned}$$

But,

$$\begin{aligned} (4.5.14)\quad \sup_{0\le t\le 1} |F\hat{H}_n^{-1}(t) - t| &\le n^{-1} + \sup_y |\hat{H}_n(y) - F(y)| \\ &= o_p(1), \end{aligned}$$

by (4.5.9) and the Glivenko - Cantelli Lemma. Hence, q being uniformly continuous, we obtain

$$\sup_{0\le z\le a, \varphi\in\mathcal{C}} |R_{n3}(n^{-1/2}z)| = o_p(1).$$

This together with (4.5.11) - (4.5.14) completes the proof of (4.5.7) whereas that of (4.5.8) follows from (4.5.7) and the fact

$$\sup_{0\le z\le a, \varphi\in\mathcal{C}} |n^{1/2}p(n^{-1/2}z) - 2zQ(f)| \to 0,$$

which in turn follows from the uniform continuity of f. □

Lemma 4.5.2 *Under the assumptions of Theorem 4.5.1,* $\forall y \ge 0$,

$$\sup_{\varphi\in\mathcal{C}} |\hat{p}_n(y) - p(y)| = o_p(1).$$

Proof. Proceed as in the proof of the previous lemma to rewrite

$$\hat{p}_n(y) - p(y) = \Lambda_{n1}(y) + \Lambda_{n2}(y) + \Lambda_{n3}(y) + u_p(1),$$

where $\Lambda_{nj} = n^{-1/2}R_{nj}$, $j = 1, 2, 3$, with R_{nj} defined at (4.5.11).

By Corollary 2.2.2 applied to the quantities given at (4.5.9), $\|Y_1\|_\infty = O_p(1)$ and hence f, φ bounded trivially imply that

$$\sup_{\varphi\in\mathcal{C}, y\ge 0} |\Lambda_{nj}(y)| = o_p(1), \quad j = 1, 2.$$

Now, rewrite

$$\begin{aligned}
&\Lambda_{n3}(y) \\
&= \Big[\int F(y+x)d\varphi(\hat{H}_n(x)) - \int F(y+x)d\varphi(F(x))\Big] \\
&\qquad - \Big[\int F(-y+x)d\varphi(\hat{H}_n(x)) - \int F(-y+x)d\varphi(F(x))\Big] \\
&= \Lambda_n(y) + \Lambda_n(-y), \quad \text{say}
\end{aligned}$$

But, $\forall\, y \in \mathbb{R}$,

$$\begin{aligned}
\Lambda_n(y) &= \int_0^1 \Big\{F(y + F^{-1}(F\hat{H}_n^{-1}(t))) - F(y + F^{-1}(t))\Big\}\, d\varphi(t) \\
&= o_p(1),
\end{aligned}$$

because of (4.5.14) and because, by (**F1**) and (**F2**), $\forall\, y \geq 0, F(y + F^{-1}(t))$ is uniformly continuous function of $t \in [0,1]$. □

Lemma 4.5.3 *Under the conditions of Theorem 4.5.1,* $\forall\, \epsilon > 0$,

$$P\left(|t_n^\alpha - t^\alpha| \leq \epsilon t^\alpha, \quad \forall\, \varphi \in \mathcal{C}\right) \longrightarrow 1.$$

Proof. Observe that the event

$$\begin{aligned}
&[\hat{p}_n(1-\epsilon)t^\alpha) < \alpha \leq \hat{p}_n(1+\epsilon)t^\alpha)] \\
&\qquad\qquad \subseteq \; [(1-\epsilon)t^\alpha \leq t_n^\alpha \leq (1+\epsilon)t^\alpha].
\end{aligned}$$

Hence, by two applications of Lemma 4.5.2, once with $y = (1+\epsilon)t^\alpha$, and once with $y = (1-\epsilon)t^\alpha$, we obtain that

$$\begin{aligned}
&\liminf_n P(|t_n^\alpha - t^\alpha| \leq \epsilon t^\alpha, \; \forall\, \varphi \in \mathcal{C}) \\
&\geq \; P(p(1-\epsilon)t^\alpha) < \alpha \leq p((1+\epsilon)t^\alpha), \; \forall \varphi \in \mathcal{C}) \\
&= \; 1.
\end{aligned}$$

□

Proof of Theorem 4.5.1. Clearly, $\forall\, \varphi \in \mathcal{C}$,

$$|\mathcal{Q}_n^\alpha - \mathcal{Q}(f)| = (2t_n^\alpha)^{-1}\left|n^{1/2}\hat{p}_n(n^{-1/2}t_n^\alpha) - 2t_n^\alpha \mathcal{Q}(f)\right|.$$

By Lemma 4.5.3, $\forall\, \epsilon > 0$,

$$P(0 < t_n^\alpha \leq (1+\epsilon)t^\alpha, \; \forall\, \varphi \in \mathcal{C}) \longrightarrow 1.$$

Hence (4.5.6) follows from (4.5.8) applied with $a = (1+\epsilon)t^\alpha$, Lemma 4.5.3 and the Slutsky Theorem. □

Remark 4.5.1 The estimator $\mathcal{Q}_n^\alpha$ shifts the burden of choosing the window width to the choice of α. There does not seem to be an easy way to recommend a universal α. In an empirical study done in Koul, Sievers and McKean (1987) that investigated level and power of some rank tests in the linear regression setting, $\alpha = 0.8$ was found to be most desirable. □

Remark 4.5.2 It is an interesting theoretical exercise to see if, for some $0 < \delta < 1$, the processes $\{n^{1/2}(\mathcal{Q}_n^\alpha - Q(f)), \delta \leq \alpha \leq 1 - \delta\}$ converge weakly to a Gaussian process. In the case $\varphi(t) \equiv t$, Thewarapperuma (1987) has proved under (**F1**), (**F2**), (**NX**), and (4.3.15), that $\forall$ fixed $0 < \alpha < 1$, $n^{1/2}(\mathcal{Q}_n^\alpha - Q(f)) \rightarrow_d \mathcal{N}(0, \sigma^2)$, where $\sigma^2 = 16\{\int f^3(x)dx - (\int f^2(x)dx)^2\}$. □

Remark 4.5.3 As mentioned earlier, $\{t_n^\alpha, \varphi \in \mathcal{C}\}$ provides a class of scale estimators for the class of scale parameters $\{t^\alpha, \varphi \in \mathcal{C}\}$. Recall that s_1 and s_2 of (4.3.5) are special cases of these estimators. The former is obtained by taking $\varphi(u) \equiv I(u \geq 0.5)$ and the latter by taking $\varphi(u) \equiv u$. For general interest we state a theorem below, giving asymptotic normality of these estimators. The details of proof are similar to those of Theorem 4.3.1. To state this theorem we need to introduce appropriately modified analogues of the entities defined at (4.3.13):

$$\begin{aligned} K_1(y) &:= \int [Y_1^0(y+x) - Y_1^0(-y+x)]d\varphi(F(x)), \\ K_2(y) &:= \int Y_1^0(x)[f(y+x) - f(-y+x)]\{f(x)\}^{-1}d\varphi(F(x)), \\ K(y) &:= K_1(y) - K_2(y), \qquad y \geq 0, \end{aligned}$$

where Y_1^0 is as (4.3.13) adapted to the i.i.d. errors setup. It is easy to check that $K(t^\alpha)$ is $n^{-1/2}\times$ { a sum of i.i.d. r.v.'s} with $EK(t^\alpha) \equiv 0$ and $0 < (\sigma^\alpha)^2 := Var(K(t^\alpha)) < \infty$, not depending on n. □

Theorem 4.5.2 *In addition to the conditions of Theorem 4.5.1, assume that either $\varphi(t) = I(t \geq u), 0 < u < 1$, fixed or φ is uniformly* *differentiable on* $[0,1]$. *Then,* $\forall\ 0<\alpha<1$,

$$n^{1/2}(t_n^\alpha - t^\alpha) \to_d \mathcal{N}(0, (\nu^\alpha)^2),$$

$$(\nu^\alpha)^2 := (\sigma^\alpha)^2 \left\{ t^\alpha \int [f(t^\alpha + x) + f(-t^\alpha + x)] d\varphi(F(x)) \right\}^{-2}. \quad \square$$

We now turn to the **arbitrary window width and kernel-type estimators** of Q. Accordingly, let K be a probability density on $\mathbb{R}$, h_n be a sequence of positive numbers and $\hat{\beta}$ and $\{\hat{e}_i\}$ be as before. Define, for $x \in \mathbb{R}$,

$$\hat{f}_n(x) := \frac{1}{nh_n}\sum_{i=1}^n K\Big(\frac{x-\hat{e}_i}{h_n}\Big), \quad f_n(x) := \frac{1}{nh_n}\sum_{i=1}^n K\Big(\frac{x-e_i}{h_n}\Big),$$

$$\hat{Q}_n := \int \hat{f}_n(x) d\varphi(\hat{H}_n(x)).$$

Theorem 4.5.3 *Assume that the model (1.1.1) with* $F_{ni} \equiv F$ *holds. In addition, assume that* (**F1**), (**F2**), (**NX**) *and (4.3.15) and the following hold:*

(i) $h_n > 0,\ h_n \to 0,\ n^{1/2}h_n \to \infty$.

(ii) K *is absolutely continuous with its a.e. derivative* $\dot{K}$ *satisfying* $\int |\dot{K}| < \infty$.

Then,

$$\sup_{\varphi \in \mathcal{C}} |\hat{Q}_n - Q(f)| = o_p(1). \tag{4.5.15}$$

Proof. First we show $\hat{f}_n$ approximates f. This is done in several steps. To begin with, summation by parts shows that

$$\hat{f}_n(x) - f_n(x) = -h_n^{-1}\int [\hat{H}_n(x - h_n z) - H_n(x - h_n z)]\dot{K}(z)dz$$

so that

$$\|\hat{f}_n - f_n\|_\infty \le (n^{1/2}h_n)^{-1} \cdot \|n^{1/2}(\hat{H}_n - H_n)\|_\infty \cdot \int |\dot{K}|.$$

Hence, by (4.5.9) and the fact that $|\mathbf{b}_n'\mathbf{v}| = O_p(1)$ guaranteed by (4.3.15), it readily follows that

$$\|\hat{f}_n - f_n\|_\infty = 0_p((n^{1/2}h_n)^{-1}) = o_p(1).$$

Now, let

$$\overline{f}_n(x) := h_n^{-1} \int K((x-y)/h_n) f(y)dy.$$

Note that integration by parts shows that

$$\overline{f}_n(x) = -h_n^{-1} \int \dot{K}(z) F(x - h_n(z) dz$$

so that

$$\begin{aligned} (4.5.16) \qquad & \|f_n - \overline{f}_n\|_\infty \\ & \le (n^{1/2}h_n)^{-1} \|n^{1/2}[H_n - F]\|_\infty \int |\dot{K}| \\ & = o_p(1), \end{aligned}$$

by (i) and by the fact that $\|n^{1/2}(H_n - F)\|_\infty = O_p(1)$. Moreover,

$$\|\overline{f}_n - f\|_\infty \le \sup_{|y-x|\le h_n} |f(y) - f(x)| = o(1), \qquad \text{by } (\mathbf{F1}).$$

Now, consider the difference

$$\begin{aligned} \hat{Q}_n - Q(f) &= \int (\hat{f}_n - f) d\varphi(\hat{H}_n) + \int f d[\varphi(\hat{H}_n) - \varphi(F)] \\ &= D_{n1} + D_{n2}, \quad \text{say}. \end{aligned}$$

Let $q(t) = f(F^{-1}(t))$. Then

$$\sup_{\varphi \in C} |D_{n2}| \le \sup_{0 \le t \le 1} |q(F(\hat{H}_n^{-1}(t))) - q(t)| = o_p(1)$$

by the uniform continuity of q and (4.5.14). Also, from the above bounds we obtain

$$\sup_{\varphi \in C} |D_{n1}| \le \|\hat{f}_n - f\|_\infty = o_p(1),$$

thereby proving (4.5.15).

5

Minimum Distance Estimators

5.1 Introduction

The practice of obtaining estimators of parameters by minimizing a certain distance between some functions of observations and parameters has long been present in statistics. The classical examples of this method are the Least Square and the minimum Chi Square estimators.

The minimum distance (m.d.) estimation method, where one obtains an estimator of a parameter by minimizing some distance between the empirical d.f. and the modeled d.f., was elevated to a general method of estimation by Wolfowitz (1953, 1954, 1957). In these papers he demonstrated that compared to the maximum likelihood estimation method, the m.d. estimation method yielded consistent estimators rather cheaply in several problems of varied levels of difficulty.

This methodology saw increasing research activity from the mid 1970's when many authors demonstrated various robustness properties of certain m.d. estimators. Beran (1977) showed that in the i.i.d. setup the minimum Hellinger distance estimators, obtained by minimizing the Hellinger distance between the modeled parametric density and an empirical density estimate, are asymptotically efficient at the true model and robust against small departures from the model, where the smallness is being measured in terms of the

Hellinger metric. Beran (1978) demonstrated the powerfulness of minimum Hellinger distance estimators in the one sample location model by showing that the estimators obtained by minimizing the Hellinger distance between an estimator of the density of the residual and an estimator of the density of the negative residual are qualitatively robust and adaptive for all those symmetric error distributions that have finite Fisher information.

Parr and Schucany (1979) empirically demonstrated that in certain location models several minimum distance estimators (where several comes from the type of distances chosen) are robust. Millar (1981, 1982, 1984) proved local asymptotic minimaxity of a fairly large class of m.d. estimators, using Cramér - von Mises type distance, in the i.i.d. setup. Donoho and Liu (1988 a, b) demonstrated certain further finite sample robustness properties of a large class of m.d. estimators and certain additional advantages of using Cramér - von Mises and Hellinger distances. All of these authors restrict their attention to the one sample models or to the two sample location model. See Parr (1981) for additional bibliography on m.d.e. through 1980.

Little was known till the late 1970's about how to extend the above methodology to one of the most applied models, viz., the multiple linear regression model (1.1.1). Given the above optimality properties in the one and two-sample location models, it became even more desirable to extend this methodology to this model. Only after realizing that one should use the weighted, rather than the ordinary, empiricals of the residuals to define m.d. estimators was it possible to extend this methodology satisfactorily to the model (1.1.1).

The main focus of this chapter is the m.d. estimators of $\boldsymbol{\beta}$ by minimizing the Cramér - von Mises type distances involving various W.E.P.'s. Some m.d. estimators involving the supremum distance are also discussed. Most of the estimators provide appropriate extension of their counterparts in the one - and two - sample location models.

Section 5.2 contains definitions of several m.d. estimators. Their finite sample properties and asymptotic distributions are discussed in Sections 5.3, 5.5, respectively. Section 5.4 discusses an asymptotic theory about general minimum dispersion estimators that is of broad and independent interest. It is a self contained section. Asymptotic relative efficiency and qualitative robustness of some of the m.d. estimators of Section 5.2 are discussed in Section 5.6. Some of the proposed m.d. functionals are Hellinger differentiable in the sense of Beran (1982) as is shown in Section 5.6. Consequently they are locally asymptotically minimax (LAM) in the sense of Hájek - Le Cam.

5.2 Definitions of M.D. Estimators

To motivate the following definitions of m.d. estimators of $\boldsymbol{\beta}$ of (1.1.1) first consider the one sample location model where $Y_1 - \theta, \cdots, Y_n - \theta$ are i.i.d. F, F a *known* d.f. Let

$$F_n(y) := n^{-1} \sum_{i=1}^{n} I(Y_i \le y), \qquad y \in \mathbb{R}. \tag{5.2.1}$$

If θ is true then $EF_n(y + \theta) = F(y), \forall y \in \mathbb{R}$. This motivates one to define m.d. estimator $\hat{\theta}$ of θ by the relation

$$\hat{\theta} = \text{argmin}\{T(t); t \in \mathbb{R}\} \tag{5.2.2}$$

where, for a $G \in \mathbb{DI}(\mathbb{R})$,

$$T(t) := n \int [F_n(y + t) - F(y)]^2 dG(y), \qquad t \in \mathbb{R}. \tag{5.2.3}$$

Observe that (5.2.2) and (5.2.3) actually define a class of estimators $\hat{\theta}$, one corresponding to each G.

Now suppose that in (1.1.1) we model the d.f. of e_{ni} to be a *known* d.f. H_{ni}, which may be different from the actual d.f. $F_{ni}, 1 \le i \le n$. How should one define a m.d. estimator of $\boldsymbol{\beta}$? Any definition should reduce to $\hat{\theta}$ when (1.1.1) is reduced to the one sample location model. One possible extension is to define

$$\hat{\boldsymbol{\beta}}_1 = \text{argmin}\{K_1(\mathbf{t}); \mathbf{t} \in \mathbb{R}^p\}, \tag{5.2.4}$$

where, for a $\mathbf{t} \in \mathbb{R}^p$,

$$K_1(\mathbf{t}) = n^{-1} \int \Big[\sum_{i=1}^{n} \{I(Y_{ni} \le y + \mathbf{x}_{ni}'\mathbf{t}) - H_{ni}(y)\}\Big]^2 dG(y).$$

If in (1.1.1) we take $p = 1$, $x_{ni1} \equiv 1$ and $H_{ni} \equiv F$ then clearly it reduces to the one sample location model and $\hat{\beta}_1$ coincides with $\hat{\theta}$ of (5.2.2). But this is also true for the estimator $\hat{\beta}_{\mathbf{X}}$ defined as follows. Recall the definition of $\{V_j\}$ from (1.1.2). Let, for $y \in \mathbb{R}, \mathbf{t} \in \mathbb{R}^p$,

$$(5.2.5) \qquad \mathcal{Z}_j(y, \mathbf{t}) := V_j(y, \mathbf{t}) - \sum_{i=1}^{n} x_{nij} H_{ni}(y), \quad 1 \le j \le p,$$
$$K_{\mathbf{X}}(\mathbf{t}) := \int \mathcal{Z}'(y, \mathbf{t})(\mathbf{X}'\mathbf{X})^{-1} \mathcal{Z}(y, \mathbf{t}) dG(y),$$

where $\mathcal{Z}' := (\mathcal{Z}_1, \cdots, \mathcal{Z}_p)$ and define

$$(5.2.6) \qquad \hat{\beta}_{\mathbf{X}} = \operatorname{argmin}\{K_{\mathbf{X}}(\mathbf{t}), \mathbf{t} \in \mathbb{R}^p\}.$$

Which of the two estimators is the right extension of $\hat{\theta}$? Since $\{V_j, 1 \le j \le p\}$ summarize the data in (1.1.1) with probability one under the continuity assumption of $\{e_{ni}, 1 \le i \le n\}$, $\hat{\beta}_{\mathbf{X}}$ should be considered the right extension of $\hat{\theta}$. In Section 5.6 we shall see that $\hat{\beta}_{\mathbf{X}}$ is asymptotically efficient among a class of estimators $\{\hat{\beta}_{\mathbf{D}}\}$ defined as follows.

Let $\mathbf{D} = ((d_{nij}))\,, 1 \le i \le n,\ 1 \le j \le p$, be an $n \times p$ real matrix and define, for $y \in \mathbb{R},\ \mathbf{t} \in \mathbb{R}^p$,

$$(5.2.7) \quad V_{jd}(y, \mathbf{t}) := \sum_{i=1}^{n} d_{nij} I(Y_{ni} \le y + \mathbf{x}'_{ni}\mathbf{t}), \quad 1 \le j \le p,$$
$$K_{\mathbf{D}}(\mathbf{t}) := \sum_{j=1}^{p} \int \Big[V_{jd}(y, \mathbf{t}) - \sum_{i=1}^{n} d_{nij} H_{ni}(y) \Big]^2 dG(y).$$

Define a class of estimators, one for each $\mathbf{D}$,

$$(5.2.8) \qquad \hat{\beta}_{\mathbf{D}} := \operatorname{argmin}\{K_{\mathbf{D}}(\mathbf{t}), \mathbf{t} \in \mathbb{R}^p\}.$$

If $\mathbf{D} = n^{-1/2}[\mathbf{1}, \mathbf{0}, \cdots, \mathbf{0}]_{n \times p}$ then $\hat{\beta}_{\mathbf{D}} = \hat{\beta}_1$ and if $\mathbf{D} = \mathbf{X}\mathbf{A}$ then $\hat{\beta}_{\mathbf{D}} = \hat{\beta}_{\mathbf{X}}$, where $\mathbf{A}$ is as in (2.3.30). The above mentioned optimality of $\hat{\beta}_{\mathbf{X}}$ is stated and proved in Theorem 5.6a.1 below.

Another way to define m.d. estimators in the case the *modeled error d.f.'s are known* is as follows. Let, for $s \in 0,1]$, $y \in \mathbb{R}$, $\mathbf{t} \in \mathbb{R}^p$,

$$(5.2.9) \quad M(s,y,\mathbf{t}) := n^{-1/2}\sum_{i=1}^{ns}\{I(Y_{ni} \le y) - H_{ni}(y - \mathbf{x}_{ni}'\mathbf{t})\},$$

$$Q(\mathbf{t}) := \int_0^1 \int \{M(s,y,\mathbf{t})\}^2 dG(y)dL(s),$$

where L is a d.f. on $[0,1]$. Define

$$(5.2.10) \qquad \bar{\boldsymbol{\beta}} = \text{argmin}\{Q(\mathbf{t}), \mathbf{t} \in \mathbb{R}^p\}.$$

The estimator $\bar{\boldsymbol{\beta}}$ with $L(s) \equiv s$ is essentially Millar's (1982) proposal.

Now suppose $\{H_{ni}\}$ are *unknown*. How should one define m.d. of $\boldsymbol{\beta}$ in this case.? Again, let us examine the one sample location model. In this case θ can not be identified unless the errors are symmetric about 0. Suppose that is the case. Then the r.v.'s $\{Y_i - \theta, 1 \le i \le n\}$ have the same distribution as $\{-Y_i + \theta, 1 \le i \le n\}$. A m.d. estimator θ^+ of θ is thus defined by the relation

$$(5.2.11) \qquad \theta^+ = \text{argmin}\{T^+(t), t \in \mathbb{R}\}$$

where

$$T^+(t) := n^{-1}\int \left[\sum_{i=1}^{n}\{I(Y_i \le y + t) - I(-Y_i < y - t)\}\right]^2 dG(y).$$

An extension of θ^+ to the model (1.1.1) is $\boldsymbol{\beta}_{\mathbf{X}}^+$ defined as follows: Let, for $y \in \mathbb{R}$, $\mathbf{t} \in \mathbb{R}^p$, $1 \le j \le p$,

$$(5.2.12) \quad \mathbf{V}_j^+(y,\mathbf{t}) := \sum_{i=1}^{n} x_{nij}\Big\{I(Y_{ni} \le y + \mathbf{x}_{ni}'\mathbf{t}) - I(-Y_{ni} < y - \mathbf{x}_{ni}'\mathbf{t})\Big\},$$

$$\mathbf{V}^+ := (V_1^+, \cdots, V_p^+)',$$

$$K_{\mathbf{X}}^+(\mathbf{t}) := \int \mathbf{V}^+(y,\mathbf{t})(\mathbf{X}'\mathbf{X})^{-1}\mathbf{V}^+(y,\mathbf{t})\, dG(y),$$

and define the estimator

$$(5.2.13) \qquad \boldsymbol{\beta}_{\mathbf{X}}^+ = \text{argmin}\{K_{\mathbf{X}}^+(\mathbf{t}), \mathbf{t} \in \mathbb{R}^p\}.$$

More generally, a class of m.d. estimators of $\boldsymbol{\beta}$ can be defined as follows. Let $\mathbf{D}$ be as before. Define, for $y \in \mathbb{R}$, $\mathbf{t} \in \mathbb{R}^p$, $1 \le j \le p$,

$$
\begin{aligned}
(5.2.14)\quad Y_j^+(y,\mathbf{t}) &:= \sum_{i=1}^n d_{nij}\Big\{I(Y_{ni} \le y + \mathbf{x}_{ni}'\mathbf{t}) \\
&\qquad\qquad -I(-Y_{ni} < y - \mathbf{x}_{ni}'\mathbf{t})\Big\}, \\
\mathbf{Y}_{\mathbf{D}}^+ &= (Y_1^+, \cdots, Y_p^+)', \\
K_{\mathbf{D}}^+(\mathbf{t}) &:= \int \mathbf{Y}_{\mathbf{D}}^{+'}(y,\mathbf{t})\mathbf{Y}_{\mathbf{D}}^{+'}(y,\mathbf{t})\, dG(y).
\end{aligned}
$$

and $\boldsymbol{\beta}_{\mathbf{D}}^+$ by the relation

$$
(5.2.15)\qquad \boldsymbol{\beta}_{\mathbf{D}}^+ = \operatorname{argmin}\{K_{\mathbf{D}}^+(\mathbf{t}), \mathbf{t} \in \mathbb{R}^p\}.
$$

Note that $\boldsymbol{\beta}_{\mathbf{X}}^+$ is $\boldsymbol{\beta}_{\mathbf{D}}^+$ with $\mathbf{D} = \mathbf{XA}$.

Next, suppose that the errors in (1.1.1) are modeled to be i.i.d., i.e. $H_{ni} \equiv F$ and F is *unknown* and *not necessarily symmetric*. Here, of course the location parameter can not be estimated. However, the regression parameter vector $\boldsymbol{\beta}$ can be estimated provided the rank of $\mathbf{X}_c$ is p, where $\mathbf{X}_c$ is defined at (4.2.11). In this case a class of m.d. estimators of $\boldsymbol{\beta}$ is defined by $\hat{\boldsymbol{\beta}}_{\mathbf{D}}$ of (5.2.8) provided we assume that

$$
(5.2.16)\qquad \sum_{i=1}^n d_{nij} = 0, \qquad 1 \le j \le p.
$$

An interesting member of this class is obtained upon taking $\mathbf{D} = \mathbf{X}_c\mathbf{A}_1$, $\mathbf{A}_1$ as in (4.2.12).

Another way to define m.d. estimators here is via the ranks of the residuals. With R_{it} as in (3.1.1), let

$$
\begin{aligned}
(5.2.17)\quad T_j(s,\mathbf{t}) &:= \sum_{i=1}^n d_{nij} I(R_{it} \le ns), \quad s \in [0,1],\ 1 \le j \le p, \\
K_{\mathbf{D}}^*(\mathbf{t}) &:= \int \mathbf{T}_{\mathbf{D}}'(s,\mathbf{t})\, \mathbf{T}_{\mathbf{D}}(s,\mathbf{t})\, dL(s), \quad \mathbf{t} \in \mathbb{R},
\end{aligned}
$$

where $\mathbf{T}_{\mathbf{D}}' = (T_1, \cdots, T_p)$ and L is a d.f. on $[0,1]$. Assume that $\mathbf{D}$ satisfies (5.2.16). Define

$$
(5.2.18)\qquad \boldsymbol{\beta}_{\mathbf{D}}^* = \operatorname{argmin}\{K_{\mathbf{D}}^*(\mathbf{t}),\ \mathbf{t} \in \mathbb{R}^p\}.
$$

Observe that $\{\hat{\boldsymbol{\beta}}_{\mathbf{D}}\}, \{\boldsymbol{\beta}_{\mathbf{D}}^{+}\}$ and $\{\overline{\boldsymbol{\beta}}\}$ are not scale invariant in the sense of (4.3.2). One way to make them so is to modify their definitions as follows. Define, for $\mathbf{t} \in \mathbb{R}^p$, $a \geq 0$,

$$
\begin{aligned}
(5.2.19)\quad K_{\mathbf{D}}(a, \mathbf{t}) &:= \sum_{j=1}^{p} \int [V_{jd}(ay, \mathbf{t}) - \sum_{i=1}^{n} d_{nij} H_{ni}(y)]^2 dG(y), \\
K_{\mathbf{D}}^{+}(a, \mathbf{t}) &:= \int \mathbf{Y}_{\mathbf{D}}^{+'}(ay, \mathbf{t}) \mathbf{Y}_{\mathbf{D}}^{+}(ay, \mathbf{t}) dG(y).
\end{aligned}
$$

Now, scale invariant analogues of $\hat{\boldsymbol{\beta}}_{\mathbf{D}}$ and $\boldsymbol{\beta}_{\mathbf{D}}^{+}$ are defined as

$$
\begin{aligned}
(5.2.20)\qquad \hat{\boldsymbol{\beta}}_{\mathbf{D}}^{0} &:= \operatorname{argmin}\{K_{\mathbf{D}}(s, \mathbf{t}), \mathbf{t} \in \mathbb{R}^p\}, \\
\boldsymbol{\beta}_{\mathbf{D}}^{+0} &:= \operatorname{argmin}\{K_{\mathbf{D}}(s, \mathbf{t}), \mathbf{t} \in \mathbb{R}^p\},
\end{aligned}
$$

where s is a scale estimator satisfying (4.3.3) and (4.3.4). One can modify $\{\hat{\boldsymbol{\beta}}\}$ in a similar fashion to make it scale invariant. The class of estimators $\{\boldsymbol{\beta}_{\mathbf{D}}^{*}\}$ is scale invariant because the ranks are.

Now we define a m.d. estimator based on the *supremum distance* in the case the errors are correctly modeled to be i.i.d. F, F an arbitrary d.f. . Here we shall *restrict* ourselves only to the *case of* $p = 1$. Define

$$
\begin{aligned}
(5.2.21)\quad V_c(y, t) &:= \sum_{i=1}^{n} (x_i - \overline{x}) I(Y_i \leq y + t x_i), \qquad t,\ y \in \mathbb{R}, \\
\mathcal{D}_n^{+}(t) &:= \sup\{V_c(y, t); y \in \mathbb{R}\}, \\
\mathcal{D}_n^{-}(t) &:= -\inf\{V_c(y, t); y \in \mathbb{R}\}, \\
\mathcal{D}_n(t) &:= \max\{\mathcal{D}_n^{+}(t), \mathcal{D}_n^{-}(t), \mathcal{D}_n^{-}(t)\} \\
&= \sup\{|V_c(y, t)|; y \in \mathbb{R}\}, \qquad t \in \mathbb{R}.
\end{aligned}
$$

Finally, define the m.d. estimator

$$
(5.2.22)\qquad \hat{\beta}_s := \operatorname{argmin}\{\mathcal{D}_n(t);\ \ t \in \mathbb{R}\}.
$$

Section 5.3 discusses some computational aspects including the existence and some finite sample properties of the above estimators. Section 5.5 proves the uniform asymptotic quadraticity of $K_{\mathbf{D}}, K_{\mathbf{D}}^{+}, K_{\mathbf{D}}^{*}$ and Q as processes in $\mathbf{t}$. These results are used in Section 5.6 to study the asymptotic distributions and robustness of the above defined estimators.

5.3 Finite Sample Properties

The purpose here is to discuss some computational aspects, the existence and the finite sample properties of the *four* classes of estimators introduced in the previous section. To facilitate this the dependence of these estimators and their defining statistics on the weight matrix $\mathbf{D}$ will not be exhibited in this section.

We first turn to *some computational aspects* of these estimators. To begin with, suppose that $p = 1$ and $G(y) = y$ in (5.2.7) and (5.2.8). Write $\hat{\beta}, x_i, d_i$ for $\hat{\boldsymbol{\beta}}, x_{i1}, d_{i1}$, respectively, $1 \le i \le n$. Then

$$\text{(5.3.1)} \quad K(t) = \int \left[\sum_{i=1}^n d_i \{ I(Y_i \le y + x_i t) - H_i(y) \} \right]^2 dy$$
$$= \sum_{i=1}^n \sum_{j=1}^n d_i d_j \int \Big\{ I(Y_i \le y + x_i t) - H_i(y) \Big\} \times \Big\{ I(Y_j \le y + x_j t) - H_j(y) \Big\} dy.$$

No further simplification of this occurs except for some special cases. One of them is the case of the one sample location model where $x_i \equiv 1$ and $H_i \equiv F$, in which case

$$K(t) = \int \left[\sum_{i=1}^n d_i \{ I(Y_i \le y) - F(y - t) \} \right]^2 dy.$$

Differentiating under the integral sign w.r.t. t (which can be justified under the sole assumption: F has a density f w.r.t. λ) one obtains

$$\dot{K}(t) = 2 \int \sum_{i=1}^n d_i \{ I(Y_i \le y + t) - F(y) \} dF(y)$$
$$= -2 \sum_{i=1}^n d_i \{ F(Y_i - t) - 1/2 \}.$$

Upon taking $d_i \equiv n^{-1/2}$ one sees that in the one sample location model $\hat{\theta}$ of (5.2.2) corresponding to $G(y) = y$ is given as a solution of

$$\text{(5.3.2)} \qquad \sum_{i=1}^n F(Y_i - \hat{\theta}) = n/2.$$

Note that this $\hat{\theta}$ is precisely the m.l.e. of θ when $F(x) \equiv \{1 + \exp(-x)\}^{-1}$, i.e., when the errors have logistic distribution!

Another simplification of (5.3.1) occurs when we assume

$$\sum_{i=1}^{n} d_i = 0, \qquad \text{and} \qquad H_i \equiv F.$$

Fix a $t \in \mathbb{R}$ and let $c := \max\{Y_i - x_i t; 1 \le i \le n\}$. Then

$$\begin{aligned}
(5.3.3)\quad K(t) &= \int \left[\sum_{i=1}^{n} d_i I(Y_i \le y + x_i t)\right]^2 dy \\
&= \sum_{i=1}^{n}\sum_{j=1}^{n} d_i d_j \int I\Big[\max(Y_j - x_j t,\, Y_i - x_i t) \\
&\qquad\qquad \le y < c\Big]\, dy \\
&= -\sum_{i=1}^{n}\sum_{j=1}^{n} d_i\, d_j \max(Y_j - x_j t,\, Y_i - x_i t).
\end{aligned}$$

Using the relationship

$$(5.3.4)\qquad 2\max(a, b) = a + b + |a - b|, \qquad a, b \in \mathbb{R},$$

and the assumption $\sum_{i=1}^{n} d_i = 0$, one obtains

$$(5.3.5)\qquad K(t) \equiv -2 \sum\sum_{1 \le i < j \le n} d_i d_j |Y_j - Y_i - (x_j - x_i)t|.$$

If $d_i = x_i - \overline{x}$ in (5.3.5), then the corresponding $\hat{\beta}$ is asymptotically equivalent to the Wilcoxon type R-estimator of β as was shown by Williamson (1979). This result will also follow from the general asymptotic theory of Sections 5.5 and 5.6 below.

If $d_i = x_i - \overline{x}, 1 \le i \le n$, and $x_i = 0, 1 \le i \le r; x_i = 1, r + 1 \le i \le n$ then (1.1.1) becomes the two sample location model and

$$K(t) = -2 \sum_{i=1}^{r} \sum_{j=r+1}^{n} |Y_j - Y_i - t| + \text{ a r.v. constant in } t.$$

Consequently here $\hat{\beta} = med\{|Y_j - Y_i|, r + 1 \le j \le n, 1 \le i \le r\}$, the usual Hodges - Lehmann estimator. The fact that in the two sample location model the Cramér - von Mises type m.d. estimator of the location parameter is the Hodges - Lehmann estimator was first noted by Fine (1966).

Note that a relation like (5.3.5) is true for general p and G. That is, suppose that $p \geq 1$, $G \in \mathcal{DI}(\mathbb{R})$ and (5.2.16) holds, then $\forall\, \mathbf{t} \in \mathbb{R}^p$,

$$(5.3.6) \quad K(\mathbf{t}) = -2\sum_{j=1}^{p}\sum_{1\le i}\sum_{<k\le n} d_{ij}\, d_{kj}|G((Y_k - \mathbf{x}_k'\mathbf{t})_-) - G((Y_i - \mathbf{x}_i'\mathbf{t})_-)|.$$

To prove this proceed as in (5.3.3) to conclude first that

$$K(\mathbf{t}) = -2\sum_{j=1}^{p}\sum_{1\le i}\sum_{<k\le n} d_{ij}d_{kj}G(\max(Y_k - \mathbf{x}_k'\mathbf{t}, Y_i - \mathbf{x}_i'\mathbf{t})_-)$$

Now use the fact that $G(a\vee b)_-) = G(a_-)\vee G(b_-)$, (5.2.16) and (5.3.4) to obtain (5.3.6). Clearly, this formula can be used to compute $\hat{\boldsymbol{\beta}}$ in general.

Next, consider K^+. To simplify the exposition, fix a $\mathbf{t} \in \mathbb{R}^p$ and let $r_i := Y_i - \mathbf{x}_i'\mathbf{t}$, $1 \le i \le n$; $b := \max\{r_i - r_i; 1 \le i \le n\}$. Then from (5.2.14) we obtain

$$K^+(\mathbf{t}) = \sum_{j=1}^{p}\int\left[\sum_{i=1}^{n} d_{ij}\{I(r_i \le y) - I(-r_i < y)\}\right]^2 dG(y).$$

Observe that the integrand is zero for $y > b$. Now expand the quadratic and integrate term by term, noting that G may have jumps, to obtain

$$\begin{aligned} K^+(\mathbf{t}) \;=\; & \sum_{j=1}^{p}\sum_{i=1}^{n}\sum_{k=1}^{n} d_{ij}d_{kj}\Big\{2G(r_i \vee -r_k)\} - 2J(r_i) \\ & -G((r_i \vee r_k)_-) - G(-r_i \vee -r_k)\Big\}, \end{aligned}$$

where $J(y) := G(y) - G(y_-)$, the jump in G at $y \in \mathbb{R}$. Once again use the fact that $G(a \vee b) = G(a) \vee G(b)$, (5.3.4), the invariance of the double sum under permutation and the definition of $\{r_i\}$ to conclude that

$$(5.3.7)\quad K^+(t) = \sum_{j=1}^{p}\sum_{i=1}^{n}\sum_{k=1}^{n} d_{ij}d_{kj}\Big[\big|G(Y_i - \mathbf{x}_i'\mathbf{t}) - G(-Y_k + \mathbf{x}_k'\mathbf{t})\big| - \frac{1}{2}\Big\{\big|G(Y_i - \mathbf{x}_i'\mathbf{t})_-) - G(Y_k - \mathbf{x}_k'\mathbf{t})_-)\big| + \big|G(-Y_i + \mathbf{x}_i'\mathbf{t}) - G(-Y_k + \mathbf{x}_k'\mathbf{t})\big|\Big\} - J(Y_i - \mathbf{x}_i'\mathbf{t})\Big].$$

Before proceeding further it is convenient to recall at this time the definition of symmetry for a $G \in \mathcal{DI}(\mathbb{R})$.

Definition 5.3.1. An arbitrary $G \in \mathcal{DI}(\mathbb{R})$, including a σ - finite measure on the Borel line $(\mathbb{R}, \mathcal{B})$, is said to be *symmetric* around 0 if

$$(5.3.8)\quad |G(y) - G(x)| = |G(-x_-) - G(-y_-)|, \qquad \forall\, x,\, y \in \mathbb{R}.$$

Or

$$(5.3.9)\qquad dG(y) = -dG(-y), \qquad \forall\, y \in \mathbb{R}.$$

If G is continuous then (5.3.8) is equivalent to

$$(5.3.10)\qquad |G(y) - G(x)| = |G(-x) - G(-y)|, \qquad \forall\, x, y \in \mathbb{R}.$$

Conversely, if (5.3.10) holds then G is symmetric around 0 and continuous.

Now suppose that G satisfies (5.3.8). Then (5.3.7) simplifies to

$$(5.3.11)\quad K^+(\mathbf{t}) = \sum_{j=1}^{p}\sum_{i=1}^{n}\sum_{k=1}^{n} d_{ij}d_{kj}\Big[\big|G(Y_i - \mathbf{x}_i'\mathbf{t}) - G(-Y_k + \mathbf{x}_k'\mathbf{t})\big| - \big|G(-Y_i + \mathbf{x}_i'\mathbf{t}) - G(-Y_k + \mathbf{x}_k'\mathbf{t})\big| - J(Y_i - \mathbf{x}_i'\mathbf{t})\Big].$$

And if G satisfies (5.3.10) then we obtain the relatively simpler expression

$$(5.3.12)\quad K^{+}(\mathbf{t}) = \sum_{j=1}^{p}\sum_{i=1}^{n}\sum_{k=1}^{n} d_{ij}d_{kj}\Big[\Big|G(Y_i - \mathbf{x}_i'\mathbf{t}) - G(-Y_k + \mathbf{x}_k'\mathbf{t})\Big| - \Big|G(Y_i - \mathbf{x}_i'\mathbf{t}) - G(Y_k - \mathbf{x}_k'\mathbf{t})\Big|\Big].$$

Upon specializing (5.3.12) to the case $G(y) = y$, $p = 1$, $d_i \equiv n^{-1/2}$ and $x_i \equiv 1$ we obtain

$$K^{+}(t) = n^{-1}\sum_{i=1}^{n}\sum_{k=1}^{n}\Big\{|Y_i + Y_k - 2t| - |Y_i - Y_k|\Big\}$$

and the corresponding minimizer is the well celebrated median of the pairwise means $\{(Y_i + Y_j)/2; 1 \le i \le j \le n\}$.

Suppose we specialize (1.1.1) to a completely randomized design with p treatments, i.e., take

$$\begin{aligned} x_{ij} &= 1, \qquad m_{j-1} + 1 \le i \le m_j, \\ &= 0, \qquad \text{otherwise}, \end{aligned}$$

where $1 \le n_j \le n$ is the jth sample size, $m_0 = 0, m_j = n_1 + \cdots + n_j, 1 \le j \le p, m_p = n$. Then, upon taking $G(y) \equiv y, d_{ij} \equiv x_{ij}$ in (5.3.12), we obtain

$$K^{+}(\mathbf{t}) = \sum_{j=1}^{p}\sum_{i=1}^{n_j}\sum_{k=1}^{n_j}\Big\{|Y_{ij} + Y_{kj} - 2t_j| - |Y_{ij} - Y_{kj}|\Big\}, \quad \mathbf{t} \in \mathbb{R}^p,$$

where Y_{ij} = the ith observation from the jth treatment, $1 \le j \le p$. Consequently, $\boldsymbol{\beta}^{+} = (\beta_1^{+}, \cdots, \beta_p^{+})$, where $\beta_j^{+} = med\{(Y_{ij} + Y_{kj})2^{-1}, 1 \le i \le k \le n_j\}, 1 \le j \le p$. That is, in a completely randomized design with p treatments, $\boldsymbol{\beta}^{+}$ corresponding to the weights $\mathbf{d}_i = \mathbf{x}_i$ and $G(y) \equiv y$ is the vector of Hodges - Lehmann estimators. Similar remark applies to the randomized block, factorial and other similar designs.

The class of estimators $\boldsymbol{\beta}^{+}$ also includes the well celebrated *least absolute deviation* (LAD) estimator. To see this, assume that *the errors are continuous.* Choose $G = \delta_0$ - the measure degenerate at 0 in K^+, to obtain

$$\begin{aligned} K^+(\mathbf{t}) & \\ &= \sum_{j=1}^{p} \Big[\sum_{i=1}^{n} d_{ij} \{ I(Y_i - \mathbf{x}_i'\mathbf{t} \le 0) - I(Y_i - \mathbf{x}_i'\mathbf{t} > 0) \} \Big]^2 \\ &= \sum_{j=1}^{p} \Big(\sum_{i=1}^{n} d_{ij} \mathrm{sgn}(Y_i - \mathbf{x}_i'\mathbf{t}) \Big)^2, \quad w.p.1, \quad \forall\, \mathbf{t} \in \mathbb{R}^p. \end{aligned}$$

Upon choosing $\mathbf{d}_i \equiv \mathbf{x}_i$, one sees that this expression is precisely the square of the norm of a.e. differential of the sum of absolute deviations $\mathcal{D}(\mathbf{t}) := \sum_{i=1}^{n} |Y_i - \mathbf{x}_i'\mathbf{t}|, \mathbf{t}|, \mathbf{t} \in \mathbb{R}^p$. Clearly the minimizer of $\mathcal{D}(\mathbf{t})$ is also a minimizer of the above $K^+(\mathbf{t})$.

Any one of the expressions among (5.3.7), (5.3.11) or (5.3.12) may be used to compute $\boldsymbol{\beta}^+$ for a general G. It is also apparent from the above discussion that both classes $\{\hat{\boldsymbol{\beta}}\}$ and $\{\boldsymbol{\beta}^+\}$ include rather interesting estimators. On the one hand we have a smooth unbounded G, v.i.z., $G(y) \equiv y$, giving rise to Hodges - Lehmann type estimators and on the other hand a highly discrete G, v.i.z., $G = \delta_0$ giving rise to the LAD estimator. Any distribution theory should be general enough to cover both of these cases.

We now address the question of the *existence* of these estimators in the case $p = 1$. As before when $p = 1$, we write unbold letters for scalars and d_i, x_i for d_{i1}, x_{i1}, $1 \le i \le n$. Before stating the result we need to define

$$\Gamma(y) := \sum_{i=1}^{n} I(x_i = 0)\, d_i \,\{ I(Y_i \le y) - I(-Y_i < y) \}, \qquad y \in \mathbb{R}.$$

Arguing as for (5.3.7) we obtain, with $b = \max\{Y_i, -Y_i;\, 1 \le i \le n\}$,

$$\begin{aligned} (5.3.13) \int |\Gamma| dG & \\ &\le \sum_{i=1}^{n} I(x_i = 0)\, |d_i| \, [G(b_-) - G(Y_{i-}) + G(b_-) - G(-Y_i)] \\ &< \infty. \end{aligned}$$

Moreover, directly from (5.3.7) we can conclude that

$$\int \Gamma^2 dG < \infty. \tag{5.3.14}$$

Both (5.3.13) and (5.3.14) hold for all $n \geq 1$, for every sample $\{Y_i\}$ and for all real numbers $\{d_i\}$.

Lemma 5.3.1. *Assume that* (1.1.1) *with* $p = 1$ *holds. In addition, assume that either*

$$d_i x_i \geq 0, \qquad \forall\, 1 \leq i \leq n, \tag{5.3.15}$$

or

$$d_i x_i \leq 0, \qquad \forall\, 1 \leq i \leq n. \tag{5.3.16}$$

Then a minimizer of K^+ *exists if either Case 1:* $G(\mathbb{R}) = \infty$, *or Case 2:* $G(\mathbb{R}) < \infty$ *and* $d_i = 0$ *whenever* $x_i = 0, 1 \leq i \leq n$.

If G *is continuous then a minimizer is measurable.*

Proof. The proof uses Fatou's Lemma and the D.C.T. Specialize (5.2.14) to the case $p = 1$ to obtain

$$K^+(t) = \int \left[\sum_{i=1}^n d_i\{I(Y_i \leq y + x_i t) - I(-Y_i < y - x_i t)\}\right]^2 dG(y).$$

Let $\mathcal{K}^+(y,t)$ denote the integrand without the square. Then

$$\mathcal{K}^+(y,t) = \Gamma(y) + \mathcal{K}^*(y,t),$$

where

$$\begin{aligned}
&\mathcal{K}^*(y,t)\\
&\quad = \sum_{i=1}^n I(x_i > 0) d_i\{I(Y_i \leq y + x_i t) - I(-Y_i < y - x_i t)\}\\
&\qquad + \sum_{i=1}^n I(x_i < 0) d_i\{I(Y_i \leq y + x_i t) - I(-Y_i < y - x_i t)\}.
\end{aligned}$$

Clearly, $\forall\ y, t \in \mathbb{R}$,

$$|\mathcal{K}^*(y,t)| \leq \sum_{i=1}^n I(x_i \neq 0)|d_i| =: \alpha, \text{ say.}$$

Hence

$$\Gamma(y) - \alpha \leq \mathcal{K}^+(y,t) \leq \Gamma(y) + \alpha, \quad \forall\, y,\ t \in \mathbb{R}. \tag{5.3.17}$$

Suppose that **(5.3.15) holds**. Then, it follows that $\forall\, y \in \mathbb{R}$,

$$\mathcal{K}^*(y,t) \to \pm\alpha \ \text{ as } \ t \to \pm\infty,$$

so that $\forall\ \ y \in \mathbb{R}$,

$$\mathcal{K}^+(y,t) \to \Gamma(y) \pm \alpha, \quad \text{as} \ \ t \to \pm\infty. \tag{5.3.18}$$

Now consider **Case 1**. If $\alpha = 0$ then either all $x_i \equiv 0$ or $d_0 = 0$ for those i for which $x_i \neq 0$. In either case one obtains from (5.3.14) and (5.3.17) that $\forall t \in \mathbb{R}, K^+(t) = \int \Gamma^2 dG < \infty$ and hence a minimizer trivially exists.

If $\alpha > 0$ then, from (5.3.13) and (5.3.14) it follows that

$$\int (\Gamma(y) \pm \alpha)^2 dG(y) = \infty,$$

and by (5.3.17) and the Fatou Lemma,

$$\liminf_{t \to \pm\infty} K^+(t) = \infty.$$

On the other hand by (5.3.7), $K^+(t)$ is a finite number for every real t, and hence a minimizer exists.

Next, consider **Case 2**. Here, clearly $\Gamma \equiv 0$. From (5.3.17), we obtain

$$\{\mathcal{K}^+(y,t)\}^2 \leq \alpha^2, \quad \forall\ \ y, t\ \in \mathbb{R},$$

and hence

$$K^+(t) \leq \alpha^2 G(\mathbb{R}), \quad \forall t \in \mathbb{R}.$$

By (5.3.18), $\mathcal{K}^+(y,t) \to \pm\alpha$, as $t \to \pm\infty$. By the D.C.T. we obtain

$$K^+(t) \to \alpha^2 G(\mathbb{R}), \ \text{ as } \ |t| \to \infty,$$

thereby proving the existence of a minimizer of K^+ in Case 2.

The continuity of G together with (5.3.12) shows that K^+ is a continuous function on $\mathbb{R}$ thereby ensuring the measurability of a minimizer, by Corollary 2.1 of Brown and Purves (1973). This completes the proof in the case of (5.3.15). It is exactly similar when (5.3.16) holds, hence no details will be given for that case. □

Remark 5.3.1. Observe that in some cases minimizers of K^+ could be measurable even if G is not continuous. For example, in the case of LAD estimator, G is degenerate at 0 yet a measurable minimizer exists.

Also, note that (5.3.15) is a *priori* satisfied by the weights $d_i \equiv x_i$.

The above proof is essentially due to Dhar (1991a). Dhar (1991b) gives proofs of the existence of classes of estimators $\{\hat{\beta}\}$ and $\{\beta^+\}$ of (5.2.8) and (5.2.15) for $p \geq 1$, among other results. These proofs are somewhat complicated and will not be reproduced here. In both of these papers Dhar carries out some finite sample simulation studies and concludes that both, $\hat{\beta}$ and β^+ corresponding to $G(y) \equiv y$, show some superiority over some of the well known estimators. □

Now we discuss $\bar{\beta}$ of (5.2.10). Rewrite

$$\begin{aligned} Q(\mathbf{t}) &= n^{-1}\sum_{i=1}^{n}\sum_{j=1}^{n} L_{ij} \int \Big\{I(Y_i \leq y) - H_i(y - \mathbf{x}_i'\mathbf{t})\Big\} \\ &\qquad\qquad \times\Big\{I(Y_j \leq y) - H_j(y - \mathbf{x}_j'\mathbf{t})\Big\}\, dG(y) \end{aligned}$$

where $L_{ij} = 1 - L(i \vee j)n^{-1})$, $1 \leq i, j \leq n$. Differentiating Q w.r.t. $\mathbf{t}$ under the integral sign (which can be justified assuming $\{H_i\}$ have Lebesgue densities $\{h_i\}$ and some other mild conditions) we obtain

$$\begin{aligned} \dot{\mathbf{Q}}(\mathbf{t}) &= 2n^{-1}\sum_{i=1}^{n}\sum_{j=1}^{n} L_{ij} \int \Big\{I(Y_i \leq y) - (H_i(y - \mathbf{x}_i'\mathbf{t})\Big\} \\ &\qquad\qquad \times h_i(y - \mathbf{x}_j'\mathbf{t})\, dG(y)\, \mathbf{x}_j. \end{aligned}$$

Specialize this to the case $G(y) \equiv y$, $L(s) \equiv s$, $p = 1, x_i \equiv 1$ and integrate by parts, to obtain

$$\begin{aligned} \dot{Q}(t) &= -2n^{-2}\sum_{i=1}^{n}\sum_{j=1}^{n} \min(n-i, n-j)\,\{H_i(Y_i - t) - 1/2\} \\ &= -n^{-2}\sum_{i=1}^{n}(n-i)(n+i-1)\{H_i(Y_i - t) - 1/2\}. \end{aligned}$$

Now suppose further that $H_i \equiv F$. Then $\bar{\beta}$ is a solution t of

$$(5.3.19) \qquad \sum_{i=1}^{n}(n-i)(n+i-1)\{F(Y_i - t) - 1/2\} = 0.$$

Compare this $\bar{\beta}$ with $\hat{\theta}$ of (5.3.2). Clearly $\bar{\beta}$ given by (5.3.19) is a weighted M-estimator of the location parameter whereas $\hat{\theta}$ given by (5.3.2) is an ordinary M-estimator. Of course, if we choose $L(s) = I(s \geq 1), p = 1, x_i \equiv 1, G(y) = y$ then $\hat{\theta} = \bar{\boldsymbol{\beta}}$. In general $\bar{\boldsymbol{\beta}}$ may be obtained as a solution of $\dot{\mathbf{Q}}(\mathbf{t}) = \mathbf{0}$.

Next, consider $\boldsymbol{\beta}^*$ of (5.2.18). For the time being focus on the case $p = 1$ and $d_i \equiv x_i - \bar{x}$. Assume, without loss of generality, that the data is so arranged that $x_1 \leq x_2 \leq \cdots \leq x_n$. Let $\mathcal{Y} := \{(Y_j - Y_i)/(x_j - x_i);\, i < j,\, x_i < x_j\}$, $t_0 := \min\{t; t \in \mathcal{Y}\}$ and $t_1 := \max\{t; t \in \mathcal{Y}\}$. Then for $x_i < x_j$, $t < t_0$ implies $t < (Y_j - Y_i)/(x_j - x_i)$ so that $R_{it} < R_{jt}$. In other words the residuals $\{Y_j - tx_j; 1 \leq j \leq n\}$ are naturally ordered for all $t < t_0$, w.p.1., assuming the continuity of the errors. Hence, with $T(s,t)$ denoting the $T_{1d}(s,t)$ of (5.2.17), we obtain for $t < t_0$,

$$\begin{aligned} T(s,t) &= \textstyle\sum_{i=1}^{k} d_i, \qquad k/n \leq s < (k+1)/n,\ 1 \leq k \leq n-1, \\ &= 0, \qquad\qquad\qquad 0 \leq s < 1/n,\ s = 1. \end{aligned}$$

Hence,

$$K^*(t) = \sum_{k=1}^{n-1} \omega_k \left\{ \sum_{i=1}^{k} d_i \right\}^2.$$

Similarly using the fact $\sum_{i=1}^{n} d_i = 0$, one obtains

$$K^*(t) = \sum_{k=1}^{n-1} \omega_k \left\{ \sum_{i=1}^{k} d_i \right\}^2 = K^*(t_{1+}),\ t > t_1.$$

As t crosses over t_0 only one pair of adjacent residuals change their ranks. Let $x_j < x_{j+1}$ denote their respective regression constants. Then

$$\begin{aligned}
K^*(t_{0-}) - K^*(t_{0+}) &= \sum_{k=1}^{n-1} \omega_k \left\{ \sum_{i=1}^{k} d_i \right\}^2 - \sum_{k=1, k\neq j}^{n-1} \omega_k \left\{ \sum_{i=1}^{k} d_i \right\}^2 \\
&\quad - \omega_j \left\{ d_{j+1} + \sum_{i=1}^{j-1} d_i \right\}^2 \\
&= \omega_j \left[\left\{ \sum_{i=1}^{j} d_i \right\}^2 - \left\{ d_{j+1} + \sum_{i=1}^{j-1} d_i \right\}^2 \right].
\end{aligned}$$

But $x_1 \leq x_2 \leq \cdots \leq x_n, x_j < x_{j+1}$ and $\sum_{i=1}^n d_i = 0$ imply

$$\sum_{i=1}^{j} d_i < d_{j+1} + \sum_{i=1}^{j-1} d_i \leq 0.$$

Hence $K^*(t_{0-}) > K^*(t_{0+})$. Similarly it follows that $K^*(t_{1+}) > K^*(t_{1-})$. Consequently, β_1 and β_2 are finite, where

$$\begin{aligned}
\beta_1 &:= \min\{t \in \mathcal{Y}_1 K^*(t_+) = \inf_{\Delta \in \mathcal{Y}^c} K^*(\Delta)\}, \\
\beta_2 &:= \max\{t \in \mathcal{Y}, K^*(t_-) = \inf_{\Delta \in \mathcal{Y}^c} K^*(\Delta)\},
\end{aligned}$$

and where $\mathcal{Y}^c$ denotes the complement of $\mathcal{Y}$. Then β^* can be uniquely defined by the relation $\beta^* = (\beta_1 + \beta_2)/2$.

This β^* corresponding to $L(a) \equiv s$ was studied by Williamson (1979, 1982). In general this estimator is asymptotically relatively more efficient than Wilcoxon type R-estimators as will be seen later on in Section 5.6.

There does not seem to be such a nice characterization for $p \geq 1$ and general $\mathbf{D}$ satisfying (5.2.16). However, proceeding as in the derivation of (5.3.6), a computational formula for K^* of (5.2.17) can be obtained to be

(5.3.20) $K^*(t)$

$$= -2 \sum_{k=1}^{p} \sum_{i=1}^{n} \sum_{j=1}^{n} d_{ik} d_{jk} \left| L\left(\frac{R_{it}}{n} -\right) - L\left(\frac{R_{jt}}{n} -\right) \right|.$$

This formula is valid for a general σ - finite measure L and can be used to compute $\boldsymbol{\beta}^*$.

We now turn to the m.d. estimator defined at (5.2.21) and (5.2.22). Let $d_i \equiv x_i - \bar{x}$. The first observation one makes is that for $t \in \mathbb{R}$,

$$\mathcal{D}_n(t) \;:=\; \sup_{y\in\mathbb{R}} \left|\sum_{i=1}^{n} d_i I(Y_i \le y + td_i)\right| = \sup_{0\le s\le 1} \left|\sum_{i=1}^{n} d_i I(R_{it} \le ns)\right|.$$

Proceedings as in the above discussion pertaining to β^*, assume, without loss of generality, that the data is so arranged that $x_1 \le x_2 \le \cdots \le x_n$ so that $d_1 \le d_2 \le \cdots \le d_n$. Let $\mathcal{Y}_1 := \{(Y_j - Y_i)/(d_j - d_i); d_i < 0, d_j \ge 0, 1 \le i < j \le n\}$. It can be proved that $\mathcal{D}_n^+(D_n^-)$ is a left continuous non-decreasing (right continuous non-increasing) step function on $\mathbb{R}$ whose points of discontinuity are a subset of $\mathcal{Y}_1$. Moreover, if $-\infty = t_0 < t_1 \le t_2 \cdots \le t_m < t_{m+1} = \infty$ denote the ordered members of $\mathcal{Y}_1$ then $\mathcal{D}_n^+(t_{1-}) = 0 = \mathcal{D}_n^-(t_{m+})$ and $\mathcal{D}_n^+(t_{m+}) = \sum_{i=1}^n d_i^+ = \mathcal{D}_n^-(t_{1-})$, where $d_i^+ \equiv \max(d_i, 0)$. Consequently, the following entities are finite:

$$\begin{aligned}\beta_{s1} &:= \inf\{t \in \mathbb{R}; \mathcal{D}_n^+(t) \ge \mathcal{D}_n^-(t)\},\\ \beta_{s2} &:= \sup\{t \in \mathbb{R}; \mathcal{D}_n^+(t) \le \mathcal{D}_n^-(t)\}.\end{aligned}$$

Note that $\beta_{s2} \ge \beta_{s1}$ w.p.1.. One can now take $\beta_s = (\beta_1 + \beta_{s2})/2$.

Williamson (1979) provides the proofs of the above claims and obtains the asymptotic distribution of β_s. This estimator is the precise generalization of the m.d. estimator of the two sample location parameter of Rao, Schuster and Littell (1975). Its asymptotic distribution is the same as that of their estimator.

We shall now discuss some additional distributional properties of the above m.d. estimators. To facilitate this discussion let $\tilde{\boldsymbol{\beta}}$ denote any one of the estimators defined at (5.2.8), (5.2.15), (5.2.18) and (5.2.22). As in Section 4.3, we shall write $\tilde{\boldsymbol{\beta}}(\mathbf{X}, \mathbf{Y})$ to emphasize the dependence on the data $\{(\mathbf{x}_i', \mathbf{Y}_i); 1 \le i \le n\}$. It also helps to think of the defining distances K, K^+, etc. as functions of residuals. Thus we shall some times write $K(\mathbf{Y} - \mathbf{Xt})$ etc. for $K(\mathbf{t})$. Let $\tilde{K}$ stand for either K or K^+ or K^* of (5.2.7), (5.2.14) and (5.2.17). To begin with, observe that

$$\tilde{K}(\mathbf{t} - \mathbf{b}) = \tilde{K}(\mathbf{Y} + \mathbf{Xb} - \mathbf{Xt}), \ \forall \mathbf{t}, \mathbf{b} \in \mathbb{R}^p, \tag{5.3.21}$$

so that

$$\tilde{\boldsymbol{\beta}}(\mathbf{X}, \mathbf{Y} + \mathbf{X}\mathbf{b}) = \tilde{\boldsymbol{\beta}}(\mathbf{X}, \mathbf{Y}) + \mathbf{b}, \ \forall \, \mathbf{b} \in \mathbb{R}^p. \tag{5.3.22}$$

Consequently, the distribution of $\tilde{\boldsymbol{\beta}} - \boldsymbol{\beta}$ does not depend on $\boldsymbol{\beta}$.

The distance measure Q of (5.2.9) does not satisfy (5.3.21) and hence the distribution of $\bar{\boldsymbol{\beta}} - \boldsymbol{\beta}$ will generally depend on $\boldsymbol{\beta}$.

In general, the classes of estimators $\{\hat{\boldsymbol{\beta}}\}$ and $\{\boldsymbol{\beta}^+\}$ are not scale invariant. However, as can be readily seen from (5.3.6) and (5.3.7), the class $\{\hat{\boldsymbol{\beta}}\}$ corresponding to $G(y) \equiv y$, $H_i \equiv F$ and those $\{\mathbf{D}\}$ that satisfy (5.2.16) and the class $\{\boldsymbol{\beta}^+\}$ corresponding to $G(y) \equiv y$ and general $\{\mathbf{D}\}$ are scale invariant in the sense of (4.3.2).

An interesting property of all of the above m.d. estimators is that they are invariant under nonsingular transformation of the design matrix $\mathbf{X}$. That is,

$$\tilde{\boldsymbol{\beta}}(\mathbf{XB}, \mathbf{Y}) = \mathbf{B}^{-1}\tilde{\boldsymbol{\beta}}(\mathbf{X}, \mathbf{Y}) \quad \forall \ p \times p \text{ nonsingular matrix } \ \mathbf{B}.$$

A similar statement holds for $\bar{\boldsymbol{\beta}}$.

We shall end this section by discussing the *symmetry* property of these estimators. In the following lemma it is implicitly assumed that all integrals involved are finite. Some sufficient conditions for that to happen will unfold as we proceed in this chapter.

Lemma 5.3.2. *Let* (1.1.1) *hold with the actual and the modeled d.f. of* e_i *equal to* $H_i, 1 \le i \le n$.

(i) *If either*

(ia) $\{H_i, 1 \le i \le n\}$ *and* G *are symmetric around* 0 *and* $\{H_i, 1 \le i \le n\}$ *are continuous,*

or

(ib) $d_{ij} = -d_{n-i+1,j}$, $x_{ij} = -x_{n-i+1,j}$ *and* $H_i \equiv F \ \forall \ 1 \le i \le n, 1 \le j \le p$,

then

$\hat{\boldsymbol{\beta}}$ *and* $\boldsymbol{\beta}^*$ *are symmetrically distributed around* $\boldsymbol{\beta}$, *whenever they exist uniquely.*

(ii) *If* $\{H_i, 1 \le i \le n\}$ *and* G *are symmetric around* 0 *and either* $\{H_i, 1 \le i \le n\}$ *are continuous or* G *is continuous,*

then

$\boldsymbol{\beta}^+$ is symmetrically distributed around $\boldsymbol{\beta}$, whenever it exists uniquely.

Proof. In view of (5.3.22) there is no loss of generality in assuming that the true $\boldsymbol{\beta}$ is $\mathbf{0}$.

Suppose that **(ia) holds**. Then $\hat{\boldsymbol{\beta}}(\mathbf{X},\mathbf{Y}) =_d \hat{\boldsymbol{\beta}}(\mathbf{X},-\mathbf{Y})$. But, by definition (5.2.8), $\hat{\boldsymbol{\beta}}(\mathbf{X},-\mathbf{Y})$ is the minimizer of $K(-\mathbf{Y}-\mathbf{X}\mathbf{t})$ w.r.t. $\mathbf{t}$. Observe that $\forall\ \mathbf{t} \in \mathbb{R}^p$,

$$
\begin{aligned}
&K(-\mathbf{Y}-\mathbf{X}\mathbf{t})\\
&\quad= \sum_{j=1}^{p}\int\Big[\sum_{i=1}^{n} d_{ij}\Big\{I(-Y_i \le y + \mathbf{x}_i'\mathbf{t}) - H_i(y)\Big\}\Big]^2 dG(y)\\
&\quad= \sum_{j=1}^{p}\int\Big[\sum_{i=1}^{n} d_{ij}\Big\{1 - I(Y_i < -y - \mathbf{x}_i'\mathbf{t}) - \mathbf{H}_i(y)\Big\}\Big]^2 dG(y)\\
&\quad= \sum_{j=1}^{p}\int\Big[\sum_{i=1}^{n} d_{ij}\Big\{I(Y_i < y - \mathbf{x}_i'\mathbf{t}) - H_i(y_-)\Big\}\Big]^2 dG(y)
\end{aligned}
$$

by the symmetry of $\{H_i\}$ and G. Now use the continuity of $\{H_i\}$ to conclude that, w.p.1.,

$$
K(-\mathbf{Y}-\mathbf{X}\mathbf{t}) = K(\mathbf{Y}+\mathbf{X}\mathbf{t}), \quad \forall\ \mathbf{t} \in \mathbb{R}^p,
$$

so that $\hat{\boldsymbol{\beta}}(\mathbf{X},-\mathbf{Y}) = -\hat{\boldsymbol{\beta}}(\mathbf{X},\mathbf{Y})$, w.p.1, and the claim follows because $-\hat{\boldsymbol{\beta}}(\mathbf{X},\mathbf{Y}) = \operatorname{argmin}\{K(\mathbf{Y}+\mathbf{X}\mathbf{t});\ \mathbf{t} \in \mathbb{R}^p\}$.

Now suppose that **(ib)** holds. Then

$$
\begin{aligned}
&K(\mathbf{Y}+\mathbf{X}\mathbf{t})\\
&\quad= \sum_{j=1}^{p}\int\Big[\sum_{i=1}^{n} d_{n-i+1,j}\Big\{I(Y_i \le y + \mathbf{x}_{n-i+1}'\mathbf{t}) - F(y)\Big\}\Big]^2 dG(y)\\
&\quad=_d \sum_{j=1}^{p}\int\Big[\sum_{i=1}^{n} d_{n-i+1,j}\Big\{I(Y_{n-i+1} \le y + \mathbf{x}_{n-i+1}'\mathbf{t}) - F(y)\Big\}\Big]^2\\
&\qquad\qquad\qquad\qquad\qquad\qquad\qquad\qquad \times dG(y)\\
&\quad= K(\mathbf{Y}-\mathbf{X}\mathbf{t}), \quad \forall\ \mathbf{t} \in \mathbb{R}^p.
\end{aligned}
$$

This shows that $-\hat{\boldsymbol{\beta}}(\mathbf{X},\mathbf{Y}) =_d \hat{\boldsymbol{\beta}}(\mathbf{X},\mathbf{Y})$ as required. The proof for $\boldsymbol{\beta}^*$ is similar.

Proof of (ii). Again, $\boldsymbol{\beta}^+(\mathbf{X}, \mathbf{Y}) =_d \boldsymbol{\beta}^+(\mathbf{X}, -\mathbf{Y})$, because of the symmetry of $\{H_i\}$. But,

$$\begin{aligned} K^+(-\mathbf{Y} - \mathbf{Xt}) &= \sum_{j=1}^{p} \int \Big[\sum_{i=1}^{n} d_{ij} \Big\{ I(-Y_i \le y + \mathbf{x}_i'\mathbf{t}) - 1 \\ &\qquad\qquad + I(-Y_i \le -y + \mathbf{x}_i'\mathbf{t}) \Big\} \Big]^2 dG(y) \\ &= \sum_{j=1}^{p} \int \Big[\sum_{i=1}^{n} d_{ij} \Big\{ I(Y_i < y + \mathbf{x}_i'\mathbf{t}) - 1 \\ &\qquad\qquad + I(Y_i < -y + \mathbf{x}_i'\mathbf{t}) \Big\} \Big]^2 dG(y) \\ &= K^+(\mathbf{Y} + \mathbf{Xt}), \quad \forall\, \mathbf{t} \in \mathbb{R}^p, \end{aligned}$$

w.p.1, if either $\{H_i\}$ or G are continuous. □

5.4 A General M. D. Estimator

This section gives a general overview of an asymptotic theory useful in inference based on minimizing an objective function of the data and parameter in general models. It is a self contained section of broad interest.

In an inferential problem consisting of a vector of n observations $\boldsymbol{\zeta}_n = (\zeta_{n1}, \cdots, \zeta_{nn})'$, not necessarily independent, and a p - dimensional parameter $\boldsymbol{\theta} \in \mathbb{R}^p$, an estimator of $\boldsymbol{\theta}$ is often based on an objective function $M_n(\boldsymbol{\zeta}_n, \boldsymbol{\theta})$, herein called *dispersion*. In this section an estimator of $\boldsymbol{\theta}$ obtained by minimizing $M_n(\boldsymbol{\zeta}_n, \cdot)$ will be called *minimum dispersion estimator*.

Typically the sequence of dispersion M_n admits the following approximate quadratic structure. Writing $M_n(\boldsymbol{\theta})$ for $M_n(\boldsymbol{\zeta}_n, \boldsymbol{\theta})$, often it turns out that $M_n(\boldsymbol{\theta}) - M_n(\boldsymbol{\theta}_0)$, under $\boldsymbol{\theta}_0$, is asymptotically like a quadratic form in $(\boldsymbol{\theta} - \boldsymbol{\theta}_0)$, for $\boldsymbol{\theta}$ close to $\boldsymbol{\theta}_0$ in a certain sense, with the coefficient of the linear term equal to a random vector which is typically asymptotically normally distributed. This approximation in turn is used to obtain the asymptotic distribution of the corresponding minimum dispersion estimators.

The two classical examples of the above type are Gauss's least square and Fisher's maximum likelihood estimators. In the former the dispersion M_n is the error sum of squares while in the latter M_n equals $-\log L_n$, L_n denoting the likelihood function of $\boldsymbol{\theta}$ based on $\boldsymbol{\zeta}_n$. In the least squares method, $M_n(\boldsymbol{\theta}) - M_n(\boldsymbol{\theta}_0)$ is exactly quadratic in $(\boldsymbol{\theta} - \boldsymbol{\theta}_0)$, uniformly in $\boldsymbol{\theta}$ and $\boldsymbol{\theta}_0$. The random vector appearing in the linear term is typically asymptotically normally distributed. In the likelihood method, the well celebrated locally asymptotically normal (LAN) models of Le Cam (1960, 1986) obey the above type of approximate quadratic structure. Other well known examples include the least absolute deviation and the minimum chi-square estimators.

The main purpose of this section is to unify the basic structure of asymptotics underlying the minimum dispersion estimators by exploiting the above type of common asymptotic quadratic structure inherent in most of the dispersions. The formulation given below is general enough to cover some irregular cases where the coefficient of linear term may not be asymptotically normally distributed.

We now formulate general conditions for a given dispersion to be uniformly locally asymptotically quadratic (ULAQ). Accordingly, let Ω be an open subset of $\mathbb{R}^p$ and $M_n, n \geq 1$, be a sequence of real valued functions defined on $\mathbb{R}^n \times \Omega$ such that $M_n(\cdot, \boldsymbol{\theta})$ is measurable for each $\boldsymbol{\theta}$. We shall often suppress the $\boldsymbol{\zeta}_n$ coordinate in M_n and write $M_n(\boldsymbol{\theta})$ for $M_n(\boldsymbol{\zeta}_n, \boldsymbol{\theta})$.

In order to state general conditions we need to define a sequence of neighborhoods $\mathcal{N}_n(\boldsymbol{\theta}_0) := \{\boldsymbol{\theta} \in \Omega, |\boldsymbol{\delta}_n(\boldsymbol{\theta}_0)(\boldsymbol{\theta} - \boldsymbol{\theta}_0)| \leq b\}$, where $\boldsymbol{\theta}_0$ is a fixed parameter value in Ω, b is a finite number and $\{\boldsymbol{\delta}_n(\boldsymbol{\theta}_0)\}$ is a sequence $p \times p$ symmetric positive definite matrices with norms $\|\boldsymbol{\delta}_n(\boldsymbol{\theta}_0)\|$ tending to infinity. Since $\boldsymbol{\theta}_0$ is fixed, write $\boldsymbol{\delta}_n$, $\mathcal{N}_n$ for $\boldsymbol{\delta}_n(\boldsymbol{\theta}_0)$, $\mathcal{N}_n(\boldsymbol{\theta}_0)$, respectively. Similarly, let P_n denote the probability distribution of $\boldsymbol{\zeta}_n$ when $\boldsymbol{\theta} = \boldsymbol{\theta}_0$.

Definition 5.4.1. A sequence of dispersions $\{M_n(\boldsymbol{\theta}), \boldsymbol{\theta} \in \mathcal{N}_n\}$, $n \geq 1$, satisfying condition $(\tilde{A1})$ - $(\tilde{A3})$ given below is said to be *uniformly locally asymptotically quadratic* (ULAQ).

$(\tilde{A1})$ There exist a sequence of $p \times 1$ random vector $\mathbf{S}_n(\boldsymbol{\theta}_0)$ and a sequence of $p \times p$, possibly random, matrices $\mathbf{W}_n(\boldsymbol{\theta}_0)$, such that, for every $0 < b < \infty$,

$$M_n(\boldsymbol{\theta}) \\ = M_n(\boldsymbol{\theta}_0) + (\boldsymbol{\theta} - \boldsymbol{\theta}_0)'\mathbf{S}_n(\boldsymbol{\theta}_0) + \frac{1}{2}(\boldsymbol{\theta} - \boldsymbol{\theta}_0)'\mathbf{W}_n(\boldsymbol{\theta}_0)(\boldsymbol{\theta} - \boldsymbol{\theta}_0) \\ + u_p(1),$$

where "$u_p(1)$" is a sequence of stochastic processes in $\boldsymbol{\theta}$ converging to zero, *uniformly in* $\boldsymbol{\theta} \in \mathcal{N}_n$, in P_n - probability.

($\tilde{A2}$) There exists a $p \times p$ non-singular, possibly random, matrix $\mathbf{W}(\boldsymbol{\theta}_0)$ such that

$$\boldsymbol{\delta}_n^{-1}\mathbf{W}_n(\boldsymbol{\theta}_0)\boldsymbol{\delta}_n^{-1} = \mathbf{W}(\boldsymbol{\theta}_0) + o_p(1), \quad (P_n).$$

($\tilde{A3}$) There exists a $p \times 1$ r.v. $\mathbf{Y}(\boldsymbol{\theta}_0)$ such that

$$\mathcal{L}_n\Big(\boldsymbol{\delta}_n^{-1}\mathbf{S}_n(\boldsymbol{\theta}_0), \boldsymbol{\delta}_n^{-1}\mathbf{W}_n(\boldsymbol{\theta}_0)\boldsymbol{\delta}_n^{-1}\Big) \longrightarrow_d \mathcal{L}\Big(\mathbf{Y}(\boldsymbol{\theta}_0), \mathbf{W}(\boldsymbol{\theta}_0)\Big)$$

where $\mathcal{L}_n$, $\mathcal{L}$ denote joint probability distributions under P_n and in the limit, respectively.

Denote the conditions ($\tilde{A1}$), ($\tilde{A2}$) by (A1) and (A2), respectively, whenever $\mathbf{W}$ is *non-random* in these conditions. A sequence of dispersions $\{M_n\}$ is called *uniformly locally asymptotically* **normal** *quadratic* (ULANQ) if (A1), (A2) hold and if (A3), instead of ($\tilde{A3}$), holds, where (A3) is as follows:

(A3) There exists a positive definite $p \times p$ matrix $\boldsymbol{\Sigma}(\boldsymbol{\theta}_0)$ such that

$$\boldsymbol{\delta}_n^{-1}\mathbf{S}_n(\boldsymbol{\theta}_0) \longrightarrow_d \mathcal{N}(\mathbf{0}, \boldsymbol{\Sigma}(\boldsymbol{\theta}_0)), \quad (P_n).$$

If ($\tilde{A1}$) holds *without* the *uniformity* requirement and ($\tilde{A2}$), ($\tilde{A3}$) hold then we call the given sequence M_n LAQ (*locally asymptotically quadratic*). If (A1) holds *without* the *uniformity* requirement and (A2), (A3) hold then the given sequence M_n is called *locally asymptotically normal quadratic* (LANQ).

In the case $M_n(\boldsymbol{\theta}) = -\ell n L_n(\boldsymbol{\theta})$, the condition non-uniform (A1), (A2), (A3) with $\|\delta_n\| = O(n^{1/2})$, determine the well celebrated LAN models of Le Cam (1960, 1986). For this particular case, $\mathbf{W}(\boldsymbol{\theta}_0)$, $\boldsymbol{\Sigma}(\boldsymbol{\theta}_0)$ and the limiting Fisher information matrix, $\mathbf{F}(\boldsymbol{\theta}_0)$, whenever it exists, are the same.

In the above general formulation, M_n is an arbitrary dispersion satisfying ($\tilde{A1}$) - ($\tilde{A3}$) or (A1) - (A3). In the latter the three matrices $\mathbf{W}(\boldsymbol{\theta}_0)$, $\boldsymbol{\Sigma}(\boldsymbol{\theta}_0)$ and $\mathbf{F}(\boldsymbol{\theta}_0)$ are not necessarily identical. The LANQ dispersions can thus be viewed as a generalization of the LAN models.

Typically in the classical i.i.d. setup the normalizing matrix $\boldsymbol{\delta}_n$ is of the order square root of n whereas in the linear regression model (1.1.1) it is of the order $(\mathbf{X}'\mathbf{X})^{1/2}$. In general $\boldsymbol{\delta}_n$ will depend on $\boldsymbol{\theta}_0$ and is determined by the order of the asymptotic magnitude of $\mathbf{S}_n(\boldsymbol{\theta}_0)$.

We now turn to the asymptotic distribution of the minimum dispersion estimators. Let $\{M_n\}$ be a sequence of ULAQ dispersions. Define

$$\hat{\boldsymbol{\theta}}_n = \operatorname{argmin}\{M_n(\mathbf{t}),\ \mathbf{t} \in \Omega\}. \tag{5.4.1}$$

Our aim is to investigate the asymptotic behavior of $\hat{\boldsymbol{\theta}}_n$ and $M_n(\hat{\boldsymbol{\theta}}_n)$. Akin to the study of the asymptotic distribution of m.l.e.'s, we must first ensure that there is a $\hat{\boldsymbol{\theta}}_n$ satisfying (5.4.1) such that

$$|\boldsymbol{\delta}_n(\hat{\boldsymbol{\theta}}_n - \boldsymbol{\theta}_0)| = O_p(1). \tag{5.4.2}$$

Unfortunately the ULAQ assumptions are not enough to guarantee (5.4.2). One set of additional assumptions that ensures this is the following.

(A4) $\forall\ \epsilon > 0\ \exists$ a $0 < z_\epsilon < \infty$ and $N_{1\epsilon}$ such that

$$P_n(|M_n(\boldsymbol{\theta}_0)| \le z_\epsilon) \ge 1 - \epsilon, \qquad \forall\ n \ge N_{1\epsilon}.$$

(A5) $\forall\ \epsilon > 0$ and $0 < \alpha < \infty, \exists$ an $N_{2\epsilon}$ and a b (depending on ϵ and α) such that

$$P_n\Big(\inf_{\|\boldsymbol{\delta}_n(\boldsymbol{\theta}-\boldsymbol{\theta}_0)\|>b} M_n(\boldsymbol{\theta}) \ge \alpha\Big) \ge 1 - \epsilon, \qquad \forall\, n \ge N_{2\epsilon}.$$

It is convenient to let, for a $\boldsymbol{\theta} \in \mathbb{R}^p$,

$$Q_n(\boldsymbol{\theta}, \boldsymbol{\theta}_0) := (\boldsymbol{\theta} - \boldsymbol{\theta}_0)'\mathbf{S}_n(\boldsymbol{\theta}_0) + \frac{1}{2}(\boldsymbol{\theta} - \boldsymbol{\theta}_0)'\mathbf{W}_n(\boldsymbol{\theta}_0)(\boldsymbol{\theta} - \boldsymbol{\theta}_0),$$

and $\tilde{\boldsymbol{\theta}}_n := \operatorname{argmin}\{Q_n(\boldsymbol{\theta}, \boldsymbol{\theta}_0),\ \boldsymbol{\theta} \in \mathbb{R}^p\}$. Clearly, $\tilde{\boldsymbol{\theta}}_n$ must satisfy the relation

$$\boldsymbol{\mathcal{B}}_n\boldsymbol{\delta}_n(\tilde{\boldsymbol{\theta}}_n - \boldsymbol{\theta}_0) = -\boldsymbol{\delta}_n^{-1}\mathbf{S}_n(\boldsymbol{\theta}_0) \tag{5.4.3}$$

where $\mathcal{B}_n := \boldsymbol{\delta}_n^{-1}\mathbf{W}_n\boldsymbol{\delta}_n^{-1}$, where $\mathbf{W}_n = \mathbf{W}_n(\boldsymbol{\theta}_0)$.

Some generality is achieved by making the following assumption.

(A6) $\quad \|\boldsymbol{\delta}_n(\tilde{\boldsymbol{\theta}}_n - \boldsymbol{\theta}_0)\| = O_p(1)$.

Note that ($\tilde{\text{A2}}$) and ($\tilde{\text{A3}}$) imply (A6). We now state and prove

Theorem 5.4.1 *Let the dispersions M_n satisfy* ($\tilde{\text{A1}}$), (A4) - (A6). *Then, under P_n,*

$$|(\hat{\boldsymbol{\theta}}_n - \tilde{\boldsymbol{\theta}}_n)'\boldsymbol{\delta}_n\mathcal{B}_n\boldsymbol{\delta}_n(\hat{\boldsymbol{\theta}}_n - \tilde{\boldsymbol{\theta}}_n)| = o_p(1), \tag{5.4.4}$$

$$\begin{aligned}&\inf_{\boldsymbol{\theta}\in\Omega} M_n(\boldsymbol{\theta}) - M_n(\boldsymbol{\theta}_0) \\ &= -(1/2)(\tilde{\boldsymbol{\theta}}_n - \boldsymbol{\theta}_0)'\mathbf{W}_n(\tilde{\boldsymbol{\theta}}_n - \boldsymbol{\theta}_0) + o_p(1).\end{aligned} \tag{5.4.5}$$

Consequently, if (A6) *is replaced by* ($\tilde{\text{A2}}$) *and* ($\tilde{\text{A3}}$), *then*

$$\boldsymbol{\delta}_n(\hat{\boldsymbol{\theta}}_n - \boldsymbol{\theta}_0) \longrightarrow_d \{\mathbf{W}(\boldsymbol{\theta}_0)\}^{-1}\mathbf{Y}(\boldsymbol{\theta}_0), \tag{5.4.6}$$

$$\begin{aligned}&\inf_{\boldsymbol{\theta}\in\Omega} M_n(\boldsymbol{\theta}) - M_n(\boldsymbol{\theta}_0) \\ &= -(1/2)S_n'(\boldsymbol{\theta}_0)\boldsymbol{\delta}_n^{-1}\mathcal{B}_n^{-1}\boldsymbol{\delta}_n^{-1}S_n(\boldsymbol{\theta}_0) + o_p(1).\end{aligned} \tag{5.4.7}$$

If, instead of ($\tilde{\text{A1}}$) - ($\tilde{\text{A3}}$), *M_n satisfies* (A1) - (A3), *and if* (A4) *and* (A5) *hold then also* (5.4.4) - (5.4.7) *hold and*

$$\delta_n(\hat{\boldsymbol{\theta}}_n - \boldsymbol{\theta}_0) \longrightarrow_d \mathcal{N}(\mathbf{0}, \boldsymbol{\Gamma}(\boldsymbol{\theta}_0)), \tag{5.4.8}$$

where

$$\boldsymbol{\Gamma}(\boldsymbol{\theta}_0) = \{\mathbf{W}(\boldsymbol{\theta}_0)\}^{-1}\boldsymbol{\Sigma}(\boldsymbol{\theta}_0)\{\mathbf{W}(\boldsymbol{\theta}_0)\}^{-1}.$$

Proof. Let z_ϵ be as in (A4). Choose an $\alpha > z_\epsilon$ in (A5). Then

$$\begin{aligned}&\Big[|M_n(\boldsymbol{\theta}_0)| \le z_\epsilon, \inf_{\|\mathbf{h}\|>b} M_n(\boldsymbol{\theta}_0 + \boldsymbol{\delta}_n^{-1}\mathbf{h}) \ge \alpha\Big] \\ &\subset \Big[\inf_{\|\mathbf{h}\|\le b} M_n(\boldsymbol{\theta}_0 + \boldsymbol{\delta}_n^{-1}\mathbf{h}) \le z_\epsilon, \inf_{\|\mathbf{h}\|>b} M_n(\boldsymbol{\theta}_0 + \boldsymbol{\delta}_n^{-1}\mathbf{h}) \ge \alpha\Big] \\ &\subset \Big[\inf_{\|\mathbf{h}\|>b} M_n(\boldsymbol{\theta}_0 + \boldsymbol{\delta}_n^{-1}\mathbf{h}) > \inf_{\|\mathbf{h}\|\le b} M_n(\boldsymbol{\theta}_0 + \boldsymbol{\delta}_n^{-1}\mathbf{h})\Big].\end{aligned}$$

Hence by (A4) and (A5), for any $\epsilon > 0$ there exists a $0 < b < \infty$ (now depending only on ϵ) such that $\forall\, n \ge N_{1\epsilon} \vee N_{2\epsilon}$,

$$P_n\left(\inf_{\|\mathbf{h}\|>b} M_n(\boldsymbol{\theta}_0 + \boldsymbol{\delta}_n^{-1}\mathbf{h}) > \inf_{\|\mathbf{h}\|\le b} M_n(\boldsymbol{\theta}_0 + \boldsymbol{\delta}_n^{-1}\mathbf{h})\right) \ge 1 - \epsilon.$$

This in turn ensures the validity of (5.4.2), which in turn together with (A$\tilde{1}$) yields

$$(5.4.9) \qquad M_n(\hat{\boldsymbol{\theta}}_n) = M_n(\boldsymbol{\theta}_0) + Q_n(\hat{\boldsymbol{\theta}}_n, \boldsymbol{\theta}_0) + o_p(1), \quad (P_n).$$

From (A6), the inequality

$$\begin{aligned} \Big| \inf_{\boldsymbol{\theta}\in\mathcal{N}_n} M_n(\boldsymbol{\theta}) \; &- \inf_{\boldsymbol{\theta}\in\mathcal{N}_n} [M_n(\boldsymbol{\theta}_0) + Q_n(\boldsymbol{\theta}, \boldsymbol{\theta}_0)] \Big| \\ &\le \sup_{\boldsymbol{\theta}\in\mathcal{N}_n} \Big| M_n(\boldsymbol{\theta}) - [M_n(\boldsymbol{\theta}_0) + Q_n(\boldsymbol{\theta}, \boldsymbol{\theta}_0)] \Big| \end{aligned}$$

and (A$\tilde{1}$), we obtain

$$(5.4.10) \qquad M_n(\hat{\boldsymbol{\theta}}_n) = M_n(\boldsymbol{\theta}_0) + Q_n(\tilde{\boldsymbol{\theta}}_n, \boldsymbol{\theta}_0) + o_p(1), \quad (P_n).$$

Now, (5.4.9) and (5.4.10) readily yield

$$Q_n(\hat{\boldsymbol{\theta}}_n, \boldsymbol{\theta}_0) = Q_n(\tilde{\boldsymbol{\theta}}_n, \boldsymbol{\theta}_0) + o_p(1), \quad (P_n),$$

which is precisely equivalent to the statement (5.4.4). The claim (5.4.5) follows from (5.4.3) and (5.4.10). The rest is obvious. □

Remark 5.4.1. Roughly speaking, the assumption (A5) assures that the smallest value of $M_n(\boldsymbol{\theta})$ for $\boldsymbol{\theta}$ outside of $\mathcal{N}_n$ can be made asymptotically arbitrarily large with arbitrarily large probability. The assumption (A4) means that the sequence of r.v.'s $\{M_n(\boldsymbol{\theta}_0)\}$ is bounded in probability. This assumption is usually verified by an application of the Markov inequality in the case $E_n|M_n(\boldsymbol{\theta}_0)| = O(1)$, where E_n denotes the expectation under P_n. In some applications $M_n(\boldsymbol{\theta}_0)$ converges weakly to a r.v. which also implies (A4). Often the verification of (A5) is rendered easy by an application of a variant of the C-S inequality. Examples of this appear in the next section when dealing with m.d. estimators of the previous section.

We now discuss the *minimum dispersion tests of simple hypothesis*, briefly *without many details.* Consider the simple hypothesis $H_0 : \boldsymbol{\theta} = \boldsymbol{\theta}_0$. In the special case when M_n is $-\ell n L_n$, the likelihood ratio statistic for testing H_0 is given by $-2\inf\{M_n(\boldsymbol{\theta})-M_n(\boldsymbol{\theta}_0); \boldsymbol{\theta} \in \Omega\}$. Thus, given a general dispersion function M_n, we are motivated to base a test of H_0 on the statistic

$$T_n = 2\inf\{M_n(\boldsymbol{\theta}) - M_n(\boldsymbol{\theta}_0);\ \boldsymbol{\theta} \in \Omega\},$$

with large values of T_n being significant.

To study the asymptotic null distribution of T_n, note that by (5.4.7), $T_n = \mathbf{S}_n' \boldsymbol{\delta}_n^{-1} \boldsymbol{\mathcal{B}}_n^{-1} \boldsymbol{\delta}_n^{-1} \mathbf{S}_n + o_p(1)$, (P_n). Let $\mathbf{Y}$, $\mathbf{W}$ etc. stand for $\mathbf{Y}(\boldsymbol{\theta}_0)$, $\mathbf{W}(\boldsymbol{\theta}_0)$, etc.

Proposition 5.4.1. *Under* $(\tilde{\text{A1}})$ - $(\tilde{\text{A3}})$, (A4), (A5), *the asymptotic null distribution of* T_n *is the same as that of* $\mathbf{Y'W^{-1}Y}$.

Under (A1) - (A5), *the asymptotic null distribution of* T_n *is the same as that of* $\mathbf{Z'BZ}$ *where* $\mathbf{Z}$ *is a* $\mathcal{N}(\mathbf{0}, \mathbf{I}_{p\times p})$ *r.v. and* $\mathbf{B} = \mathbf{\Sigma}^{1/2}\mathbf{W}^{-1}\mathbf{\Sigma}^{1/2}$. □

Remark 5.4.2. Clearly if $\mathbf{W}(\boldsymbol{\theta}_0) = \mathbf{\Sigma}(\boldsymbol{\theta}_0)$, then the asymptotic null distribution of T_n is χ^2_p. However, if $\mathbf{W}(\boldsymbol{\theta}_0) \neq \mathbf{\Sigma}(\boldsymbol{\theta}_0)$, the limit distribution of T_n is not a chi-square. We shall not discuss the distribution of T_n under alternatives. □

Here we shall now discuss a few examples illustrating the above theory.

EXAMPLE 1: THE ONE SAMPLE I.I.D. SETUP. Let Θ be an open interval in $\mathbb{R}$ and $\{F_\theta;\ \theta \in \Theta\}$ be a family of distribution functions on $\mathbb{R}$. Let g be real valued function on $\mathbb{R} \times \Theta$ such that $g(\cdot, \theta)$ is measurable for each $\theta \in \Theta$. Let $X_1, \cdots, X_n$ be i.i.d. from an F_{θ_0}, where θ_0 is fixed value of θ in Θ. Define

$$M_n(\theta) := \sum_{i=1}^n g(X_i,\ \theta), \qquad \theta \in \Theta.$$

We make the following additional assumptions: Let A be an open neighborhood of θ_0.

a. $g(x,\ \theta)$ is absolutely continuous in $\theta \in A$ and for almost all x (F_{θ_0}), with $\dot{g}(x, \cdot)$ denoting the a.e. derivative of $g(x, \cdot)$ satisfying

$$\text{(5.4.11)} \qquad \int \dot{g}^2(x, \theta)\, dF_{\theta_0}(x) < \infty, \qquad \theta \in A.$$

b. The function $\lambda(\theta) := \int \dot{g}(x, \theta)\, dF_{\theta_0}(x)$ is differentiable on A and the derivative $\dot{\lambda}$ satisfies

$$\text{(5.4.12)} \qquad \lim_{\epsilon \to 0} \epsilon^{-1} \int_0^\epsilon |\dot{\lambda}(\theta_0 + s) - \dot{\lambda}(\theta_0)|\, ds = 0.$$

c. The function $s \mapsto \int [\dot{g}(x, s) - \dot{g}(x, \theta_0)]^2\, dF_{\theta_0}(x)$ is continuous at θ_0.

Under the above assumptions, for every $0 < b < \infty$ and for every sequence of real numbers $\theta_n \in \mathcal{N}_n := \{t : t \in \Theta, n^{1/2}|t - \theta_0| \le b\}$,

$$(5.4.13) \quad M_n(\theta_n) = M_n(\theta_0) - (\theta_n - \theta_0)\sum_{i=1}^n \dot g(X_i, \theta_0) - \frac{(\theta_n - \theta_0)^2}{2}\, n\dot\lambda(\theta_0) + o_p(1), \quad (P_n).$$

Proof. Let D_n denote the difference between $M_n(\theta_n)$ and the leading r.v.'s on the right hand side of (5.4.13), $F \equiv F_{\theta_0}$, E, V denote the expectation and variance with respect to F, $u_n := \theta_n - \theta_0$, and let $t_n := n^{1/2}\, u_n$. To prove (5.4.13), it suffices to prove

$$(5.4.14) \qquad ED_n \to 0, \qquad V(D_n) \to 0.$$

First, suppose $u_n > 0$. Consider

$$\begin{aligned}
&E[M_n(\theta_n) - M_n(\theta_0) - (\theta_n - \theta_0)\sum_{i=1}^n \dot g(X_i, \theta_0)] \\
&= n\int \Big[g(x,\theta_n) - g(x,\theta_0) - (\theta_n - \theta_0)\,\dot g(x,\theta_0)\Big]\, dF(x) \\
&= n\int\int_0^{u_n} \Big[\dot g(x,\theta_0 + s) - \dot g(x,\theta_0)\Big]\, ds\, dF(x), \\
&= n\int_0^{u_n}\int \Big[\dot g(x,\theta_0 + s) - \dot g(x,\theta_0)\Big]\, dF(x)\, ds, \\
&= n\int_0^{u_n} \Big[\lambda(\theta_0 + s) - \lambda(\theta_0)\Big]\, ds \\
&= n^{1/2}\int_0^{t_n} \Big[\lambda(\theta_0 + sn^{-1/2}) - \lambda(\theta_0)\Big]\, ds.
\end{aligned}$$

In the above, the second equality follows from **a** and (5.4.11) while the third uses Fubini. Note that $t_n^2/2 = \int_0^{t_n} s\, ds$. Hence

$$\begin{aligned}
|ED_n| &= \Big|\int_0^{t_n} \Big[n^{1/2}\{\lambda(\theta_0 + tn^{-1/2}) - \lambda(\theta_0)\} - t\dot\lambda(\theta_0)\Big]\, dt\Big| \\
&= \Big|\int_0^{t_n} n^{1/2}\int_0^{tn^{-1/2}} \{\dot\lambda(\theta_0 + s) - \dot\lambda(\theta_0)\}\, ds dt\Big|
\end{aligned}$$

$$\begin{aligned} &\le b^2(bn^{-1/2})^{-1}\int_0^{bn^{-1/2}} |\dot{\lambda}(\theta_0+s)-\dot{\lambda}(\theta_0)|\,ds \\ &\to 0, \end{aligned}$$

by (5.4.12), thereby proving the first part of (5.4.14) when $u_n > 0$. To prove the second part in this case, similarly we have

$$\begin{aligned} V(D_n) &= nV\Big[g(X,\theta_n)-g(X,\theta_0)-(\theta_n-\theta_0)\dot{g}(X,\theta_0)\Big] \\ &\le nE\Big[g(X,\theta_n)-g(X,\theta_0)-(\theta_n-\theta_0)\dot{g}(X,\theta_0)\Big]^2 \\ &= n\int\Big[\int_0^{u_n}\{\dot{g}(x,s+\theta_0)-\dot{g}(x,\theta_0)\}\,ds\Big]^2\,dF(x) \\ &\le n\,u_n\int\int_0^{u_n}\Big\{\dot{g}(x,s+\theta_0)-\dot{g}(x,\theta_0)\Big\}^2 ds dF(x) \\ &= \tau_n n^{1/2}\int_0^{u_n}\int\Big\{\dot{g}(x,s+\theta_0)-\dot{g}(x,\theta_0)\Big\}^2 dF(x)ds \\ &\le b^2(bn^{-1/2})^{-1} \\ &\qquad\times\int_0^{bn^{-1/2}}\int\{\dot{g}(x,s+\theta_0)-\dot{g}(x,\theta_0)\}^2 dF(x) \\ &\to 0, \qquad\qquad \text{by the assumption } \mathbf{c}. \end{aligned}$$

This completes the proof of (5.4.13) in the case $\theta_n > \theta_0$. The proof is similar in the opposite case, thereby completing the proof of (5.4.13).

Actually we can prove more under the same assumptions. Let

$$\begin{aligned} D_n(t) &:= M_n(\theta_0+tn^{-1/2})-M_n(\theta_0) \\ &\qquad - tn^{-1/2}\sum_{i=1}^{n}\dot{g}(X_i,\theta_0)-\frac{t^2}{2}\,\dot{\lambda}(\theta_0),\ t\in\mathbb{R}. \end{aligned}$$

We shall now show, under the same assumptions as above, that for every $0 < b < \infty$,

$$E\Big(\sup_{|t|\le b}|D_n(t)|\Big)^2 \to 0. \tag{5.4.15}$$

To prove this, because of **a**, we can rewrite with probability 1 for all $t \ge 0$ and for all $n \ge 1$,

$$
\begin{aligned}
D_n(t) &= \sum_{i=1}^{n} \big[g(X_i, \theta_0 + tn^{-1/2}) - g(X_i, \theta_0) \\
&\qquad - tn^{-1/2}\dot{g}(X_i, \theta_0) - \frac{t^2}{2n}\dot{\lambda}(\theta_0)\big] \\
&= \int_0^t n^{-1/2} \sum_{i=1}^{n} \big[\, \dot{g}(X_i, \theta_0 + sn^{-1/2}) - \dot{g}(X_i, \theta_0) \\
&\qquad - sn^{-1/2}\dot{\lambda}(\theta_0) \,\big]\, ds \\
&= \int_0^t n^{-1/2} \sum_{i=1}^{n} \big[\, \dot{g}(X_i, \theta_0 + sn^{-1/2}) - \dot{g}(X_i, \theta_0) \\
&\qquad - \lambda(\theta_0 + sn^{-1/2}) + \lambda(\theta_0) \,\big]\, ds \\
&\quad + \int_0^t \big[\, n^{1/2}[\lambda(\theta_0 + sn^{-1/2}) - \lambda(\theta_0) \\
&\qquad - sn^{-1/2}\dot{\lambda}(\theta_0) \,\big]\, ds \\
&= D_{n1}(t) + D_{n2}(t), \qquad \text{say.}
\end{aligned}
$$

Hence,

$$
\begin{aligned}
&E\Big(\sup_{0\le t\le b} |D_{n1}(t)|^2\Big) \\
&\le E\Big(\int_0^b \Big| n^{-1/2} \sum_{i=1}^{n} \big[\, \dot{g}(X_i, \theta_0 + sn^{-1/2}) - \dot{g}(X_i, \theta_0) \\
&\qquad - \lambda(\theta_0 + sn^{-1/2}) + \lambda(\theta_0) \,\big]\Big|\, ds\Big)^2 \\
&\le b \int_0^b E\Big(n^{-1/2} \sum_{i=1}^{n} \big[\, \dot{g}(X_i, \theta_0 + sn^{-1/2}) - \dot{g}(X_i, \theta_0) \\
&\qquad - \lambda(\theta_0 + sn^{-1/2}) + \lambda(\theta_0) \,\big]\Big)^2 ds \\
&= b \int_0^b V\Big(\dot{g}(X, \theta_0 + sn^{-1/2}) - \dot{g}(X_i, \theta_0) \\
&\qquad - \lambda(\theta_0 + sn^{-1/2}) + \lambda(\theta_0)\Big)\, ds \\
&\le b^2 (bn^{-1/2})^{-1} \int_0^{bn^{-1/2}} E\big[\dot{g}(X, \theta_0 + u) - \dot{g}(X, \theta_0)\big]^2 du
\end{aligned}
$$

$$\rightarrow \quad 0, \qquad \text{by } \mathbf{c}.$$

Similarly,

$$\begin{aligned}\sup_{0 \le t \le b} |D_{n2}(t)| &\le b^2 (bn^{-1/2})^{-1} \int_0^{bn^{-1/2}} |\dot\lambda(\theta_o + s) - \dot\lambda(\theta_0)|\, ds \\ &= o(1), \qquad \text{by } \mathbf{b}.\end{aligned}$$

This completes the proof of (5.4.15) for $t \ge 0$. The details are exactly similar for the case $t \le 0$. Hence (5.4.15) is proved, which in turn together with the Markov inequality implies that for every $0 < b < \infty$, $\sup_{|t| \le b} |D_n(t)| = o_p(1)$.

EXAMPLE 2. Now consider a more general setup where the parameter $\boldsymbol{\theta}$ is p-dimensional real vector, the observations X_{ni}, $1 \le i \le n$, are independent r.v.'s, each having possibly a different distribution depending on the true parameter value $\boldsymbol{\theta}_0$ under a sequence of probability measures $P_n := \prod_{i=1}^n F_{ni}$. Here, F_{ni} denotes the d.f. of X_{ni}, $1 \le i \le n$, that will typically depend on $\boldsymbol{\theta}_0$. Also, let $\mathbf{c}_{ni}$, $1 \le i \le n$, be arrays of p-vectors in $\mathbb{R}^p$, $\Theta \subseteq \mathbb{R}^p$, and

$$M_n(\boldsymbol{\theta}) := \sum_{i=1}^n g(X_{ni}, \mathbf{c}_{ni}'\boldsymbol{\theta}), \qquad \boldsymbol{\theta} \in \Theta.$$

Let $\mathbf{C}$ denote the $n \times p$ matrix whose i^{th} row is $\mathbf{c}_{ni}'$, $1 \le i \le n$, and $\lambda_{ni}(t) := E_n g(X_{ni}, t)$, $t \in \mathbb{R}$, where E_n is the expectation under P_n. About the constants $\mathbf{c}_{ni}$, assume the following:

d. The vectors $\mathbf{c}_{ni}$ are such that $\max_{1 \le i \le n} \|\mathbf{c}_{ni}\| = o(1)$, and $\mathbf{C}'\mathbf{C} = \mathbf{I}_{p \times p}$.

e. The function g is as in **a** satisfying

$$\sum_{i=1}^n \|\mathbf{c}_{ni}\|^2 \int g^2(x, \mathbf{c}_{ni}'\boldsymbol{\theta}_0)\, dF_{ni}(x) = O(1), \tag{5.4.16}$$

$$\lim_{s \to 0} \limsup_{n \to \infty} \max_{1 \le i \le n} \int \Big\{ \dot g(x, \mathbf{c}_{ni}'\boldsymbol{\theta}_0 + s) - \dot g(x, \mathbf{c}_{ni}'\boldsymbol{\theta}_0) \Big\}^2 dF_{ni}(x) = 0. \tag{5.4.17}$$

f. The functions $\{\lambda_{ni}\}$ are absolutely continuous with their a.e.

derivatives $\{\dot{\lambda}_{ni}\}$ satisfy the following: For any sequence a_{ni} of positive numbers with $\max_{1\le i\le n} a_{ni} = o(1)$,

$$(5.4.18)\quad \max_{1\le i\le n} a_{ni}^{-1}\int_0^{a_{ni}} |\dot{\lambda}_{ni}(\mathbf{c}_{ni}'\boldsymbol{\theta}_0 + s) - \dot{\lambda}_{ni}(\mathbf{c}_{ni}'\boldsymbol{\theta}_0)|\, ds = o(1).$$

Using the arguments of the type used in Example 1 above, one can prove that under **d** - **f** the following holds: For every $0 < b < \infty$,

$$(5.4.19)\quad M_n(\boldsymbol{\theta}) = M_n(\boldsymbol{\theta}_0) - (\boldsymbol{\theta}-\boldsymbol{\theta}_0)'\sum_{i=1}^n \mathbf{c}_{ni}\dot{g}(X_{ni}, \mathbf{c}_{ni}'\boldsymbol{\theta}_0) - \frac{1}{2}(\boldsymbol{\theta}-\boldsymbol{\theta}_0)'\sum_{i=1}^n \mathbf{c}_{ni}\mathbf{c}_{ni}'\dot{\lambda}_{ni}(\mathbf{c}_{ni}'\boldsymbol{\theta}_0)(\boldsymbol{\theta}-\boldsymbol{\theta}_0) + u_p(1), \qquad (P_n),$$

where $u_p(1)$ is a sequence of processes in θ that converges to zero uniformly over the set $\|\boldsymbol{\theta}-\boldsymbol{\theta}_0\| \le b$, in probability.

Consider as a special case the M-dispersion used by Huber (1981) to construct M-estimators in the linear regression model (1.1.1). Here one takes $\mathbf{c}_{ni} \equiv \mathbf{x}_{ni}(\mathbf{X}'\mathbf{X})^{-1/2}$, and

$$\mathcal{M}_n(\boldsymbol{\beta}) = \sum_{i=1}^n \rho(Y_{ni} - \mathbf{x}_{ni}'\boldsymbol{\beta}) = \sum_{i=1}^n \rho(e_{ni} - \mathbf{c}_{ni}(\mathbf{X}'\mathbf{X})^{1/2}(\boldsymbol{\beta}-\boldsymbol{\beta}_0)),$$

where ρ is measurable function from $\mathbb{R}$ to $\mathbb{R}$. Upon identifying $X_{ni} \equiv e_{ni}$, $\boldsymbol{\theta} = (\mathbf{X}'\mathbf{X})^{1/2}(\boldsymbol{\beta}-\boldsymbol{\beta}_0)$, $\boldsymbol{\theta}_0 = \mathbf{0}$, one readily obtains the following:

Suppose (1.1.1) holds with the errors e_{ni} i.i.d. F; ρ is absolutely continuous with a.e. derivative ψ such that

$$\int |\psi(x+y)|dF(x) < \infty, \forall\, y \in \mathbb{R}, \qquad \int \psi dF = 0, \int \psi^2 dF < \infty.$$

Let $\gamma_r(t) := \int[\psi(x-t) - \psi(x)]^r dF(x), t \in \mathbb{R}, r = 1, 2$. Note that upon identifying $g(y,t)$ of the EXAMPLE 1 with $\rho(y-t)$, one sees that $\dot{g}(y,t) \equiv -\psi(y-t)$ and that $\gamma_1 = \lambda(0) - \lambda(t)$. Suppose, additionally, that γ_1 is continuously differentiable at 0 and that γ_2 is continuous at 0. Then, under (**NX**), it follows that Huber's dispersion $\mathcal{M}_n$ is

ULANQ with

$$\begin{aligned}
\boldsymbol{\theta}_0 &= \mathbf{0}, \quad \boldsymbol{\delta}_n = (\mathbf{X}'\mathbf{X})^{1/2}, \quad \mathbf{S}_n(\boldsymbol{\beta}) = -\sum_{i=1}^{n} \mathbf{x}_i \psi(Y_i - \mathbf{x}_i'\boldsymbol{\beta}), \\
\mathbf{W}_n(\boldsymbol{\beta}) &= \dot{\lambda}(0)\mathbf{X}'\mathbf{X}, \quad \mathbf{W} = \dot{\lambda}(0)\mathbf{I}_{p\times p}, \quad \boldsymbol{\Sigma} = \int \psi^2 dF\, \mathbf{I}_{p\times p}.
\end{aligned}$$

If ρ is additionally assumed to be convex, then using a result in Rockafeller (1970), to prove the ULANQ, it suffices to prove that the score statistic $S_n(\boldsymbol{\beta})$ is asymptotically linear in $(\mathbf{X}'\mathbf{X})^{1/2}(\boldsymbol{\beta} - \boldsymbol{\beta}_0)$, for $\boldsymbol{\beta}$ in the set $\|(\mathbf{X}'\mathbf{X})^{1/2}(\boldsymbol{\beta} - \boldsymbol{\beta}_0)\| \le b$, in probability. See Heiler and Weiler (1988) and Pollard (1991) for details.

For $\rho(x) = |x|$ and F continuous, $\psi(x) = sgn(x)$ and $\gamma_r(t) = 2^r|F(t) - F(0)|$. The condition on γ_1 now translates to the usual condition on F in terms of the density f at 0. For $\rho(x) = x^2$, $\psi(x) \equiv 2x$, $\gamma_1(t) \equiv 2t$, so that γ_1 is trivially continuously differentiable with $\dot{\gamma}_1(0) = 2$. Note that in general $\mathbf{W} \neq \boldsymbol{\Sigma}$ unless $\dot{\lambda}(0) = \int \psi^2 dF$ which is the case when ψ is related to the likelihood scores. □

EXAMPLE 3. An example where the full strength of $(\tilde{A1})$ - $(\tilde{A3})$ is realized is obtained by considering the least square dispersion in an explosive AR(1) process where $X_0 = 0$ and for some $|\theta| > 1$,

$$X_i = \theta X_{i-1} + \varepsilon_i, \qquad i = 1, 2, \cdots,$$

where it is assumed that ε_i are mean zero finite variance i.i.d. r.v.'s. Let

$$M_n(t) := \frac{1}{2}\sum_{i=1}^{n} (X_i - tX_{i-1})^2, \qquad |t| > 1.$$

Then the least square estimator of θ is $\hat{\theta}_n := argmin_t M_n(t)$.

We now verify $(\tilde{A1}) - (\tilde{A3})$ for this dispersion. Let θ_0 denote the true value. Rewrite, for a $|\theta| > 1$,

$$\begin{aligned}
M_n(\theta) &= \frac{1}{2}\sum_{i=1}^{n} (\varepsilon_i - (\theta - \theta_0)X_{i-1})^2 \\
&= \frac{1}{2}\sum_{i=1}^{n} \varepsilon_i^2 - (\theta - \theta_0)\sum_{i=1}^{n} \varepsilon_i X_{i-1} + \frac{1}{2}(\theta - \theta_0)^2 \sum_{i=1}^{n} X_{i-1}^2
\end{aligned}$$

$$= \quad M_n(\theta_0) - (\theta - \theta_0)S_n + \frac{1}{2}(\theta - \theta_0)^2 W_n, \quad \text{say.}$$

So one readily sees that what S_n and W_n should be. Now the question is what should be δ_n, W and Y. First note that one can rewrite

$$X_j = \sum_{i=1}^{j} \theta_0^{j-i} \varepsilon_i, \qquad j \geq 1,$$

so that

$$\theta_0^{-j} X_j = \sum_{i=1}^{j} \theta_0^{-i} \varepsilon_i$$

is a mean zero square integrable martingale. Note also that, with $\sigma^2 = E\varepsilon_1^2$, and because $\theta_0^2 > 1$, as $j \to \infty$,

$$\begin{aligned} E\left(\frac{X_j}{\theta_0^j}\right)^2 &= \sigma^2 \sum_{i=1}^{j} \theta_0^{-2i} = \sigma^2 \frac{\theta_0^{-2} - \theta_0^{-2(j+1)}}{1 - \theta_0^{-2}} \\ &\to \frac{\sigma^2}{\theta_0^2 - 1} =: \tau^2, \qquad \text{say.} \end{aligned}$$

The above derivations also show that $\sup_j E(\frac{X_j}{\theta_0^j})^2 = \tau^2 < \infty$. Hence, by the martingale convergence theorem, there exists a r.v. Z with mean zero and variance τ^2 such that

$$\frac{X_j}{\theta_0^j} \to Z, \qquad \text{a.s. and in } L_2, \text{ as } j \to \infty.$$

This implies that $|X_j|$ explodes to infinity as $j \to \infty$. Also note that because $\sum_{i=1}^{j} \theta_0^{-i}\varepsilon_i$ is a sum of i.i.d. mean zero finite variance r.v.'s, by the CLT, Z is a $\mathcal{N}(0, (\theta_0^2 - 1)^{-1})$ r.v. Now let $Z_i := \theta_0^{-i} X_i$. Then

$$\theta_0^{-2n} W_n = \sum_{i=1}^{n} \theta_0^{-2i} Z_{n-i}^2 \to Z^2/(\theta_0^2 - 1), \quad a.s.$$

This suggests that the $\delta_n = \theta_0^n$ and $W = Z^2/(\theta_0^2 - 1)$. Now consider

$$\theta_0^{-n} S_n = \sum_{i=1}^{n} \theta_0^{-i} Z_{n-i}\, \varepsilon_{n-i+1}.$$

Because of the independence of ε_i from X_{i-1}, S_n can be verified to be a means zero square integrable martingale. By the martingale central limit theorem it follows that

$$(\theta_0^{-n} S_n,\, \theta_0^{-2n} W_n) \Longrightarrow (|Z|Z_1,\, Z),$$

where Z_1 is a $\mathcal{N}(0, (\theta_0^2 - 1)^{-1})$ r.v., independent of Z. So here $Y = |Z|Z_1$. □

The next section is devoted to verifying (A1) - (A5) for various dispersion introduce in Section 5.2.

5.5 Asymptotic Uniform Quadraticity

In this section we shall give sufficient conditions under which $K_{\mathbf{D}}, K_{\mathbf{D}}^+$ of Section 5.2 will satisfy (5.4.A1), (5.4.A4), (5.4.A5) and $K_{\mathbf{D}}^*$ and Q of Section 5.2 will satisfy (5.4.A1). As is seen from the previous section this will bring us a step closer to obtaining the asymptotic distributions of various m.d. estimators introduced in Section 5.2.

To begin with we shall focus on (5.4.A1) for $K_{\mathbf{D}}, K_{\mathbf{D}}^+$ and $K_{\mathbf{D}}^*$. Our basic objective is to study the asymptotic distribution of $\hat{\boldsymbol{\beta}}_{\mathbf{D}}$ when the actual d.f.'s of $\{e_{ni}, 1 \le i \le n\}$ are $\{F_{ni}, 1 \le i \le n\}$ but we model them to be $\{H_{ni}, 1 \le i \le n\}$. Similarly, we wish to study the asymptotic distribution of $\boldsymbol{\beta}_{\mathbf{D}}^+$ when actually the errors may not be symmetric but we model them to be so. To achieve these objectives it is necessary to obtain the asymptotic results under as general a setting as possible. This of course makes the exposition that follows look somewhat complicated. The results thus obtained will enable us to study not only the asymptotic distributions of these estimators at the true model but also some of their robustness properties. With this in mind we proceed to state our assumptions.

(*a*) $\mathbf{X}$ satisfies (**NX**).

(*b*) With $\mathbf{d}_{(j)}$ denoting the j-th column of $\mathbf{D}$, $\|d_{(j)}\|^2 > 0$ for at least one j; $\|\mathbf{d}_{(j)}\|^2 = 1$ for all those j for which $\|\mathbf{d}_{(j)}\|^2 > 0$, $1 \le j \le p$.

(*c*) $\{F_{ni}, 1 \le i \le n\}$ admit densities $\{f_{ni},\, 1 \le i \le n\}$ w.r.t. λ.

(*d*) $\{G_n\}$ is a sequence in $\mathcal{DI}(\mathbb{R})$.

(*e*) With $\mathbf{d}'_{ni} = (d_{ni1}, \cdots, d_{nip})$, the i-th row of $\mathbf{D}, 1 \leq i \leq n$,

$$\int \sum_{i=1}^{n} \|\mathbf{d}_{ni}\|^2 F_{ni}(1 - F_{ni}) dG_n = O(1).$$

(*f*) With $\gamma_n := \sum_{i=1}^{n} \|\mathbf{d}_{ni}\|^2 f_{ni}$,

$$\limsup_n \int_{a_n}^{b_n} \int \gamma_n(y + x) dG_n(y) dx = 0$$

for any real sequences $\{a_n\}, \{b_n\}, a_n < b_n, b_n - a_n \to 0$.

(*g*) With $d_{nij} = d_{nij}^{+} - d_{nij}^{-}$, $1 \leq j \leq p$; $\mathbf{c}_{ni} = \mathbf{A}\mathbf{x}_{ni}, \kappa_{ni} := \|\mathbf{c}_{ni}\|$, $1 \leq i \leq n$, $\forall\, \delta > 0$, $\forall\, \|\mathbf{v}\| \leq b$,

$$\limsup_n \sum_{j=1}^{p} \int \Big[\sum_{i=1}^{n} d_{nij}^{\pm}\{F_{ni}(y + \mathbf{v}'\mathbf{c}_{ni} + \delta\kappa_{ni}) - F_{ni}(y + \mathbf{v}'\mathbf{c}_{ni} - \delta\kappa_{ni})\}\Big]^2 dG_n(y) \leq k\delta^2,$$

where k is a constant not depending on $\mathbf{v}$ and δ.

(*h*). With $\mathbf{R}_{nj} := \sum_{i=1}^{n} d_{nij}\mathbf{x}_{ni} f_{ni}$, $\boldsymbol{\nu}_{nj} := \mathbf{A}\mathbf{R}_{nj}$, $1 \leq j \leq p$,

$$\sum_{j=1}^{p} \int \|\boldsymbol{\nu}_{nj}\|^2 dG_n = O(1).$$

(*i*) With $\mu_{nj}^0(y, \mathbf{u}) := \sum_{i=1}^{n} d_{nij} F_{ni}(y + \mathbf{c}'_{ni}\mathbf{u})$, for each $\mathbf{u} \in \mathbb{R}^p$,

$$\sum_{j=1}^{p} \int [\mu_{nj}^0(y, \mathbf{u}) - \mu_{nj}^0(y, \mathbf{0}) - \mathbf{u}'\boldsymbol{\nu}_{nj}(y)]^2 dG_n(y) = o(1).$$

(*j*) With $m_{nj} := \sum_{i=1}^{n} d_{nij}[F_{ni} - H_{ni}]$, $1 \leq j \leq p$; $\mathbf{m}'_{\mathbf{D}} = (m_{n1}, \cdots, m_{np})$,

$$\int \|\mathbf{m}_{\mathbf{D}}\|^2 dG_n = O(1).$$

(k) With $\mathbf{\Gamma}'_n(y) := (\boldsymbol{\nu}_{n1}(y), \cdots, \boldsymbol{\nu}_{np}(y)) = \mathbf{D}'\mathbf{\Lambda}^*(y)\mathbf{X}\mathbf{A}$, where $\mathbf{\Lambda}^*$ is defined at (4.2.1), and with $\bar{\mathbf{\Gamma}}_n := \int \mathbf{\Gamma}_n g_n dG_n$, where $g_n \in L_r(G_n)$, $r = 1, 2$, $n \geq 1$, is such that $g_n > 0$,

$$0 < \liminf_n \int g_n^2 dG_n \leq \limsup_n \int g_n^2 dG_n < \infty,$$

and such that there exists an $\alpha > 0$ satisfying

$$\liminf_n \inf \left\{ \mathbf{e}'\bar{\mathbf{\Gamma}}_n \mathbf{e};\ \mathbf{e} \in \mathbb{R}^p, \|\mathbf{e}\| = 1 \right\} \geq \alpha.$$

(l) Either

(1) $\mathbf{e}'\mathbf{d}_{ni}\mathbf{x}'_{ni}\mathbf{A}\mathbf{e} \geq 0, \quad \forall\ 1 \leq i \leq n$ and $\forall\ \mathbf{e} \in \mathbb{R}^p, \quad \|\mathbf{e}\| = 1.$

Or

(2) $\mathbf{e}'\mathbf{d}_{ni}\mathbf{x}'_{ni}\mathbf{A}\mathbf{e} \leq 0, \quad \forall\ 1 \leq i \leq n$ and $\forall\ \mathbf{e} \in \mathbb{R}^p, \quad \|\mathbf{e}\| = 1.$

In most of the subsequent applications of the results obtained in this section, the sequence of integrating measures $\{G_n\}$ will be a fixed G. However, we formulate the results of this section in terms of sequences $\{G_n\}$ to allow extra generality. Note that if $G_n \equiv G$, $G \in \mathcal{DI}(\mathbb{R})$, then there always exist a a $g \in L_r(G)$, $r = 1, 2$, such that $g > 0$, $0 < \int g^2 dG < \infty$.

Define, for $y \in \mathbb{R}, \mathbf{u} \in \mathbb{R}^p, 1 \leq j \leq p$,

$$\begin{aligned} S_j^0(y, \mathbf{u}) &:= V_{jd}(y, \mathbf{A}\mathbf{u}), \\ Y_j^0(y, \mathbf{u}) &:= S_j^0(y, \mathbf{u}) - \mu_j^0(y, \mathbf{u}). \end{aligned} \tag{5.5.1}$$

Note that for each j, S_j^0, μ_j^0, Y_j^0 are the same as in (2.3.2) applied to $X_{ni} = Y_{ni}$, $\mathbf{c}_{ni} = \mathbf{A}\mathbf{x}_{ni}$ and $d_{ni} = d_{nij}$, $1 \leq i \leq n$, $1 \leq j \leq p$.

Notation. For any functions g, $h : \mathbb{R}^{p+1} \to \mathbb{R}$,

$$|g_{\mathbf{u}} - h_{\mathbf{v}}|_n^2 := \int \{g(y, \mathbf{u}) - h(y, \mathbf{v})\}^2 dG_n(y).$$

Occasionally we shall write $|g|_n^2$ for $|g_{\mathbf{0}}|_n^2$.

Lemma 5.5.1. *Let $Y_{n1}, \cdots, Y_{nn}$ be independent r.v.'s with respective* d.f.'s $F_{n1}, \cdots, F_{nn}$. *Then (e) implies*

$$E \sum_{j=1}^{p} |Y_{j\mathbf{0}}^0|_n^2 = O(1). \tag{5.5.2}$$

Proof. By Fubini's Theorem,

$$E\sum_{j=1}^{p}|Y_{j\mathbf{o}}|_n^2 = \int\sum_{i=1}^{n}\|\mathbf{d}_i\|^2 F_i(1-F_i)dG_n$$

and hence (e) implies the Lemma. □

Lemma 5.5.2. *Let $\{Y_{ni}\}$ be as in Lemma 5.5.1. Then assumptions (a) - (d), (f) - (j) imply that, for every $0 < b < \infty$,*

$$E\sup_{\|\mathbf{u}\|\le b}\sum_{j=1}^{p}|Y_{j\mathbf{u}}^0 - Y_{j\mathbf{0}}^0|_n^2 = o(1). \tag{5.5.3}$$

Proof. Let $\mathcal{N}(b) := \{\mathbf{u} \in \mathbb{R}^p;\ \|\mathbf{u}\| \le b\}$. By Fubini's Theorem, $\forall\ \mathbf{u} \in \mathcal{N}(b)$,

$$\begin{aligned}
E&\sum_{j=1}^{p}|Y_{j\mathbf{u}}^0 - Y_{j\mathbf{0}}^0|_n^2\\
&\le \int\sum_{i=1}^{n}\|\mathbf{d}_i\|^2|F_i(y+\mathbf{c}_i'\mathbf{u}) - F_i(y)|dG_n\\
&\le \int_{-b_n}^{b_n}\Big(\int\gamma_n(y+x)dG_n(y)\Big)dx
\end{aligned}$$

where $b_n = b\max_i\kappa_i$, γ_n as in (f). Therefore, by the assumption (f),

$$E\sum_{j=1}^{p}\left|Y_{j\mathbf{u}}^0 - Y_{j\mathbf{0}}^0\right|_n^2 = o(1), \quad \forall\ \mathbf{u} \in \mathbb{R}^p. \tag{5.5.4}$$

To complete the proof of (5.5.3), because of the compactness of $\mathcal{N}(b)$, it suffices to show that $\forall\ \epsilon > 0$, $\exists$ a $\delta > 0$ such that $\forall\ \mathbf{v} \in \mathcal{N}(b)$,

$$\limsup_n E\sup_{\|\mathbf{u}-\mathbf{v}\|\le\delta}\sum_{j=1}^{p}|L_{j\mathbf{u}} - L_{j\mathbf{v}}| \le \epsilon, \tag{5.5.5}$$

where

$$L_{j\mathbf{u}} := |Y_{j\mathbf{u}}^0 - Y_{j\mathbf{0}}^0|_n^2, \qquad \mathbf{u} \in \mathbb{R}^p,\ \ 1 \le j \le p.$$

Expand the quadratic, apply the C-S inequality to the cross product terms to obtain, for all $1 \le j \le p$,

$$|L_{j\mathbf{u}} - L_{j\mathbf{v}}| \le |Y_{j\mathbf{u}}^0 - Y_{j\mathbf{v}}^0|_n^2 + 2|Y_{j\mathbf{u}}^0 - Y_{j\mathbf{v}}^0|_n\,|Y_{j\mathbf{v}}^0 - Y_{j\mathbf{0}}^0|_n.$$

Moreover, for all $1 \le j \le p$,

$$
\begin{aligned}
(5.5.6) \quad |Y^0_{j\mathbf{u}} - Y^0_{j\mathbf{v}}|^2_n &\le 2\{|S^0_{j\mathbf{u}} - S^0_{j\mathbf{v}}|^2_n + |\mu^0_{j\mathbf{u}} - \mu^0_{j\mathbf{v}}|^2_n\}, \\
|S^0_{j\mathbf{u}} - S^0_{j\mathbf{v}}|^2_n &\le 2\{|S^+_{j\mathbf{u}} - S^+_{j\mathbf{v}}|^2_n + |S^-_{j\mathbf{u}} - S^-_{j\mathbf{v}}|^2_n\}, \\
|\mu^0_{j\mathbf{u}} - \mu^0_{j\mathbf{v}}|^2_n &\le 2\{|\mu^+_{j\mathbf{u}} - \mu^+_{j\mathbf{v}}|^2_n + |\mu^-_{j\mathbf{u}} - \mu^-_{j\mathbf{v}}|^2_n\},
\end{aligned}
$$

where $S^\pm_j$, $\mu^\pm_j$ are the S^0_j, μ^0_j with d_{ij} replaced by $d^\pm_{ij} := \max(0, d_{ij})$, $d^-_{ij} := d^+_{ij} - d_{ij}$, $1 \le i \le n$, $1 \le j \le p$.

Now, $\|\mathbf{u} - \mathbf{v}\| \le \delta$, nonnegativity of $\{d^\pm_{ij}\}$, and the monotonicity of $\{F_i\}$ yields (use (2.3.17) here), that for all $1 \le j \le p$,

$$
\begin{aligned}
&|\mu^\pm_{j\mathbf{u}} - \mu^\pm_{j\mathbf{v}}|^2_n \\
&\le \int \left[\sum_{i=1}^n d^\pm_{ij}\left\{F_i(y + \mathbf{c}_i'\mathbf{v} + \delta\kappa_i) - F_i(y + \mathbf{c}_i'\mathbf{v} - \delta\kappa_i)\right\}\right]^2 dG_n(y).
\end{aligned}
$$

Therefore, by assumption (g),

$$
(5.5.7) \qquad \limsup_n \sup_{\|\mathbf{u}-\mathbf{v}\| \le \delta} \sum_{j=1}^p |\mu^0_{j\mathbf{u}} - \mu^0_{j\mathbf{v}}|^2_n \le 4k\delta^2.
$$

By the monotonicity of $S^\pm_j$ and (2.3.17), $\|\mathbf{u}-\mathbf{v}\| \le \delta$ implies that for all $1 \le j \le p, y \in \mathbb{R}$,

$$
\begin{aligned}
-\sum_{i=1}^n & d^\pm_{ij} I(-\delta\kappa_i < Y_i - \mathbf{v}'c_i - y \le 0) \\
&\le S^\pm_j(y, \mathbf{u}) - S^\pm_j(y, \mathbf{v}) \\
&\qquad \le \sum_{i=1}^n d^\pm_{ij} I(0 < Y_i - \mathbf{v}'\mathbf{c}_i - y \le \delta\kappa_i).
\end{aligned}
$$

This in turn implies (using the fact that $a \le b \le c$ implies $b^2 \le a^2 + c^2$ for any reals a, b, c)

$$
\begin{aligned}
&\left\{S^\pm_j(y, \mathbf{u}) - S^\pm_j(y, \mathbf{v})\right\}^2 \\
&\le \left\{\sum_{i=1}^n d^\pm_{ij} I(0 < Y_i - y - \mathbf{v}'\mathbf{c}_i \le \delta\kappa_i)\right\}^2 \\
&\qquad + \left\{\sum_{i=1}^n d^\pm_{ij} I(-\delta\kappa_i < Y_i - y - \mathbf{v}'\mathbf{c}_i \le 0)\right\}^2
\end{aligned}
$$

$$\le \; 2\left\{\sum_{i=1}^{n} d_{ij}^{\pm} I\left(-\delta\kappa_i < Y_i - y - \mathbf{v}'\mathbf{c}_i \le \delta\kappa_i\right)\right\}^2,$$

for all $1 \le j \le p$ and all $y \in \mathbb{R}$. Now use the fact that for a, b real, $(a+b)^2 \le 2a^2 + 2b^2$ to conclude that, for all $1 \le j \le p$,

$$\begin{aligned}
&|S_{j\mathbf{u}}^{\pm} - S_{j\mathbf{v}}^{\pm}|_n^2 \\
&\le \; 4\int \left\{\sum_{i=1}^{n} d_{ij}^{\pm}\left[I(-\delta\kappa_i < Y_i - y - \mathbf{v}'c_i \le \delta\kappa_i) - p_i(y, \mathbf{v}, \delta)\right]\right\}^2 \\
&\qquad\qquad \times dG_n(y) \\
&\qquad + 4\int \left\{\sum_{i=1}^{n} d_{ij}^{\pm} p_i(y, \mathbf{v}, \delta)\right\}^2 dG_n(y) \\
&= \; 4\{I_j + II_j\}, \qquad \text{(say)},
\end{aligned}$$

where $p_i(y, \mathbf{v}, \delta) \equiv F_i(y + \mathbf{v}'c_i\delta\kappa_i) - F_i(y + \mathbf{v}'c_i - \delta\kappa_i)$.

But $(d_{ij}^{\pm})^2 \le d_{ij}^2$ for all i and j implies that

$$\begin{aligned}
E\sum_{j=1}^{p} I_j &= \sum_{j=1}^{p}\int\sum_{i=1}^{n}(d_{ij}^{\pm})^2 p_i(y, \mathbf{v}, \delta)(1 - p_i(y, \mathbf{v}, \delta))dG_n(y) \\
&\le \int\sum_{i=1}^{n}\|\mathbf{d}_i\|^2 p_i(y, \mathbf{v}, \delta)dG_n(y) \\
&\le \int_{a_n}^{b_n}\left(\int \gamma_n(y+s)dG_n(y)\right)ds,
\end{aligned}$$

by (c) and Fubini, where $a_n = (-b-\delta)\max_i \kappa_i$, $b_n = (b+\delta)\max_i \kappa_i$ and where γ_n is defined in (f). Therefore, by the assumption (f),

$$E\sum_{j=1}^{p} I_j = o(1).$$

From the definition of II_j above and the assumption (g),

$$\limsup_n \sum_{j=1}^{p} II_j \le k\delta^2.$$

From the above bounds, we readily obtain

$$\limsup_n E \sup_{\|\mathbf{u}-\mathbf{v}\|\le\delta} \sum_{j=1}^{p} |Y_{j\mathbf{u}}^0 - Y_{j\mathbf{v}}^0|_n^2 \le 40k\delta^2. \tag{5.5.8}$$

Thus if we choose $0 < \delta \le (\epsilon/40k)^{1/2}$, then (5.5.5) will follow from (5.5.8), (5.5.7), (5.5.6) and (5.5.4). This also completes the proof of (5.5.3). □

To state the next theorem we need to introduce

$$\begin{aligned}&\hat{K}_{\mathbf{D}}(\mathbf{t})\\ &:= \sum_{j=1}^{p} \int \Big\{Y_j^0(y,\mathbf{0}) + \mathbf{t}'\mathbf{R}_j(y) + m_j(y)\Big\}^2 \, dG_n(y).\end{aligned}$$

In (5.5.9) below, the G in $K_{\mathbf{D}}$ is assumed to have been replaced by the sequence G_n, just for extra generality.

Theorem 5.5.1 *Let $Y_{n1}, \cdots, Y_{nn}$ be independent r.v.'s with respective d.f.'s $F_{n1}, \cdots, F_{nn}$. Suppose that $\{\mathbf{X}, F_{ni}, H_{ni}, \mathbf{D}, G_n\}$ satisfy (a) - (j). Then, for every $0 < b < \infty$,*

$$E \sup_{\|\mathbf{u}\| \le b} |K_{\mathbf{D}}(\mathbf{Au}) - \hat{K}_{\mathbf{D}}(\mathbf{Au})| = o(1). \tag{5.5.9}$$

Proof. Write $K, \hat{K}$ etc. for $K_{\mathbf{D}}, \hat{K}_{\mathbf{D}}$ etc. Note that

$$\begin{aligned}&K(\mathbf{Au})\\ &= \sum_{j=1}^{p} \int \Big[S_j^0(y,\mathbf{u}) - \mu_j^0(y) + m_j(y)\Big]^2 dG_n(y)\\ &= \sum_{j=1}^{p} \int \Big[Y_j^0(y,\mathbf{u}) - Y_j^0(y) + Y_j^0(y) + \mathbf{u}'\boldsymbol{\nu}_j(y) + m_j(y)\\ &\qquad + \mu_j^0(y,\mathbf{u}) - \mu_j^0(y) - \mathbf{u}'\boldsymbol{\nu}_j(y)\Big]^2 dG_n(y)\end{aligned}$$

where $Y_j^0(y) \equiv Y_j^0(y,\mathbf{0})$, $\mu_j^0(y) \equiv \mu_j^0(y,\mathbf{0})$. Expand the quadratic and use the C-S inequality on the cross product terms to obtain

$$\begin{aligned}&|K(\mathbf{Au}) - \hat{K}(\mathbf{Au})|\\ &\le \sum_{j=1}^{p} \Big\{|Y_{j\mathbf{u}}^0 - Y_{j\mathbf{v}}^0|_n^2 + |\mu_{j\mathbf{u}}^0 - \mu_j^0 - \mathbf{u}'\boldsymbol{\nu}_j|_n^2\\ &\qquad + 2\,|Y_{j\mathbf{u}}^0 - Y_{j\mathbf{v}}^0|_n \Big[|Y_j^0 + \mathbf{u}'\boldsymbol{\nu}_j + m_j|_n\\ &\qquad\qquad + |\mu_{j\mathbf{u}}^0 - \mu_j^0 - \mathbf{u}'\boldsymbol{\nu}_j|_n\Big]\\ &\qquad + 2\,|Y_j^0 + \mathbf{u}'\boldsymbol{\nu}_j + m_j|_n \cdot |\mu_{j\mathbf{u}}^0 - \mu_j^0 - \mathbf{u}'\boldsymbol{\nu}_j|_n\Big\}.\end{aligned} \tag{5.5.10}$$

In view of Lemmas 5.5.1, 5.5.2 and assumptions (h) and (j), (5.5.9) will follow from the above inequality, if we prove

$$(5.5.11) \qquad \sup_{\|\mathbf{u}\|\le \mathbf{B}} \sum_{j=1}^{p} |\mu_{j\mathbf{u}}^0 - \mu_j^0 - \mathbf{u}'\boldsymbol{\nu}_j|_n^2 = o(1).$$

Let $\xi_{j\mathbf{u}} := |\mu_{j\mathbf{u}}^0 - \mu_j^0 - \mathbf{u}'\boldsymbol{\nu}_j|_n^2$, $1 \le j \le p$, $\mathbf{u} \in \mathbb{R}^p$. In view of the compactness of $\mathcal{N}(b)$ and the assumption (i), it suffices to prove that $\forall\ \epsilon > 0, \exists\ a\ \delta > 0 \ni \forall\ \mathbf{v} \in \mathcal{N}(b)$,

$$(5.5.12) \qquad \limsup_n \sup_{\|\mathbf{u}-\mathbf{v}\|\le\delta} \sup_{j=1}^{P} \xi_{j\mathbf{u}} - \xi_{j\mathbf{v}}| \le \epsilon.$$

But

$$\begin{aligned} |\xi_{j\mathbf{u}} - \xi_{j\mathbf{v}}| \;\le\; & 2\Big\{|\mu_{j\mathbf{u}}^0 - \mu_{j\mathbf{v}}^0|_n^2 + \|\mathbf{u}-\mathbf{v}\|^2\,\|\boldsymbol{\nu}_j\|_n^2 \\ & + \xi_{jv}^{1/2}\Big[|\mu_{j\mathbf{u}}^0 - \mu_{j\mathbf{v}}^0|_n + \|\mathbf{u}-\mathbf{v}\|\,\|\boldsymbol{\nu}_j\|_n\Big] \\ & + |\mu_{j\mathbf{u}}^0 - \mu_{j\mathbf{v}}^0|_n\|\mathbf{u}-\mathbf{v}\|\,\|\boldsymbol{\nu}_j\|_n\Big\}. \end{aligned}$$

Hence, from (5.5.7) and the assumption (i), the left hand side of (5.5.12) is bounded above by

$$2\left\{4k\delta^2 + \delta^2(a + 2k^{1/2}a^{1/2})\right\} = k_1\delta^2$$

where $a = \limsup_n \sum_{j=1}^{p} \|\boldsymbol{\nu}_j\|_n^2$. Thus upon choosing $\delta^2 \le \epsilon/k_1$ one obtains (5.5.12), hence (5.5.11) and therefore the Theorem. □

Our next goal is to obtain an analogue of (5.5.9) for $K_{\mathbf{D}}^+$. Before stating it rigorously, it helps to rewrite $K_{\mathbf{D}}^+$ in terms of standardized processes $\{Y_j^0\}$ and $\{\mu_j^0\}$ defined at (5.5.1). In fact, we have

$$\begin{aligned} K_{\mathbf{D}}^+(\mathbf{Au}) &= \sum_{j=1}^{p} \int \Big[S_j^0(y,\mathbf{u}) - \sum_{i=1}^{n} d_{ij} + S_j^0(-y,\mathbf{u})\Big]^2 dG_n(y) \\ &= \sum_{j=1}^{p} \int \Big[Y_j^0(y,\mathbf{u}) - Y_j^0(y) + Y_j^0(-y,\mathbf{u}) - Y_j^0(-y) \\ &\qquad + \mu_j^0(y,\mathbf{u}) - \mu_j^0(y) - \mathbf{u}'\boldsymbol{\nu}_j(y) \\ &\qquad + \mu_j^0(-y,\mathbf{u}) - \mu_j^0(-y) - \mathbf{u}'\boldsymbol{\nu}_j(y) \\ &\qquad + \mathbf{u}'\boldsymbol{\nu}_j^+(y) + W_j^+(y) + m_j^+(y)\Big]^2 dG_n(y), \end{aligned}$$

where

$$\begin{aligned}
W_j^+(y) &:= Y_j^0(y) + Y_j^0(-y), \qquad \boldsymbol{\nu}_j^+(y) = \boldsymbol{\nu}_j(y) + \boldsymbol{\nu}_j(-y), \\
m_j^+(y) &:= \sum_{i=1}^n d_{ij}\{F_i(y) - 1 + F_i(-y)\} \\
&= \mu_j^0(y) + \mu_j^0(-y) - \sum_{i=1}^n d_{ij}, \qquad y \in \mathbb{R}, 1 \le j \le p.
\end{aligned}$$

Let

$$\hat{K}_{\mathbf{D}}^+(\mathbf{Au}) = \sum_{j=1}^p \int [W_j^+ + m_j^+ + \mathbf{u}'\boldsymbol{\nu}_j^+]^2 dG_n, \quad \mathbf{u} \in \mathbb{R}^p. \tag{5.5.13}$$

Now proceeding as in (5.5.10), one obtains a similar upper bound for $|K_{\mathbf{D}}^+(\mathbf{Au}) - \hat{K}_{\mathbf{D}}^+(\mathbf{Au})|$ involving terms like those in R.H.S. of (5.5.10) and the terms like $|Y_{j\mathbf{u}}^0 - Y_{j\mathbf{0}}^0|_{-n}$, $|\mu_{j\mathbf{u}}^0 - \mu_j^0 - \mathbf{u}'\boldsymbol{\nu}_j|_{-n}$, $\|\boldsymbol{\nu}_j\|_{-n}$, $|Y_{j\mathbf{0}}|_{-n}$, where for any function $h : \mathbb{R}^{p+1} \to \mathbb{R}$,

$$|h_{\mathbf{u}}|_{-n}^2 := \int h^2(-y, \mathbf{u}) dG_n(y).$$

It thus becomes apparent that one needs an analogue of Lemmas 5.5.1 and 5.5.2 with $G_n(\cdot)$ replaced by $G_n(-\cdot)$. That is, if the conditions (*e*) - (*j*) are also assumed to hold for measures $\{G_n(-\cdot)\}$ then obviously analogues of these lemmas will hold. Alternatively, the statement of the following theorem and the details of its proof are considerably simplified if one assumes G_n to be symmetric around zero, as we shall do for convenience. Before stating the theorem, we state

Lemma 5.5.3. *Let* $Y_{n1}, \cdots, Y_{nn}$ *be independent r.v.'s with respective d.f.'s* $F_{n1}, \cdots, F_{nn}$. *Assume* (*a*) - (*d*), (*f*), (*g*) *hold,* $\{G_n\}$ *satisfies* (5.3.8) *and that* (5.5.14) *hold, where*

$$\int \sum_{i=1}^n \|\mathbf{d}_{ni}\|^2 \{F_{ni}(-y) + 1 - F_{ni}(y)\} dG_n(y) = O(1). \tag{5.5.14}$$

Then,

$$E \sum_{j=1}^p |Y_{j\mathbf{0}}^0|_{-n}^2 = O(1), \tag{5.5.15}$$

and $\forall\ 0 < b < \infty$,

$$(5.5.16)\qquad E \sup_{\|\mathbf{u}\| \le b} \sum_{j=1}^{p} |Y_{j\mathbf{u}}^0 - Y_{j\mathbf{0}}^0|_{-n}^2 = o(1). \qquad \square$$

This lemma follows from Lemmas 5.5.1 and 5.5.2, since under (5.3.8), LHS's of (5.5.15) and (5.5.16) are equal to those of (5.5.2) and (5.5.3), respectively. The proof of the following theorem is similar to that of Theorem 5.5.1.

Theorem 5.5.2 *Let* $Y_{n1}, \cdots, Y_{nn}$ *be independent r.v.'s with respective d.f.'s* $F_{n1}, \cdots, F_{nn}$. *Suppose that* $\{\mathbf{X}, F_{ni}, \mathbf{D}, G_n\}$ *satisfy* (*a*) - (*d*), (*f*) - (*i*), (5.3.8) *for all* $n \ge 1$, (5.5.14) *and that*

$$(5.5.17)\qquad \sum_{j=1}^{p} \int \{m_j^+(y)\}^2 dG_n(y) = O(1),$$

Then, $\forall\ 0 < b < \infty$,

$$(5.5.18)\qquad E \sup_{\|\mathbf{u}\} \le b} |K_{\mathbf{D}}^+(\mathbf{Au}) - \hat{K}_{\mathbf{D}}^+(\mathbf{Au})| = o(1).$$

Remark 5.5.1. Recall that we are interested in the asymptotic distribution of $\mathbf{A}^{-1}(\hat{\boldsymbol{\beta}}_{\mathbf{D}} - \boldsymbol{\beta})$ which is a minimizer of $K_{\mathbf{D}}(\boldsymbol{\beta} + \mathbf{Au})$ w.r.t. $\mathbf{u}$. Since $\hat{\boldsymbol{\beta}}_{\mathbf{D}}$ satisfies (5.3.22), there is no loss of generality in taking the true $\boldsymbol{\beta}$ equal to $\mathbf{0}$. Then (5.5.9) asserts that $(1/2)K_{\mathbf{D}}$ satisfies (5.4.A1) with

$$(5.5.19)\qquad \begin{aligned} \boldsymbol{\theta}_0 &= \mathbf{0},\ \boldsymbol{\delta}_n = \mathbf{A}^{-1}, \\ \mathbf{S}_n &= \mathbf{A}^{-1}\mathcal{J}_n, \mathbf{W}_n = \mathbf{A}\mathcal{B}_n\mathbf{A}, \\ \mathcal{J}_n &:= -\int \boldsymbol{\Gamma}_n(y)\{\mathbf{Y}_{\mathbf{D}}^0(y) + \mathbf{m}_{\mathbf{D}}(y)\} dG_n(y), \\ \mathcal{B}_n &:= \int \boldsymbol{\Gamma}_n(y)\boldsymbol{\Gamma}_n'(y) dG_n(y), \end{aligned}$$

where $\boldsymbol{\Gamma}_n(y) = \mathbf{AX}'\boldsymbol{\Lambda}^*(y)\mathbf{D}, \boldsymbol{\Lambda}^*$ as in (4.2.1), $\mathbf{Y}_{\mathbf{D}}^{0'} := (Y_1^0, \cdots, Y_p^0)$ and $\mathbf{m}_{\mathbf{D}}' := (m_1, \cdots, m_p)$.

In view of Lemma 5.5.1, the assumptions (*e*) and (*h*) of this section imply that $EK_{\mathbf{D}}(\mathbf{0}) = O(1)$, thereby ensuring the validity of (5.4.A4).

Similarly, (5.5.18) asserts that $(1/2)K_{\mathbf{D}}^{+}$ satisfies (5.4.A1) with

$$
\begin{aligned}
(5.5.20)\qquad \boldsymbol{\theta}_0 &= \mathbf{0},\ \boldsymbol{\delta}_n = \mathbf{A}^{-1},\\
\mathbf{S}_n &= \mathbf{A}^{-1}\mathcal{J}_n^{+},\quad \mathbf{W}_n = \mathbf{A}\mathcal{B}_n^{+}\mathbf{A},\\
\mathcal{J}_n^{+} &:= -\int \boldsymbol{\Gamma}_n^{+}(y)\{\mathbf{W}_{\mathbf{D}}^{0}(y) + \mathbf{m}_{\mathbf{D}}(y)\}dG_n(y),\\
\mathcal{B}_n^{+} &:= \int \boldsymbol{\Gamma}_n(y)\boldsymbol{\Gamma}_n^{+'}(y)dG_n(Y),
\end{aligned}
$$

where $\boldsymbol{\Gamma}_n^{+}(y) := \mathbf{A}\mathbf{X}'\boldsymbol{\Lambda}^{+}(y)\mathbf{D}, \boldsymbol{\Lambda}^{+}(y) := \boldsymbol{\Lambda}^{*}(y) + \boldsymbol{\Lambda}^{*}(-y), y \in \mathbb{R}^p$, $\mathbf{W}_{\mathbf{D}}^{+'} := (W_1^{+}, \cdots, W_p^{+})$ and $\mathbf{m}_{\mathbf{D}}^{+'} := (m_1^{+}, \cdots, m_p^{+})$.

In view of (5.5.12), (5.5.14) and (5.3.8) it follows that (5.4.A4) is satisfied by $K_{\mathbf{D}}^{+}(\mathbf{0})$.

Theorem 5.4.1 enables one to study the asymptotic distribution of $\hat{\boldsymbol{\beta}}_{\mathbf{D}}$ when in (1.1.1) the actual error d.f. F_{ni} is not necessarily equal to the modeled d.f. $H_{ni}, 1 \leq i \leq n$. Theorem 5.4.2 enables one to study the asymptotic distribution of $\boldsymbol{\beta}_{\mathbf{D}}^{+}$ when in (1.1.1) the error d.f. F_{ni} is not necessarily symmetric around $\mathbf{0}$, but we model it to be so, $1 \leq i \leq n$. □

So far we have not used the assumptions (k) and (l). They will be now used to obtain (5.4.A5) for $K_{\mathbf{D}}$ and $K_{\mathbf{D}}^{+}$.

Lemma 5.5.4. *In addition to the assumptions of Theorem* 5.5.1 *assume that* (k) *and* (l) *hold. Then,* $\forall \epsilon > 0, 0 < z < \infty, \exists N$ *(depending only on* ϵ*) and a* $b \in (0, \infty)$ *(depending on* ϵ, z*)* such that,

$$(5.5.21)\qquad P(\inf_{\|\mathbf{u}\|>b} K_{\mathbf{D}}(\mathbf{A}\mathbf{u}) \geq z) \ \geq\ 1 - \epsilon,\quad \forall n \geq N,$$

$$(5.5.22)\qquad P(\inf_{\|\mathbf{u}\|>b} \hat{K}_{\mathbf{D}}(\mathbf{A}\mathbf{u}) \geq z) \ \geq\ 1 - \epsilon,\quad \forall n \geq N,$$

Proof. As before write K, $\hat{K}$ etc. for $K_{\mathbf{D}}, \hat{K}_{\mathbf{D}}$ etc. Recall the definition of $\overline{\boldsymbol{\Gamma}}_n$ from (k). Let $k_n(\boldsymbol{e}) := \boldsymbol{e}'\overline{\boldsymbol{\Gamma}}_n\boldsymbol{e}$, $\boldsymbol{e} \in \mathbb{R}^p$. By the C-S inequality and (k),

$$(5.5.23)\qquad \sup_{\|\boldsymbol{e}\|=1} |k_n(\boldsymbol{e})|^2 \leq \|\overline{\boldsymbol{\Gamma}}_n\|^2 \leq \sum_{j=1}^{p} \|\boldsymbol{\nu}_j\|_n^2 \cdot |g_n|_n^2 = O(1).$$

Fix an $\epsilon > 0$ and a $z \in (0, \infty)$. Define, for $\mathbf{t} \in \mathbb{R}^p, 1 \le j \le p$,

$$\begin{aligned} \hat{V}_j(\mathbf{t}) &:= \int \{Y_j^0 + \mathbf{t}'\mathbf{R}_j + m_j\} g_n dG_n, \\ V_j(\mathbf{t}) &:= \int [V_{jd}(y, \mathbf{t}) - \sum_{i=1}^n d_{nij} H_{ni}(y)]\ g_n(y) dG_n(y). \end{aligned}$$

Also, let

$$\hat{\mathbf{V}}' := (\hat{V}_1, \cdots, \hat{V}_p),\ \mathbf{V}' := (V_1, \cdots, V_p),\ \gamma_n := |g_n|_n^2,\ \gamma := \limsup_n \gamma_n.$$

Write a $\mathbf{u} \in \mathbb{R}^p$ with $\|\mathbf{u}\| > b$ as $\mathbf{u} = r\mathbf{e}$, $|r| > b$, $\|\mathbf{e}\| = 1$. Then, by the C-S inequality,

$$\begin{aligned} \inf_{\|\mathbf{u}\|>b} K(\mathbf{A}\mathbf{u}) &\ge \inf_{|r|>b, \|\mathbf{e}\|=1} (\mathbf{e}'\mathbf{V}(r\mathbf{A}\mathbf{e}))^2/\gamma_n, \\ \inf_{\|\mathbf{u}\|>b} \hat{K}(\mathbf{A}\mathbf{u}) &\ge \inf_{|r|>b, \|\mathbf{e}\|=1} (\mathbf{e}'\hat{\mathbf{V}}(r\mathbf{A}\mathbf{e}))^2/\gamma_n, \end{aligned}$$

It thus suffices to show that $\exists$ a $b \in (0, \infty)$ and $N \ni$, $\forall\ n \ge N$,

$$(5.5.24) \quad P\Big(\inf_{|r|>b, \|\mathbf{e}\|=1} (\mathbf{e}'\mathbf{V}(r\mathbf{A}\mathbf{e}))^2/\gamma_n \ge z\Big) \ \ge\ 1 - \epsilon,$$

$$(5.5.25) \quad P\Big(\inf_{|r|>b, \|\mathbf{e}\|=1} (\mathbf{e}'\hat{\mathbf{V}}(r\mathbf{A}\mathbf{e}))^2/\gamma_n \ge z\Big) \ \ge\ 1 - \epsilon.$$

But $\forall\ \mathbf{u} \in \mathbb{R}^p$,

$$\begin{aligned} &\|\mathbf{V}(\mathbf{A}\mathbf{u}) - \hat{\mathbf{V}}(\mathbf{A}\mathbf{u})\| \\ &\le\ 2\gamma_n \sum_{j=1}^p \{|Y_{j\mathbf{u}}^0 - Y_{j0}^0|_n^2 + |\mu_{j\mathbf{u}}^0 - \mu_j^0 - \mathbf{u}'\boldsymbol{\nu}_j|_n^2\}. \end{aligned}$$

Thus, from (k), (5.5.3) and (5.5.10), it follows that $\forall\ B \in (0, \infty)$,

$$(5.5.26) \qquad \sup_{\|\mathbf{u}\| \le b} \|\mathbf{V}(\mathbf{A}\mathbf{u}) - \hat{\mathbf{V}}(\mathbf{A}\mathbf{u})\| = o_p(1).$$

Now, let $T_j := \int \{Y_j^0 + m_j\} g_n dG_n,\ 1 \le j \le p;\ \mathbf{T}' := (T_1, \cdots, T_p)$, and rewrite

$$\mathbf{e}'\hat{\mathbf{V}}(r\mathbf{A}\mathbf{e}) = \mathbf{e}'\mathbf{T} + r\mathbf{k}_n(\mathbf{e}).$$

Again, by the C-S inequality, Fubini, (5.5.3) and the assumptions (j) and (k) it follows that $\exists N_1$ and b_1, possibly both depending on ϵ, such that

$$(5.5.27) \qquad P(\|\mathbf{T}\| \le b_1) \ge 1 - (\epsilon/2), \qquad \forall\ n \ge N_1.$$

Choose b such that

$$(5.5.28) \qquad b \ge (b_1 + (z\gamma)^{1/2})\alpha^{-1},$$

where α is as in (k). Then, with $\alpha_n := \inf\{|k_n(\mathbf{e})|;\ \|\mathbf{e}\| = 1\}$,

$$\begin{aligned}
(5.5.29)\ & P\Big(\inf_{|r|=b, \|\mathbf{e}\|=1} \frac{(\mathbf{e}'\hat{\mathbf{V}}(r\mathbf{Ae}))^2}{\gamma_n} \ge z\Big) \\
&= P\Big(|\mathbf{e}'\hat{\mathbf{V}}(r\mathbf{Ae})| \ge (z\gamma_n)^{1/2},\ \forall\ \|\mathbf{e}\| = 1, |r| = b\Big) \\
&\ge P\Big(\Big|\,|\mathbf{e}'\mathbf{T}| - |r||k_n(\mathbf{e})|\,\Big| \ge (z\gamma_n)^{1/2}, \\
&\qquad\qquad\qquad \forall\, \|\mathbf{e}\| = 1, |r| = b\Big) \\
&\ge P\Big(\|\mathbf{T}| \le -(z\gamma_n)^{1/2} + b\alpha_n\Big) \\
&\ge P\Big(\|\mathbf{T}\| \le -(z\gamma)^{1/2} + b\alpha\Big) \\
&\ge P\Big(\|\mathbf{T}\| \le b_1\Big) \ge 1 - (\epsilon/2), \qquad\qquad \forall\, n \ge N_1.
\end{aligned}$$

In the above, the first inequality follows from the fact that $|\,|d| - |c|\,| \le |d+c|$, d, c real numbers; the second uses the fact that $|\mathbf{e}'\mathbf{T}| \le \|\mathbf{T}\|$ for all $\|\mathbf{e}\| = 1$; the third uses the relation $(-\infty, -(z\gamma)^{1/2} + b\alpha) \subset (-\infty, -(z\gamma_n)^{1/2} + b\alpha_n)$; while the last inequality follows from (5.5.27) and (5.5.28).

Observe that $\mathbf{e}'\hat{\mathbf{V}}(r\mathbf{Ae})$ is monotonic in r for every $\|\mathbf{e}\| = 1$. Therefore, (5.5.29) implies (5.5.25) and hence (5.5.22) in a straight forward fashion.

Next, consider $\mathbf{e}'\mathbf{V}(r\mathbf{Ae})$. Rewrite

$$\begin{aligned}
&\mathbf{e}'\mathbf{V}(r\mathbf{Ae}) \\
&\quad = \int \sum_{i=1}^{n} (\mathbf{e}'\mathbf{d}_i)[I(Y_{ni} \le y + r\mathbf{x}_{ni}'\mathbf{Ae})) - H_{ni}(y)]g_n(y)dG_n(y)
\end{aligned}$$

which, in view of the assumption (l), shows that $\mathbf{e}'\mathbf{V}(r\mathbf{A}\mathbf{e})$ is monotonic in r for every $\|\mathbf{e}\| = 1$. Therefore, by (5.5.26) $\exists\, N_2$, depending on $\epsilon, \ni$

$$\begin{aligned}
P\left(\inf_{|r|>b,\|\mathbf{e}\|=1} \frac{(\mathbf{e}'\mathbf{V}(r\mathbf{A}\mathbf{e}))^2}{\gamma_n} \geq z\right) \\
\geq\ P\left(\inf_{|r|=b,\|\mathbf{e}\|=1} \frac{(\mathbf{e}'\mathbf{V}(r\mathbf{A}\mathbf{e}))^2}{\gamma_n} \geq z\right) \\
\geq\ P\left(\inf_{|r|=b,\|\mathbf{e}\|=1} \frac{(\mathbf{e}'\hat{\mathbf{V}}(r\mathbf{A}\mathbf{e}))^2}{\gamma_n} \geq z\right) - (\epsilon/2), \qquad \forall n \geq N_2, \\
\geq\ 1-\epsilon, \qquad \forall n \geq N_2 \vee N_1,
\end{aligned}$$

by (5.5.29). This proves (5.5.24) and hence (5.5.21). □

The next lemma gives an analogue of the previous lemma for $K^+_{\mathbf{D}}$. Since the proof is quite similar no details will be given.

Lemma 5.5.5. *In addition to the assumptions of Theorem* 5.5.2 *assume that* (k^+) *and* (l) *hold, where* (k^+) *is the condition* (k) *with* $\mathbf{\Gamma}_n$ *replaced by* $\mathbf{\Gamma}_n^+ := (\boldsymbol{\nu}_1^+, \cdots, \boldsymbol{\nu}_p^+)$ *and where* $\{\boldsymbol{\nu}_j^+\}$ *are defined just above* (5.5.13).

Then, $\forall \epsilon > 0$, $0 < z < \infty$, $\exists N$ *(depending only on* ϵ*) and a* b *(depending on* ϵ, z*)* $\ni\ \forall\, n \geq N$,

$$\begin{aligned}
P\left(\inf_{\|\mathbf{u}\|>b} K^+_{\mathbf{D}}(\mathbf{A}\mathbf{u}) \geq z\right) \ &\geq\ 1-\epsilon, \\
P(\inf_{\|\mathbf{u}\|>b} \hat{K}^+_{\mathbf{D}}(\mathbf{A}\mathbf{u}) \geq z) \ &\geq\ 1-\epsilon.
\end{aligned}$$

□

The above two lemmas verify (5.4.A5) for the two dispersions K and K^+. Also note that (5.5.22) together with (e) and (j) imply that $\|\mathbf{A}^{-1}(\hat{\mathbf{\Delta}} - \boldsymbol{\beta})\| = O_p(1)$, where $\hat{\mathbf{\Delta}}$ is defined at (5.5.31) below. Similarly, Lemma 5.5.5, (e), (5.5.17) and the symmetry assumption (5.3.8) about $\{G_n\}$ imply that $\|\mathbf{A}^{-1}(\mathbf{\Delta}^+ - \boldsymbol{\beta})\| = O_p(1)$, where $\mathbf{\Delta}^+$ is defined at (5.5.35) below. The proofs of these facts are exactly similar to that of (5.4.2) given in the proof of Theorem 5.4.1.

In view of Remark 5.5.1 and Theorem 5.4.1, we have now proved the following theorems.

Theorem 5.5.3 *Assume that* (1.1.1) *holds with the modeled and actual d.f.'s of the errors* $\{e_{ni}, 1 \leq i \leq n\}$ *equal to* $\{H_{ni}, 1 \leq i \leq n\}$

and $\{F_{ni}, 1 \le i \le n\}$, *respectively. In addition, suppose that* (a) - (l) *hold. Then*

$$(5.5.30) \qquad (\hat{\boldsymbol{\beta}}_{\mathbf{D}} - \hat{\boldsymbol{\Delta}})'\mathbf{A}^{-1}\boldsymbol{\mathcal{B}}_n\mathbf{A}^{-1}(\hat{\boldsymbol{\beta}}_{\mathbf{D}} - \hat{\boldsymbol{\Delta}})o_p(1),$$

where $\hat{\boldsymbol{\Delta}}$ *satisfies the equation*

$$(5.5.31) \qquad \boldsymbol{\mathcal{B}}_n\mathbf{A}^{-1}(\hat{\boldsymbol{\Delta}} - \boldsymbol{\beta}) = \boldsymbol{\mathcal{J}}_n.$$

If in addition,

$$(5.5.32) \qquad \boldsymbol{\mathcal{B}}_n{}^{-1} \text{exists for all } n \ge p,$$

then,

$$(5.5.33) \qquad \mathbf{A}^{-1}(\hat{\boldsymbol{\beta}}_{\mathbf{D}} - \boldsymbol{\beta}) = \boldsymbol{\mathcal{B}}_n^{-1}\boldsymbol{\mathcal{J}}_n + o_p(1),$$

where $\boldsymbol{\mathcal{J}}_n$ *and* $\boldsymbol{\mathcal{B}}_n$ *are defined at* (5.5.19). □

Theorem 5.5.4 *Assume that* (1.1.1) *holds with the actual d.f.'s of the errors* $\{e_{ni}, 1 \le i \le n\}$ *equal to* $\{F_{ni}, 1 \le i \le n\}$. *In addition, suppose that* $\{\mathbf{X}, F_{ni}, \mathbf{D}, G_n\}$ *satisfy* (a) - (d), (f) - (i), (5.3.8) *for all* $n \ge 1$, (k), (l) *and* (5.5.14). *Then,*

$$(5.5.34) \qquad (\boldsymbol{\beta}^+_{\mathbf{D}} - \boldsymbol{\Delta}^+)'\mathbf{A}^{-1}\boldsymbol{\mathcal{B}}^+_n\mathbf{A}^{-1}(\boldsymbol{\beta}^+_{\mathbf{D}} - \boldsymbol{\Delta}^+) = o_p(1),$$

where $\boldsymbol{\Delta}^+$ *satisfies the equation*

$$(5.5.35) \qquad \boldsymbol{\mathcal{B}}^+_n\mathbf{A}^{-1}(\boldsymbol{\Delta}^+ - \boldsymbol{\beta}) = \boldsymbol{\mathcal{J}}^+_n.$$

If, in addition,

$$(5.5.36) \qquad (\boldsymbol{\mathcal{B}}^+_n)^{-1} \; exists \; for \; all \; n \ge p,$$

then,

$$(5.5.37) \qquad \mathbf{A}^{-1}(\boldsymbol{\beta}^+_{\mathbf{D}} - \boldsymbol{\beta}) = (\boldsymbol{\mathcal{B}}^+_n)^{-1}\boldsymbol{\mathcal{J}}^+_n + o_p(1),$$

where $\boldsymbol{\mathcal{J}}^+_n$ *and* $\boldsymbol{\mathcal{B}}^+_n$ *are defined at* (5.5.19). □

Remark 5.5.2. If $\{F_i\}$ are symmetric about zero then $\mathbf{m}^+_{\mathbf{D}} \equiv \mathbf{0}$ and $\boldsymbol{\beta}^+_{\mathbf{D}}$ is consistent for $\boldsymbol{\beta}$ even if the errors are not identically distributed. On the other hand, if the errors are identically distributed, but not symmetrically, then $\boldsymbol{\beta}_{\mathbf{D}}^{+}$ will be asymptotically biased. This is not surprising because here the symmetry, rather than the identically distributed nature of the errors is relevant.

If $\{F_i\}$ are symmetric about an unknown common point then that point can be also estimated by the above m.d. method by simply augmenting the design matrix to include the column **1**, if not present already. □

Next we turn to the $K_{\mathbf{D}}^{*}$ and $\beta_{\mathbf{D}}^{*}$ (5.2.17) and (5.2.18). First we state a theorem giving an analogue of (5.5.9) for $K_{\mathbf{D}}^{*}$. Let Y_j, μ_j be Y_d, μ_d of (2.3.1) with $\{d_{ni}\}$ replaced by $\{d_{nij}\}$, $j = 1, \cdots, p$, X_{ni} replaced by Y_{ni} and $\mathbf{c}_{ni} = \mathbf{A}_1(\mathbf{x}_{ni} - \overline{\mathbf{x}}_n), 1 \leq i \leq n$, where $\mathbf{A}_1$ and $\overline{\mathbf{x}}_n$ are defined at (4.3.10). Set

$$\mathbf{R}_j^*(s) := \sum_{i=1}^{n} (d_{nij} - \tilde{d}_{nj}(s))(\mathbf{x}_{ni} - \overline{\mathbf{x}}_n) q_{ni}(s),$$

where, for $1 \leq j \leq p$, $\tilde{d}_{nj}(s) := n^{-1} \sum_{i=1}^{n} d_{nij} \ell_{ni}(s)$, $0 \leq s \leq 1$, with $\{l_{ni}\}$ as in (3.2.34) and $q_{ni} \equiv f_{ni}(H^{-1}), 1 \leq i \leq n$. Let

$$\hat{K}_{\mathbf{D}}^{*}(\mathbf{t}) \;\; := \;\; \sum_{j=1}^{p} \int_0^1 \{Y_j(s, \mathbf{0}) - \mathbf{t}'\mathbf{R}_j^*(s) + \mu_j(s, \mathbf{0})\}^2 dL_n(s).$$

In the assumptions of theorem below, L in $K_{\mathbf{D}}^{+}$ is supposed to have been replaced by L_n.

Theorem 5.5.5. *Let $Y_{ni}, \cdots, Y_{nn}$ be independent r.v.'s with respective d.f.'s $F_{n1}, \cdots, F_{nn}$. Assume $\{\mathbf{D}, \mathbf{X}, F_{ni}\}$ satisfy* (a)-(c), (2.3.5), (3.2.12), (3.2.34) *and* (3.2.35) *with $w_i = d_{ij}$, $1 \leq j \leq p$, $1 \leq i \leq n$. Let $\{L_n\}$ be a sequence of d.f.'s on $[0,1]$ and assume that*

$$\sum_{j=1}^{p} \int_0^1 \mu_j^2(s, \mathbf{0}) dL_n(s) = O(1). \tag{5.5.38}$$

Then, for every $0 < b < \infty$,

$$\sup_{\|\mathbf{u}\| \leq b} |K_{\mathbf{D}}^{*}(\mathbf{A}\mathbf{u}) - \hat{K}_{\mathbf{D}}^{*}(\mathbf{A}\mathbf{u})| = o_p(1).$$

Proof. The proof uses the AUL result of Theorems 3.2.1 and 3.2.4. Details are left out as an exercise. □

This result shows that the dispersion $K_{\mathbf{D}}^*$ satisfies (5.4.A1) with

$$\begin{aligned}
(5.5.39) \qquad \boldsymbol{\theta}_0 &= \mathbf{0}, \;\; \boldsymbol{\delta}_n = \mathbf{A}_1^{-1}, \;\; \mathbf{S}_n = \mathbf{A}_1^{-1}\mathcal{J}_n^*, \\
\mathbf{W}_n &= \mathbf{A}_1'\mathcal{B}_n^*\mathbf{A}_1, \\
\mathcal{J}_n^* &:= -\int_0^1 \boldsymbol{\Gamma}_n^*(s)\{\mathbf{Y}_{\mathbf{D}}(s) + \mu_{\mathbf{D}}(s)\}dL_n(s), \\
\mathcal{B}_n^* &:= \int_0^1 \boldsymbol{\Gamma}_n^*(s)\boldsymbol{\Gamma}_n^{*'}(s)dL_n(s),
\end{aligned}$$

where $\boldsymbol{\Gamma}_n^*(s) = \mathbf{A}_1'\mathbf{X}_c\boldsymbol{\Lambda}(s)\mathbf{D}(s)$, $\mathbf{D}(s) := ((d_{nij} - \tilde{d}_{nj}(s)),\ 1 \le i \le n,\ 1 \le j \le p$; $\boldsymbol{\Lambda}(s)$ as in (2.3.30), $0 \le s \le 1; \mathbf{X}_c$ as in (4.2.11); $\mathbf{Y}_{\mathbf{D}} := (Y_1, \cdots, Y_p), \mu_{\mathbf{D}}' = (\mu_1, \cdots, \mu_p)$ *with* $Y_j(s) \equiv Y_j(s, \mathbf{0}), \mu_j(s) \equiv \mu_j(s, \mathbf{0})$.

Call the condition (k) by the name of (k^*) if it holds when (Γ_n, G_n) is replaced by (Γ_n^*, L_n). Analogous to Theorem 5.5.4 we have

Theorem 5.5.6 *Assume that* (1.1.1) *holds with the actual d.f.'s of the errors* $\{e_{ni}, 1 \le i \le n\}$ *equal to* $\{F_{ni}, 1 \le i \le n\}$. *In addition, assume that* $\{\mathbf{D}, \mathbf{X}, F_{ni}\}$ *satisfy* ($\mathbf{NX}^*$), (b), (c), (2.3.5), (3.2.12), (3.2.34), (3.2.35) *with* $w_i = d_{ij}, 1 \le j \le p, 1 \le i \le n$, (k^*) *and* (l). *Let* $\{L_n\}$ *be a sequence of d.f.'s on* $[0, 1]$ *satisfying* (5.5.38). *Then*

$$(\boldsymbol{\beta}_{\mathbf{D}}^* - \boldsymbol{\Delta}^*)'\mathbf{A}^{-1}\mathcal{B}_n^*\mathbf{A}^{-1}(\boldsymbol{\beta}_{\mathbf{D}}^* - \boldsymbol{\Delta}^*) = o_p(1),$$

where $\boldsymbol{\Delta}^*$ *satisfies the equation*

$$(5.5.40) \qquad \mathcal{B}_n^*\mathbf{A}^{-1}(\boldsymbol{\Delta}^* - \boldsymbol{\beta}) = \mathcal{J}_n^*.$$

If, in addition,

$$(5.5.41) \qquad (\mathcal{B}_n^*)^{-1} \text{ exists for } n \ge p,$$

then,

$$\mathbf{A}^{-1}(\boldsymbol{\beta}_{\mathbf{D}}^* - \boldsymbol{\beta}) = (\mathcal{B}_n^*)^{-1}\mathcal{J}_n^* + o_p(1).$$

The proof of this theorem is similar to that of Theorem 5.5.3. The details are left out for interested readers. □

Remark 5.5.3. Discussion of the assumptions (a) - (j). Among all of these assumptions, (g) and (i) are relatively harder to verify.

First, we shall give some sufficient conditions that will imply (g), (i) and the other assumptions. Then, we shall discuss these assumptions in detail for three cases, v.i.z., the case when the errors are correctly modeled to be i.i.d. F, F a known d.f., the case when we model the errors to be i.i.d. F but they actually have heteroscedastic gross errors distributions, and finally, the case when the errors are modeled to be i.i.d. F but they actually are heteroscedastic due to difference in scales.

To begin with consider the following assumptions.

(5.5.42) For any sequence of numbers $\{a_{ni}, b_{ni}\}, a_{ni} < b_{ni}$, with

$$\max_{1\le i\le n}(b_{ni} - a_{ni}) \to 0,$$

$$\limsup_n \max_{1\le i\le n} \frac{1}{b_{ni} - a_{ni}} \int_{a_{ni}}^{b_{ni}} \int \{f_{ni}(y+z) - f_{ni}(y)\}^2 \times dG_n(y)dz = 0.$$

(5.5.43)
$$\max_{1\le i\le n} \int f_{ni}^2 dG_n = O(1).$$

Claim 5.5.1. *Assumptions* (a) - (d), (5.5.42), (5.5.43) *imply* (g) *and* (i).

Proof. Use the C-S inequality twice, the fact that $(d_{ij}^{\pm})^2 \le d_{ij}^2$ for all i, j and (b) to obtain

$$\begin{aligned}
&\sum_{j=1}^{p} \int \Big[\sum_{i=1}^{n} d_{ij}^{\pm}\{F_i(y + \mathbf{c}_i'\mathbf{v} + \delta\kappa_i) - F_i(y + \mathbf{c}_i\mathbf{v} - \delta\kappa_i)\}\Big]^2 dG_n(y) \\
&\le 2\sum_{i=1}^{n} \|\mathbf{d}_i\|^2 \int \sum_{i=1}^{n} \delta\kappa_i \int_{a_i}^{b_i} f_i^2(y+z)dzdG_n(y) \\
&\le 4p^2\delta^2 \max_i (2\delta\kappa_i)^{-1} \int_{a_i}^{b_i} \int f_i^2(y+z)dG_n(y)dz, \quad \text{(by Fubini)},
\end{aligned}$$

where $a_i = -\kappa_i\delta + \mathbf{c}_i'\mathbf{v}, b_i = \kappa_i\delta + \mathbf{c}_i'\mathbf{v}, 1 \le i \le n$. Therefore, by (5.5.42), (5.5.43) and (a), the

$$\text{L.H.S. } (g) \le 4p^2\delta^2 k, \qquad (k = \limsup_n \max_i |f_i|_n^2),$$

which shows that (g) holds.

Next, by (b) and two applications of the C-S inequality, the L.H.S. (i)

$$
\begin{aligned}
&= \sum_{j=1}^{p} \int \Big[\sum_{i=1}^{n} d_{ij} \{ F_i(y + \mathbf{c}_i'\mathbf{u}) - F_i(y) - \mathbf{c}_i'\mathbf{u} f_i(y) \} \Big]^2 dG_n(y) \\
&\le p \int \sum_{i=1}^{n} \Big\{ F_i(y + \mathbf{c}_i'\mathbf{u}) - F_i(y) - \mathbf{c}_i'\mathbf{u} f_i(y) \Big\}^2 dG_n(y) \\
&= p \Big\{ \int \Sigma_i^+ \Big[\int_0^{\mathbf{c}_i'\mathbf{u}} (f_i(y+z) - f_i(y)) dz \Big]^2 dG_n(y) \\
&\qquad + \int \Sigma_i^- \Big[- \int_{\mathbf{c}_i'\mathbf{u}}^{0} (f_i(y+z) - f_i(y)) dz \Big]^2 dG_n(y) \Big\} \\
&\le 2p \int \Big[\Sigma_i^+ \mathbf{c}_i'\mathbf{u} \int^{\mathbf{c}_i'\mathbf{u}} \{ f_i(y+z) - f_i(y) \}^2 dz \\
&\qquad + \Sigma_i^- (-\mathbf{c}_i'\mathbf{u}) \int_{\mathbf{c}_i'\mathbf{u}}^{0} \int \{ f_i(y+z) - f_i(y) \}^2 dz \Big] dG_n(y) \\
&\le \Big[\max_i \frac{2}{|\mathbf{c}_i'\mathbf{u}|} \int_{-|\mathbf{c}_i'\mathbf{u}|}^{|\mathbf{c}_i'\mathbf{u}|} \int \{ f_i(y+z) - f_i(y) \}^2 dG_n(y) dz \Big] \\
&\qquad\qquad \times 4p \sum_{i=1}^{n} (\mathbf{c}_i'\mathbf{u})^2,
\end{aligned}
$$

where $\Sigma_i^+ (\Sigma_i^-)$ is the sum over those i for which $\mathbf{c}_i'\mathbf{u} \ge 0 (\mathbf{c}_i'\mathbf{u} < 0)$. Since $\sum_{i=1}^{n} (\mathbf{c}_i'\mathbf{u})^2 \le pB$ for all $\mathbf{u} \in \mathcal{N}(b)$, ($i$) now follows from (5.5.42) and (a). □

Now we consider the three special cases mentioned above.

Case 5.5.1. *Correctly modeled i.i.d. errors:* $F_{ni} \equiv F \equiv H_{ni}$, $G_n \equiv G$. Suppose that F has a density f w.r.t. λ. Assume that

(5.5.44) $\quad$ (a) $0 < \int f dG < \infty$, $\quad$ (b) $0 < \int f^2 dG < \infty$.

(5.5.45) $\quad \int F(1-F) dG < \infty$.

(5.5.46) $\quad$ (a) $\lim_{z \to 0} \int f(y+z) dG(y) = \int f dG$,

$\quad$ (b) $\lim_{z \to 0} \int f^2(y+z) dG)y) = \int f^2 dG$.

Claim 5.5.2. *Assumptions* (a), (b), (d) *with* $G_n \equiv G$, (5.5.44) - (5.5.46) *imply* (a) - (j) *with* $G_n \equiv G$.

This is easy to see. In fact here (e) and (f) are equivalent to (5.5.44a), (5.5.45) and (5.5.46a); (5.5.42) and (5.5.43) are equivalent to (5.5.44b) and (5.5.46b). The LHS (j) is equal to 0.

Note that if G is absolutely continuous then (5.5.44) implies (g). If G is purely discrete and f continuous at the points of jumps of G then (5.5.46) holds. In particular if $G = \delta_0$, i.e. if G is degenerate at 0, $\infty > f(0) > 0$ and f is continuous at 0 then (5.5.44), (5.5.46) are trivially satisfied. If $G(y) \equiv y$, (5.5.44a) and (5.5.46a) are *a priori* satisfied while (5.5.45) is equivalent to assuming that $E|e_1 - e_2| < \infty$, e_1, e_2 i.i.d. F.

If $dG = \{F(1-F)\}^{-1}dF$, the so called Darling - Anderson measure, then (5.5.44) - (5.5.46) are satisfied by a class of d.f.'s that includes normal, logistic and double exponential distributions.

Case 5.5.2. *Heteroscedastic gross errors* : $H_{ni} \equiv F$, $F_{ni} \equiv (1 - \delta_{ni})F + \delta_{ni}F_0$. We shall also assume that $G_n \equiv G$. Let f and f_0 be continuous densities of F and F_0. Then $\{F_{ni}\}$ have densities $f_{ni} = f + \delta_{ni}(f_0 - f)$, $1 \le i \le n$. Hence (c) is satisfied. Consider the assumption

$$0 \le \delta_{ni} \le 1, \quad \max_i \delta_{ni} \to 0, \tag{5.5.47}$$

$$\int |F_0 - F| dG < \infty. \tag{5.5.48}$$

Claim 5.5.3. *Suppose that* f_0, f, F *satisfy* (5.5.44) (5.5.46), *and* (5.5.45), *and suppose that* (a), (b) *and* (d) *hold. Then* (5.5.47) *and* (5.5.48) *imply* (e) - (i).

Proof. The relation $f_i \equiv f + \delta_i(f_0 - f)$ implies that

$$\boldsymbol{\nu}_j - \sum_{i=1}^n d_{ij}\mathbf{c}_i f = \sum_{i=1}^n d_{ij}\mathbf{c}_i\delta_i(f_0 - f), \qquad 1 \le j \le p,$$

and

$$\gamma_n - \sum_{i=1}^n \|\mathbf{d}_i\|^2 f = \sum_{i=1}^n \|\mathbf{d}_i\|^2 \delta_i(f_0 - f).$$

Because $\sum_{i=1}^n \|\mathbf{d}_i\|^2 \le p$, $\sum_{i=1}^n \|\mathbf{c}_i\|^2 = p$, we obtain $\forall\, x \in \mathbb{R}$,

$$\begin{aligned}&\left|\int [\gamma_n(y+x) - \sum_{i=1}^n \|\mathbf{d}_i\|^2 f(y+x)]dG(y)\right| \\ &\le \; p \max_i \delta_i \left|\int [f_0(y+x) - f(y+x)]dG(y)\right|,\end{aligned}$$

Therefore, by (5.5.47), (5.5.44a) and (5.5.46a), it follows that (f) is satisfied. Similarly, the inequality

$$\begin{aligned}&\sum_{j=1}^p \int \left\| \boldsymbol{\nu}_j - \sum_{i=1}^n d_{ij}\mathbf{c}_i f \right\|^2 dG \\ &\qquad\qquad \le 2p^2 \max_i \delta_i^2 \left\{ \int f_0^2 dG + \int f^2 dG \right\}\end{aligned}$$

ensures the satisfaction of (h). The inequality

$$\begin{aligned}&\left| \int \sum_{i=1}^n \|\mathbf{d}_i\|^2 \{F_i(1-F_i) - F(1-F)\} dG \right| \\ &\qquad\qquad \le 2p \max_i \delta_i \int |F_0 - F| dG,\end{aligned}$$

(5.5.45), (5.5.47) and (5.5.48) imply (e). Next,

$$\begin{aligned}&\int \{f_i(y+x) - f_i(y)\}^2 dG(y) \\ &\le \; 2(1+2\delta_i^2) \int \{f(y+x) - f(y)\}^2 dG(y) \\ &\qquad +4\delta_i^2 \int \{f_0(y+x) - f_0(y)\}^2 dG(y).\end{aligned}$$

Note that (5.5.44b), (5.5.46b) and the continuity of f imply that

$$\lim_{x \to 0} \int \{f(y+x) - f(y)\}^2 dG(y) = 0$$

and a similar result for f_0. Therefore from the above inequality, (5.5.46) and (5.5.47) we see that (5.5.42) and (5.5.43) are satisfied. By Claim 5.5.1, it follows that (g) and (i) are satisfied. □

Suppose that G is a *finite measure.* Then (**F1**) implies (5.5.44) - (5.5.46) and (5.5.48). In particular these assumptions are satisfied by all those f's that have finite Fisher information.

The assumption (j), in view of (5.5.48), amounts to requiring that

$$\sum_{j=1}^{p}\left(\sum_{i=1}^{n} d_{ij}\delta_i\right)^2 = O(1). \tag{5.5.49}$$

But

$$\sum_{j=1}^{p}(\sum_{i=1}^{n} d_{ij}\delta_i)^2 = \sum_{i=1}^{n}\sum_{k=1}^{n} \mathbf{d}_i'\delta_i\mathbf{d}_k\delta_k \leq (\sum_{i=1}^{n} \|\mathbf{d}_i\|\delta_i)^2.$$

This and (b) suggest a choice of $\delta_i \equiv p^{-1/2}\|\mathbf{d}_i\|$ will satisfy (5.5.49). Note that if $\mathbf{D} = \mathbf{XA}$ then $\|\mathbf{d}\|^2 \equiv \mathbf{x}_i'(\mathbf{X}'\mathbf{X})^{-1}\mathbf{x}_i$.

When studying the robustness of $\hat{\beta}_{\mathbf{X}}$ in the following section, $\delta_i^2 \equiv p^{-1}\mathbf{x}_i'(\mathbf{X}'\mathbf{X})^{-1}\mathbf{x}_i$ is a natural choice to use. It is an analogue of $n^{-1/2}-$ contamination in the i.i.d. setup. □

Case 5.5.3. *Heteroscedastic scale errors* : $H_{ni} \equiv F$, $F_{ni}(y) \equiv F(\tau_{ni}y)$, $G_n \equiv G$. Let F have continuous density f. Consider the conditions

$$\tau_{ni} \equiv \sigma_{ni} + 1; \quad \sigma_{ni} > 0, \quad 1 \leq i \leq n; \quad \max_i \sigma_{ni} \to 0. \tag{5.5.50}$$

$$\lim_{s\to 1} \int |y|^j f^k(sy)dG(y) = \int |y|^j f^k(y)dG(y), \qquad j = 1, k = 1, \ j = 0, k = 1, 2. \tag{5.5.51}$$

Claim 5.5.4. *Under* (a), (b), (d) *with* $G_n \equiv G$, (5.5.44) - (5.5.46), (5.5.50) *and* (5.5.51), *the assumptions* (e) - (i) *are satisfied.*

Proof. By the Lemmas 9.1.5 and 9.1.6 of the Appendix below, and by (5.5.23), (5.5.27), and (5.5.31),

$$\lim_{x\to 0} \limsup_n \max_i \int |f(\tau_1(y+x)) - f(y+x)|^r dG(y) = 0, \tag{5.5.52}$$

$$\lim_{x\to 0} \int |f(y+x) - f(y)|^r dG(y) = 0, \quad r = 1, 2.$$

Now,

$$\begin{aligned}
&\left|\int \sum_{i=1}^{n} \|\mathbf{d}\|^2 \{F_i(1-F_i) - F(1-F)\} dG\right| \\
&\leq \; 2p \max_i \int |F(\tau_1 y) - F(y)| dG(y) \\
&\leq \; 2p \max_i \int_1^{\tau_i} \int |y| f(sy) dG(y) ds = o(1),
\end{aligned}$$

by (5.5.30) and (5.5.31) with $j = 1, r = 1$.

Hence (5.5.45) implies (e). Next,

$$\begin{aligned}
&\left|\int \gamma_n(y+x) dG(y) - \sum_{i=1}^{n} \|\mathbf{d}_i\|^2 \int f dG\right| \\
&\leq \; \sum_{i=1}^{n} \|d_i\|^2 \tau_i \int \Big\{ |f(\tau_i(y+x)) - f(y+x)| \\
&\qquad\qquad + |f(y+x) - f(y)| \Big\} dG(y) + \max_i \sigma_i p \int f dG.
\end{aligned}$$

Therefore, in view of (5.5.30), (5.5.52) and (5.5.44) we obtain (f).

Next, consider

$$\begin{aligned}
&\int \{f_i(y+x) - f_i(y)\}^2 dG(y) \\
&\leq \; 4\tau_i^2 \int \Big\{ |f(\tau_i(y+x)) - f(y+x)]^2 \\
&\qquad\qquad + [f(y+x) - f(y)]^2 + [f(\tau_i y) - f(y)]^2 \Big\} dG(y)
\end{aligned}$$

Therefore, (5.5.50) and (5.5.32) imply (5.5.21), and hence (g) and (i) by Claim 5.5.1. Note that (5.5.32) and (5.5.23b) imply (5.5.22). Finally,

$$\begin{aligned}
&\sum_{j=1}^{p} \int \|\boldsymbol{\nu}_j - \sum_{i=1}^{n} d_{ij} \mathbf{c}_i f\|^2 dG \\
&\leq \; p^2 \max_i \int \{\tau_i f(\tau_i y) - f(y)\}^2 dG(y) \\
&\leq \; 2p^2 \max_i \tau_i^2 Big[\int \{f(\tau_i y) - f(y)\}^2 dG(y) + \int f^2 dG \Big] = o(1),
\end{aligned}$$

by (5.5.50), (5.5.46b), (5.5.52). Hence (5.5.46b) and the fact that

$$\sum_{j=1}^{p} \Big\| \sum_{i=1}^{n} d_{ij} \mathbf{c}_i \Big\|^2 \le p^2$$

implies (h). □

Here, the assumption (j) is equivalent to having

$$\sum_{j=1}^{p} \int [\sum_{i=1}^{n} d_{ij}\{F(\tau_i y) - F(y)\}]^2 dG(y) = O(1).$$

One sufficient condition for this, besides requiring F to have density f satisfying

$$\lim_{s \to 1} \int (yf(sy))^2 dG(y) = \int (yf(y))^2 dG(y) < \infty, \tag{5.5.53}$$

is to have

$$\sum_{i=1}^{n} \sigma_i^2 = O(1). \tag{5.5.54}$$

One choice of $\{\sigma_i\}$ satisfying (5.5.54) is $\sigma_i^2 \equiv n^{-1/2}$ and the other choice is $\sigma_i^2 = \mathbf{x}_i'(\mathbf{X}'\mathbf{X})^{-1}\mathbf{x}_i$, $1 \le i \le n$.

Again, if f satisfies (**F1**), (**F2**) and G is a finite measure then (5.5.44), (5.5.46), (5.5.51) and (5.5.53) are a *priori* satisfied.

Now we shall give a set of sufficient conditions that will yield (5.4.A1) for the Q of (5.2.9). Since Q does not satisfy (5.3.21), the distribution of Q under (1.1.1) is not independent of $\boldsymbol{\beta}$. Therefore care has to be taken to exhibit this dependence clearly when formulating a theorem pertaining to Q. This of course complicates the presentation somewhat. As before with $\{H_{ni}\}, \{F_{ni}\}$ denoting the modeled and the actual d.f.'s of $\{e_{ni}\}$, define for $0 \le s \le 1, y \in \mathbb{R}, \mathbf{t} \in \mathbb{R}^p$,

$$\begin{aligned}
m_n(s,y) &:= n^{-1/2} \sum_{i=1}^{ns} \{F_{ni}(y - \mathbf{x}_{ni}'\boldsymbol{\beta}) - H_{ni}(y - \mathbf{x}_{ni}'\boldsymbol{\beta})\}, \\
\overline{H}_n(s,y,\mathbf{t}) &:= n^{-1} \sum_{i=1}^{ns} H_{ni}(y - \mathbf{x}_{ni}'\mathbf{t}), \\
M_{1n}(s,y) &:= n^{-1/2} \sum_{i=1}^{ns} \{I(Y_{ni} \le y) - F_{ni}(y - \mathbf{x}_{ni}'\boldsymbol{\beta})\}, \\
d\alpha_n(s,y) &:= dL_{n^*}(s) dG_n(y)
\end{aligned}$$

Observe that

$$Q(\mathbf{t}) = \int [M_{1n}(s,y) + m_n(s,y) - n^{1/2}\{\overline{H}_n(s,y,\mathbf{t}) - \overline{H}_n(s,y,\boldsymbol{\beta})\}]^2 d\alpha_n(s,y),$$

where the integration is over the set $[0,1] \times \mathbb{R}$.

Assume that $\{H_{ni}\}$ have densities $\{h_{ni}\}$ w.r.t. λ and set, for $s \in [0,1]$, $y \in \mathbb{R}$,

$$\text{(5.5.55)} \qquad \overline{\mathbf{R}}_n(s,y) := n^{-1/2}\sum_{i=1}^{ns} \mathbf{x}_{ni} h_{ni}(y - \mathbf{x}_{ni}'\boldsymbol{\beta}),$$

$$\overline{h_n^2}(y) := n^{-1}\sum_{i=1}^{n} h_{ni}^2(y - \mathbf{x}_{ni}'\boldsymbol{\beta}),$$

$$\overline{\boldsymbol{\nu}}_n := \mathbf{A}\overline{\mathbf{R}}_n, \quad \mathcal{B}_{in} := \int \overline{\boldsymbol{\nu}}_n \overline{\boldsymbol{\nu}}_{n'} d\alpha_n.$$

Finally define, for $\mathbf{t} \in \mathbb{R}^p$,

$$\hat{Q}(\mathbf{t}) := \int [M_{1n}(s,y) + m_n(s,y) + \mathbf{t}'\overline{\mathbf{R}}_n(s,y)]^2 d\alpha_n(s,y).$$

Theorem 5.5.7. *Assume that* (1.1.1) *holds with the actual and the modeled d.f.'s of the errors* $\{e_{ni}, 1 \le i \le n\}$ *equal to* $\{F_{ni}, 1 \le i \le n\}$ *and* $\{H_{ni}, 1 \le i \le n\}$, *respectively. In addition, assume that* (a) *holds,* $\{H_{ni}, 1 \le i \le n\}$ *have densities* $\{h_{ni}, 1 \le i \le n\}$ *w.r.t.* λ, *and the following hold.*

$$\text{(5.5.56)} \qquad |\overline{h_n^2}|_n = O(1).$$

$\forall\, \mathbf{v} \in \mathcal{N}(b)$, $\forall\, \delta > 0$,

$$\limsup_n \max_{1 \le i \le n} \frac{1}{2\delta\kappa_{ni}} \int_{a_{ni}}^{b_{ni}} \int h_{ni}^2(y - \mathbf{x}_{ni}'\boldsymbol{\beta} + z)dG_n(y)dz = \limsup_n \max_{1 \le i \le n} \int h_{ni}^2(y - \mathbf{x}_{ni}'\boldsymbol{\beta})dG_n(y) < \infty,$$

where $a_{ni} = -\delta\kappa_{ni} - \mathbf{c}_{ni}'\mathbf{v}$, $b_{ni} = \delta\kappa_{ni} - \mathbf{c}_{ni}'\mathbf{v}$, $\kappa_{ni} = \|\mathbf{c}_{ni}\|$, $\mathbf{c}_{ni} = \mathbf{A}\mathbf{x}_{ni}$, $1 \le i \le n$.

$\forall\, \mathbf{u} \in \mathcal{N}(b)$,

$$\int \Big\{ n^{1/2}[\overline{H}_n(s,y,\boldsymbol{\beta}+\mathbf{A}\mathbf{u}) - \overline{H}_n(s,y,\boldsymbol{\beta})] + \mathbf{u}'\overline{\boldsymbol{\nu}}_n \Big\}^2 d\alpha_n(s,y) = o(1).$$

$$\int n^{-1}\sum_{i=1}^n F_{ni}(y - \mathbf{x}_{ni}'\boldsymbol{\beta})(1 - F_{ni}(y - \mathbf{x}_{ni}'\boldsymbol{\beta}))dG_n(y) = O(1).$$

$$\int m_n^2(s,y)d\alpha_n(s,y) = O(1).$$

Then $\forall\ 0 < b < \infty$,

$$E \sup_{\|\mathbf{u}\|\le b} |Q(\boldsymbol{\beta}+\mathbf{A}\mathbf{u}) - \hat{Q}(\mathbf{A}\mathbf{u})| = o(1).$$

The details of the proof are similar to those of Theorem 5.5.1 and are left out as an exercise for interested readers.

An analogue of (5.5.34) of $\overline{\boldsymbol{\beta}}$ will appear in the next section as Theorem 5.6a.3. Its asymptotic distribution in the case when the errors are correctly modeled to be i.i.d. will be also discussed here.

We shall end this section by stating analogues of some of the above results that will be useful when an unknown scale is also being estimated. To begin with, consider $K_{\mathbf{D}}$ of (5.2.19). To simplify writing, let

$$K_{\mathbf{D}}^0(s,\mathbf{u}) := K_{\mathbf{D}}((1+sn^{-1/2}), \mathbf{A}\mathbf{u}), \quad s \in \mathbb{R}, \mathbf{u} \in \mathbb{R}^p.$$

Write $a_s := (1+sn^{-1/2})$. From (5.2.19),

$$K_{\mathbf{D}}^0(s,\mathbf{u}) = \sum_{j=1}^p \int \Big\{ Y_j^0(ya_s,\mathbf{u}) + \mu_j^0(ya_s,\mathbf{u}) - \sum_{i=1}^n d_{ij}H_i(y) \Big\}^2 dG_n(y)$$

where H_i is the d.f. of e_i, $1 \le i \le n$, and where μ_j^0, Y_j^0 are as in (i) and (5.5.1)), respectively. Writing $\mu_j^0(y), Y_j^0(y)$ etc. for $\mu_j^0(y,\mathbf{0}), Y_j^0(y,\mathbf{0})$

etc., rewrite

$$
\begin{aligned}
&K_{\mathbf{D}}^{0}(s,\mathbf{u}) \\
&= \sum_{j=1}^{p} \int \Big\{ Y_j^0(ya_s,\mathbf{u}) - Y_j^0(y) + \mu_j^0(ya_s) - \mu_j^0(y) - sy\nu_j^*(y) \\
&\qquad + Y_j^0(y) + \mathbf{u}'\boldsymbol{\nu}_j(y) + sy\nu_j^*(y) + m_j(y) \\
&\qquad + \mu_j^0(ya_s,\mathbf{u}) - \mu_j^0(ya_s) - \mu'\boldsymbol{\nu}_j(ya_s) \\
&\qquad + \mathbf{u}'[\boldsymbol{\nu}_j(ya_s) - \boldsymbol{\nu}_j(y)]\Big\}^2 dG_n(y)
\end{aligned}
$$

where $\boldsymbol{\nu}_j$ are as in (h) and $\nu_j^*(y) := n^{-1/2}\sum_{i=1}^n d_{nij} f_{ni}(y)$, $1 \le j \le p$. This representation suggests the following approximating candidate:

$$
\hat{K}_{\mathbf{D}}^{0}(s,\mathbf{u}) := \sum_{j=1}^{p} \int \{Y_j^0 + \mathbf{u}'\nu_j + sy\nu_j^* + m_j\}^2 dG_n.
$$

We now state

Lemma 5.5.5. *With γ_n as in (f), assume that $\forall\ s \in \mathbb{R}$,*

$$
\begin{aligned}
(5.5.57)\qquad \lim_{x\to 0}\limsup_n \int \gamma_n((1+sn^{-1/2})y + x)dG_n(y) \\
= \limsup_n \int \gamma_n(y)dG_n(y) < \infty,
\end{aligned}
$$

and

$$
\begin{aligned}
(5.5.58)\qquad \lim_{s\to 0}\limsup_n \int |y|\gamma_n(y+zy)dG_n(y) \\
= \limsup_n \int |y|\gamma_n(y)dG_n(y) < \infty
\end{aligned}
$$

Moreover, assume that $\forall (s,\mathbf{v}) \in [-b,b] \times \mathcal{N}(b) =: \mathcal{C}_1$, $0 < b < \infty$, and $\forall \delta > 0$

$$
\begin{aligned}
&\limsup_n \sum_{j=1}^{p} \int \Big[\sum_{i=1}^{n} d_{nij}^{\pm} \Big\{ F_{ni}(ya_s + \mathbf{c}_{ni}'\mathbf{v} + \delta(n^{-1/2}|y| + \kappa_{ni})) \\
(5.5.59)\qquad &- F_{ni}(ya_s + \mathbf{c}_{ni}\mathbf{v} - \delta(n^{-1/2}|y| + \kappa_{ni}))\Big\}\Big]^2 dG_n(y) \le k\delta^2,
\end{aligned}
$$

for some k not depending on $(s,\mathbf{v})$ and δ.

Then, $\forall\, 0 < b < \infty$,

$$(5.5.60) \qquad E \sup \sum_{j=1}^{p} \int \Big\{ Y_j^0 (1 + sn^{-1/2}) y, \mathbf{u}) - Y_j^0(y) \Big\}^2 dG_n(y) = o(1),$$

where the supremum is taken over $(s, \mathbf{u}) \in C_1$.

Proof. For each $(s, \mathbf{u}) \in C_1$, with $a_s = 1 + sn^{-1/2}$,

$$\begin{aligned} & E \sum_{j=1}^{p} \int \{Y_j^0(ya_s, \mathbf{u}) - Y_j^0(y)\}^2 dG_n(y) \\ & \leq \int_{-B_n}^{B_n} \int \gamma_n(ya_s + z) dG_n(y) dz \\ & \qquad + \int_{-b_n}^{b_n} \int |y| \gamma_n(y + zy) dG_n(y) dz \end{aligned}$$

where $B_n = b \max_i \|\kappa_i\|$, $b_n = bn^{-1/2}$. Therefore, by (5.5.57) and (5.5.58), for every $(s, \mathbf{u}) \in C_1$,

$$E \sum_{j=1}^{p} \int \{Y_j^0(ya_s, \mathbf{u}) - Y_j^0(y)\}^2 dG_n(y) = o(1).$$

Now proceed as in the proof of (5.5.3), using the monotonicity of $V_{jd}(a, \mathbf{t})$, $\mu_j^0(a, \mathbf{u})$ and the compactness of C_1 to conclude (5.5.60). Use (5.5.59) in place of (g). The details are left out as an exercise. □

The proof of the following lemma is quite similar to that of (5.5.10).

Lemma 5.5.6. *Let* $G_n^*(y) = G_n(y/a_\tau)$. *Assume that for each fixed* $(\tau, \mathbf{u}) \in C_1$, (h) *and* (i) *hold with* G_n *replaced by* G_n^*. *In addition, assume the following:*

$$\sum_{j=1}^{p} \int (y\nu_j^*(y))^2 dG_n(y) = O(1).$$

$$\sum_{j=1}^{p} \int \{\mu_j^0(ya_s) - \mu_j^0(y) - \tau y \nu_j^*(y)\}^2 dG_n(y) = o(1), \quad \forall\, |s| \leq b.$$

Then, $\forall\ 0 < b < \infty$,

$$\sup_{(s,\mathbf{u})\in\mathcal{C}_1} \sum_{j=1}^{p} \int \Big\{ \mu_j^0(ya_s, \mathbf{u}) - \mu_j^0(a\tau y) - \mathbf{u}'\nu_j(a\tau y) \Big\}^2 dG_n(y) = o(1),$$

$$\sup_{|s|\le b} \sum_{j=1}^{p} \int \{\mu_j^0(ya_s) - \mu_j^0(y) - \tau y \nu_j^*(y)\}^2 dG_n(y) = o(1).$$

Theorem 5.5.8 *Let* $Y_{n1}, \cdots, Y_{nn}$ *be independent r.v.'s with respective d.f.'s* $F_{n1}, \cdots, F_{nn}$. *Assume* (a) - (e), (h), (j), (5.5.57) - (5.5.59) *and the conditions of Lemma* 5.5.6 *hold. Moreover assume that for each* $|s| \le b$

$$\sum_{j=1}^{p} \int \|\boldsymbol{\nu}_j(ya_s) - \boldsymbol{\nu}_j(y)\|^2 dG_n(y) = o(1).$$

Then, $\forall\ 0 < b < \infty$,

$$E \sup |K_{\mathbf{D}}^0(\tau, \mathbf{u}) - \hat{K}_{\mathbf{D}}^0(\tau, \mathbf{u}) = o(1).$$

where the supremum is taken over $(s, \mathbf{u}) \in \mathcal{C}_1$.

The proof of this theorem is quite similar to that of Theorem 5.5.1. □

5.6 Asymptotic Distributions, Efficiency & Robustness

5.6.1 Asymptotic distributions and efficiency

To begin with consider the *Case* 5.5.1 *and the class of estimators* $\{\hat{\boldsymbol{\beta}}_{\mathbf{D}}\}$. Recall that in this case the errors $\{e_{ni}\}$ of (1.1.1) are correctly modeled to be i.i.d. F, i.e., $H_{ni} \equiv F \equiv F_{ni}$. We shall also take $G_n \equiv G, G \in \mathbb{DI}(\mathbb{R})$. Assume that (5.5.44) - (5.5.46) hold. The various quantities appearing in (5.5.19) and Theorem 5.5.3 now take the following simpler forms.

$$(5.6.1) \qquad \boldsymbol{\Gamma}_n(y) = \mathbf{AX'D}f(y), \ y \in \mathbb{R},$$

$$\mathcal{B}_n = \mathbf{AX'DD'XA} \int f dG, \quad \mathcal{J}_n = -\mathbf{AX'D} \int \mathbf{Y}_{\mathbf{D}}^0 f dG.$$

Note that $\mathcal{B}_n^{-1}$ will exist if and only if the rank of $\mathbf{D}$ is p. Note also that

$$(5.6.2)\quad \mathcal{B}_n^{-1}\mathcal{J}_n = -(\mathbf{D}'\mathbf{X}\mathbf{A})^{-1}\int \mathbf{Y}_\mathbf{D}^0 f dG/(\int f^2 dG)^{-1}$$
$$= (\mathbf{D}'\mathbf{X}\mathbf{A}\int f^2 dG)^{-1}\sum_{i=1}^n \mathbf{d}_i[\psi(e_i) - E\psi(e_i)],$$

where $\psi(y) = \int_{-\infty}^y f dG,\ y \in \mathbb{R}$.

Because $G_n \equiv G \in \mathbb{DI}(\mathbb{R})$, there always exists a $g \in L_r^2(G)$ such that $g > 0$, and $0 < \int g^2 dG < \infty$. The condition (5.5.k) with $g_n \equiv g$ becomes

$$(5.6.3)\quad \liminf_n \inf_{\|\mathbf{e}\|=1} |\mathbf{e}'\mathbf{D}'\mathbf{X}\mathbf{A}\mathbf{e}| \ge \alpha,\quad \text{for some } \alpha > 0.$$

Condition (5.5.l) implies that $\mathbf{e}'\mathbf{D}'\mathbf{X}\mathbf{A}\mathbf{e} \ge 0$ or $\mathbf{e}'\mathbf{D}'\mathbf{X}\mathbf{A}\mathbf{e} \le 0,\ \forall \|\mathbf{e}\| = 1$ and $\forall\ n \ge 1$. It need not imply (5.6.3). The above discussion together with the L-F Cramér-Wold Theorem leads to

Corollary 5.6a.1. *Assume that* (1.1.1) *holds with the error r.v.'s correctly modeled to be i.i.d.* F, F *known. In addition, assume that* (5.5.a), (5.5.b), (5.5.l), (5.5.44) - (5.5.46), (5.6.3) *and* (5.6.4) *hold, where*

$$(5.6.4)\quad (\mathbf{D}'\mathbf{X}\mathbf{A})^{-1} \text{ exists for all } n \ge p.$$

Then,

$$(5.6.5)\quad \mathbf{A}^{-1}(\hat{\boldsymbol{\beta}}_\mathbf{D} - \boldsymbol{\beta})$$
$$= (\mathbf{D}'\mathbf{X}\mathbf{A}\int f^2 dG)^{-1}\sum_{i=1}^n \mathbf{d}_{ni}[\psi(e_{ni}) - E\psi(e_{ni})] + o_p(1).$$

If, in addition, we assume

$$(5.6.6)\quad \max_{1\le i\le n} \|\mathbf{d}_{ni}\|^2 = o(1),$$

then

$$(5.6.7)\quad \boldsymbol{\Sigma}_\mathbf{D}^{-1}\mathbf{A}^{-1}(\hat{\boldsymbol{\beta}}_\mathbf{D} - \boldsymbol{\beta}) \longrightarrow_d \mathcal{N}(\mathbf{0}, \tau^2\mathbf{I}_{p\times p})$$

where

$$\mathbf{\Sigma_D} \;:=\; (\mathbf{D'XA})^{-1}\mathbf{D'D}(\mathbf{AX'D})^{-1}, \;\; \tau^2 = \frac{Var\psi(e_1)}{(\int f^2 dG)^2}. \quad \square$$

For any two square matrices $\mathbf{L}_1$ and $\mathbf{L}_2$ of the same order, by $\mathbf{L}_1 \geq \mathbf{L}_2$ we mean that $\mathbf{L}_1 - \mathbf{L}_2$ is non-negative definite. Let $\mathbf{L}$ and $\mathbf{J}$ be two $p \times n$ matrices such that $(\mathbf{LL'})^{-1}$ exists. The C-S inequality for matrices states that

$$\text{(5.6.8)} \qquad \mathbf{JJ'} \geq \mathbf{JL'}(\mathbf{LL'})^{-1}\mathbf{LJ'}$$

with equality if and only if $\mathbf{J} \propto \mathbf{L}$.

Now note that if $\mathbf{D} = \mathbf{XA}$ then $\mathbf{\Sigma_D} = \mathbf{I}_{p\times p}$. In general, upon choosing $\mathbf{J} = \mathbf{D'}, \mathbf{L} = \mathbf{AX'}$ in this inequality, we obtain

$$\mathbf{D'D} \geq \mathbf{D'XA} \cdot \mathbf{AX'D} \quad \text{or} \quad \mathbf{\Sigma_D} \geq \mathbf{I}_{p\times p}$$

with equality if and only if $\mathbf{D} \propto \mathbf{XA}$. From these observations we deduce

Theorem 5.6a.1 (*Optimality of* $\hat{\boldsymbol{\beta}}_{\mathbf{X}}$). *Suppose that* (1.1.1) *holds with the error r.v.'s correctly modeled to be i.i.d. F. In addition, assume that* (5.5.a), (5.5.d) *with* $G_n \equiv G$, (5.5.44) - (5.5.46) *hold. Then, among the class of estimators* $\{\hat{\boldsymbol{\beta}}_{\mathbf{D}}; \mathbf{D}$ *satisfying* (5.5.b), (5.5.l), (5.6.3), (5.6.4) *and* (5.6.6) $\}$, *the estimator that minimizes the asymptotic variance of* $\mathbf{b'A}^{-1}(\hat{\boldsymbol{\beta}}_{\mathbf{D}} - \boldsymbol{\beta})$, *for every* $\mathbf{b} \in \mathbb{R}^p$, *is* $\hat{\boldsymbol{\beta}}_{\mathbf{X}}-$ *the* $\hat{\boldsymbol{\beta}}_{\mathbf{D}}$ *with* $\mathbf{D} = \mathbf{XA}$. $\square$

Observe that under (5.5.a), $\mathbf{D} = \mathbf{XA}$ *a priori* satisfies (5.5.b), (5.6.3), (5.6.4) and (5.6.6). Consequently we obtain

Corollary 5.6a.2. (*Asymptotic normality of* $\hat{\boldsymbol{\beta}}_{\mathbf{X}}$).
Assume that (1.1.1) *holds with the error r.v.'s correctly modeled to be i.i.d. F. In addition, assume that* (5.5.a) *and* (5.5.44) - (5.5.46) *hold. Then,*

$$\mathbf{A}^{-1}(\hat{\boldsymbol{\beta}}_{\mathbf{X}} - \boldsymbol{\beta}) \longrightarrow_d \mathcal{N}(\mathbf{0}, \tau^2 \mathbf{I}_{p\times p}). \quad \square$$

Remark 5.6a.1. Write $\hat{\boldsymbol{\beta}}_{\mathbf{D}}(G)$ for $\hat{\boldsymbol{\beta}}_{\mathbf{D}}$ to emphasize the dependence on G. The above theorem proves the optimality of $\hat{\boldsymbol{\beta}}_{\mathbf{X}}(G)$ among a class of estimators $\{\hat{\boldsymbol{\beta}}_{\mathbf{D}}(G)$, as $\mathbf{D}$ varies$\}$. To obtain an asymptotically efficient estimator at a given F among the class of estimators

$\{\hat{\boldsymbol{\beta}}_{\mathbf{X}}(G), G$ varies $\}$ one must have F and G satisfy the following relation. Assume that F satisfies (3.2.a) of Theorem 3.2.3 and all of the derivatives that occur below make sense and that (5.5.44) hold. Then, a G that will give asymptotically efficient $\hat{\boldsymbol{\beta}}_{\mathbf{X}}(G)$ must satisfy the relation

$$-f dG = (1/I(f) \cdot d(\dot{f}/f), \quad I(f) := \int (\dot{f}/f)^2 dF.$$

From this it follows that the m.d. estimators $\hat{\boldsymbol{\beta}}_{\mathbf{X}}(G)$, for G satisfying the relations $dG(y) = (2/3)dy$ and $dG(y) = 4d\delta_0(y)$, are asymptotically efficient at logistic and double exponential error d.f.'s, respectively.

For $\hat{\boldsymbol{\beta}}_{\mathbf{X}}(G)$ to be asymptotically efficient at $\mathcal{N}(0,1)$ errors, G would have to satisfy $f(y)dG(y) = dy$. But such a G does not satisfy (5.5.58). Consequently, under the current art of affairs, one can not estimate $\boldsymbol{\beta}$ asymptotically efficiently at the $\mathcal{N}(0,1)$ error d.f. by using a $\hat{\boldsymbol{\beta}}_{\mathbf{X}}(G)$. This naturally leaves one open problem, v.i.z., *Is the conclusion of Corollary* 5.6a.2 *true without requiring* $\int f dG < \infty, 0 < \int f^2 dG < \infty$? □

Observe that Theorem 5.6a.1 does not include the estimator $\hat{\boldsymbol{\beta}}_1$ - the $\hat{\boldsymbol{\beta}}_{\mathbf{D}}$ when $\mathbf{D} = n^{1/2}[1, 0, \cdots, 0]_{n \times p}$ i.e., the m.d. estimator defined at (5.2.4), (5.2.5) after H_{ni} is replaced by F in there. The main reason for this being that the given $\mathbf{D}$ does not satisfy (5.6.4). However, Theorem 5.5.3 is general enough to cover this case also. Upon specializing that theorem and applying (5.5.31) one obtains the following

Theorem 5.6a.2. *Assume that* (1.1.1) *holds with the errors correctly modeled to be i.i.d. F. In addition, assume that* (5.5.*a*), (5.5.44) - (5.5.46) *and the following hold.*

(5.6.9) *Either*

$$n^{-1/2} e_1 \mathbf{x}'_{ni} \mathbf{A} \mathbf{e} \geq 0 \text{ for all } 1 \leq i \leq n, \text{ all } \|\mathbf{e}\| = 1,$$

or

$$n^{-1/2} e_1 \mathbf{x}'_{ni} \mathbf{A} \mathbf{e} \leq 0 \text{ for all } 1 \leq i \leq n, \text{ all } \|\mathbf{e}\| = 1.$$

(5.6.10) $$\liminf_n \inf_{\|\mathbf{e}\|=1} |n^{1/2} e_1 \bar{\mathbf{x}}'_n \mathbf{A} \mathbf{e}| \geq \alpha > 0,$$

where $\overline{\mathbf{x}}_n$ *is as in* (4.2a.11) *and* θ_1 *is the first coordinate of* $\boldsymbol{\theta}$. *Then*

$$(5.6.11) \qquad n^{1/2}\overline{\mathbf{x}}_n'\mathbf{A}\cdot\mathbf{A}^{-1}(\hat{\boldsymbol{\beta}}_1-\boldsymbol{\beta}) = \frac{Z_n}{\int f^2 dG} + o_p(1),$$

where

$$Z_n = n^{-1/2}\sum_{i=1}^{n}\{\psi(e_{ni}) - E\psi(e_{ni})\}, \text{ with } \psi \text{ as in (5.6.2)}.$$

Consequently, $n^{1/2}\overline{\mathbf{x}}_n'(\hat{\boldsymbol{\beta}}_1-\boldsymbol{\beta})$ *is asymptotically a* $\mathcal{N}(0,\tau^2)$ *r.v.* □

Next, we focus on the class of estimators $\{\boldsymbol{\beta}_{\mathbf{D}}^+\}$ and the *case of i.i.d. symmetric errors.* An analogue of Corollary 5.6a.1 is obtained with the help of Theorem 5.5.4 instead of Theorem 5.5.3 and is given in Corollary 5.6a.3. The details of its proof are similar to those of Corollary 5.6a.1.

Corollary 5.6a.3. *Assume that* (1.1.1) *holds with the errors correctly modeled to be i.i.d. symmetric around* 0. *In addition, assume that* (5.3.8), (5.5.*a*), (5.5.*b*), (5.5.*d*) *with* $G_n \equiv G$, (5.5.44), (5.5.46), (5.6.3), (5.6.4) *and* (5.6.12) *hold, where*

$$(5.6.12) \qquad \int_0^\infty (1-F)dG < \infty$$

Then,

$$(5.6.13) \qquad \mathbf{A}^{-1}(\boldsymbol{\beta}_{\mathbf{D}}^+ - \boldsymbol{\beta})$$
$$= -\{2\mathbf{AX'D}\int f^2 dG\}^{-1}\int \mathbf{W}^+(y)f^+(y)dG(y) + o_p(1),$$

where $f^+(y) := f(y) + f(-y)$ *and* $\mathbf{W}^+(y)$ *is* $\mathbf{W}^+(y,\mathbf{0})$ of (5.5.13). *If, in addition,* (5.6.6) *holds, then*

$$(5.6.14) \qquad \boldsymbol{\Sigma}_{\mathbf{D}}^{-1}\mathbf{A}^{-1}(\boldsymbol{\beta}_{\mathbf{D}}^+ - \boldsymbol{\beta}) \longrightarrow_d \mathcal{N}(\mathbf{0},\tau^2\mathbf{I}_{p\times p}). \quad \square$$

Consequently, an analogue of Theorem 5.6a.1 holds for $\boldsymbol{\beta}_{\mathbf{X}}^+$ also and Remark 5.6a.1 applies equally to the class of estimators $\{\boldsymbol{\beta}_{\mathbf{X}}^+(G),$ G varies$\}$, assuming that the errors are symmetric around 0. We leave it to interested readers to state and prove an analogue of Theorem 5.6a.2 for $\boldsymbol{\beta}_1^+$.

Now consider the class of estimators $\{\boldsymbol{\beta}_{\mathbf{D}}^*\}$ of (5.2.18). Recall the notation in (5.5.39) and Theorem 5.5.6. The distributions of these estimators will be discussed when the errors in (1.1.1) are correctly modeled to be i.i.d. F, F an arbitrary d.f. and when $L_n \equiv L$. In this case various entities of Theorem 5.5.6 acquire the following forms:

$$\begin{aligned}
\boldsymbol{\mu}_{\mathbf{D}} &\equiv \mathbf{0}; \quad \ell_{ni}(s) \equiv 1; \quad \mathbf{D}(s) \equiv \mathbf{D}, \quad \text{under (5.2.21)}; \\
\boldsymbol{\Gamma}_n^*(s) &\equiv \mathbf{A}_1 \mathbf{X}_c \mathbf{D} q(s), \quad q = f(F^{-1}); \\
\mathcal{J}_n^* &= -\mathbf{A}_1 \mathbf{X}_c' \mathbf{D} \int \mathbf{Y}_{\mathbf{D}} q dL = \mathbf{A}_1 \mathbf{X}_c' \mathbf{D} \sum_{i=1}^n \mathbf{d}_{ni} \varphi_0(F(e_{ni})); \\
\mathcal{B}_n^* &= (\mathbf{A}_1 \mathbf{X}_c' \mathbf{D} \mathbf{D}' \mathbf{X}_c \mathbf{A}_1) \int q^2 dL,
\end{aligned}$$

where $\mathbf{X}_c$ and $\mathbf{A}_1$ are defined at (4.2.11) and where

(5.6.15) $$\varphi_0(u) := \int_0^u q(s) dL(s), \quad 0 \le u \le 1.$$

Arguing as for Corollary 5.6a.1, one obtains the following

Corollary 5.6a.4. *Assume that* (1.1.1) *holds with the errors correctly modeled to be i.i.d.* F *and that* L *is a d.f. In addition, assume that* (**F1**), ($\mathbf{NX}_c$), (5.2.16), (5.5.b), *and the following hold.*

(5.6.16) $$\liminf_n \inf_{\|\mathbf{e}\|=1} |\mathbf{e}' \mathbf{D}' \mathbf{X}_c \mathbf{A}_1 \mathbf{e}| \ge \alpha > 0.$$

(5.6.17) *Either*

$$\mathbf{e}' \mathbf{d}_{ni} (\mathbf{x}_{ni} - \bar{\mathbf{x}}_n)' \mathbf{A}_1 \mathbf{e} \ge 0, \ \forall\, 1 \le i \le n, \ \ \forall\, \|\mathbf{e}\| = 1,$$

or

$$\mathbf{e}' \mathbf{d}_{ni} (x_{ni} - \bar{\mathbf{x}}_n)' \mathbf{A}_1 \mathbf{e} \le 0, \ \forall\, 1 \le i \le n, \ \forall\, \|\mathbf{e}\| = 1.$$

(5.6.18) $$\Big(\mathbf{D}' \mathbf{X}_c \mathbf{A}_1\Big)^{-1} \ \text{exists for all} \ \ n \ge p.$$

Then,

(5.6.19) $$\begin{aligned} &\mathbf{A}_1^{-1} (\boldsymbol{\beta}_{\mathbf{D}}^* - \boldsymbol{\beta}) \\ &= (\mathbf{D}' \mathbf{X}_c \mathbf{A}_1 \int_0^1 q^2 dL)^{-1} \sum_{i=1}^n \mathbf{d}_{ni} \varphi(F(e_{ni})) + o_p(1). \end{aligned}$$

If, in addition, (5.6.6) *holds, then*

$$(\Sigma^*_{\mathbf{D}})^{-1}\mathbf{A}_1^{-1}(\boldsymbol{\beta}^*_{\mathbf{D}} - \boldsymbol{\beta}) \longrightarrow_d \mathcal{N}(\mathbf{0}, \sigma_0^2\mathbf{I}_{p\times p}) \tag{5.6.20}$$

where $\Sigma^*_{\mathbf{D}} = (\mathbf{D}'\mathbf{X}_c\mathbf{A}_1)^{-1}\mathbf{D}'\mathbf{D}(\mathbf{A}_1\mathbf{X}_c'\mathbf{D})^{-1}$, and

$$\sigma_0^2 = \frac{Var\varphi_0(F(e_1))}{(\int_0^1 q^2 dL)^2}, \tag{5.6.21}$$

with φ_0 as in (5.6.15).

Consequently,

$$\mathbf{A}_1^{-1}(\boldsymbol{\beta}^*_{\mathbf{X}_c} - \boldsymbol{\beta}) \longrightarrow_d \mathcal{N}(\mathbf{0}, \sigma_0^2\mathbf{I}_{p\times p}) \tag{5.6.22}$$

and $\{\boldsymbol{\beta}^*_{\mathbf{X}_c}\}$ *is asymptotically efficient among all* $\{\boldsymbol{\beta}^*_{\mathbf{D}}, \mathbf{D}$ *satisfying above conditions .* □

Consider the case when $L(s) \equiv s$. Then

$$\sigma_0^2 = \frac{\int\int[F(x \wedge y) - F(x)F(y)]f^2(x)f^2(y)dxdy}{\left(\int f^3(x)dx\right)^2}.$$

It is interesting to make a numerical comparison of this variance with that of some other well celebrated estimators. Let $\sigma_w^2, \sigma_{lad}^2, \sigma_{ls}^2$ and σ_{ns}^2 denote the respective asymptotic variances of the Wilcoxon rank, the least absolute deviation, the least square and the normal scores estimators of $\boldsymbol{\beta}$. Recall, from Chapter 4 that

$$\begin{aligned} \sigma_w^2 &= \left\{12\left(\int f^2(x)dx\right)^2\right\}^{-1}; \quad \sigma_{lad}^2 = (2f(0))^{-2}; \\ \sigma_{ls}^2 &= \sigma^2; \quad \sigma_{ns}^2 = \left\{[\int f^2(x)/\varphi(\Phi^{-1}(F))]dx\right\}^{-2}; \end{aligned}$$

where σ^2 is the error variance. Using these we obtain the following table.

Table 1

F	σ_0^2	σ_w^2	σ_{lad}^2	σ_{ns}^2	σ^2
Double Exp.	1.2	1.333	1	$\pi 2$	2
Logistic	3.0357	3	4	π	$\pi^2/3$
Normal	1.0946	$\pi/3$	$\pi/2$	1	1
Cauchy	2.5739	3.2899	2.46		∞

It thus follows that the m.d. estimator $\beta^*_{\mathbf{X}_c}(L)$, with $L(s) \equiv s$, is superior to the Wilcoxon rank estimator and the LAD estimator at double exponential and logistic errors, respectively. At normal errors, it has smaller variance than the LAD estimator and compares favorably with the optimal estimator. The same is true for the m.d. estimator $\hat{\beta}_{\mathbf{X}}(F)$.

Next, we shall discuss $\overline{\beta}$. In the following theorem the framework is the same as in Theorem 5.5.7. Also see (5.5.82) for the definitions of $\overline{\boldsymbol{\nu}}_n, \mathcal{B}_{1n}$ etc.

Theorem 5.6a.3. *In addition to the assumptions of Theorem* 5.5.7 *assume that*

$$(5.6.23)\quad \liminf_n \inf_{\|\boldsymbol{\theta}\|=1} \Big|\int \overline{\boldsymbol{\nu}}_n' d\alpha_n \boldsymbol{\theta}\Big| \geq \alpha, \quad \text{for some} \quad \alpha > 0.$$

Moreover, assume that (5.6.9)) *holds and that*

$$(5.6.24)\qquad \mathcal{B}_{1n}^{-1} \text{ exists for all } n \geq p.$$

Then,

$$\begin{aligned}(5.6.25)\quad & \mathbf{A}^{-1}(\overline{\beta} - \beta) \\ &= -\mathcal{B}_{1n}^{-1} \int\int \overline{\boldsymbol{\nu}}_n(s,y)\{\mathcal{M}_{1n}(s,y) + m_n(s,y)\}d\alpha_n(s,y) \\ &\qquad +o_p(1).\end{aligned}$$

Proof. The proof of (5.6.25) is similar to that of (5.5.33), hence no details are given. □

Corollary 5.6a.5. *Suppose the conditions of Theorem* 5.6a.3 *are satisfied by* $F_{ni} \equiv F \equiv H_{ni}$, $G_n \equiv G$, $L_n \equiv L$, *where* F *is supposed to have continuous density* f. *Let*

$$\begin{aligned}\mathbf{C} &= \int\int\int_0^1\int_0^1 \Big[\Big\{\mathbf{A}n^{-1}\sum_{i=1}^{ns}\sum_{j=1}^{nt} \mathbf{x}_i\mathbf{x}_j' f_i(y) f_j(y)\mathbf{A}\Big\}(s \wedge t) \\ &\qquad \times \Big(F(y\wedge z) - F(y)F(z)\Big)\Big] d\alpha(s,y)d\alpha(t,z),\end{aligned}$$

where $f_i(y) = f(y - \mathbf{x}_i'\beta)$, *and* $d\alpha(s,y) = dL(s)dG(y)$. *Then the asymptotic distribution of* $\mathbf{A}^{-1}(\overline{\beta}-\beta)$ *is* $\mathcal{N}(\mathbf{0}, \boldsymbol{\Sigma}_0(\beta))$ *where* $\boldsymbol{\Sigma}_0(\beta) = \mathcal{B}_{1n}^{-1}\mathbf{C}\mathcal{B}_{1n}^{-1}$. □

Because of the dependence of $\mathbf{\Sigma}_0$ on $\boldsymbol{\beta}$, no clear cut comparison between $\overline{\boldsymbol{\beta}}$ and $\hat{\boldsymbol{\beta}}_{\mathbf{X}}$ in terms of their asymptotic covariance matrices seems to be feasible. However, some comparison at a given $\boldsymbol{\beta}$ can be made. To demonstrate this, consider the case when $L(s) = s$, $p = 1$ and $\beta_1 = 0$. Write x_i for x_{i1} etc.

Note that here, with $\tau_x^2 = \sum_{i=1}^n x_i^2$,

$$\begin{aligned}
\mathcal{B}_{1n} &= \tau_x^{-2} \int_n^1 n^{-1} \sum_{i=1}^{ns} x_i \sum_{j=1}^{ns} x_j ds \cdot \int f^2 dG, \\
C &= \tau_x^{-2} \int_0^1 \int_0^1 n^{-1} \sum_{i=1}^{ns} x_i \sum_{j=1}^{nt} x_j (s \wedge t) ds dt \\
&\quad \times \int \int [F(y \wedge z) - F(y)F(z)] d\psi(y) d\psi(z).
\end{aligned}$$

Consequently

$$\begin{aligned}
\Sigma_0(0) &= \frac{\tau_x^{-2} \int_0^1 \int_0^1 (s \wedge t) n^{-1} \sum_{i=1}^{ns} x_i \sum_{j=1}^{nt} x_j ds dt}{(\tau_x^{-2} \int_0^1 n^{-1} \sum_{i=1}^{ns} x_i \sum_{j=1}^{ns} x_j (ds)^2} \times \tau^2 \\
&= r_n \times \tau^2, \qquad \text{say.}
\end{aligned}$$

Recall that τ^2 is the asymptotic variance of $\tau_x(\hat{\beta}_x - \beta)$. Direct integration shows that in the cases $x_i \equiv 1$ and $x_i \equiv i$, $r_n \to \frac{18}{15}$ and $\frac{50}{21}$, respectively. Thus, in the cases of the one sample location model and the first degree polynomial through the origin, in terms of the asymptotic variance, $\hat{\beta}_x$ dominates $\overline{\beta}$ with $L(s) = s$ at $\beta = 0$. □

5.6.2 Robustness

In a linear regression setup an estimator needs to be robust against departures in the assumed design variables and the error distributions. As seen in the previous section, one purpose of having general weights $\mathbf{D}$ in $\hat{\boldsymbol{\beta}}_{\mathbf{D}}$ was to prove that $\hat{\boldsymbol{\beta}}_{\mathbf{X}}$ is asymptotically efficient among a certain class of m.d. estimators $\{\hat{\boldsymbol{\beta}}_{\mathbf{D}}, \mathbf{D} \text{ varies}\}$. Another purpose is to robustify these estimators against the extremes in the design by choosing $\mathbf{D}$ to be a bounded function of $\mathbf{X}$ that satisfies all other conditions of Theorem 5.6a.1. Then the corresponding $\hat{\boldsymbol{\beta}}_{\mathbf{D}}$ would be asymptotically normal and robust against the extremes in the design, but not as efficient as $\hat{\boldsymbol{\beta}}_{\mathbf{X}}$. This gives another example of the phenomenon that compromises efficiency in return for robustness. A similar remark applies to $\{\boldsymbol{\beta}^+_{\mathbf{D}}\}$ and $\{\boldsymbol{\beta}^*_{\mathbf{D}}\}$.

We shall now focus on the *qualitative robustness* (see Definition 4.4.1) of $\hat{\boldsymbol{\beta}}_{\mathbf{X}}$ and $\boldsymbol{\beta}^+_{\mathbf{X}}$. For simplicity, we shall write $\hat{\boldsymbol{\beta}}$ and $\boldsymbol{\beta}^+$ for $\hat{\boldsymbol{\beta}}_{\mathbf{X}}$ and $\boldsymbol{\beta}^+_{\mathbf{X}}$, respectively, in the rest of this section. To begin with consider $\hat{\boldsymbol{\beta}}$. Recall Theorem 5.5.3 and the notation of (5.5.19). We need to apply these to the case when the errors in (1.1.1) are modeled to be i.i.d. F, but their actual d.f.'s are $\{F_{ni}\}$, $\mathbf{D} = \mathbf{XA}$ and $G_n = G$. Then various quantities in (5.5.19) acquire the following form.

$$\begin{aligned}
\boldsymbol{\Gamma}_n(y) &= \mathbf{AX}'\boldsymbol{\Lambda}^*(y)\mathbf{XA}, \qquad \mathcal{B}_n = \mathbf{AX}'\int \boldsymbol{\Lambda}^*\boldsymbol{\Pi}\boldsymbol{\Lambda}^* dG\mathbf{XA},\\
\mathcal{J}_n &= \int \boldsymbol{\Gamma}_n(y)\mathbf{AX}'[\boldsymbol{\alpha}_n(y) + \boldsymbol{\Delta}_n(y)]dG(y) = \mathbf{Z}_n + \mathbf{b}_n, \quad \text{say},
\end{aligned}$$

where for $1 \le i \le n,\ y \in \mathbb{R}$,

$$\begin{aligned}
\alpha_{ni}(y) &:= I(e_{ni} \le y) - F_{ni}(y), \quad \Delta_{ni}(y) := F_{ni}(y) - F(y);\\
\boldsymbol{\alpha}'_n &:= (\alpha_{n1}, \alpha_{n2}, \cdots, \alpha_{nn}), \quad \boldsymbol{\Delta}'_n := (\Delta_{n1}, \Delta_{n2}, \cdots, \Delta_{nn});\\
\boldsymbol{\Pi} &:= \mathbf{X}(\mathbf{X}'\mathbf{X})^{-1}\mathbf{X}'; \qquad \mathbf{b}_n := \int \boldsymbol{\Gamma}_n(y)\mathbf{AX}'\boldsymbol{\Delta}_n(y)dG(y).
\end{aligned}$$

The assumption (5.5.a) ensures that the design matrix $\mathbf{X}$ is of the full rank p. This in turn implies the existence of $\mathcal{B}_n^{-1}$ and the satisfaction of (5.5.b), (5.5.l) in the present case. Moreover, because $G_n \equiv G$, (5.5.k) now becomes

$$\text{(5.6.26)} \qquad \liminf_n \inf_{\|\mathbf{e}\|=1} k_n(\mathbf{e}) \ge \gamma, \quad \text{for some } \gamma > 0,$$

where

$$k_n(\mathbf{e}) := \mathbf{e}'\mathbf{AX}'\int \boldsymbol{\Lambda}^* g dG\,\mathbf{XAe}, \qquad \|\mathbf{e}\| = 1,$$

and where g is a function from $\mathbb{R}$ to $[0, \infty], 0 < \int g^r dG < \infty, r = 1, 2$. Because G is a σ - finite measure, such a g always exists.

Upon specializing Theorem 5.5.3 to the present case, we readily obtain

Corollary 5.6b.1. *Assume that in* (1.1.1) *the actual and modeled d.f.'s of the errors* $\{e_{ni}, 1 \le i \le n\}$ *are* $\{F_{ni}, 1 \le i \le n\}$ *and* F, *respectively. In addition, assume that* (5.5.a), (5.5.c) - (5.5.j) *with* $\mathbf{D} = \mathbf{XA}$, $H_{ni} \equiv F$, $G_n \equiv G$, *and* (5.6.26) *hold. Then*

$$\mathbf{A}^{-1}(\hat{\boldsymbol{\beta}} - \boldsymbol{\beta}) = -\boldsymbol{\mathcal{B}}_n^{-1}\{\mathbf{Z}_n + \mathbf{b}_n\} + o_p(1). \qquad \Box$$

Observe that $\boldsymbol{\mathcal{B}}_n^{-1}\mathbf{b}_n$ measures the amount of the asymptotic bias in the estimator $\hat{\boldsymbol{\beta}}$ when $F_{ni} \neq F$. Our goal here is to obtain the asymptotic distribution of $\mathbf{A}^{-1}(\hat{\boldsymbol{\beta}} - \boldsymbol{\beta})$ when $\{F_{ni}\}$ converge to F in a certain sense. The achievement of this goal is facilitated by the following lemma. Recall that for any square matrix $\mathbf{L}$, $\|\mathbf{L}\|_\infty = \sup\{\|\mathbf{t}'\mathbf{L}\| \leq 1\}$. Also recall the fact that

$$\|\mathbf{L}\|_\infty \leq \{tr.\mathbf{L}\mathbf{L}'\}^{1/2}, \tag{5.6.27}$$

where tr. denotes the trace operator.

Lemma 5.6b.1. *Let F and G satisfy* (5.5.44). *Assume that* (5.5.e) *and* (5.5.j) *are satisfied by* $G_n \equiv G, \{F_{ni}\}, H_{ni} \equiv F$ *and* $\mathbf{D} = \mathbf{XA}$. *Moreover assume that* (5.5.c) *holds and that*

$$\rho_n := \int \Big(\sum_{i=1}^n \|\mathbf{x}_{ni}' A\|^2 |f_{ni} - f|\Big)^2 dG = o(1). \tag{5.6.28}$$

Then with $\mathbf{I} = \mathbf{I}_{p\times p}$,

(i) $\|\boldsymbol{\mathcal{B}}_n - \mathbf{I}\int f^2 dG\|_\infty = o(1)$.

(ii) $\|\boldsymbol{\mathcal{B}}_n^{-1} - \mathbf{I}(\int f^2 dG)^{-1}\|_\infty = o(1)$.

(iii) $|tr.\boldsymbol{\mathcal{B}}_n - p\int f^2 dG| = o(1)$.

(iv) $|\sum_{j=1}^p \int \|\boldsymbol{\nu}_j\|^2 dG - p\int f^2 dG| = o(1)$.

(v) $\|\mathbf{b}_n - \int \mathbf{AX}'\boldsymbol{\Delta}_n(y) f(y) dG(y)\} = o(1)$.

(vi) $\|\mathbf{Z}_n - \int \mathbf{AX}'\boldsymbol{\alpha}_n(y) f(y) dG(y)\| = o_p(1)$.

(vii) $\sup_{\|\mathbf{e}\|=1} |k_n(\mathbf{e}) - \int f g dG| = o(1)$.

Remark 5.6b.1. Note that the condition (5.5.j) with $\mathbf{D} = \mathbf{XA}$, $G_n \equiv G$ now becomes

$$\int \|\mathbf{AX}'\boldsymbol{\Delta}_n\|^2 dG = O(1). \tag{5.6.29}$$

Proof. To begin with, because $\mathbf{AX}'\mathbf{XA} \equiv \mathbf{I}$, we obtain the relation

$$\begin{aligned}
\boldsymbol{\Gamma}_n(y)\boldsymbol{\Gamma}_n'(y) - f^2(y)\mathbf{I} &= \mathbf{AX}'[\boldsymbol{\Lambda}^*(y) - f(y)\mathbf{I}]\mathbf{XA} \\
&\qquad \cdot \mathbf{AX}'[\boldsymbol{\Lambda}^*(y) - f(y)\mathbf{I}]\mathbf{XA} \\
&= \mathbf{AX}'\boldsymbol{\mathcal{C}}(y)\mathbf{XA} \cdot \mathbf{AX}'\boldsymbol{\mathcal{C}}(y)\mathbf{XA} \\
&= \boldsymbol{\mathcal{D}}(y)\boldsymbol{\mathcal{D}}'(y), \qquad y \in \mathbb{R},
\end{aligned}$$

where $\mathcal{C}(y) := \mathbf{\Lambda}^*(y) - \mathbf{I}f(y)$, $\mathcal{D}(y) := \mathbf{AX}'\mathcal{C}(y)\mathbf{XA}, y \in \mathbb{R}$. Therefore,

$$\text{(5.6.30)} \quad \|\mathcal{B}_n - \mathbf{I}\int f^2 dG\|_\infty \le \sup_{\|\mathbf{t}\|\le 1} \int \|\mathbf{t}'\mathcal{D}(y)\mathcal{D}'(y)\|dG(y) \le \int \{tr.\mathbf{LL}'\}^{1/2} dG$$

where $\mathbf{L} = \mathcal{D}\mathcal{D}'$. Note that, by the C-S inequality,

$$tr.\mathbf{LL}' = tr.\mathcal{D}\mathcal{D}'\mathcal{D}'\mathcal{D} \le \{tr.\mathcal{D}\mathcal{D}'\}^2.$$

Let $\delta_i = f_i - f$, $1 \le i \le n$. Then

$$\begin{aligned}
\text{(5.6.31)} \quad |tr.\mathcal{D}\mathcal{D}'| &= \left| tr. \sum_{i=1}^n \sum_{j=1}^n \mathbf{Ax}_i\mathbf{x}_i'\mathbf{A} \cdot \mathbf{Ax}_j\mathbf{x}_j'\mathbf{A} \cdot \delta_i\delta_j \right| \\
&= \left| \sum_{i=1}^n \sum_{j=1}^n \delta_i\delta_j (\mathbf{x}_j'\mathbf{AAx}_i)^2 \right| \\
&\le \sum_{i=1}^n \sum_{j=1}^n |\delta_i\delta_j| \cdot \|\mathbf{x}_i'\mathbf{A}\|^2 \cdot \|\mathbf{x}_j'\mathbf{A}\|^2 \\
&= \left(\sum_{i=1}^n \|\mathbf{Ax}_i\|^2 |\delta_i| \right)^2 = \rho_n.
\end{aligned}$$

Consequently, from (5.6.30) and (5.6.31),

$$\|\mathcal{B}_n - \mathbf{I}\int f^2 dG\|_\infty \le \int \left(\sum_{i=1}^n \|\mathbf{Ax}_i\| \right)^2 |f_i - f|^2 dG = o(1),$$

by (5.6.28). This proves (i) while (ii) follows from (i) by using the determinant and co-factor formula for the inverses.

Next, (iii) follows from (5.6.28) and the fact that

$$|tr.\mathcal{B}_n - p\int f^2 dG| = |\int tr.\mathcal{D}\mathcal{D}' dG| \le \rho_n, \quad \text{by (5.6.31)}.$$

To prove (iv), note that with $\mathbf{D} = \mathbf{XA}$,

$$\begin{aligned}
&\sum_{j=1}^p \int \|\boldsymbol{\nu}_j\|^2 dG \\
&= \sum_{i=1}^n \sum_{k=1}^n \int \mathbf{x}_i'\mathbf{AAx}_k\mathbf{x}_k'\mathbf{AAx}_i f_i(y) f_k(y) dG(y).
\end{aligned}$$

Note that the R.H.S. is $p\int f^2 dG$ in the case $f_i \equiv f$. Thus

$$\left|\sum_{j=1}^{p}\int \|\boldsymbol{\nu}_j\|^2 dG - p\int f^2 dG\right| = \left|\int tr.\mathcal{D}\mathcal{D}' dG\right| \le \rho_n.$$

This and (5.6.28) prove (iv).

Similarly, with $\mathbf{d}_j'(y)$ denoting the j-th row of $\mathcal{D}(y)$, $1 \le j \le p$,

$$\begin{aligned}
\left\|\mathbf{b}_n - \int \mathbf{A}\mathbf{X}'\boldsymbol{\Delta}_n f dG\right\|^2 &= \|\int \mathcal{D}\mathbf{A}\mathbf{X}\boldsymbol{\Delta} dG\|^2 \\
&= \sum_{j=1}^{p}\Big\{\int \mathbf{d}_j'(y)\mathbf{A}\mathbf{X}'\boldsymbol{\Delta}_n(y)dG(y)\Big\}^2 \\
&\le \rho_n \int \|\mathbf{A}\mathbf{X}'\boldsymbol{\Delta}_n(y)\|^2 dG(y) \qquad (5.6.32)
\end{aligned}$$

and

$$\begin{aligned}
&\|\mathbf{Z}_n - \int \mathbf{A}\mathbf{X}'\boldsymbol{\alpha}_n(y)f(y)dG(y)\|^2 \qquad (5.6.33)\\
&\le \rho_n \int \|\mathbf{A}\mathbf{X}'\boldsymbol{\alpha}_n\|^2 dG.
\end{aligned}$$

Moreover,

$$E\int \|\mathbf{A}\mathbf{X}'\alpha_n\|^2 dG = \int \sum_{i=1}^{n}\|\mathbf{x}_i'\mathbf{A}\|^2 F_i(1-F_i)dG. \qquad (5.6.34)$$

Consequently, (v) follows from (5.6.28), (5.6.29) and (5.6.32). The claim (vi) follows from (5.5.e), (5.6.28), (5.6.33) and (5.6.34). Finally, with $\mathcal{D}^{1/2} = \mathbf{A}\mathbf{X}'\mathcal{C}^{1/2}$, $\forall\ \boldsymbol{e} \in \mathbb{R}^p$,

$$\left|k_n(\boldsymbol{e}) - \int fgdG\right| = \left|\boldsymbol{e}'\int \mathcal{D}gdG\ \boldsymbol{e}\right| = \int \|\boldsymbol{e}'\mathcal{D}^{1/2}\|^2 gdG.$$

Therefore,

$$\begin{aligned}
\sup_{\|\boldsymbol{e}\|=1}\left|k_n(\boldsymbol{e}) - \int fgdG\right| &\le \int\left\{\sum_{i=1}^{n}\|\mathbf{A}\mathbf{x}_i\|^2|f_1 - f|\right\}gdG \\
&\le \rho_n\left\{\int g^2 dG\right\}^{1/2} = o(1). \text{ by } (6). \quad \square
\end{aligned}$$

Corollary 5.6b.2. *Assume that* (1.1.1) *holds with the actual and the modeled d.f.'s of* $\{e_{ni}, 1 \le i \le n\}$ *equal to* $\{F_{ni}, 1 \le i \le n\}$ *and* F, *respectively. In addition, assume that* (5.5.*a*), (5.5.*c*) - (5.5.*g*), (5.5.*i*), (5.5.*j*) *with* $\mathbf{D} = \mathbf{XA}$, $H_{ni} \equiv F$, $G_n \equiv G$; (5.5.44) *and* (5.6.28) *hold.*

Then, (5.5.*h*) *and* (5.5.*b*) *are satisfied and*

$$\mathbf{A}^{-1}(\hat{\boldsymbol{\beta}} - \boldsymbol{\beta}) = -\Big(\int f^2 dG\Big)^{-1}\{\hat{\mathbf{Z}}_n + \hat{\mathbf{b}}_n\} + o_p(1) \tag{5.6.35}$$

where

$$\begin{aligned}
\hat{\mathbf{Z}}_n &:= \int \mathbf{A}\mathbf{X}'\boldsymbol{\alpha}_n(y)d\psi(y) \\
&= \mathbf{A}\sum_{i=1}^n \mathbf{x}_{ni}[\psi(e_{ni}) - \int \psi(x)dF_{ni}(x)], \\
\hat{\mathbf{b}}_n &:= \int \mathbf{A}\mathbf{X}'\boldsymbol{\Delta}_n(y)d\psi(y) = \int \sum_{i=1}^n \mathbf{A}\mathbf{x}_{ni}[F_{ni} - F]d\psi,
\end{aligned}$$

with ψ *as in* (5.6.2). □

Consider $\hat{\mathbf{Z}}_n$. Note that with $\sigma_{ni}^2 := Var\{\psi(e_{ni})|F_{ni}\}$, $1 \le i \le n$,

$$E\hat{\mathbf{Z}}_n\hat{\mathbf{Z}}_n' = \sum_{i=1}^n \mathbf{A}\mathbf{x}_{ni}\mathbf{x}_{ni}'\mathbf{A}\cdot\sigma_{ni}^2.$$

One can rewrite

$$\sigma_{ni}^2 = \int\int [F_{ni}(x \wedge y) - F_{ni}(x)F_{ni}(y)]\, d\psi(x)d\psi(y), \quad 1 \le i \le n.$$

By (5.5.44), ψ is nondecreasing and bounded. Hence $\max_i \|F_{ni} - F\|_\infty \to 0$ readily implies that $\max_i \sigma_{ni}^2 \to \sigma^2, \sigma^2 := Var\{\psi(e)|F\}$. Moreover, we have the inequality

$$|e\hat{\mathbf{Z}}_n\hat{\mathbf{Z}}_n' - \sigma^2\mathbf{I}_{p\times p}| \le \sum_{i=1}^n \|\mathbf{A}\mathbf{x}_{ni}\|^2|\sigma_{ni}^2 - \sigma^2|.$$

It thus readily follows from the L-F CLT that (5.5.*a*) implies that $\hat{\mathbf{Z}}_n \to_d \mathcal{N}(\mathbf{0}, \sigma^2\mathbf{I}_{p\times p})$, if $\max_i \|F_{ni} - F\|_\infty \to 0$. Consequently, we have

Theorem 5.6b.1. **(Qualitative Robustness).** *Assume the same setup and conditions as in Corollary* 5.6b.2. *In addition, suppose that*

$$\max_i \|F_{ni} - F\|_\infty = o(1), \tag{5.6.36}$$

$$\|\mathbf{A}\|_\infty = o(1). \tag{5.6.37}$$

Then, the distribution of $\hat{\boldsymbol\beta}$ *under* $\prod_{i=1}^n F_{ni}$ *converges weakly to the degenerate distribution, degenerate at* $\boldsymbol\beta$.

Proof. It suffices to show that the asymptotic bias is bounded. To that effect we have the inequality

$$\left\|\left(\int f^2 dG\right)^{-1}\hat{\mathbf{b}}_n\right\|^2 \le \int \|AX'\Delta\|^2 dG < \infty, \quad \text{by (7).}$$

From this, (5.6.35), and the above discussion about $\{\hat{\mathbf{Z}}_n\}$, we obtain that $\forall\ \eta > 0$, $\exists\ K_\eta$ such that $P^n(E_\eta) \to 1$ where P^n denotes the probability under $\prod_{i=1}^n F_{ni}$ and $E_\eta = \{\|\mathbf{A}^{-1}(\hat{\boldsymbol\beta} - \boldsymbol\beta)\| \le K_\eta\}$. Theorem now follows from this and the elementary inequality $\|\hat{\boldsymbol\beta} - \boldsymbol\beta\| \le \|\mathbf{A}\|_\infty \|\mathbf{A}^{-1}(\hat{\boldsymbol\beta} - \boldsymbol\beta)\|$. □

Remark 5.6b.2. The conditions (5.6.28) and (5.6.36) together need not imply (5.5.g), (5.5.i) and (5.5.j). The condition (5.5.j) is heavily dependent on the rate of convergence in (5.6.36). Note that

$$\|\hat{\mathbf{b}}_n\|^2 \le \min\Big\{\psi(\infty)\int \|\mathbf{AX}'\boldsymbol\Delta\|^2 d\psi, \left(\int f^2 dG\right)\int \|\mathbf{AX}'\boldsymbol\Delta\|^2 dG\Big\}. \tag{5.6.38}$$

This inequality shows that because of (5.5.44), it is possible to have $\|\hat{\mathbf{b}}_n\|^2 = O(1)$ even if (5.6.29) (or (5.5.j) with $\mathbf{D} = \mathbf{XA}$) may not be satisfied. However, our general theory requires (5.6.29) any way.

Now, with $\varphi = \psi$ or G,

$$\begin{aligned} \int \|\mathbf{AX}'\boldsymbol\Delta\|^2 d\varphi &= \int \sum_{i=1}^n \sum_{j=1}^n \mathbf{x}_i'\mathbf{A}\mathbf{A}\mathbf{x}_j \Delta_j \Delta_j d\varphi \\ &\le \int \left(\sum_{i=1}^n \|\mathbf{Ax}_i\| |\Delta_i|\right)^2 d\varphi. \end{aligned} \tag{5.6.39}$$

Thus, if

$$(5.6.40)\qquad \sum_{i=1}^{n} \|\mathbf{Ax}_i\| |F_i(y) - F)y)| \le k\Delta_n^*(y), \quad y \in]R,$$

where k is a constant and Δ_n^* is a function such that

$$(5.6.41)\qquad \limsup_n \int (\Delta_n^*)^2 d\varphi < \infty,$$

then (5.6.29) would be satisfied and in view of (5.6.38), $\|\hat{\mathbf{b}}_n\| = O(1)$.

The inequality (5.6.40) clearly shows that not every sequence $\{F_{ni}\}$ satisfying (5.6.28), (5.6.36) and (5.5.c) - (5.5.i) with $\mathbf{D} = \mathbf{XA}$ will satisfy (5.6.29). The rate at which $F_{ni} \Longrightarrow F$ is crucial for the validity of (5.6.29) or (5.6.40). □

We now discuss two interesting examples.

Example 5.6b.1. $F_{ni} = (1 - \delta_{ni})F + \delta_{ni}F_0, 1 \le i \le n$. This is the Case 5.5.2. From the Claim 5.5.3, (5.5.e) - (5.5.i) are satisfied by this model as long as (5.5.44) - (5.5.46) and (5.5.a) hold. To see if (5.6.28) and (5.6.29) are satisfied, note that here

$$\begin{aligned} \rho_n &= \int \Big(\sum_{i=1}^{n} \|\mathbf{Ax}_i\|\Big)^2 \delta_i |f - f_0\|^2 dG \\ &\le 2 \max_i \delta_i^2 p^2 \cdot \int (f^2 + f_0^2) dG, \end{aligned}$$

and

$$\sum_{i=1}^{n} \|\mathbf{Ax}_i\| |F_i - F| = \sum_{i=1}^{n} \|\mathbf{Ax}_i\| \delta_i |F - F_0|.$$

Consequently, here (5.6.28) is implied by (5.5.44) for (f, G), (f_0, G) and by (5.5.47), while (5.6.29) follows from (5.5.48), (5.6.39) to (5.6.41) upon taking $\Delta_n^* \equiv |F - F_0|$, provided we additionally assume that

$$(5.6.42)\qquad \sum_{i=1}^{n} \|\mathbf{Ax}_i\| \delta_i = O(1).$$

There are two obvious choices of $\{\delta_i\}$ that satisfy (5.6.42). They are:

(5.6.43) (a) $\delta_{ni} = n^{-1/2}$

or

(b) $\delta_{ni} = p^{-1/2} \|\mathbf{Ax}_{ni}\|, \quad 1 \le i \le n.$

The gross error models with $\{\delta_i\}$ given by (5.6.43b) are more natural than those given by (5.6.43a) to linear regression models with unbounded designs. We suggest that in these models, a proportion of contamination one can allow for the ith observation is $p^{-1/2}\|\mathbf{A}\mathbf{x}_i\|$. If δ_i is larger than this in the sense that $\sum_{i=1}^{n}\|\mathbf{A}\mathbf{x}_i\|\delta_i \to \infty$ then the bias of $\hat{\boldsymbol{\beta}}$ blows up.

Note that if G is a finite measure, f uniformly continuous and $\{\delta_i\}$ are given by (5.6.43b) then all the conditions of the above theorem are satisfied by the above $\{F_i\}$ and F. Thus we have

Corollary 5.6b.3. *Every $\hat{\boldsymbol{\beta}}$ corresponding to a finite measure G is qualitatively robust for $\boldsymbol{\beta}$ against heteroscedastic gross errors at all those F's which have uniformly continuous densities provided $\{\delta_i\}$ are given by* (5.6.43b) *and provided* (5.5.a) *and* (5.6.37) *hold.*

Example 5.6b.2. Here we consider $\{F_{ni}\}$ given in the Case 5.5.3. We leave it to the reader to verify that one choice of $\{\sigma_{ni}\}$ that implies (5.6.29) is to take $\sigma_{ni} = \|\mathbf{A}\mathbf{x}_{ni}\|$, $1 \le i \le n$. One can also verify that in this case, (5.5.44) - (5.5.46), (5.5.50) and (5.5.51) entail the satisfaction of all the conditions of Theorem 5.6b.1. Again, the following corollary holds.

Corollary 5.6b.4. *Every $\hat{\boldsymbol{\beta}}$ corresponding to a finite measure G is qualitatively robust for $\boldsymbol{\beta}$ against heteroscedastic scale errors at all those F's which have uniformly continuous densities provided $\{\sigma_{ni}\} = \|\mathbf{A}\mathbf{x}_{ni}\|$, $1 \le i \le n$, and provided* (5.5.a) *and* (5.6.37) *hold.*

As an example of a σ - finite G with $G(\mathbb{R}) = \infty$ that yields a robust estimator, consider $G(y) \equiv (2/3)y$. Assume that the following hold.

(*i*) F, F_0 have continuous densities f, f_0;

$$0 < \int f^2 d\lambda, \ \int f_0^2 d\lambda < \infty.$$

$$(ii) \quad \int F(1-F)d\lambda < \infty. \qquad\qquad (iii) \quad \int |F - F_0| d\lambda < \infty.$$

Then the corresponding $\hat{\boldsymbol{\beta}}$ is qualitatively robust at F against the heteroscedastic gross errors of Example 5.6a.1 with $\{\delta_{ni}\}$ given by (5.6.43b).

Recall, from Remark 5.6a.1, that this $\hat{\boldsymbol{\beta}}$ is also asymptotically efficient at logistic errors. Thus we have a m.d. estimator $\hat{\boldsymbol{\beta}}$ that is asymptotically efficient *and* qualitatively robust at logistic error d.f. against the above gross errors models!!

We leave it to an interested reader to obtain analogues of the above results for $\boldsymbol{\beta}^+$ and $\boldsymbol{\beta}^*$. The reader will find Theorems 5.5.4 and 5.5.6 useful here. □

5.6.3 Locally asymptotically minimax property

In this subsection we shall show that the class of m.d. estimators $\{\boldsymbol{\beta}^+\}$ are locally asymptotically minimax (LAM) in the Hájek - Le Cam sense (Hájek (1972), Le Cam (1972)). In order to achieve this goal we need to recall an inequality from Beran (1982) that gives a lower bound on the local asymptotic minimax risk for estimators of Hellinger differentiable functionals on the class of product probability measures. Accordingly, let Q_{ni}, P_{ni} be probability measures on $(\mathbb{R}, \mathcal{B})$, μ_{ni}, ν_{ni} be a σ-finite measures on $(\mathbb{R}, \mathcal{B})$ with ν_{ni} dominating Q_{ni}, P_{ni}; $q_{ni} := dQ_{ni}/d\nu_{ni}$, $p_{ni} := dP_{ni}/d\nu_{ni}$; $1 \le i \le n$. Let $Q^n = Q_{n1} \times \cdots \times Q_{nn}$ and $P^n = P_{n1} \times \cdots \times P_{nn}$ and let Π^n denote the class of n-fold product probability measures $\{Q^n\}$ on $(\mathbb{R}^n, \mathcal{B}^n)$.

Define, for a $c > 0$ and for sequences $0 < \eta_{n1} \to 0$, $0 < \eta_{n2} \to 0$,

$$\mathcal{H}_n(P^n, c) = \Big\{Q^n \in \Pi^n; \sum_{i=1}^{n} \int (q_{ni}^{1/2} - p_{ni}^{1/2})^2 d\nu_{ni} \le c^2\Big\},$$

$$\mathcal{K}_n(P^n, c, \eta_n) = \Big\{Q^n \in \Pi^n; Q^n \in \mathcal{H}_n(P^n, c),$$
$$\max_i \int (q_{ni} - p_{ni})^2 d\mu_{ni} \le \eta_{n1},$$
$$\max_i \int (q_{ni}^{1/2} - p_{ni}^{1/2})^2 d\nu_{ni} \le \eta_{n2}\Big\},$$

where $\eta_n' := (\eta_{n1}, \eta_{n2})$.

Definition 5.6c.1. A sequence of vector valued functionals $\{\mathbf{S}_n : \Pi^n \to \mathbb{R}^p, n \ge 1\}$ is Hellinger - (H -) differentiable at $\{P^n \in \Pi^n\}$ if there exists a triangular array of $p \times 1$ random vectors $\{\boldsymbol{\xi}_{ni}, 1 \le i \le n\}$ and a sequence of $p \times p$ matrices $\{\mathbf{A}_n, n \ge 1\}$ having the following properties:

(1) $\int \boldsymbol{\xi}_{ni} dP_{ni} = 0, \quad \int \|\boldsymbol{\xi}_{ni}\|^2 dP_{ni} < \infty, \quad 1 \le i \le n;$
$\sum_{i=1}^n \int \boldsymbol{\xi}_{ni} \boldsymbol{\xi}_{ni}' dP_{ni} \equiv \mathbf{I}_{p \times p}.$

(2) For every $0 < c < \infty$, every sequence $\eta_n \to 0$,

$$\sup \Bigg\| \mathbf{A}_n\{\mathbf{S}_n(Q^n) - \mathbf{S}_n(P^n)\} - 2\sum_{i=1}^{n} \int \boldsymbol{\xi}_{ni}(p_{ni}^{1/2} - p_{ni}^{1/2}) d\nu_{ni} \Bigg\| = o(1),$$

where the supremum is over all $Q^n \in \mathcal{H}_n(P^n, c, \eta_n)$.

(3) For every $\epsilon > 0$ and every $\boldsymbol{\alpha} \in \mathbb{R}^p$ with $\|\boldsymbol{\alpha}\| = 1$,

$$\sum_{i=1}^{n} \int (\boldsymbol{\alpha}'\boldsymbol{\xi}_{ni})^2 I(|\boldsymbol{\alpha}'\boldsymbol{\xi}_{ni}| > \epsilon) dP_{ni} = o(1).$$

Now, let $X_{n1}, \cdots, X_{nn}$ be independent r.v.'s with $Q_{n1}, \cdots, Q_{nn}$ denoting their respective distributions and $\hat{\mathbf{S}}_n = \hat{\mathbf{S}}_n(X_{n1}, \cdots, X_{nn})$ be an estimator of $\mathbf{S}_n(Q^n)$. Let $\mathcal{U}$ be a nondecreasing bounded function on $[0, \infty]$ to $[0, \infty)$ and define the risk of estimating $\mathbf{S}_n$ by $\hat{\mathbf{S}}_n$ to be

$$R_n(\hat{\mathbf{S}}_n, Q^n) = E^n \left[\mathcal{U}(\|\mathbf{A}_n\{\hat{\mathbf{S}}_n - \mathbf{S}_n(Q^n)\}\|) \right],$$

where E^n is the expectation under Q^n.

Theorem 5.6c.1. *Suppose that $\{\mathbf{S}_n : \Pi^n \to \mathbb{R}^p, n \geq 1\}$ is a sequence of H-differentiable functionals and that the sequence $\{P^n \in \prod^n\}$ is such that*

$$\max_i \int p_{ni}^2 d\mu_{ni} = O(1). \tag{5.6.44}$$

Then,

$$\lim_{c \to 0} \liminf_{n} \inf_{\hat{\mathbf{S}}_n} \sup_{Q^n \in \mathcal{K}_n(P^n, c, \eta_n)} R_n(\hat{\mathbf{S}}_n, Q^n) \geq E\mathcal{U}(\|\mathbf{Z}\|) \tag{5.6.45}$$

where $\mathbf{Z}$ is a $\mathcal{N}(\mathbf{0}, \mathbf{I}_{p \times p})$ r.v..

Sketch of a proof. This is a reformulation of a result of Beran (1982), pp. 425-426. He actually proved (5.6.45) with $\mathcal{K}_n(P^n, c, \eta_n)$ replaced by $\mathcal{H}_n(P^n, c)$ and without requiring (5.6.44). The assumption (5.6.44) is an assumption on the fixed sequence $\{P^n\}$ of probability measures. Beran's proof proceeds as follows:

Under (1) and (2) of the Definition 5.6c.1, there exists a sequence of probability measures $\{Q^n(\mathbf{h})\}$ such that for every $0 < b < \infty$,

$$\sup_{\|\mathbf{h}\| \le b} \sum_{i=1}^{n} \int \left\{ q_{ni}^{1/2}(\mathbf{h}) - p_{ni}^{1/2} - (1.2)\mathbf{h}'\boldsymbol{\xi}_{ni} p_{ni}^{1/2} \right\}^2 d\nu_{ni} = o(1).$$

Consequently,

$$\lim_n \sup_{\|\mathbf{h}\| \le b} \sum_{i=1}^{n} \int \{q_{ni}^{1/2}(\mathbf{h}) - p_{ni}^{1/2}\}^2 d\nu_{ni} = 4^{-1} b^2,$$

and for n sufficiently large, the family $\{Q^n(\mathbf{h}), \|\mathbf{h}\} \le b, \mathbf{h} \in \mathbb{R}^p\}$ is a subset of $\mathcal{H}_n(P^n, (b/2))$. Hence, $\forall\ c > 0, \forall$ sequence of statistics $\{\hat{\mathbf{S}}_n\}$,

$$\text{(5.6.46)} \qquad \begin{aligned} &\liminf_n \inf_{\hat{\mathbf{S}}_n} \sup_{Q^n \in \mathcal{H}_n(P^n, c)} R_n(\hat{\mathbf{S}}_n, Q^n) \\ &\ge \liminf_n \inf_{\hat{\mathbf{S}}_n} \sup_{\|\mathbf{h}\| \le 2c} R_n(\hat{\mathbf{S}}_n, Q^n(\mathbf{h})) \end{aligned}$$

Then the proof proceeds as in Hájek - Le Cam setup for the parametric family $\{Q^n(\mathbf{h}), \|\mathbf{h}\| \le b\}$, under the LAN property of the family $\{Q^n(\mathbf{h}), \|\mathbf{h}\| \le b\}$ with $b = 2c$.

Thus (5.6.45) will be proved if we verify (5.6.46) with $\mathcal{H}_n(P^n, c)$ replaced by $\mathcal{H}_n(P^n, c, \eta_n)$ under the additional assumption (5.6.44). That is, we have to show that there exist sequences $0 < \eta_{n1} \to 0, 0 < \eta_{n2} \to 0$ such that the above family $\{Q^n(\mathbf{h}), \|\mathbf{h}\| \le b\}$ is a subset of $\mathcal{H}_n(P^n, (b/2), \eta_n)$ for sufficiently large n. To that effect we recall the family $\{Q^n(\mathbf{h})\}$ from Beran. With $\boldsymbol{\xi}_{ni}$ as in the Definition 5.6c.1, let ξ_{nij} denote the j-th component of $\xi_{ni}, 1 \le j \le p, 1 \le i \le n$. By (3), there exist a sequence $\epsilon_n > 0, \epsilon_n \downarrow 0$ such that

$$\max_{1 \le j \le p} \sum_{i=1}^{n} \int \xi_{nij}^2 \mathbf{I}(|\xi_{nij}| > \epsilon_n) dP_{ni} = o(1).$$

Now, define, for $1 \le i \le n,\ 1 \le j \le p$,

$$\begin{aligned} \xi_{nij}^* &:= \xi_{nij} \mathbf{I}(|\xi_{nij}| \le \epsilon_n), \qquad \bar{\xi}_{nij} := \xi_{nij}^* - \int \xi_{nij}^* dP_{ni}, \\ \bar{\boldsymbol{\xi}}_{ni} &:= (\bar{\xi}_{ni1}, \cdots, \bar{\xi}_{nip})'. \end{aligned}$$

Note that

$$(5.6.47) \qquad \|\bar{\boldsymbol{\xi}}_{ni}\| \le 2p\epsilon_n, \quad \int \bar{\boldsymbol{\xi}}_{ni} dP_{ni} = 0, \quad 1 \le i \le n.$$

For a $0 < b < \infty, \|\mathbf{h}\| \le b, 1 \le i \le n$, define

$$\begin{aligned} q_{ni}(\mathbf{h}) &= (1 + \mathbf{h}'\bar{\boldsymbol{\xi}}_{ni})p_{ni}, & \epsilon_n < (2bp)^{-1}, \\ &= p_{ni}, & \epsilon_n \ge (2bp)^{-1}. \end{aligned}$$

Because of (5.6.47), $\{q_{ni}(\mathbf{h}), \|h\| \le b, 1 \le i \le n\}$ are probability density functions. Let $\{Q_{ni}(\mathbf{h}); \|\mathbf{h}\| \le b, 1 \le i \le n\}$ denote the corresponding probability measures and $Q^n(\mathbf{h}) = Q_{n1}(\mathbf{h}) \times \cdots \times Q_{nn}(\mathbf{h})$. Now, note that for $\|\mathbf{h}\| \le b$, $1 \le i \le n$,

$$\begin{aligned} \int (q_{ni}(\mathbf{h}) - p_{ni})^2 d\mu_{ni} &= 0, & \epsilon_n \ge (2bp)^{-1}, \\ &= \int (\mathbf{h}'\bar{\boldsymbol{\xi}}_{ni})^2 p_{ni}^2 d\mu_{ni}, & \epsilon_n < (2bp)^{-1}. \end{aligned}$$

Consequently, since $\epsilon_n \downarrow 0, \epsilon_n < (2bp)^{-1}$ eventually, and

$$\begin{aligned} &\sup_{\|\mathbf{h}\| \le b} \max_i \int (q_{ni}(\mathbf{h}) - p_{ni})^2 d\mu_{ni} \\ &\quad \le \; (2p\epsilon_n)^2 b^2 \max_i \int p_{ni}^2 d\mu_{ni} =: \eta_{n1}. \end{aligned}$$

Similarly, for a sufficiently large n,

$$\sup_{\|\mathbf{h}\| \le b} \max_i \int (q_{ni}^{1/2}(\mathbf{h}) - p_{ni}^{1/2})^2 d\nu_{ni} \le 2bp\epsilon_n =: \eta_{n2}.$$

Because of (5.6.44) and because $\epsilon_n \downarrow 0, \max\{\eta_{n1}, \eta_{n2}\} \to 0$.

Consequently, for every $b > 0$ and for n sufficiently large, $\{Q^n(\mathbf{h}), \|\mathbf{h}\| \le b\}$ is a subset of $\mathcal{H}_n(P^n, (b/2), \eta_n)$ with the above η_{n1}, η_{n2} and an analogue of (5.6.46) with $\mathcal{H}_n(P^n, c)$ replaced by $\mathcal{K}_n(P^n, (b/2), \eta_n)$ holds. The rest is the same as in Beran. □

We shall now show that $\boldsymbol{\beta}^+$ achieves the lower bound in (5.6.45). Fix a $\boldsymbol{\beta} \in \mathbb{R}^p$ and consider the model (1.1.1). As before, let F_{ni} be the actual d.f. of $e_{ni}, 1 \le i \le n$, and suppose we model the errors to be i.i.d. F, F symmetric around zero. The d.f. F need not be known. Then the actual and the modeled d.f. of Y_{ni} of (1.1.1) is $F_{ni}(\cdot - \mathbf{x}_{ni}'\boldsymbol{\beta}), F(\cdot - \mathbf{x}_{ni}'\boldsymbol{\beta})$, respectively.

In Theorem 5.6c.1 take $X_{ni} \equiv Y_{ni}$ and $\{Q_{ni}, P_{ni}, \nu_{ni}\}$ as follows:

$$\begin{aligned} Q_{ni}^{\boldsymbol{\beta}}(Y_{ni} \le \cdot) &= F_{ni}(\cdot - \mathbf{x}_{ni}'\boldsymbol{\beta}), \quad P_{ni}^{\boldsymbol{\beta}}(Y_{ni} \le \cdot) = F(\cdot - \mathbf{x}_{ni}'\boldsymbol{\beta}), \\ x\mu_{ni}^{\boldsymbol{\beta}}(\cdot) &= G(\cdot - \mathbf{x}_{ni}'\boldsymbol{\beta}), \qquad \nu_{ni} \equiv \lambda, \ \ 1 \le i \le n. \end{aligned}$$

Also, let $Q_{\boldsymbol{\beta}}^n = Q_{n1}^{\boldsymbol{\beta}} \times \cdots \times Q_{nn}^{\boldsymbol{\beta}}; P_{\boldsymbol{\beta}}^n = P_{n1}^{\boldsymbol{\beta}} \times \cdots \times P_{nn}^{\boldsymbol{\beta}}$. The absence of $\boldsymbol{\beta}$ from the sub - or the super - script of a probability measure indicates that the measure is being evaluated at $\boldsymbol{\beta} = \mathbf{0}$. Thus, for example we write Q^n for $Q_{\mathbf{0}}^n (= \prod_{i=1}^n F_{ni})$ and P^n for $P_{\mathbf{0}}^n$, etc. Also for an integrable function g write $\int g$ for $\int g d\lambda$.

Let f_{ni}, f denote the respective densities of F_{ni}, F, w.r.t. λ. Then $q_{ni}^{\boldsymbol{\beta}}(\cdot) = f_{ni}(\cdot - \mathbf{x}_{ni}'\boldsymbol{\beta})$, $p_{ni}^{\boldsymbol{\beta}}(\cdot) = f(\cdot - \mathbf{x}_{ni}'\boldsymbol{\beta})$ and, because of the translation invariance of the Lebesgue measure,

$$\begin{aligned} \mathcal{H}_n(P_{\boldsymbol{\beta}}^n, c) &= \left\{ Q_{\boldsymbol{\beta}}^n \in \Pi^n; \sum_{i=1}^n \int \{(q_{ni}^{\boldsymbol{\beta}})^{1/2} - (p_{ni}^{\boldsymbol{\beta}})^{1/2}\}^2 \le c^2 \right\} \\ &= \left\{ Q^n \in \Pi^n; \sum_{i=1}^n \int (f_{ni}^{1/2} - f^{1/2})^2 \le c^2 \right\} \\ &= \mathcal{H}_n(P^n, c). \end{aligned}$$

That is the set $\mathcal{H}_n(P_{\boldsymbol{\beta}}^n, c)$ does not depend on $\boldsymbol{\beta}$. Similarly,

$$\begin{aligned} \mathcal{K}_n(P_{\boldsymbol{\beta}}^n, c, \eta_n) &= \Big\{ Q^n \in \Pi^n; Q^n \in \mathcal{H}_n(P^n, c), \\ &\qquad \max_i \int (f_{ni} - f)^2 dG \le \eta_{n1}, \\ &\qquad \max_i \int (f_{ni}^{1/2} - f^{1/2}) \le \eta_{n2} \Big\} \\ &= \mathcal{K}_n(P^n, c, \eta_n). \end{aligned}$$

Next we need to define the relevant functionals. For $\mathbf{t} \in \mathbb{R}^p, y \in$

$\mathbb{R}, 1 \le i \le n$, define

$$\begin{aligned} m_{ni}^{+}(y,\mathbf{t}) &= F_{ni}(y+\mathbf{x}_{ni}'(\mathbf{t}-\boldsymbol{\beta})) - 1 + F_{ni}(-y+\mathbf{x}_{ni}'(\mathbf{t}-\boldsymbol{\beta})), \\ \mathbf{b}_n(y,\mathbf{t}) &:= \sum_{i=1}^{n} \mathbf{A}\mathbf{x}_{ni} m_{ni}^{+}(y,\mathbf{t}), \\ \mu_n(\mathbf{t}, Q_{\boldsymbol{\beta}}^n) &\equiv \mu_n(\mathbf{t},\mathbf{F}) := \int \|\mathbf{b}_n(y,\mathbf{t})\|^2 dG(y), \\ \mathbf{F}' &:= (F_{ni}, \cdots, F_{nn}). \end{aligned}$$

Now, recall the definition of ψ from (5.6.2) and let $\mathbf{T}_n(\boldsymbol{\beta}, Q_{\boldsymbol{\beta}}^n) \equiv \mathbf{T}_n(\boldsymbol{\beta},\mathbf{F})$ be defined by the relation

$$\begin{aligned} \mathbf{T}_n(\boldsymbol{\beta},\mathbf{F}) &:= \boldsymbol{\beta} + (\mathbf{X}'\mathbf{X}\int f^2 dG)^{-1} \\ &\quad \times \int \sum_{i=1}^{n} \mathbf{x}_{ni}[F_{ni}(y) - 1 + F_{ni}(-y)]d\psi(y). \end{aligned}$$

Note that, with $\mathbf{b}_n(y) \equiv \mathbf{b}_n(y,\boldsymbol{\beta})$,

$$\text{(5.6.48)} \quad \mathbf{A}^{-1}(\mathbf{T}_n(\boldsymbol{\beta},\mathbf{F}) - \boldsymbol{\beta}) = \Big(\int f^2 dG\Big)^{-1} \int \mathbf{b}_n(y) d\psi(y).$$

Some times we shall write $\mathbf{T}_n(\mathbf{F})$ for $\mathbf{T}_n(\boldsymbol{\beta},\mathbf{F})$.

Observe that if $\{F_{ni}\}$ are symmetric around 0, then $\mathbf{T}_n(\boldsymbol{\beta},\mathbf{F}) = \boldsymbol{\beta} = \mathbf{T}_n(\boldsymbol{\beta}, P_{\boldsymbol{\beta}}^n)$. In general, the quantity $\mathbf{A}^{-1}(\mathbf{T}_n(\mathbf{F}) - \boldsymbol{\beta})$ measures the asymptotic bias in $\boldsymbol{\beta}^+$ due to the asymmetry of the errors.

We shall prove the LAM property of $\boldsymbol{\beta}^+$ by showing that $\mathbf{T}_n$ is H-differentiable and that $\boldsymbol{\beta}^+$ is an estimator of $\mathbf{T}_n$ that achieves the lower bounds in (5.6.45). To that effect we first state a lemma. Its proof follows from Theorem 5.5.4 in the same fashion as that of Lemma 5.6b.1 and Corollary 5.6b.2 from Theorem 5.5.3. Observe that the conditions (5.5.17) and (5.5.k^+) with $\mathbf{D} = \mathbf{XA}$, respectively, become

$$\text{(5.6.49)} \quad \int \|\mathbf{b}_n(y)\|^2 dG(y) = O(1),$$

$$\text{(5.6.50)} \quad \liminf_n \inf_{\|\mathbf{e}\|=1} \mathbf{e}'\mathbf{AX}' \int \boldsymbol{\Lambda}^+ g dG\, \mathbf{XA}\,\mathbf{e} \ge \alpha, \text{ for an } \alpha > 0,$$

where $\boldsymbol{\Lambda}^+$ is defined at (5.5.20) and g is as in (5.6.3).

Lemma 5.6c.1. *Assume that (1.1.1) holds with the actual d.f.'s of* $\{e_{ni}, 1 \le i \le n\}$ *equal to* $\{F_{ni}, 1 \le i \le n\}$ *and suppose that we model the errors to be i.i.d. F, F symmetric around zero. In addition, assume that* (5.3.8); (5.5.*a*), (5.5.*c*), (5.5.*d*), (5.5.*f*), (5.5.*g*), (5.5.*i*) *with* $\mathbf{D} = \mathbf{XA}, G_n \equiv G$; (5.5.44), (5.6.12), (5.6.28) *and* (5.6.49) *hold. Then* (5.5.*h*) *and its variant where the argument y in the integrand is replaced by* $-y$, (5.5.14), (5.6.50) *and the following hold.*

$$\text{(5.6.51)} \qquad \mathbf{A}^{-1}(\boldsymbol{\beta}^+ - \mathbf{T}_n(\mathbf{F})) = -\Big\{2\int f^2 dG\Big\}^{-1}\mathbf{Z}_n^+ + o_p(1), \quad \textit{under } \{Q^n\}.$$

where

$$\mathbf{Z}_n^+ = \sum_{i=1}^n \mathbf{A}\mathbf{x}_{ni}\Big\{\psi(-e_{ni}) - \psi(e_{ni}) - \int m_{ni}^+(y)dG(y)\Big\},$$

with $m_{ni}^+(y) \equiv m_{ni}^+(y, \boldsymbol{\beta})$ *and* ψ *as in* (5.6.2). □

Now, define, for an $0 < a < \infty$,

$$\begin{aligned}
&\mathcal{M}_n(P^n, a) \\
&= \Big\{Q^n \in \Pi^n; Q^n = \prod_{i=1}^n F_{ni}, \ \max_i \|F_{ni} - F\|_\infty \to 0, \\
&\qquad \max_i \int |f_{ni} - f|^r dG \to 0, \ r = 1, 2 \\
&\qquad \int \Big[\sum_{i=1}^n \|\mathbf{A}\mathbf{x}_{ni}\| |F_{ni} - F|\Big]^2 dG \le a^2\Big\}.
\end{aligned}$$

Lemma 5.6c.2. *Assume that (1.1.1) holds with the actual d.f.'s of* $\{e_{ni}, 1 \le i \le n\}$ *equal to* $\{F_{ni}, 1 \le i \le n\}$ *and suppose that we model the errors to be i.i.d. F, F symmetric around zero. In addition, assume that* (5.3.8), (5.5.*a*), (5.5.44) *and the following hold.*

$$\text{(5.6.52)} \qquad G \textit{is a finite measure.}$$

Then, for every $0 < a < \infty$ *and sufficiently large n,*

$$\text{(5.6.53)} \qquad \mathcal{M}_n(P^n, a) \supset \mathcal{H}_n(P^n, b_a, \eta_n),$$

where $b_a := (4p\alpha)^{-1/2}a$, $\alpha := G(\mathbb{R})$. *Moreover, all assumptions of Lemma* 5.6c.1 *are satisfied.*

Proof. Fix an $0 < a < \infty$. It suffices to show that

$$\sum_{i=1}^{n} \int (f_{ni}^{1/2} - f^{1/2})^2 \le b_a^2, \quad n \ge 1, \tag{5.6.54}$$

and

$$\text{(a)} \quad \max_i \int (f_{ni} - f)^2 dG \le \eta_{n1}, \quad n \ge 1, \tag{5.6.55}$$

$$\text{(b)} \quad \max_i \int (f_{ni}^{1/2} - f^{1/2})^2 \le \eta_{n2}, \quad n \ge 1,$$

imply all the conditions describing $\mathcal{M}_n(P^n, a)$.

Claim. (5.6.54) implies

$$\int \left[\sum_{i=1}^{n} \|\mathbf{A}\mathbf{x}_{ni}\| |F_{ni} - F| \right]^2 dG \le a^2, \quad n \ge 1$$

By the C-S inequality, $\forall\ 1 \le i \le n, x \in \mathbb{R}$,

$$\begin{aligned} |F_{ni}(x) - F(x)|^2 &= \left| \int_{-\infty}^{x} (f_{ni} - f) \right|^2 \\ &\le \int_{-\infty}^{x} (f_{ni}^{1/2} - f^{1/2})^2 \int_{-\infty}^{x} (f_{ni}^{1/2} + f^{1/2})^2 \\ &\le 4 \int (f_{ni}^{1/2} - f^{1/2})^2. \end{aligned}$$

Hence,

$$\begin{aligned} &\int \Big[\sum_{i=1}^{n} \|\mathbf{A}\mathbf{x}_{ni}\| |F_{ni} - F| \Big]^2 dG \\ &\le \sum_{i=1}^{n} \|\mathbf{A}\mathbf{x}_{ni}\|^2 \cdot \sum_{i=1}^{n} \int (F_{ni} - F)^2 dG \\ &\le 4p\alpha \cdot \sum_{i=1}^{n} \int (f_{ni}^{1/2} - f^{1/2})^2, \end{aligned}$$

which proves the Claim.

The above bound, the finiteness of G and (5.6.55b) with $\eta_{n2} \to 0$ imply that $\max_i \|F_{ni} - F\|_\infty \to 0$ in a routine fashion. The rest uses (5.5.42), (5.5.43) and details are straightforward.

Now let $\varphi(y) = \psi(-y) - \psi(y)$, $y \in \mathbb{R}$. Note that $d\psi(-y) \equiv -d\psi(y)$, $d\varphi \equiv -2d\psi$, $d\psi = f dG$ and because F is symmetric around 0, $\int \varphi f = 0$. Let

$$\begin{aligned} \sigma^2 &= Var\{\psi(e)|F\}, \quad \tau = \int f^2 dG, \quad \rho = \varphi/\sigma, \\ \boldsymbol{\xi}_{ni} &\equiv \xi_{ni}(Y_{ni}, \boldsymbol{\beta}) \equiv \mathbf{A}\mathbf{x}_{ni}\rho(e_{ni}). \end{aligned}$$

Use the above facts to obtain

$$\begin{aligned} 2\sum_{i=1}^n \int \boldsymbol{\xi}_{ni}(y,\boldsymbol{\beta})(p_{ni}^{\boldsymbol{\beta}}(y))^{1/2}&\Big\{\left(q_{ni}^{\boldsymbol{\beta}}(y)\right)^{1/2} - (p_{ni}^{\boldsymbol{\beta}}(y))^{1/2}\Big\}^2 dy \\ &= 2\sum_{i=1}^n \mathbf{A}\mathbf{x}_{ni}\int \rho f^{1/2}\left(f_{ni}^{1/2} - f^{1/2}\right) \\ &= \sum_{i=1}^n \mathbf{A}\mathbf{x}_{ni}\Big\{\int \rho f_{ni} - \int \rho(f_{ni}^{1/2} - f^{1/2})^2\Big\} \\ &= -\sigma^{-1}\sum_{i=1}^n \mathbf{A}\mathbf{x}_{ni}\Big\{\int [F_{ni} - F]d\varphi - \int \rho(f_{ni}^{1/2} - f^{1/2})^2\Big\} \\ &= \sigma^{-1}\sum_{i=1}^n \mathbf{A}\mathbf{x}_{ni}\Big\{2\int [F_{ni} - F]f dG \\ &\qquad - \int \rho\left(f_{ni}^{1/2} - f^{1/2}\right)^2\Big\}. \end{aligned} \tag{5.6.56}$$

The last but one equality follows from integrating the first term by parts.

Now consider the R.H.S. of (5.6.48). Note that because F and G are symmetric around 0,

$$\begin{aligned} \int \mathbf{b}_n f dG &= \int \sum_{i=1}^n \mathbf{A}\mathbf{x}_{ni}[F_{ni}(y) - 1 + F_{ni}(-y)]d\psi(y) \\ &= \int \sum_{i=1}^n \mathbf{A}\mathbf{x}_{ni}[F_{ni}(y) - F(y) \\ &\qquad + F_{ni}(-y) - F(-y)]d\psi(y) \end{aligned} \tag{5.6.57}$$

$$= \; 2\int \sum_{i=1}^{n} \mathbf{A}\mathbf{x}_{ni}[F_{ni} - F] f dG.$$

Recall that by definition $\mathbf{T}_n(\boldsymbol{\beta}, P^n_{\boldsymbol{\beta}}) \equiv \boldsymbol{\beta}$. Now take $\mathbf{A}_n$ of (2) of the Definition 5.6c.1 to be $\mathbf{A}^{-1}\tau\sigma^{-1}$ and conclude from (5.6.54), (5.6.56), (5.6.57), that

$$\begin{aligned}
&\Bigg\| \mathbf{A}_n\{\mathbf{T}_n(\boldsymbol{\beta}, Q^n_{\boldsymbol{\beta}}) - \mathbf{T}_n(\boldsymbol{\beta}, P^n_{\boldsymbol{\beta}})\} \\
&\qquad - 2\sum_{i=1}^{n} \int \boldsymbol{\xi}_{ni}(y, \boldsymbol{\beta}) \left(p^{\boldsymbol{\beta}}_{ni}(y)\right)^{1/2} \\
&\qquad\qquad \times \left\{(q^{\boldsymbol{\beta}}_{ni}(y))^{1/2} - (p^{\boldsymbol{\beta}}_{ni}(y))^{1/2}\right\}^2 dy \Bigg\| \\
&\le \; \left\| \sum_{i=1}^{n} \mathbf{A}\mathbf{x}_{ni} \int \int \rho (f_{ni}^{1/2} - f^{1/2})^2 \right\| \\
&\le \; \max_i \|\mathbf{A}\mathbf{x}_{ni}\} \cdot \|\rho\|_\infty \cdot b_a^2 = o(1),
\end{aligned}$$

uniformly for $\{Q^n\} \in \mathcal{H}_n(P^n, b_a, \eta_n)$.

This proves that the requirement (2) of the Definition 5.6c.1 is satisfied by the functional $\mathbf{T}_n$ with the $\{\boldsymbol{\xi}_{ni}\}$ given as above. The fact that these $\{\boldsymbol{\xi}_{ni}\}$ satisfy (1) and (3) of the Definition 5.6c.1 follows from (5.3.8), (5.5.a), (5.6.52), (5.6.53) and the symmetry of F. This then verifies the H-differentiability of the above m.d. functional $\mathbf{T}_n$.

We shall now derive the asymptotic distribution of $\boldsymbol{\beta}^+$ under any sequence $\{Q^n\} \in \mathcal{M}_n(P^n, a)$ under the conditions of Lemma 5.6c.2. For that reason consider the above $\mathbf{Z}_n^+$. Note that under $Q^n, (1/2)\mathbf{Z}_n^+$ is the sum of independent centered triangular random arrays and the boundedness of ψ and (5.5.a), imply, via the L-F CLT, that $\mathbf{C}_n^{-1/2}\mathbf{Z}_n^+ \longrightarrow_d \mathcal{N}(\mathbf{0}, \mathbf{I}_{\times p})$, where

$$\begin{aligned}
\mathbf{C}_n \;&=\; 4^{-1} E\mathbf{Z}_n^+\mathbf{Z}_n^{+'} = \sum_{i=1}^{n} \mathbf{A}\mathbf{x}_{ni}\mathbf{x}'_{ni}\mathbf{A}\sigma^2_{ni} \\
\sigma^2_{ni} \;&=\; Var\{\psi(e_{ni})|F_{ni}\}, \quad 1 \le i \le n.
\end{aligned}$$

But the boundedness of ψ implies that $\max_i |\sigma^2_{ni} - \sigma^2| \to 0$, for every $Q^n \in \mathcal{M}_n(P^n, a)$, where $\sigma^2 = Var\{\psi(e_1)|F\}$. Therefore $\sigma^{-1}\mathbf{Z}_n^+ \to_d \mathcal{N}(\mathbf{0}, \mathbf{I}_{p\times p})$.

Consequently, from Lemma 5.6c.1,

$$\lim_{c\to 0}\limsup_{n}\sup_{\boldsymbol{\beta}}\sup_{Q^n\in\mathcal{K}_n(P^n_{\boldsymbol{\beta}},c,\eta_n)} E\Big\{\mathcal{U}(\|\mathbf{A}_n(\boldsymbol{\beta}^+ - \mathbf{T}_n(\boldsymbol{\beta},Q^n_{\boldsymbol{\beta}}))\|\Big|Q^n_{\boldsymbol{\beta}}\Big\}$$
$$= E\mathcal{U}(\|\mathbf{Z}\|),$$

for every bounded nondecreasing function $\mathcal{U}$, where $\mathbf{Z}$ is a $\mathcal{N}(\mathbf{0},\mathbf{I}_{p\times p})$ r.v. This and Lemma 5.6c.2 shows that the sequence of the m.d. estimators $\{\boldsymbol{\beta}^+\}$ achieves the lower bound of (5.6.45) and hence is LAM. □

Remark 5.6c.1. It is an *interesting problem* to see if one can remove the requirement of the finiteness of the integrating measure G in the above LAM result. The LAM property of $\{\hat{\boldsymbol{\beta}}\}$ can be obtained in a similar fashion. For an alternative definition of LAM see Millar (1984) where, among other things, he proves the LAM property, in his sense, of $\{\hat{\boldsymbol{\beta}}\}$ for $p = 1$.

A Problem : To this date an appropriate extension of Beran (1978) to the model (1.1.1) does not seem to be available. Such an extension would provide asymptotically fully efficient estimators at every symmetric density with finite Fisher information and would also be LAM. □

Note : The contents of this chapter are based on the works of Koul (1979, 1980, 1984, 1985 a,b), Williamson (1979, 1982), Koul and DeWet (1983), Basawa and Koul (1988) and Dhar (1991a, b).

6

Goodness-of-fit Tests in Regression

6.1 Introduction

In this chapter we shall discuss the two problems of the goodness-of-fit. The first one pertains to the error d.f. of the linear model (1.1.1) and the second one pertains to fitting a parametric regression model to a regression function. The proposed tests will be based on certain residual weighted empiricals for the first problem and a partial sum process of the residuals for the second problem. The first five sections of this chapter deal with the first problem and Section 6.6, with several subsections, discusses the second problem. To begin with we shall focus on the first problem.

Consider the model (1.1.1) and the goodness-of-fit hypothesis

$$H_0 : F_{ni} \equiv F_0, \quad F_0 \text{ a known continuous d.f..} \tag{6.1.1}$$

This is a classical problem yet not much is readily available in literature. Observe that even if F_0 is known, having an unknown $\boldsymbol{\beta}$ in the model poses a problem in constructing tests of H_0 that would be implementable, at least asymptotically.

One test of H_0 could be based on $\hat{D}_1$ of (1.3.3). This test statistic is suggested by looking at the estimated residuals and mimicking the one sample location model technique. In general, its large sample distribution depends on the design matrix. In addition, it does not reduce to the Kiefer (1959) tests of goodness-of-fit in the k-sample location problem when (1.1.1) is reduced to this model. The test statistics that overcome these deficiencies are those that are based on the W.E.P.'s $\mathbf{V}$ of (1.1.2). For example, the two candidates that will be considered in this chapter are

$$\hat{D}_2 := \sup_y |\mathbf{W}^0(y, \hat{\beta})|, \quad \hat{D}_3 := \|\mathbf{W}^0(y, \hat{\beta})\|, \tag{6.1.2}$$

where $\hat{\beta}$ is an estimator of β and, for $y \in \mathbb{R}$, $\mathbf{t} \in \mathbb{R}^p$,

$$\mathbf{W}^0(y, \mathbf{t}) := (\mathbf{X}'\mathbf{X})^{-1/2}\{\mathbf{V}(y, \mathbf{t}) - \mathbf{X}'\mathbf{1}F_0(y)\}, \tag{6.1.3}$$

with $\mathbf{1}' := (1, \cdots, 1)_{1\times n}$, and where $|\mathbf{x}| = \max\{|x_j|; 1 \le j \le p\}$, for any $\mathbf{x} \in \mathbb{R}^p$. Other classes of tests are based on $K^0_{\mathbf{X}}(\hat{\beta}_{\mathbf{X}})$ and $\inf\{K^0_{\mathbf{X}}(\mathbf{t}), \mathbf{t} \in \mathbb{R}^p\}$, where $K^0_{\mathbf{X}}$ is equals to the $K_{\mathbf{X}}$ of (1.3.2) with $\mathbf{W}$ replaced by $\mathbf{W}^0$ in there.

Section 6.2.1 discusses the asymptotic null distributions of the supremum distance test statistics for H_0 when β is estimated arbitrarily and asymptotically efficiently. Also discussed in this section are some asymptotically distribution free (ADF) tests for H_0. Some comments about the asymptotic power of these tests appear at the end of this section. Section 6.2.2 discusses a smooth bootstrap null distribution of $\hat{D}_3$.

Analogous results for tests of H_0 based on L_2 - distances involving the ordinary and weighted empirical processes appear in Section 6.3.

A closely related problem to H_0 is that of testing the composite hypothesis

$$H_1 : F_{ni}(\cdot) = F_0(\cdot/\sigma), \ \sigma > 0, \ F_0 \text{ a known d.f.} \tag{6.1.4}$$

Modifications of various tests of H_0 suitable for testing H_1 and their asymptotic null distributions are discussed in Section 6.4.

Another problem of interest is to test the composite hypothesis of symmetry of the errors:

$$H_s : F_{ni} = F, \ 1 \le i \le n, \ n \ge 1; \tag{6.1.5}$$

F a d.f. symmetric around 0, not necessarily known.

This is a more general hypothesis than H_0. In some situations it may be of interest to test H_s before testing, say, that the errors are normally distributed. Rejection of H_s would *a priori* exclude any possibility of normality of the errors. A test of H_s could be based on

$$\hat{D}_{1s} := \sup_y |W_1^+(y, \hat{\beta})|, \tag{6.1.6}$$

where, for $y \in \mathbb{R}$, $\mathbf{t} \in \mathbb{R}^p$,

$$(6.1.7) \quad W_1^+(y, \mathbf{t}) := n^{-1/2} \sum_{i=1}^{n} [I(Y_{ni} \leq y + \mathbf{x}_{ni}'\mathbf{t}) - I(-Y_{ni} < y - \mathbf{x}_{ni}'\mathbf{t})] = H_n(y, \mathbf{t}) - 1 + H_n(-y, \mathbf{t}),$$

with H_n as in (1.2.1). Other candidates are

$$(6.1.8) \quad \begin{aligned} \hat{D}_{2s} &:= \sup_y |\mathbf{W}^+(y, \hat{\boldsymbol{\beta}})|, \\ \hat{D}_{3s} &:= \sup_y \|\mathbf{W}^+(y, \hat{\boldsymbol{\beta}})\| \\ &= \sup_y [\mathbf{V}^{+'}(y, \hat{\boldsymbol{\beta}})(\mathbf{X}'\mathbf{X})^{-1}\mathbf{V}^+(y, \hat{\boldsymbol{\beta}})]^{1/2}, \end{aligned}$$

where $\mathbf{W}^+ := \mathbf{A}\mathbf{V}^+$, $\mathbf{V}^{+'} := (V_1^+, \cdots, V_p^+)$, with

$$V_j^+(y, \mathbf{t}) := V_j(y, \mathbf{t}) - \sum_{i=1}^{n} x_{nij} + V_j(-y, \mathbf{t}),$$

for $1 \leq j \leq p$, $y \in \mathbb{R}$, $\mathbf{t} \in \mathbb{R}^p$. Yet other tests can be obtained by considering various L_2 - norms involving W_1^+ and $\mathbf{W}^+$. The asymptotic null distribution of all these test statistics is given in Section 6.5.

It will be observed that the tests based on the vectors $\mathbf{W}^0$ and $\mathbf{W}^+$ of W.E.P.'s will have asymptotic distributions similar to their counterparts in the k-sample location models. Consequently these tests can use, at least for the large samples, the null distribution tables that are available for such problems. For the sake of the completeness some of these tables are reproduced in the following sections.

6.2 The Supremum Distance Tests

6.2.1 Asymptotic null distributions

To begin with, define, for $0 \leq t \leq 1, s \in \mathbb{R}^p$,

$$(6.2.1) \quad \begin{aligned} W_1(t, \mathbf{s}) &:= n^{1/2}\{H_n(F_0^{-1}(t), \mathbf{s}) - t\}, \\ \mathbf{W}(t, \mathbf{s}) &:= \mathbf{W}^0(F_0^{-1}(t), \mathbf{s}). \end{aligned}$$

Let, for $0 \leq t \leq 1$,

$$(6.2.2) \quad \hat{W}_1(t) := W_1(t, \hat{\boldsymbol{\beta}}), \quad \hat{\mathbf{W}}(t) := \mathbf{W}(t, \hat{\boldsymbol{\beta}}).$$

Clearly, if F_0 is continuous then the distribution of $\hat{D}_j$, $j = 1, 2, 3$, is the same as that of

$$\|\hat{W}_1\|_\infty, \quad \sup\{|\mathbf{W}(t)|;\, 0 \le t \le 1\}, \quad \sup\{\|\hat{\mathbf{W}}(t)\|;\, 0 \le t \le 1\},$$

respectively. Consequently, from Corollaries 2.3.3 and 2.3.5 one readily obtains the following Theorem 6.2.1. Recall the conditions ($\mathbf{F_0 1}$) and (**NX**) from Corollary 2.3.1 and just after Corollary 2.3.2.

Theorem 6.2.1 *Suppose that the model (1.1.1) and H_0 hold. In addition, assume that* $\mathbf{X}$ *and* F_0 *satisfy* (**NX**) *and* ($\mathbf{F_0 1}$), *and that* $\hat{\beta}$ *satisfies*

$$\|\mathbf{A}^{-1}(\hat{\beta} - \beta)\| = O_p(1). \tag{6.2.3}$$

Then

$$\sup |W_1(t, \hat{\beta}) - \{W_1(t, \beta) + q_0(t) \cdot n^{1/2} \bar{\mathbf{x}}_n' \mathbf{A} \cdot \mathbf{A}^{-1}(\hat{\beta} - \beta)\}| = o_p(1), \tag{6.2.4}$$

$$\sup \|\mathbf{W}(t, \hat{\beta}) - \{W(t, \beta) + q_0(t) \mathbf{A}^{-1}(\hat{\beta} - \beta)\}\| = o_p(1), \tag{6.2.5}$$

where $q_0 := f_0(F_0^{-1})$ *and the supremum is over* $0 \le t \le 1$. □

Write $W_1(t), \mathbf{W}(t)$ for $W_1(t, \beta), \mathbf{W}(t, \beta)$, respectively. The following lemma gives the weak limits of W_1 and W under H_0.

Lemma 6.2.1 *Suppose that the model (1.1.1) and H_0 hold. Then*

$$W_1 \Longrightarrow B, \quad B \text{ a Brownian bridge in } C[0, 1]. \tag{6.2.6}$$

In addition, if $\mathbf{X}$ *satisfies* (**NX**), *then*

$$\mathbf{W} \Longrightarrow \mathbf{B}' := (B_1, \cdots, B_p) \tag{6.2.7}$$

where $B_1, \cdots, B_p$ *are independent Brownian bridges in* $C[0, 1]$.

Proof. The result (6.2.6) is well known or may be deduced from Corollary 2.2.2. The same corollary implies (6.2.7). To see this, rewrite

$$\mathbf{W}(t) = \mathbf{A} \sum_{i=1}^n \mathbf{x}_{ni} \{I(e_{ni} \le F_0^{-1}(t)) - t\} = \mathbf{A}\mathbf{X}' \boldsymbol{\alpha}_n(t),$$

where $\boldsymbol{\alpha}_n(t) := (\alpha_{n1}(t), \cdots, \alpha_{nn}(t))'$, with

$$\alpha_{ni}(t) := \{I(e_{ni} \le F_0^{-1}(t)) - t\}, \quad 1 \le i \le n, \ 0 \le t \le 1.$$

Clearly, under H_0,

$$(6.2.8) \qquad \begin{aligned} EW &\equiv 0, \\ Cov(\mathbf{W}(s), \mathbf{W}(t)) &= (s \wedge t - st)\mathbf{I}_{p\times p}, \quad 0 \le s, t \le 1. \end{aligned}$$

Now apply Corollary 2.2.2 p times, j^{th} time to the W.E.P. with the weights and r.v.'s given as in (6.2.9) below, $1 \le j \le p$, to conclude (6.2.7).

$$(6.2.9) \qquad \begin{aligned} \mathbf{d}_{(j)} &\equiv \text{the } j^{th} \text{ column of } \mathbf{XA}, \ X_{ni} \equiv e_{ni}, \ \text{ and} \\ F &\equiv F_0, \ 1 \le j \le p, \end{aligned}$$

See (2.3.31) and (2.3.32) for ensuring the applicability of Corollary 2.2.2 to this case. □

Remark 6.2.1 From (6.2.5) it follows that if $\hat{\beta}$ is chosen so that the finite dimensional asymptotic distributions of $\{\mathbf{W}(t) + q_0(t)\mathbf{A}^{-1}(\hat{\beta} - \beta); 0 \le t \le 1\}$ do not depend on the design matrix then the asymptotic null distribution's of $\hat{D}_j, j = 2, 3$, will also not depend on the design matrix. The classes of estimators that satisfy this requirement include M-, R- and m.d. estimators. Consequently, in these cases, the asymptotic null distribution's of $\hat{D}_j, j = 2, 3$ are design free.

On the other hand, from (6.2.4), the asymptotic null distribution of $\hat{D}_1$ depends on the design matrix through $n^{1/2}\overline{\mathbf{x}}_n'\mathbf{A}$. Of course, if $\overline{\mathbf{x}}_n$ equals to zero, then this distribution is free from F_0 and the design matrix. □

Remark 6.2.2 *The effect of estimating the parameter β efficiently.* To describe this, assume that F_0 has an a.c. density f_0 with a.e. derivative $\dot{f}_0$ satisfying

$$(6.2.10) \qquad 0 < I_0 := \int (\dot{f}_0/f_0)^2 dF_0 < \infty.$$

Define

$$(6.2.11) \qquad \begin{aligned} s_{ni} &:= -\dot{f}_0(e_{ni})/f_0(e_{ni}), \quad 1 \le i \le n; \\ \mathbf{s}_n &:= (s_{n1}, \cdots, s_{nn})' \end{aligned}$$

and assume that the estimator $\hat{\beta}$ satisfies

$$(6.2.12) \qquad \mathbf{A}^{-1}(\hat{\beta} - \beta) = I_0^{-1}\mathbf{AX}'\mathbf{s}_n + o_p(1).$$

Then, the approximating processes in (6.2.4) and (6.2.5), respectively, become

$$\text{(6.2.13)} \qquad \begin{aligned} W_1(t) &:= W_1(t) + q_0(t) n^{1/2} \bar{\mathbf{x}}_n' \mathbf{A} \cdot I_0^{-1} \mathbf{A} \mathbf{X}' \mathbf{s}_n, \\ \mathbf{W}(t) &:= \mathbf{W}(t) + q_0(t) I_0^{-1} \mathbf{A} \mathbf{X}' \mathbf{s}_n, \quad 0 \le t \le 1. \end{aligned}$$

Using the independence of the errors, one directly obtains

$$\text{(6.2.14)} \qquad \begin{aligned} EW_1(s)W_1(t) &= \{s(1-t) - n\bar{\mathbf{x}}_n'(\mathbf{X}'\mathbf{X})^{-1}\bar{\mathbf{x}}_n q_0(s) q_0(t) I_0^{-1}\}, \\ E\mathbf{W}(s)\mathbf{W}'(t) &= \{s(1-t) - q_0(s) q_0(t) I_0^{-1}\}\mathbf{I}_{p\times p}, \end{aligned}$$

for $0 \le s \le t \le 1$. The calculations in (6.2.14) use the facts that $Es_n \equiv \mathbf{0}$, $E\boldsymbol{\alpha}_n(t)\mathbf{s}_n' \equiv q_0(t)\mathbf{I}_{n\times n}$.

From (6.2.14), Theorem 2.2.1(i) applied to the entities given in (6.2.9), and the uniform continuity of q_0, implied by (6.2.10) (see Claim 3.2.1 above), it readily follows that $\mathbf{W} \Rightarrow \mathbf{Z} := (Z_1, \cdots, Z_p)'$, where $Z_1, \cdots, Z_p$ are continuous independent Gaussian processes, each having the covariance function

$$\text{(6.2.15)} \qquad \rho(s,t) := s(1-t) - q_0(s)q_0(t)I_0^{-1}, \quad 0 \le s \le t \le 1.$$

Consequently,

$$\text{(6.2.16)} \qquad \begin{aligned} \hat{D}_2 &\Longrightarrow \sup\{|\mathbf{Z}(t)|; 0 \le t \le 1\}, \\ \hat{D}_3 &\Longrightarrow \sup\{\|\mathbf{Z}(t)\|; 0 \le t \le 1\}. \end{aligned}$$

This shows that the asymptotic null distribution's of $\hat{D}_j$, $j = 1, 3$ are design free when an asymptotically efficient estimator of $\boldsymbol{\beta}$ is used in constructing the residuals while the same can not be said about $\hat{D}_1$.

Moreover, recall, say from Durbin (1975), that when testing for H_0 in the one sample location model, the Gaussian process Z_1 with the covariance function ρ appears as the limiting process for the analogue of $\hat{D}_1$. Note also that in this case, $\hat{D}_1 = \hat{D}_2 = \hat{D}_3$. However, it is the test based on $\hat{D}_3$ that provides the right extension of the one sample Kolmogorov goodness-of-fit test to the linear regression model (1.1.1) for testing H_0 in the sense that it includes the k-sample goodness-of-fit Kolmogorov type test of Kiefer (1959). That is, if we specialize (1.1.1) to the k-sample location model, then $\hat{D}_3$ reduces to the T_N' of Section 2 of Kiefer modulo the fact that we have to estimate $\boldsymbol{\beta}$.

The distribution of $\sup\{|Z_1(t)|; 0 \le t \le 1\}$ has been studied by Durbin (1976) when F_0 equals $\mathcal{N}(0,1)$ and some other distributions. Consequently, one can use these results together with the independence of $\mathbf{Z}_1, \cdots, \mathbf{Z}_p$ to implement the tests based on $\hat{D}_2, \hat{D}_3$ in a routine fashion. □

Remark 6.2.3 *Asymptotically distribution free (ADF) tests.* Here we shall construct estimators of $\boldsymbol{\beta}$ such that the above tests become ADF for testing H_0. To that effect, write $\mathbf{X}_n$ and $\mathbf{A}_n$ for $\mathbf{X}$ and $\mathbf{A}$ to emphasize their dependence on n. Recall that n is the number of rows in $\mathbf{X}_n$. Let $m = m_n$ be a sequence of positive integers, $m_n \leq n$. Let X_m be $m_n \times p$ matrix obtained from some m_n rows of $\mathbf{X}_n$. A way to choose m_n and these rows will be discussed later on. Relabel the rows of X_n so that its first m_n rows are the rows of X_m and let $\{e^*_{ni}, 1 \leq i \leq m_n\}, \{Y^*_{ni}; 1 \leq i \leq m_n\}$ denote the corresponding errors and observations, respectively. Define

(6.2.17)
$$\begin{aligned} s^*_{ni} &:= -\dot{f}(e^*_{ni})/f_0(e^*_{ni}), \quad 1 \leq i \leq m_n; \\ \mathbf{s}^*_m &:= (s^*_{ni}, \ 1 \leq i \leq m_n)', \\ \mathbf{T}_m &:= I_0^{-1}\mathbf{A}_m\mathbf{X}'_m\mathbf{s}^*_m, \quad \mathbf{A}_m = (\mathbf{X}'_m\mathbf{X}_m)^{-1/2}. \end{aligned}$$

Observe that under (6.2.10) and H_0,

(6.2.18)
$$E\mathbf{T}_m = \mathbf{0}, \quad E\mathbf{T}_m\mathbf{T}'_m \equiv I_0^{-1}\mathbf{I}_{p\times p}.$$

Consider the assumption

(6.2.19)
$$\begin{gathered} m_n \leq n, \quad m_n \to \infty \ \text{ such that} \\ (\mathbf{X}'_n\mathbf{X}_n)^{1/2}(\mathbf{X}'_m\mathbf{X}_m)^{-1}(\mathbf{X}'_n\mathbf{X}_n)^{1/2} \to 2\mathbf{I}_{p\times p}. \end{gathered}$$

The assumptions (6.2.19) and (**NX**) together imply

(6.2.20)
$$\max_{1\leq i\leq m} \mathbf{x}'_{ni}\mathbf{A}_m\mathbf{A}_m\mathbf{x}_{ni} = o(1).$$

Consequently one obtains, with the aid of the Cramér - Wold LF - CLT, that

(6.2.21)
$$\mathbf{T}_m \longrightarrow_d \mathcal{N}(\mathbf{0}, I_0^{-1}\mathbf{I}_{p\times p}).$$

Now use $\{(\mathbf{x}'_{ni}, Y^*_{ni});\ 1 \leq i \leq m_n\}$ to construct an estimator $\hat{\boldsymbol{\beta}}_m$ of $\boldsymbol{\beta}$ such that

(6.2.22)
$$\mathbf{A}_m^{-1}(\hat{\boldsymbol{\beta}}_m - \boldsymbol{\beta}) = \mathbf{T}_m + o_p(1).$$

Note that, by (6.2.19) and (6.2.21), $\|\mathbf{A}_n^{-1}\mathbf{A}_m\|_\infty = O(1)$ and hence

(6.2.23)
$$\mathbf{A}_n^{-1}(\hat{\boldsymbol{\beta}}_m - \boldsymbol{\beta}) = \mathbf{A}_n^{-1}\mathbf{A}_m\mathbf{T}_m + o_p(1).$$

Therefore it follows that $\hat{\beta}_m$ satisfies (6.2.3). Define

$$\mathbf{K}^*(t) := \mathbf{W}(t) + \mathbf{A}_n^{-1}\mathbf{A}_m\mathbf{T}_m q_0(t), \quad 0 \le t \le 1.$$

From (6.2.5) and (6.2.23) it now readily follows that

$$\sup_{0\le t\le 1} \|\mathbf{W}(t,\hat{\beta}) - \mathbf{K}^*(t)\| = o_p(1). \tag{6.2.24}$$

We shall now show that

$$\mathbf{K}^* \Longrightarrow \mathbf{B} \ \text{ with } \ \mathbf{B} \ \text{ as in (6.2.7)}. \tag{6.2.25}$$

First, consider the covariance function of $\mathbf{K}^*$. By the independence of the errors and by (6.2.10) one obtains that

$$\begin{aligned} & E\Big(\{I(e_{ni} \le F_0^{-1}(t)) - t\}\frac{\dot{f}_0(e^*_{nj})}{f_0(e^*_{nj})}\Big) \\ &= 0, \qquad i \ne j,\ 1 \le i \le n,\ 1 \le j \le m_n, \\ &= q_0(t), \quad 1 \le i = j \le m_n, 0 \le t \le 1. \end{aligned}$$

Use this and direct calculations to obtain that for all $0 \le s \le t \le 1$,

$$\begin{aligned} (6.2.26)\quad & E\mathbf{K}^*(s)\mathbf{K}^*(t)' \\ &= s(1-t)\mathbf{I}_{p\times p} \\ &\quad - \Big[2\mathbf{I}_{p\times p} - (\mathbf{X}_n'\mathbf{x}_n)^{1/2}(\mathbf{X}_m'\mathbf{X}_m)^{-1}(\mathbf{X}_n'\mathbf{X}_n)^{1/2}\Big] \\ &\qquad \times I_0^{-1} q_0(s) q_0(t). \end{aligned}$$

Thus (6.2.19) implies that

$$E\mathbf{K}^*(s)\mathbf{K}^*(t) \longrightarrow s(1-t)\mathbf{I}_{p\times p}, \quad \forall\ 0 \le s \le t \le 1.$$

Because of (6.2.7) and the uniform continuity of q_0, the relative compactness of the sequence $\{\mathbf{K}^*\}$ is *a priori* established, thereby completing the proof of (6.2.25). Consequently, we obtain the following

Corollary 6.2.1 *Under (1.1.1),* H_0, (**NX**), *(6.2.10), (6.2.19) and (6.2.22),*

$$\hat{D}_{2m} \longrightarrow_d \sup_{0\le t\le 1}\max_{1\le j\le p} |B_j(t)|,$$

$$\hat{D}_{3m} \longrightarrow_d \sup_{0\le t\le 1}\Big\{\sum_{j=1}^p B_j^2(t)\Big\}^{1/2},$$

where $\hat{D}_{jm}$ *stand for the* $\hat{D}_j$ *with* $\hat{\beta} = \hat{\beta}_m$, $j = 2, 3$. □

It thus follows, from the independence of the Brownian bridges $\{B_j, 1 \leq j \leq p\}$ and Theorem V.3.6.1 of Hájek and Sĭdák (1967), that the test that rejects H_0 when $\hat{D}_{2m} \geq d$ is of the asymptotic size α, provided d is determined from the relation

$$2\sum_{j=1}^{\infty}(-1)^{j+1}e^{-2j^2d^2} = 1 - (1-\alpha)^{1/p}. \tag{6.2.27}$$

Let T_p stand for the limiting r.v. of $\hat{D}_{3m}$. The distribution of T_p has been tabulated by Kiefer (1959) for $1 \leq p \leq 5$. Delong (1983) has also computed these tables for $1 \leq p \leq 7$. The following table is obtained from Kiefer for $1 \leq p \leq 5$ and Delong for $p = 6, 7$ for the sake of completeness. The last place digit is rounded from their entries.

$\alpha \setminus p$	1	2	3	4	5	6	7
.001	1.9495	2.1516	2.3030	2.4301	2.5422	2.6437	2.7373
.005	1.7308	1.9417	2.0977	2.2280	2.3424	2.445	2.540
.01	1.6276	1.8427	2.0009	2.1326	2.2480	2.3525	2.4525
.02	1.5174	1.7370	1.8974	2.0305	2.1470	2.252	2.350
.025	1.480	1.702	1.8625	1.9961	2.116	2.217	2.315
.05	1.3581	1.5838	1.7473	1.8823	2.0001	2.1053	2.2031
.10	1.2239	1.4540	1.6196	1.7559	1.8746	1.981	2.0788
.15	1.1380	1.3703	1.5370	1.6740	1.7930	1.900	1.9977
.20	1.0728	1.3061	1.4734	1.6107	1.730	1.8352	1.9349
.25	1.0192	1.2530	1.4205	1.5579	1.6773	1.785	1.8825

Table 1: Values d such that $P(T_p \geq d) \simeq \alpha$ for $1 \leq p \leq 7$. Obtained from Kiefer (1959) *& Delong (personal communication).*

Note that for $p = 1$, $\hat{D}_{2m}$ and $\hat{D}_{3m}$ are the same tests and the d of (6.2.27) is the same as the d of column 1 of Table 1 for various values of α.

The entries in Table 1 can be used to get the asymptotic critical level of $\hat{D}_{3m}$ for $1 \leq p \leq 7$. Thus for $p = 5, \alpha = .05$, the test that rejects H_0 when $\hat{D}_{3m} \geq 2.0001$ is of the asymptotic size .05, no matter what F_0 is within the class of d.f.'s satisfying (6.2.10).

Next, to make $\hat{D}_1$ - test ADF, let $r = r_n$ be a sequence of positive integers, $r_n \leq n, r_n \to \infty$. Let $\mathbf{X}_r$ denote the $r_n \times p$ matrix obtain from some r_n rows of X_n. Relabel the rows of $\mathbf{X}_n$ so that the first r_n rows are in $\mathbf{X}_r$ and let Y_i^0, e_i^0 denote the corresponding Y_i's and e_i's. Let $\mathbf{A}_r = (\mathbf{X}_r'\mathbf{X}_r)^{-1/2}$.

Assume that

$$\text{(6.2.28)} \qquad \begin{aligned} &\text{(i)} \quad \|n^{1/2}\overline{\mathbf{x}}_n' \mathbf{A}_r\| = O(1), \quad \text{and} \\ &\text{(ii)} \quad |n\overline{\mathbf{x}}_n(\mathbf{X}_r'\mathbf{X}_r)^{-1}\overline{\mathbf{x}}_n - 2r_n\overline{\mathbf{x}}_n'(\mathbf{X}_r'\mathbf{X}_r)^{-1}\overline{\mathbf{x}}_r| = o(1). \end{aligned}$$

Let $\hat{\boldsymbol{\beta}}_r$ be an estimator of $\boldsymbol{\beta}$ based on $\{(\mathbf{x}_{ni}', Y_{ni}^0),\ 1 \le i \le r_n\}$ such that

$$\text{(6.2.29)} \qquad \mathbf{A}_r^{-1}(\hat{\boldsymbol{\beta}}_r - \boldsymbol{\beta}) = \mathbf{T}_r + o_p(1), \qquad \mathbf{T}_r := I_0^{-1}\mathbf{A}_r\mathbf{X}_r'\mathbf{s}_r^0,$$

where $s_{ni}^0 = -\dot{f}(e_{ni}^0)/f(e_{ni}^0), 1 \le i \le r_n$, and $\mathbf{s}_r^0 = (s_{ni}^0, 1 \le i \le r_n)'$. Define

$$\mathbf{K}_1^*(t) := W_1(t) + n^{1/2}\overline{\mathbf{x}}_n'\mathbf{A}_r \cdot \mathbf{T}_r q_0(t), \quad 0 \le t \le 1.$$

Similar to (6.2.26), we obtain, for $s \le t$, that

$$\begin{aligned} EK_1^*(s)\mathbf{K}_1^*(t) &= s(1-t) \\ &\quad -I_0^{-1}q_0(s)q_0(t)\{\overline{\mathbf{x}}_n'(\mathbf{X}_r'\mathbf{X}_r)^{-1}[n\overline{\mathbf{x}}_n - 2r_n\overline{\mathbf{x}}_r]\}. \end{aligned}$$

Argue as for Corollary 6.2.1 to conclude

Corollary 6.2.2 *Under (1.1.1), H_0, (**NX**), (6.2.10), (6.2.28) and (6.2.29),*

$$\text{(6.2.30)} \qquad \hat{D}_{1r} \to_d \sup_{0 \le t \le 1} |B(t)|,$$

where $\hat{D}_{1r}$ is the $\hat{D}_1$ with $\hat{\beta} = \hat{\boldsymbol{\beta}}_r$. □

Remark 6.2.4 *On assumptions (6.2.19) and (6.2.28).* To begin with note that if

$$\text{(6.2.31)} \qquad \lim_n n^{-1}(\mathbf{X}_n'\mathbf{X}_n) \text{ exists and is positive definite,}$$

then (6.2.19) is equivalent to

$$\text{(6.2.32)} \qquad nm_n^{-1} \to 2.$$

If, in addition to (6.2.31), one also assumes

$$\text{(6.2.33)} \qquad \lim_n \overline{\mathbf{x}}_n \text{ exists and is finite,}$$

then (6.2.28) is equivalent to

$$\text{(6.2.34)} \qquad nr_n^{-1} \to 2.$$

There are many designs that satisfy (6.2.31) and (6.2.33). These include the one way classification, randomized block and the factorial designs, among others.

The choice of m_n and r_n rows is, of course, crucial, and obviously, depends on the design matrix. In the one way classification design with p treatments, n_j observations from the jth treatment, it is recommended to choose the first $m_{nj} = [n_j/2]$ observations from the jth treatment, $1 \le j \le p$ to estimate $\boldsymbol{\beta}$. Here $m_n = m_{n1} + \cdots + m_{np} = [n/2]$. One chooses $r_{nj} = m_{nj}, 1 \le j \le p, r_n = \Sigma_j r_{nj} = [n/2]$. The choice of m_n and r_n is made similarly in the randomized block design and other similar designs. If one had several replications of a design, where the design matrix satisfies (6.2.31) and (6.2.33), then one could use the first half of the replications to estimate $\boldsymbol{\beta}$ and all replications to carry out the test.

Thus, in those cases where designs satisfy (6.2.31) and (6.2.33), the above construction of the ADF tests is similar to the half sample technique in the one sample problem as found in Rao (1972) or Durbin (1976).

Of course there are designs of interest where (6.2.31) and (6.2.33) do not hold. An example is $p = 1$, $x_{ni} \equiv i$. Here, $\mathbf{X}_n'\mathbf{X}_n = O(n^3)$. If one decides to choose the first $m_n(r_n)\, x_i$'s, then (6.2.19) and (6.2.28) are equivalent to requiring $(m_n/n)^3 \to 1/2$ and $(r_n/n)^2 \to 1/2$. Thus, here $\hat{\mathbf{D}}_{2m}$ or $\hat{\mathbf{D}}_{3m}$ would use 89% of the observations to estimate β while $\hat{D}_{1r}$ would use 71%. On the other hand, if one decides to use the last $m_n(r_n)\, x_i$'s, then $\hat{D}_2, \hat{D}_3$ will use the last 21% observations while $\hat{D}_1$ will use the last 29% observations to estimate $\boldsymbol{\beta}$. Of course all of these tests would be based on the entire sample.

In general, to avoid the above kind of problem, one may wish to use, from the practical point of view, some other characteristics of the design matrix in deciding which m_n, r_n rows to choose. One criterion to use may be to choose those $m_n(r_n)$ rows that will approximately maximize $(m_n/n)((r_n/n))$ subject to (6.2.19) ((6.2.28)). □

Remark 6.2.5 *Construction of* $\hat{\boldsymbol{\beta}}_m$ *and* $\hat{\boldsymbol{\beta}}_r$. If F_0 is a d.f. for which the maximum likelihood estimator (m.l.e.) of $\boldsymbol{\beta}$ has a limiting distribution under (**NX**) and (6.2.10) then one should use this estimator based on $r_n(m_n)$ observations $\{(\mathbf{x}_i', Y_i)\}$ for $\hat{D}_1(\hat{D}_2$ or $\hat{D}_3)$. For example, if F_0 is the $\mathcal{N}(0,1)$ d.f., then the obvious choice for $\hat{\boldsymbol{\beta}}_r$ and $\hat{\boldsymbol{\beta}}_m$ are the least squares estimators:

$$\hat{\boldsymbol{\beta}}_r \;:=\; (\mathbf{X}_r'\mathbf{X}_r)^{-1}\mathbf{X}_r'\mathbf{Y}_r^0; \quad \hat{\boldsymbol{\beta}}_m := (\mathbf{X}_m\mathbf{X}_m)^{-1}\mathbf{X}_m'\mathbf{Y}_m^*.$$

Of course there are many d.f.'s F_0 that satisfy the above conditions, but for which the computation of m.l.e. is not easy. One way to proceed in such cases is to use one step linear approximation. To make this precise, let $\overline{\boldsymbol{\beta}}_m$ be an estimator of $\boldsymbol{\beta}$ based on $\{(\mathbf{x}'_{ni}, Y_{ni}), 1 \le i \le m_n\}$ such that

$$(6.2.35) \qquad A_m^{-1}(\overline{\boldsymbol{\beta}}_m - \boldsymbol{\beta}) = O_p(1).$$

Define

$$(6.2.36) \qquad \begin{aligned} \psi_0(y) &:= \dot{f}_0(y)/f_0(y), \quad y \in \mathbb{R}; \\ \bar{s}_{ni} &:= \psi_0(Y_{ni} - \mathbf{x}'_{ni}\overline{\boldsymbol{\beta}}_m), 1 \le i \le m_n; \\ \bar{\mathbf{s}}_m &:= (\bar{s}_{ni}, 1 \le i \le m_n)' \\ \hat{\boldsymbol{\beta}}_m &:= \overline{\boldsymbol{\beta}}_m + I_0 \mathbf{A}_m \mathbf{A}_m \mathbf{X}'_m \mathbf{s}_m \\ \mathbf{V}^*_m(y, \mathbf{t}) &= \mathbf{A}_m \sum_{i=1}^{m_n} \mathbf{x}_{ni} \mathbf{I}(Y_{ni} \le y + \mathbf{x}'_{ni}\mathbf{t}), \\ & \qquad y \in \mathbb{R}, \ t \in \mathbb{R}^p. \end{aligned}$$

Then

$$\mathbf{A}_m \mathbf{X}'_m \bar{\mathbf{s}}_m = \int \psi_0(y) \mathbf{V}^*_m(dy, \overline{\boldsymbol{\beta}}_m).$$

From this and (2.3.37), applied to $\{(\mathbf{x}'_{ni}, Y_{ni}), 1 \le i \le m_n\}$, one readily obtains

Corollary 6.2.3 *Assume that (1.1.1) and H_0 hold. In addition, assume that F_0 is strictly increasing, satisfies (6.2.10) and is such that ψ_0 is a finite linear combination of nondecreasing bounded functions, $\mathbf{X}$ and $\{\overline{\boldsymbol{\beta}}_m\}$ satisfy (NX) and (6.2.35). Then $\{\hat{\boldsymbol{\beta}}_m\}$ of (6.2.36) satisfies (6.2.22) for any sequence $m_n \to \infty$.*

Proof. Clearly,

$$\mathbf{A}_m^{-1}(\hat{\boldsymbol{\beta}}_m - \boldsymbol{\beta}) = \mathbf{A}_m^{-1}(\overline{\boldsymbol{\beta}}_m - \boldsymbol{\beta}) + I_0^{-1} \mathbf{A}_m \mathbf{X}'_m \bar{\mathbf{s}}_m.$$

But, integration by parts and (2.3.37) yield

$$\begin{aligned} \mathbf{A}_m \mathbf{X}'_m \{\bar{\mathbf{s}}_m - \mathbf{s}_m\} &= \int \psi_0(y) \{\mathbf{V}^*_m(dy, \overline{\boldsymbol{\beta}}) - \mathbf{V}^*_m(dy, \boldsymbol{\beta})\} \\ &= -\int \{\mathbf{V}^*_m(y, \overline{\boldsymbol{\beta}}) - \mathbf{V}^*_m(y, \boldsymbol{\beta})\} d\psi_0(y) \\ &= A_m^{-1}(\overline{\boldsymbol{\beta}}_m - \boldsymbol{\beta}) \int f_0(y) d\psi_0(y) + o_p(1) \\ &= -A_m^{-1}(\overline{\boldsymbol{\beta}}_m - \boldsymbol{\beta}) I_0 + o_p(1). \qquad \square \end{aligned}$$

The above result is useful, e.g., when F_0 is logistic, Cauchy or double exponential. In the first case m.l.e. is not easy to compute but F_0 has finite second moment. So take $\overline{\boldsymbol{\beta}}_m$ to be the l.s.e. and then use (6.2.36) to obtain the final estimator to be used for testing. In the case of Cauchy, $\overline{\boldsymbol{\beta}}_m$ may be chosen to be an R-estimator.

Clearly, there is an analogue of the above corollary involving $\{\hat{\boldsymbol{\beta}}_r\}$ that would satisfy (6.2.28). □

6.2.2 Bootstrap distributions

In this subsection we shall obtain a weak convergence result about a bootstrapped W.E.P.'s and then apply this to yield bootstrap distributions of some of the above tests.

Let (1.1.1) with $e_{ni} \equiv e_i$ and H_0 hold. Let E_0 and P_0 denote the expectation and probability, respectively, under these assumptions. In addition, throughout this section we shall *assume that* $(\mathbf{F_01}), (\mathbf{F_02})$ and $(\mathbf{NX})$ hold.

Recall the definition of $\mathbf{W}, \hat{\mathbf{W}}$ from (6.2.1), (6.2.2). Let $\hat{\boldsymbol{\beta}}$ be an M-estimators of $\boldsymbol{\beta}$ corresponding to a bounded nondecreasing right continuous score function ψ such that

$$\text{(6.2.37)} \qquad \int \psi dF_0 = 0, \qquad \int f_0 d\psi > 0.$$

Upon specializing (4.2.7) to the current setup one readily obtains

$$\text{(6.2.38)} \qquad \mathbf{A}^{-1}(\hat{\boldsymbol{\beta}} - \boldsymbol{\beta}) = -\kappa \sum_{i=1}^{n} \mathbf{A}\mathbf{x}_{ni}\psi(e_i) + o_p(1), \quad (P_0).$$

where $\kappa := 1/\int f_0 d\psi$.

Let the approximating process obtained from (6.2.2) and (6.2.38) be denoted by $\overline{\mathbf{W}}$, i.e., for $0 \le t \le 1$,

$$\text{(6.2.39)} \qquad \overline{\mathbf{W}}(t) := \sum_{i=1}^{n} \mathbf{A}\mathbf{x}_{ni}\{I(e_i \le F_0^{-1}(t)) - t - \kappa q_0(t)\psi(e_i)\}.$$

Define

$$\text{(6.2.40)} \qquad \begin{aligned} \sigma^2 &:= E_0\psi^2(e_1), \\ g_0(t) &:= E_0\{I(e_1 \le F_0^{-1}(t)) - t\}\psi(e_1) \\ &= \int I(x \le F_0^{-1}(t))\psi(x)dF_0(x), \quad 0 \le t \le 1, \end{aligned}$$

and, for $0 \le t \le u \le 1$,

$$(6.2.41) \qquad \rho_0(t,u) := t(1-u) - \kappa\Big[q_0(t)g_0(u) + g_0(t)q_0(u)\Big] + \kappa^2 q_0(t)q_0(u)\sigma^2.$$

Note that

$$(6.2.42) \qquad \mathcal{C}_0(t,u) := E_0\left\{\overline{\mathbf{W}}(t)\overline{\mathbf{W}}(u)'\right\} = \rho_0(t,u)\mathbf{I}_{p\times p},$$

for all $0 \le t \le u \le 1$. Let $\mathcal{G}_0 := (\mathcal{G}_{01}, \cdots, \mathcal{G}_{0p})'$ be a p - vector of independent Gaussian processes each having the covariance function ρ_0. Thus, $\mathbf{E}\mathcal{G}_0(t)\mathcal{G}_0(u)' \equiv \mathcal{C}_0(t,u)$. Since ρ_0 is continuous, $\mathcal{G}_0 \in \{\mathcal{C}[0,1]\}^p$. Moreover, from Corollary 2.2.1 applied p time, j^{th} time to the entities $X_{ni} \equiv e_i, F_{ni} \equiv F_0$ and $d_{ni} \equiv (i,j)$-th entry of $\mathbf{AX}$, $1 \le i \le p, 1 \le i \le n$, and from the uniform continuity of q_0 it readily follows that

$$(6.2.43) \qquad \overline{\mathbf{W}} \Longrightarrow \mathcal{G}_0 \text{ in } [\{\mathcal{D}[0,1]\}^p, \rho].$$

Now, let $\hat{f}_n$ be a density estimator based on $\{\hat{e}_{ni} := Y_{ni} - \mathbf{x}'_{ni}\hat{\beta}; 1 \le i \le n\}$ and $\hat{F}_n$ be the corresponding d.f. Let $\{e^*_{ni}; 1 \le i \le n\}$ represent i.i.d. $\hat{F}_n$ r.v.'s, i.e., $\{e^*_{ni}; 1 \le i \le n\}$ is a random sample from the population $\hat{F}_n$. Because $\hat{F}_n$ is continuous, the resampling procedures based on it are usually called *smooth* bootstrap procedures. Let

$$Y^*_{ni} := \mathbf{x}'_{ni}\hat{\beta} + e^*_{ni}, \quad 1 \le i \le n.$$

Define the bootstrap estimator β^* to be a solution $\mathbf{s} \in \mathbb{R}^p$ of the equation

$$(6.2.44) \qquad \sum_{i=1}^n \mathbf{Ax}_{ni}\{\psi(Y^*_{ni} - \mathbf{x}'_{ni}\mathbf{s}) - \hat{E}_n\psi(e^*_{ni})\} = \mathbf{0}.$$

where $\hat{E}_n$ is the expectation under $\hat{F}_n$. Let $\hat{P}_n$ denote the bootstrap probability under $\hat{F}_n$. Finally, define, for $0 \le t \le 1$, $\mathbf{u} \in \mathbb{R}^p$,

$$\mathcal{S}^*(t,\mathbf{u}) := \sum_{i=1}^n \mathbf{Ax}_{ni} I(e^*_{ni} \le \hat{F}_n^{-1}(t) + \mathbf{x}'_{ni}\mathbf{Au}),$$

$$\hat{\mathbf{W}}^*(t) := \sum_{i=1}^n \mathbf{Ax}_{ni}\{I(Y^*_{ni} - \mathbf{x}'_{ni}\beta^* \le \hat{F}_n^{-1}(t)) - t\}.$$

We also need

$$(6.2.45) \qquad \mathbf{W}^*(t) := \sum_{i=1}^n \mathbf{Ax}_{ni}\{I(e^*_{ni} \le \hat{F}_n^{-1}(t)) - t\}, \quad 0 \le t \le 1.$$

Our goal is to show that $\hat{\mathbf{W}}^*$ converges weakly to $\mathcal{G}_0$ in $[\{\mathbb{D}[0,1]\}^p, \rho]$, a.s. Here a.s. refers to almost all error sequences $\{e_i; i \ge 1\}$. We in fact have the following

Theorem 6.2.2 *In addition to (1.1.1), H_0, $(\mathbf{F}_0\mathbf{1})$, $(\mathbf{F}_0\mathbf{2})$ $(\mathbf{NX})$ and (6.2.37), assume that ψ is a bounded nondecreasing right continuous score function and that the following hold.*

(6.2.46) *For almost all error sequences* $\{e_i; i \ge 1\}, \hat{f}_n(x) > 0,$ *for almost all* $x \in \mathbb{R},\ n \ge 1.$

(6.2.47)
$$\|\hat{f}_n - f_0\|_\infty \to 0, \quad a.s., \quad (P_0).$$

Then, $\forall\ 0 < b < \infty$,

(6.2.48)
$$\sup \|\mathcal{S}^*(t,\mathbf{u}) - \mathcal{S}^*(t,\mathbf{0}) - \mathbf{u}\hat{f}_n(\hat{F}_n^{-1}(t))\| = o_p(1), \ (\hat{P}_n), \ a.s.,$$

where the supremum is over $0 \le t \le 1$, $\|\mathbf{u}\| \le b$.

Moreover, for almost all error sequences $\{e_i; i \ge 1\}$,

(6.2.49)
$$\begin{aligned} &\mathbf{A}^{-1}(\beta^* - \hat{\beta}) \\ &= \hat{\kappa}_n \sum_{i=1}^n \mathbf{A}\mathbf{x}_{ni}\{\psi(e^*_{ni}) - \hat{E}_n\psi(e^*_{ni})\} + o_p(1), \quad (\hat{P}_n), \end{aligned}$$

and

(6.2.50)
$$\hat{\mathbf{W}}^* \Longrightarrow \mathcal{G}_0, \quad in\ [\{\mathcal{D}[0,1]\}^p, \rho],$$

where $\hat{\kappa}_n := 1/\int \hat{f}_n d\psi$.

Proof. Fix an error sequences $\{e_i; i \ge 1\}$ for which

(6.2.51)
$$\hat{f}_n(x) > 0, \text{for almost all } x \in \mathbb{R}, \text{ and } \|\hat{f}_n - f_0\|_\infty \to 0.$$

The following arguments are carried out conditional on this sequence.

Observe that $\mathcal{S}^*(t,\mathbf{u})$ is a p-vector of W.E.P.'s $S_d(t,\mathbf{u})$ of (2.3.1) whose j^{th} component has various underlying entities as follows:

(6.2.52)
$$\begin{aligned} X_{ni} &= e^*_{ni}, \quad F_{ni} = \hat{F}_n, \quad \mathbf{c}_{ni} = \mathbf{A}\mathbf{x}_{ni}, \\ d_{ni} &= \mathbf{a}'_{(j)}\mathbf{x}_{ni}, \quad 1 \le i \le n, \end{aligned}$$

where, as usual, $\mathbf{a}_{(j)} = j$-th column of $\mathbf{A}, 1 \le j \le p$.

Thus, (6.2.48) follows from p applications of Theorem 2.3.1, j^{th} time applied to the above entities, provided we ensure the validity of the assumptions of that theorem. But, f_0 uniformly continuous and (6.2.47) readily imply that $\{\hat{f}_n, n \geq 1\}$ satisfies (2.3.4), (2.3.5). In view of (2.3.31), (2.3.32) and (**NX**), it follows that all other assumptions of Theorem 2.3.1 are satisfied. Hence, (6.2.48) follows from (2.3.8). In view of (6.2.46) we also obtain, from (2.3.9),

$$\text{(6.2.53)} \qquad \sup \|\boldsymbol{\mathcal{S}}^{0*}(x, \mathbf{u}) - \boldsymbol{\mathcal{S}}^{0*}(x, \mathbf{0}) - \mathbf{u}\hat{f}_n(x)\| = o_p(1), \ (\hat{P}_n),$$

where $\boldsymbol{\mathcal{S}}^{0*}(x, \mathbf{u}) \equiv \boldsymbol{\mathcal{S}}^{*}(\hat{F}_n(x), \mathbf{u})$ and where the supremum is over $x \in \mathbb{R}, \|\mathbf{u}\| \leq b$. Now, (6.2.49) follows from (6.2.53) in precisely the same fashion as does (4.2.7) from (2.3.9).

From (6.2.48), (6.2.49) and (6.2.62) below, we readily obtain that, under $\hat{P}_n$,

$$\text{(6.2.54)} \qquad \hat{\mathbf{W}}^*(t) = \sum_{i=1}^{n} \mathbf{A}\mathbf{x}_{ni}\{I(e^*_{ni} \leq \hat{F}_n^{-1}(t)) - t - \hat{\kappa}_n \hat{q}_n(t) \times [\psi(e^*_{ni}) - \hat{E}_n \psi(e^*_{ni})]\} + o_p(1),$$

where $\hat{q}_n := \hat{f}_n(\hat{F}_n^{-1})$.

In analogy to (6.2.40) and (6.2.41), let $\hat{g}_n, \hat{\rho}_n$ stand for g_0, ρ_0 after F_0 is replaced by $\hat{F}_n$ in these entities. Thus

$$\begin{aligned} \hat{g}_n(t) &:= \hat{E}_n\{I(e^*_{ni} \leq \hat{F}_n^{-1}(t)) - t\}\psi(e^*_{ni}) \\ &= \int I(x \leq \hat{F}_n^{-1}(t))\psi(x) d\hat{F}_n(x), \quad 0 \leq t \leq 1, \end{aligned}$$

and, for $0 \leq t \leq u \leq 1$,

$$\begin{aligned} \hat{\rho}_n(t, u) &:= t(1-u) - \hat{\kappa}_n \Big[\hat{q}_n(t)\hat{g}_n(U) - \hat{g}_n(t)\hat{q}_n(u)\Big] \\ &\quad + \hat{\kappa}_n^2 \hat{q}_n(t)\hat{q}_n(u)\hat{\sigma}_n^2. \end{aligned}$$

where $\hat{\sigma}_n^2 := \hat{E}_n[\psi(e^*_{n1} - \hat{E}_n\psi(e^*_{n1})]^2$.

Let $\tilde{\mathbf{W}}^*(t)$ denote the leading r.v. in the R.H.S. of (6.2.54). Observe that

$$\tilde{\mathcal{C}}_n(t, u) := \hat{E}_n\{\tilde{\mathbf{W}}^*(t)\tilde{\mathbf{W}}^*(u)'\} = \hat{\rho}_n(t, u)\mathbf{I}_{p\times p},$$

for all $0 \leq t \leq u \leq 1$.

(6.2.55) **Claim**: $\hat{\rho}_n(t, u) \to \rho_0(t, u), \ \forall\, 0 \leq t \leq u \leq 1.$

To prove (6.2.55), note that (6.2.51) and Lemma 9.1.4 of the Appendix below imply that for the given error sequence $\{e_i; i \geq 1\}$,

$$\delta_n := \|\hat{F}_n - F_0\|_\infty \to 0, \tag{6.2.56}$$

which, together with the continuity of $\hat{F}_n$, yields

$$\sup_{0 \leq t \leq 1} |F_0(\hat{F}_n^{-1}(t)) - t| \to 0. \tag{6.2.57}$$

Also, observe that

$$\sup_{0 \leq t \leq 1} |\hat{f}_n(\hat{F}_n^{-1}(t)) - f_0(\hat{F}_n^{-1}(t))| \leq \|\hat{f}_n - f_0\|_\infty \to 0,$$

by (6.2.51), and that, $\forall\, 0 \leq t \leq 1$,

$$|f_0(\hat{F}_n^{-1}(t)) - f_0(F_0^{-1}(t))| \equiv |q_0(F_0(\hat{F}_n^{-1}(t))) - q_0(t)|.$$

Hence, by (6.2.57) and the uniform continuity of q_0, which is implied by ($\mathbf{F}_0\mathbf{1}$),

$$\sup_{0 \leq t \leq 1} |\hat{q}_n(t) - q_0(t)| \longrightarrow 0. \tag{6.2.58}$$

Next, let

$$g_n(t) = \int I(\hat{F}_n(x) \leq t)\psi(x) f_0(x)dx, 0 \leq t \leq 1.$$

Upon rewriting

$$\hat{g}_n(t) = \int I(\hat{F}_n(x) \leq t)\psi(x)\hat{f}_n(x)dx,$$

from (6.2.51), Lemma 9.1.4 and the boundedness of ψ, we readily obtain that

$$\sup_{0 \leq t \leq 1} |\hat{g}_n(t) - g_n(t)| \leq \int |\hat{f}_n(x) - f_0(x)|dx \to 0.$$

But, the inequality $F_0(x) - \delta_n \leq \hat{F}_n(x) \leq F_0(x) + \delta_n$ for all x, implies that

$$\begin{aligned} |g_n(t) - g_0(t)| &\leq \|\psi\|_\infty \int I(F_0(x) - \delta_n \leq t \leq F_0(x) + \delta_n)dF_0(x), \\ &\leq \|\psi\|_\infty 2\delta_n, \quad \forall\, 0 \leq t \leq 1. \end{aligned}$$

Hence, by (6.2.56),

$$\sup_{0 \leq t \leq 1} |\hat{g}_n(t) - g_0(t)| \to 0. \tag{6.2.59}$$

Again by the boundedness of ψ, (6.2.51) and (6.2.56), one readily concludes that

$$(6.2.60) \qquad \hat{\kappa}_n \to \kappa, \quad \hat{\sigma}_n^2 \to \sigma^2.$$

The claim (6.2.55) now readily follows from (6.2.58) - (6.2.60).

Now recall (6.2.45) and rewrite $\tilde{\mathbf{W}}^*$ as

$$(6.2.61) \qquad \tilde{\mathbf{W}}^*(t) = \mathbf{W}^*(t) - \hat{\kappa}_n \hat{q}_n(t) \sum_{i=1}^{n} \mathbf{A}\mathbf{x}_{ni}[\psi(e_{ni}^*) - \hat{E}_n \psi(e_{ni}^*)].$$

Observe that because

$$\hat{E}_n \Big\| \sum_{i=1}^{n} \mathbf{A}\mathbf{x}_{ni}[\psi(e_{ni}^*) - \hat{E}_n \psi(e_{ni}^*)\Big\|^2 = p\, \hat{\sigma}_n^2,$$

by (6.2.60) and the Markov inequality it follows that

$$(6.2.62) \qquad \Big\| \sum_{i=1}^{n} \mathbf{A}\mathbf{x}_{ni}[\psi(e_{ni}^*) - \hat{E}_n \psi(e_{ni}^*)]\Big\| = O_p(1), \quad (\hat{P}_n).$$

Apply Corollary 2.2.1 p times, j^{th} time to the entities given at (6.2.52), to conclude that

$$\lim_{\eta \to 0} \limsup_{n} \hat{P}_n \left(\sup_{|t-s|<\eta} |\mathbf{W}^*(t) - \mathbf{W}^*(s)| > \epsilon \right) = 0, \quad \forall\ ep > 0.$$

This together with (6.2.62), (6.2.61), (6.2.58) and the uniform continuity of F_0 implies that the sequence of processes $\{\tilde{\mathbf{W}}^*\}$ is tight in the uniform metric ρ and all its subsequential limits must be in $\{\mathcal{C}[0,1]\}^p$. Now, (6.2.50) follows from this, (6.2.54), (6.2.47), (6.2.46), (6.2.42) and the Claim (6.2.55). □

Remark 6.2.6 One of the main consequences of (6.2.50) is that one can use the bootstrap analogue of $\hat{D}_3$, v.i.z.,

$$\hat{D}_3^* := \sup\{\|\hat{\mathbf{W}}^*(t)\|, 0 \le t \le 1\}$$

to carry out the test H_0. Thus an approximation to the null distribution of $\hat{D}_3$ is obtained by the distribution of $\hat{D}_3^*$ under $\hat{P}_n$. In practice it means to obtain repeated random samples of size n from $\hat{F}_n$, compute the frequency distribution of $\hat{D}_3^*$ from these samples and use that to approximate the null distribution of $\hat{D}_3$. At least asymptotically this converges to the right distribution. Obviously the smooth bootstrap distributions for $\hat{D}_1, \hat{D}_2$ can be obtained similarly.

Reader might have realized that the conclusion (6.2.50) is true for any sequence of estimators $\{\hat{\beta}\}$, $\{\beta^*\}$ satisfying (6.2.38) and (6.2.49). □

6.3 L_2-Distance Tests

Let K_1^0 and K_2^0, respectively, stand for the K_1 and $K_{\mathbf{X}}$ of (5.2.5) and (5.2.5) after the d.f.'s $\{H_{ni}\}$ there are replaced by F_0. Thus, for $G \in \mathcal{DI}(\mathbb{R})$,

(6.3.1)
$$\begin{aligned} K_1^0(\mathbf{t}) &:= \int \{W_1^0(y,\mathbf{t})\}^2 dG(y), \\ K_2^0(\mathbf{t}) &:= \int \|\mathbf{W}^0(y,\mathbf{t})\|^2 dG(y), \quad \mathbf{t} \in \mathbb{R}^p, \end{aligned}$$

where $\mathbf{W}^0$ is as in (6.1.3) and

(6.3.2)
$$W_1^0(y,\mathbf{t}) := n^{1/2}[H_n(y,\mathbf{t}) - F_0(y)], \quad y \in \mathbb{R}, \ \mathbf{t} \in \mathbb{R}^p.$$

Let $\hat{\beta}$ be an estimator of β and define the four test statistics

(6.3.3)
$$K_j^* := \inf\{K_j(\mathbf{t}); \mathbf{t} \in \mathbb{R}^p\}, \quad \hat{K}_j := K_j(\hat{\beta}), \quad j = 1,2.$$

The large values of these statistics are significant for testing H_0.

We shall first discuss the asymptotic null distribution's of K_j^*, $j = 1,2$. Let $W_1^0(\cdot)$, $\mathbf{W}^0(\cdot)$ stand for $W_1^0(\cdot,\beta)$ and $\mathbf{W}^0(\cdot,\beta)$, respectively.

Theorem 6.3.1 *Assume that (1.1.1), H_0, (**NX**), (5.5.44) - (5.5.46) with $F \equiv F_0$ hold.*

(a) If, in addition, (5.6.9) and (5.6.10) hold, then

(6.3.4)
$$K_1^* = \int \left\{ W_1^0(y) - f_0(y) \frac{\int W_1^0 f_0 dG}{\int f_0^2 dG} \right\}^2 dG + o_p(1).$$

(b) Under no additional assumptions,

(6.3.5)
$$K_2^* = \int \left\| \mathbf{W}^0(y) - f_0(y) \frac{\int \mathbf{W}^0 f_0 dG}{\int f_0^2 dG} \right\|^2 dG + o_p(1).$$

Proof. Apply Theorems 5.5.1 and 5.5.3 twice, once with $\mathbf{D} = \mathbf{XA}$, and once with $\mathbf{D} = n^{-1/2}(\mathbf{1}, \mathbf{0}, \cdots, \mathbf{0}]$ and the rest of the entities as follows:

(6.3.6)
$$Y_{ni} \equiv e_{ni}, \ H_{ni} \equiv F_0 \equiv F_{ni}, \ \ G_n \equiv G.$$

The theorem then follows from (5.5.9), (5.6.5), (5.6.11) and some algebra. See also Claim 5.5.2. □

Remark 6.3.1 Perhaps it is worth emphasizing that (6.3.5) holds without any extra conditions on the design matrix $\mathbf{X}$. Thus, at least in this sense, K_2^* is a more natural statistic to use than K_1^* for testing H_0.

A consequence of (6.3.4) is that even if $\hat{\beta}_1$ of (5.2.4) is asymptotically non-unique, K_1^* asymptotically behaves like a unique sequence of r.v.'s. Moreover, unlike the $\hat{D}_1$ - statistic, the asymptotic null distribution of K_1^* does not depend on the design matrix among all those designs that satisfy the given conditions.

The assumptions (5.6.9) and (5.6.10) are restrictive. For example, in the case $p = 1$, (5.6.9) translates to requiring that either $x_{i1} \geq 0$ for all i or $x_{i1} \leq 0$ for all i. The assumption (5.6.10) says that $\bar{\mathbf{x}} \neq \mathbf{0}$ or can not converge to $\mathbf{0}$. Compare this with the fact that if $\bar{\mathbf{x}} \approx 0$ then the asymptotic distribution of $\hat{D}_1$ does not depend on the preliminary estimator $\hat{\beta}$. □

Next, we need a result that will be useful in deriving the limiting distributions of certain quadratic forms involving W.E.P.'s. To that effect, let $L_2^p(\mathbb{R}, G)$ be the equivalence classes of measurable functions $h : \mathbb{R}$ to $\mathbb{R}^p$ such that $|h|_G^2 := \int \|h\|^2 dG < \infty$. The equivalence classes are defined in terms of the norm $|\cdot|_G^2$. In the following lemma, $\{\mathbf{a}_i; i \geq 1\}$ is a fixed orthonormal basis in $L_2^p(\mathbb{R}, G)$.

Lemma 6.3.1 *Let $\{\mathbf{Z}_n, n \geq 1\}$ be a sequence of p - vector stochastic processes with $E\mathbf{Z}_n = \mathbf{0}$,*

$$Cov(\mathbf{Z}_n(x), \mathbf{Z}_n(y)) := \boldsymbol{K}_n(x, y) = ((K_{nij}(x, y))),$$

for $1 \leq i, j \leq p$, $x, y \in \mathbb{R}$. In addition, assume the following:

There is a covariance matrix function $\boldsymbol{K}(x, y) = ((K_{ij}(x, y)))$, and a p - vector mean zero covariance $\boldsymbol{K}$ Gaussian process $\mathbf{Z}$ such that

(i) *(a)* $\sum_{j=1}^p \int K_{njj}(x, x) dG(x) < \infty, \quad n \geq 1.$

(b) $\sum_{j=1}^p \int K_{jj}(x, x) dG(x) < \infty.$

(ii) $\sum_{j=1}^p \int K_{njj}(x, x) dG(x) \to \sum_{j=1}^p \int K_{jj}(x, x) dG(x).$

(iii) *For every $m \geq 1$,*

$$\left(\int \mathbf{Z}_n' \mathbf{a}_1 dG, \cdots, \int \mathbf{Z}_n' \mathbf{a}_m dG \right)$$
$$\to_d \left(\int \mathbf{Z}' \mathbf{a}_1 dG, \cdots, \int \mathbf{Z}' \mathbf{a}_m dG \right).$$

(iv) *For each* $i \geq 1$,

$$E\left(\int \mathbf{Z}_n'\mathbf{a}_i dG\right)^2 \to E(\int \mathbf{Z}'\mathbf{a}_i dG)^2.$$

Then, $\mathbf{Z}_n$, $\mathbf{Z}$ *belong to* $L_2^p(\mathbb{R}, G)$, *and*

$$\mathbf{Z}_n \Longrightarrow \mathbf{Z} \quad in \quad L_2^p(\mathbb{R}, G). \tag{6.3.7}$$

Proof. In view of Theorem VI.2.2 of Parthasarthy (1967) and in view of (iii), it suffices to show that for any $\epsilon > 0$, there is an $N(= N_\epsilon)$ such that

$$\sup_n E \sum_{i \geq \mathbf{N}} \left(\int \mathbf{Z}_n'\mathbf{a}_i dG\right)^2 \leq \epsilon. \tag{6.3.8}$$

Because of the properties of $\{\mathbf{a}_i\}$, Fubini and (i),

$$\sum_{j=1}^p \int K_{nij}(x,x)dG(x) = E|\mathbf{Z}_n|_G^2 = \sum_{i \geq 1} E\left(\int \mathbf{Z}_n'\mathbf{a}_i dG\right)^2, \tag{6.3.9}$$

$$\sum_{j=1}^p \int K_{jj}(x,x)dG(x) = E|\mathbf{Z}|_G^2 = \sum_{i \geq 1} E\left(\int \mathbf{Z}'\mathbf{a}_i dG\right)^2. \tag{6.3.10}$$

Thus, to prove (6.3.8), it suffices to exhibit an N such that

$$\sup_n \sum_{i \geq \mathbf{N}} E\left(\int \mathbf{Z}_n'\mathbf{a}_i dG\right)^2 \leq \epsilon. \tag{6.3.11}$$

By (ii), (6.3.9) and (6.3.10), there exists $N_{1\epsilon}$ such that

$$\sum_{i \geq 1} E\left(\int \mathbf{Z}_n'\mathbf{a}_i dG\right)^2 \leq \sum_{i \geq 1} E\left(\int \mathbf{Z}'\mathbf{a}_i dG\right)^2 + \epsilon/3, \ n \geq N_{1\epsilon}$$

By (i)(b) and (6.3.10), there exists $N(= N_\epsilon)$ such that

$$\sum_{i \geq \mathbf{N}} E\left(\int \mathbf{Z}'\mathbf{a}_i dG\right)^2 \leq \epsilon/3. \tag{6.3.12}$$

By (iv), there exists $N_{2\epsilon}$ such that

$$\sum_{i < \mathbf{N}} E\left(\int \mathbf{Z}'\mathbf{a}_i dG\right)^2 \leq \sum_{i < \mathbf{N}} E\left(\int \mathbf{Z}_n'\mathbf{a}_i dG\right)^2 + \epsilon/3, \ n \geq N_{2\epsilon}.$$

Therefore, with $N = N_\epsilon := N_{1\epsilon} \vee N_{2\epsilon}$,

$$\begin{aligned}
&\sup_{n\geq \mathrm{N}} \sum_{i<\mathrm{N}} E\Big(\int \mathbf{Z}_n'\mathbf{a}_i dG\Big)^2 \\
&\leq \sup_{n\geq \mathrm{N}} \left[\sum_{i\geq 1} E\left(\int \mathbf{Z}'\mathbf{a}_i dG\right)^2 - \sum_{i<\mathrm{N}} \left(\int \mathbf{Z}_n'\mathbf{a}_i dG\right)^2\right] + \epsilon/3 \\
&\leq \epsilon.
\end{aligned}$$

Use (i) (a) to take care of the case $n < N_\epsilon$. This proves (6.3.7). □

Remark 6.3.2 Millar (1981) contains a special case of the above lemma where $p = 1$, $\mathbf{Z}_n$ is the standardized ordinary e.p. and $\mathbf{Z}$ is the Brownian bridge. The above lemma is an extension of that result to cover more general processes like the W.E.P.'s under general independent setting. In applications of the above lemma, one may choose $\{\mathbf{a}_i\}$ to be such that the support S_i of $\mathbf{a}_i$ has $G(S_i) < \infty, i \geq 1$ and such that $\{\mathbf{a}_i\}$ are bounded. □

Corollary 6.3.1 *(a) Under the conditions of Theorem 6.3.1(a),*

$$K_1^* \longrightarrow_d \int \left\{B(F_0) - f_0 \cdot \frac{\int B(F_0) f_0 dG}{\int f_0^2 dG}\right\}^2 dG =: \overline{G}_1, \quad (say).$$

(b) Under the conditions of Theorem 6.3.1(b),

$$K_1^* \longrightarrow_d \int \left\|\mathbf{B}(F_0) - f_0 \cdot \frac{\int \mathbf{B}(F_0) f_0 dG}{\int f_0^2 dG}\right\|^2 dG =: \overline{G}_1, \quad (say).$$

Here $B, \mathbf{B}$ are as in (6.2.6), (6.2.7).

Proof. (b) Apply Lemma 6.3.1, with $\mathbf{a}_i$ as in the Remark 6.3.2 above, to

$$\begin{aligned}
\mathbf{Z}_n &= \mathbf{W}^0 - \frac{\int \mathbf{W}^0 f_0 dG}{\int f_0^2 dG} \cdot f_0, \\
\mathbf{Z} &= \mathbf{B}(F_0) - \frac{\int \mathbf{B}(F_0) f_0 dG}{\int f_0^2 dG} \cdot f_0.
\end{aligned}$$

Direct calculations show that $E\mathbf{Z}_n = \mathbf{0} = E\mathbf{Z}$, and $\forall\, x, y \in \mathbb{R}$,

$$\begin{aligned}
\boldsymbol{K}_n(x,y) := E\mathbf{Z}_n(x)\mathbf{Z}_n'(y) &= \mathbf{I}_{p\times p}\ell(x,y) \\
&= \boldsymbol{K}(x,y) =: E\mathbf{Z}(x)\mathbf{Z}'(y),
\end{aligned}$$

where, for $x, y \in \mathbb{R}$,

$$\begin{aligned}\ell(x,y) &:= k(x,y) - a^{-1}f_0(y)\int k(x,s)d\psi(s)\\&\quad -a^{-1}f_0(y)\int k(y,s)d\psi(s) + a^{-2}\int\int k(s,t)d\psi(s)d\psi(t),\\k(x,y) &:= F_0(x\wedge y) - F_0(x)F_0(y),\ \psi(x) := \int_{-\infty}^{x} f_0 dG,\ a = \psi(\infty).\end{aligned}$$

Therefore, (5.5.44), (5.5.45) imply (i), (ii) and (iv). To prove (iii), let $\lambda_1, \cdots, \lambda_m$ be real numbers. Then

$$\begin{aligned}\sum_{j=1}^{m}\lambda_j\int \mathbf{Z}_n'\mathbf{a}_j dG &= \int \mathbf{W}^{0'}\mathbf{b}dG - \frac{\int \mathbf{W}^{0'}d\psi}{\int f_0 d\psi},\\&\int \mathbf{b}d\psi =: h(\mathbf{W}^0), \text{(say)},\end{aligned}$$

where $\mathbf{b} := \sum_{j=1}^{m}\lambda_j\mathbf{a}_j$. Because ψ and $\mathbf{b}dG$ are finite measures, $h(\mathbf{W}^0)$ is a uniformly continuous function of $\mathbf{W}^0$. Thus by Lemma 6.2.1 and Theorem 5.1 of Billingsley (1968), $h(\mathbf{W}^0) \to_d h(\mathbf{B}(F_0))$, under H_0 and (**NX**). This then verifies all conditions of Lemma 6.3.1. Hence $\mathbf{Z}_n \Longrightarrow \mathbf{Z}$ in $L_2^p(\mathbb{R}, G)$. In particular $\int \|\mathbf{Z}_n\|^2 dG \to_d \int \|\mathbf{Z}\|^2 dG$. This and (6.3.5) proves part (a). The proof of part (b) is similar. □

Remark 6.3.3 The r.v. $\overline{G}_1$ can be rewritten as

$$\overline{G}_1 = \int B^2(F_0)dG - \frac{\{\int B(F_0)f_0 dG\}^2}{\int f_0^2 dG}$$

Recall that $\overline{G}_1$ is the same as the limiting r.v. obtained in the one sample location model. Its distribution for various G and F_0 has been theoretically studied by Martynov (1975). Boos (1981) has tabulated some critical values of $\overline{G}_1$ when $dG = \{F_0(1-F_0)\}^{-1}dF_0$ and F_0 = Logistic. From Anderson - Darling or Boos one obtains that in this case

$$\begin{aligned}\overline{G}_1 &= \int_0^1 B^2 * (t)(t(1-t))^{-1}dt - 6\Big(\int_0^1 B(t)dt\Big)^2\\&= \sum_{j\geq 2} N_j^2/j(j+1)\end{aligned}$$

where $\{N_j\}$ are i.i.d. $\mathcal{N}(0,1)$ r.v.'s. From Boos (Table 3), one obtains the following

Table II

α	.005	.01	.025	.05
t_α	1.710	1.505	1.240	1.046

In Table II, t_α is such that $P(\overline{G}_1 > t_\alpha) = \alpha$. For some other tables see Stephens (1979).

The r.v. $\overline{G}_2$ can be rewritten as

$$\begin{aligned}\overline{G}_2 &:= \int \|\mathbf{B}(F_0)\|^2 dG - \frac{\|\mathbf{B}(F_0) f_0 dG\|^2}{\int f_0^2 dG} \\ &= \sum_{j=1}^{p} \left[\int \mathbf{B}_j^2(F_0) dG - \frac{(\int \mathbf{B}_j(F_0) f_0 dG)^2}{\int f_0^2 dG} \right],\end{aligned}$$

which is a sum of p independent r.v.'s identically distributed as $\overline{G}_1$. The distribution of such r.v.'s does not seem to have been studied yet. Until the distribution of $\overline{G}_2$ is tabulated one could use the independence of the summands in $\overline{G}_2$ and the bounds between the sum and the maximum to obtain a crude approximation to the significance level.

For $p = 1$, the asymptotic null distribution of K_1^* and K_2^* is the same but the conditions under which the results for K_1^* hold are stronger than those for K_2^*. □

The next result gives an approximation for $\hat{K}_j, j = 1, 2$. It also follows from Theorem 5.5.1 in a fashion similar to the previous theorem, and hence no details are given.

Theorem 6.3.2 *Assume that (1.1.1), H_0, (**NX**), (5.5.44) - (5.5.46) with $F \equiv F_0$ and (6.2.3) hold. Then,*

$$(6.3.13) \qquad \hat{K}_1 = \int \left[W_1^0(y) + n^{1/2} \bar{\mathbf{x}} \mathbf{A} \cdot \mathbf{A}^{-1} (\hat{\beta} - \beta) f_0(y) \right]^2 dG(y) + o_p(1)$$

$$\hat{K}_2 = \int \left\| \mathbf{W}^0(y) + \mathbf{A}^{-1} (\hat{\beta} - \beta) f_0(y) \right\|^2 dG(y) + o_p(1).$$

From this we can obtain the asymptotic null distribution of these statistics when $\hat{\beta}$ is estimated efficiently for the large samples as follows. Recall the definition of $\{s_i\}$ from (6.2.12) and let

$$\begin{aligned}\gamma_i(y) &:= \mathbf{I}(e_i \le y) - F_0(y) + n\bar{\mathbf{x}}'(\mathbf{X}'\mathbf{X})^{-1} \mathbf{x}_i s_i I_0^{-1} f_0(y), \\ \alpha_i(y) &:= \mathbf{I}(e_i \le y) - F_0(y) + s_i I_0^{-1} f_0(y), \; 1 \le i \le n, \; y \in \mathbb{R}, \\ \boldsymbol{\alpha} &= (\alpha_1, \cdots, \alpha_n)', \quad \gamma = (\gamma_1, \cdots, \gamma_n)'.\end{aligned}$$

Also, define, for $y \in \mathbb{R}$,

$$
\begin{aligned}
Z_{n1}(y) &:= W_1^0(y) + n^{1/2}\bar{\mathbf{x}}'\mathbf{A}\mathbf{A}\mathbf{X}'\mathbf{s}I_0^{-1}f_0(y) \\
&= n^{-1/2}\sum_{i=1}^{n}\gamma_i(y), \qquad (6.3.14)\\
\mathbf{Z}_{n2}(y) &:= \mathbf{W}^0(y) + \mathbf{A}\mathbf{X}'\mathbf{s}I_0^{-1}f_0(y) = \mathbf{A}\mathbf{X}'\boldsymbol{\alpha}(y).
\end{aligned}
$$

From Theorem 6.3.2 we readily obtain the

Corollary 6.3.2 *Assume that (1.1.1), H_0 (**NX**), (5.5.44) - (5.5.46) with $F \equiv F_0$, (6.2.11) and (6.2.13) hold. Then,*

$$
(6.3.15) \qquad \hat{K}_1 = \int Z_{n1}^2 dG + o_p(1).
$$

$$
(6.3.16) \qquad \hat{K}_2 = \int \|\mathbf{Z}_{n2}\|^2 dG + o_p(1).
$$

Next, observe that for $y \le \mathbf{z}$,

$$
\begin{aligned}
K_{n1}(y,z) &:= Cov(Z_{n1}(y), Z_{n1}(z)) \\
&= F_0(y)(1-F_0(z)) - n\bar{\mathbf{x}}'(\mathbf{X}'\mathbf{X})^{-1}\bar{\mathbf{x}}\frac{f_0(y)f_0(z)}{I_0} \\
&=: \ell_{n1}(y,z), \\
\boldsymbol{K}_{n2}(y,z) &:= E\mathbf{Z}_{n2}(y)\mathbf{Z}_{n2}(z)' \\
&= \{F_0(y)(1-F_0(z)) - \frac{f_0(y)f_0(z)}{I_0}\}\mathbf{I}_{p\times p} \\
&=: r_0(y,z)\,\mathbf{I}_{p\times p}, \quad \text{say.}
\end{aligned}
$$

Now apply Lemma 6.3.1 and argue just as in the proof of Corollary 6.3.1 to conclude

Corollary 6.3.3 *(a) In addition to the conditions of Corollary 6.3.2, assume that*

$$
(6.3.17) \qquad n\,\bar{\mathbf{x}}'(\mathbf{X}'\mathbf{X})^{-1}\bar{\mathbf{x}} \to c, \quad |c| < \infty.
$$

Then,

$$
\hat{K}_1 \longrightarrow_d \int Z_1^2(y)dG(y),
$$

where Z_1 is a Gaussian process in $L_2(\mathcal{R}, G)$ with the covariance function

$$
K_1(x,y) := F_0(x)(1-F_0(y)) - c\, f_0(x)f_0(y)I_0^{-1}, \quad x \le y.
$$

(b) Under the conditions of Corollary 6.3.2,

$$\hat{K}_2 \longrightarrow_d \int \|\mathbf{Y}_0\|^2 dG, \tag{6.3.18}$$

where $\mathbf{Y}_0$ *is a vector of* p *independent Gaussian processes in* $L_2^p(\mathbb{R}, G)$ *with the covariance matrix* $r_0 \cdot \mathbf{I}_{p\times p}$. □

Remark 6.3.4 Again, observe that asymptotic null distribution of the test statistic $\hat{K}_1$ based on the ordinary empirical of the residuals is design dependent whereas that of the test based on the weighted empiricals $\hat{K}_2$ is design free. In fact, for $p = 1$, the limiting r.v. in (6.3.18) is the same as the one that appears in the one sample location model. For $G = F_0 = \mathcal{N}(0,1)$ d.f. Martynov (1976) has tabulated the distribution of this r.v.. Stephens (1976) has also tabulated the distribution of this r.v. for $G = F_0, dG = dG_0 = \{F_0(1 - F_0)\}^{-1} dF_0$ and for $F_0 = \mathcal{N}(0,1)$. For $G = F_0, F_0 = \mathcal{N}(0,1)$ d.f., Stephens and Martynov's tables generally agree up to the two decimal places, though occasionally there is an agreement up to three decimal places. In any case, for $p = 1$, one could use these tables to implement the test based on $\hat{K}_2$, at least asymptotically, whereas the test based on $\hat{K}_1$, being design dependent, can not be readily implemented. For the sake of convenience we reproduce some of the Stephens (1976, 1979) tables below.

Table III

$F_0 = \mathcal{N}(0,1)$

$\hat{K}_2 \backslash$ α	0.10	0.25	.05	.10
$\hat{K}_2(F_0)$	.237	.196	.165	.135
$\hat{K}_2(G_0)$	1.541	1.281	1.088	.897

In Table III, $\hat{K}_2(G)$ stands for the $\hat{K}_2$ with G being the integrating measure. $\hat{K}_2(G_0)$ is the $\hat{K}_2$ with the Anderson - Darling weights. Table III is, of course, useful only when $p = 1$. □

As far as the *asymptotic power* of the above L_2 - tests is concerned, it is apparent that Theorems 5.5.1, 5.5.3 and Lemma 6.3.1 can be used to deduce the asymptotic power of these tests against fairly general alternatives. Here we shall discuss the asymptotic behavior of only $K_1^*, j = 1, 2$ under the heteroscedastic gross errors alternatives. More precisely, suppose F_1 is a fixed d.f., and let $\{\delta_{ni}\}$ be numbers satisfying

$$(6.3.19) \qquad 0 \le \delta_{ni} \le 1, \quad 1 \le i \le n; \qquad \max_{1\le i\le n} \delta_{ni} \longrightarrow 0.$$

Let

$$(6.3.20) \qquad F_{ni} = (1-\delta_{ni})F_0 + \delta_{ni}F_1, \quad 1 \le i \le n,$$

$$m_1 := n^{-1/2}\sum_{i=1}^n \delta_{ni}(F_1 - F_0), \quad \mathbf{m}_2 := \sum_{i=1}^n \mathbf{A}\mathbf{x}_{ni}\delta_{ni}(F_1 - F_0).$$

Lemma 6.3.2 *Let (1.1.1) hold with e_{ni} having the d.f. F_{ni} given by (6.3.20), $1 \le i \le n$. Suppose that $\mathbf{X}$ satisfies (NX); (F_0, G) and (F_1, G) satisfy (5.5.44) - (5.5.46) and that*

$$(6.3.21) \qquad \int |F_1 - F_0| dG < \infty,$$

(a) If, in addition (5.6.9) and (5.6.10) hold, then

$$(6.3.22) \qquad K_1^* = \int \{W_1^0 + m_1 - f_0 \frac{\int (W_1^0 + m_1) f_0 dG}{\int f_0^2 dG}\}^2 dG + o_p(1)$$

provided

$$(6.3.23) \qquad n^{-1/2}\sum_{i=1}^n \delta_{ni} = O(1).$$

(b) Without any additional conditions,

$$(6.3.24) \qquad K_2^* = \int \left\| \mathbf{W}^0 + \mathbf{m}_2 - f_0 \frac{\int (\mathbf{W}^0 + \mathbf{m}_2) f_0 dG}{\int f_0^2 dG} \right\|^2 dG + o_p(1),$$

provided

$$(6.3.25) \qquad \sum_{i=1}^n \mathbf{A}\mathbf{x}_{ni}\delta_{ni} = O(1).$$

Proof. Apply Theorem 5.5.1 and (5.5.31) to $Y_{ni} \equiv e_{ni}$, $H_{ni} \equiv F_0$, $\{F_{ni}\}$ given by (6.3.20) and to $\mathbf{D} = n^{-1/2}[\mathbf{1}, \mathbf{0}, \cdots, \mathbf{0}]$, to conclude (a). Apply the same results to $D = \mathbf{AX}$ and the rest of the entities as in the proof of (a) to conclude (b). □

Now apply Lemma 6.3.1 to

$$(6.3.26) \qquad \begin{aligned} Z_n &:= W^0 + m_1 - f_0 \frac{\int (W^0 + m_1) f_0 dG}{\int f_0^2 dG} \\ Z &:= B(F_0) + a_1(F_1 - F_0) \\ &\quad - f_0 \frac{\int \{B(F_0) + a_1(F_1 - F_0)\} f_- dG}{\int f_0^2 dG} \end{aligned}$$

where $a_1 := \limsup_n n^{-1/2} \sum_{i=1}^n \delta_{ni}$, to obtain

Corollary 6.3.4 *Under the conditions of Lemma 6.3.2(a),*

$$K_1^* \to_d \int Z^2 dG, \quad \textit{where } Z \textit{ is as above.}$$

Similarly, apply Lemma 6.3.1 to

$$(6.3.27) \qquad \begin{aligned} \mathbf{Z}_n &:= \mathbf{W}^0 + \mathbf{m}_2 - f_0 \frac{\int (\mathbf{W}^0 + \mathbf{m}_2) f_0 dG}{\int f_0^2 dG}, \\ \mathbf{Z} &:= \mathbf{B}(F_0) + \mathbf{a}_2(F_1 - F_0) \\ &\quad - f_0 \frac{\int \{\mathbf{B}(F_0) + \mathbf{a}_2(F_1 - F_0)\} f_0 dG}{\int f_0^2 dG}, \end{aligned}$$

where $\mathbf{a}_2 = \limsup_n \sum_{i=1}^n \mathbf{A}\mathbf{x}_{ni}\delta_{ni}$, to obtain

Corollary 6.3.5 *Under the conditions of Lemma 6.3.2 (b),*

$$K_2^* \to_d \int \|\mathbf{Z}\|^2 dG, \quad \textit{where } \mathbf{Z} \textit{ is as in (6.3.27).}$$

An interesting choice of $\delta_{ni} = p^{-1/2}\|\mathbf{A}\mathbf{x}_{ni}\|$. Another choice is $\delta_{ni} \equiv n^{-1/2}$. Both a priori satisfy (6.3.19), (6.3.23) and (6.3.25). □

6.4 Testing with Unknown Scale

Now consider (1.1.1) and the problem of testing H_1 of (6.1.4). Here we shall discuss the modifications of $\hat{D}_j, \hat{K}_j$, $j = 1, 2$, of Sections 6.2, 6.3 that will be suitable for H_1. With W_1^0, $\mathbf{W}^0$ as before, define

$$(6.4.1) \qquad \begin{aligned} D_1(a, \mathbf{u}) &:= \sup_y |W_1^0(ay, \mathbf{u})|, \quad D_2(a, \mathbf{u}) := \sup_y \|\mathbf{W}^0(ay, \mathbf{u})\|, \\ K_1(a, \mathbf{u}) &:= \int \{W_1^0(ay, \mathbf{u})\}^2 dG(y), \\ K_2(a, \mathbf{u}) &:= \int \|\mathbf{W}^0(ay, \mathbf{u}\|^2 dG(y), \quad a > 0, \ \mathbf{u} \in \mathbb{R}^p. \end{aligned}$$

Let $(\tilde{\sigma}, \tilde{\beta})$ be estimators of $(\sigma, \beta), \tilde{D}_j$ and $\tilde{K}_j$ stand for $D_j(\tilde{\sigma}, \tilde{\beta})$ and $K_j(\tilde{\sigma}, \tilde{\beta})$, respectively, $j = 1, 2$. The following two theorems give the asymptotic null distribution's of these statistics. Theorem 6.4.1 follows from Corollary 2.3.4 in a similar fashion as does Theorem 6.2.1 from Corollaries 2.3.3 and 2.3.5. Theorem 6.4.2 follows from Theorem 5.5.8 in a similar fashion as does Theorem 6.3.2from Theorem 5.5.1. Recall the conditions $(\mathbf{F}_0\mathbf{1})$ and $(\mathbf{F}_0\mathbf{3})$ from Section 2.3.

Theorem 6.4.1 *In addition to (1.1.1) and H_1, assume that* (**NX**), $(\mathbf{F}_0\mathbf{1})$, $(\mathbf{F}_0\mathbf{3})$ *and the following hold.*

(6.4.2) *(a)* $|n^{1/2}(\tilde{\sigma} - \sigma)\sigma^{-1}| = O_p(1)$. *(b)* $\|\mathbf{A}^{-1}(\tilde{\beta} - \beta)\| = O_p(1)$.

Then,

$$\begin{aligned}
\tilde{D}_1 &= \sup \Big| W_1(t) + q_0(t)\{n^{1/2}\bar{\mathbf{x}}_n(\tilde{\beta} - \beta) \\
&\qquad + n^{1/2}(\tilde{\sigma} - \sigma)F_0^{-1}(t)\}\sigma^{-1}\Big| + o_p(1), \\
\tilde{D}_2 &= \sup \Big\| \mathbf{W}(t) + q_0(t)\{\mathbf{A}^{-1}(\tilde{\beta} - \beta) \\
&\qquad + n^{1/2}\mathbf{A}\bar{\mathbf{x}}_n \cdot n^{1/2}(\tilde{\sigma} - \sigma)F_0^{-1}(t)\}\sigma^{-1}\Big\| + o_p(1),
\end{aligned}$$

where now $W_1(\cdot) := W_1^0(\sigma F_0^{-1}(\cdot), \beta)$ *and* $\mathbf{W}(\cdot) := \mathbf{W}^0(\sigma F_0^{-1}(\cdot), \beta)$.

Theorem 6.4.2 *Suppose (1.1.1) and H_1,* (**NX**), *(6.4.2), (5.5.45) with $F = F_0$ hold. In addition, supppose F_0 has a continuous density f_0 such that*

$$\begin{aligned}
(6.4.3) \quad & 0 < \int |y|^j f_0^k(y)dG(y) < \infty, \; j = 0, \; k = 1, 2; \; j = 2, \; k = 2, \\
& \lim_{s\to 0} \limsup_n \int f_0^k(y + \tau n^{-1/2} + s)dG(y) = \int f_0^k dG(y), \\
& k = 1, 2, \; \tau \in \mathbb{R}, \\
& \lim_{s\to 0} \int |y| f_0(y(1+s)dG(y) = \int |y| f_0(y)dG(y).
\end{aligned}$$

Then,

$$\begin{aligned}
\tilde{K}_1 &= \int \Big[W_1^0(\sigma y, \beta) + f_0(y)\{n^{1/2}\bar{\mathbf{x}}_n'(\tilde{\beta} - \beta) \\
&\qquad + n^{1/2}(\tilde{\sigma} - \sigma)y\}\sigma^{-1}\Big]^2 dG(y) + o_p(1), \\
\tilde{K}_2 &= \int \Big\|\mathbf{W}^0(\sigma y, \beta) + f_0(y)\{\mathbf{A}^{-1}(\tilde{\beta} - \beta) \\
&\qquad + n^{1/2}\mathbf{A}\bar{\mathbf{x}}_n \cdot n^{1/2}(\tilde{\sigma} - \sigma)y\}\sigma^{-1}\Big\|^2 dG(y) + o_p(1).
\end{aligned}$$

Clearly, from these theorems one can obtain an analogue of Corollary 6.3.2 when $(\tilde{\sigma}, \tilde{\boldsymbol{\beta}})$ are chosen to be asymptotically efficient estimators.

As is the case in the classical least square theory or in the M-estimation methodology, neither of the two dispersions $K_1(a, \mathbf{u})$ and $K_2(a, \mathbf{u})$ can be used to satisfactorily estimate $(\sigma, \boldsymbol{\beta})$ by the simultaneous minimization process. The analogues of the m.d. goodness-of-fit tests that should be used are $\inf\{K_j(\tilde{\sigma}, \mathbf{u}); \mathbf{u} \in \mathbb{R}^p\}, j = 1, 2$. The methodology of Section 5 may be used to obtain the asymptotic distributions of these statistics in a fashion similar to the above. □

6.5 Testing for the Symmetry of the Errors

Consider the model (1.1.1) and the hypothesis H_s of symmetry of the errors specified at (6.1.5). The proposed tests are to be based on $\hat{D}_{js}$, $j = 1, 2, 3$ of (6.1.6), (6.1.7), $K_j^+(\hat{\boldsymbol{\beta}})$, and $\inf\{K_j^+(\mathbf{t}),\ \mathbf{t} \in \mathbb{R}^p\}$, $j = 1, 2$, where

$$\text{(6.5.1)} \qquad \begin{aligned} K_1^+(\mathbf{t}) &:= \int \{W_1^+(y, \mathbf{t})\}^2 dG(y), \\ K_2^+(\mathbf{t}) &:= \int \|\mathbf{W}^+(y, \mathbf{t})\|^2 dG(y), \quad \mathbf{t} \in \mathbb{R}^p, \end{aligned}$$

with W_1^+ and $\mathbf{W}^+$ as in (6.1.7) and (6.1.8). Large values of these statistics are considered to be significant for H_s.

Although the results of Chapters 2 and 5 can be used to obtain their asymptotic behavior under fairly general alternatives, here we shall focus only on the asymptotic null distribution's of these tests. To state these, we need some more notation. For a d.f. F, define

$$F_+(y) := F(y) - F(-y), \quad y \geq 0.$$

Then, with F^{-1} denoting the usual inverse of a d.f. F, we have

$$\text{(6.5.2)} \qquad \begin{aligned} F_+^{-1}(t) &= F^{-1}(1+t)/2), \\ -F_+^{-1}(t) &= F^{-1}((1-t)/2), \quad 0 \leq t \leq 1, \end{aligned}$$

for all F that are continuous and symmetric around 0. Finally, let

$$\begin{aligned} W_1^*(t) &:= W_1^+(F_+^{-1}(t), \boldsymbol{\beta}), \quad \mathbf{W}^*(t) := \mathbf{W}^+(F_+^{-1}(t), \boldsymbol{\beta}), \\ q^+(t) &:= f(F_+^{-1}(t)), \quad 0 \leq t \leq 1. \end{aligned}$$

We are now ready to state and prove

Theorem 6.5.1 *In addition to (1.1.1), H_s and (**NX**), assume that F in H_s and the estimator $\hat{\beta}$ satisfy (**F1**) and*

$$\|\mathbf{A}^{-1}(\hat{\beta}-\beta)\| = O_p(1), \quad under \ \ H_s. \tag{6.5.3}$$

Then,

$$\hat{D}_{1s} = \sup_{0\le t\le 1} |W_1^*(t) + 2q^+(t)n^{1/2}\bar{\mathbf{x}}_n'\mathbf{A}\mathbf{A}^{-1}(\hat{\beta}-\beta)| + o_p(1), \tag{6.5.4}$$

$$\hat{D}_{2s} = \sup_{0\le t\le 1} \left\|\mathbf{W}^*(t) + 2q^+(t)\mathbf{A}^{-1}(\hat{\beta}-\beta)\right\| + o_p(1), \tag{6.5.5}$$

and

$$\hat{D}_{3s} = \sup_{0\le t\le 1} \|\mathbf{W}^*(t) + 2q^+(t)\mathbf{A}^{-1}(\hat{\beta}-\beta)\| + o_p(1). \tag{6.5.6}$$

Proof. The proof follows from Theorem 2.3.1 in the following fashion. The details will be given only for (6.5.6), as they are the same for (6.5.5) and quite similar for (6.5.4). Because F is continuous and symmetric around 0 and because $\mathbf{W}^+(\cdot,\cdot) \equiv \mathbf{W}^+(-\cdot,\cdot)$, $\hat{D}_{3s} =_d \sup_{0\le t\le 1} \mathbf{W}^+(F_+^{-1}(t),\hat{\beta})$. But, from the definition (6.1.8) and (6.5.2), it follows that for a $\mathbf{v} \in \mathbb{R}^p$,

$$\begin{aligned}
\mathbf{W}^+(F_+^{-1}(t),\mathbf{v}) \quad & \\
= \sum_{i=1}^n \mathbf{A}\mathbf{x}_{ni}\Big\{&\mathbf{I}\Big(e_{ni} \le F^{-1}\Big(\frac{1+t}{2}\Big) + \mathbf{c}_{ni}'\mathbf{u}\Big) \\
&+\mathbf{I}\Big(e_{ni} \le F^{-1}\Big(\frac{1-t}{2}\Big) + \mathbf{c}_{ni}'\mathbf{u}\Big) - 1\Big\} \\
= \mathbf{S}\left(\frac{1+t}{2},\mathbf{u}\right) &+ \mathbf{S}\left(\frac{1-t}{2},\mathbf{u}\right) - \sum_{i=1}^n \mathbf{A}\mathbf{x}_{ni}, \quad 0\le t\le 1,
\end{aligned} \tag{6.5.7}$$

where

$$\mathbf{S}(t,\mathbf{u}) := \sum_{i=1}^n \mathbf{A}\mathbf{x}_{ni} I(e_{ni} \le F^{-1}(t) + \mathbf{c}_{ni}'\mathbf{u}), \quad 0\le t\le 1,$$

is a p-vector of S_d - processes of (2.3.1) with $X_{ni} \equiv e_{ni}, F_{ni} \equiv F \equiv H, \mathbf{c}_{ni} \equiv \mathbf{A}\mathbf{x}_{ni}, \mathbf{u} = \mathbf{A}^{-1}(\mathbf{v}-\beta)$ and where the jth process has the weights $\{d_{ni}\}$ given by the jth column of $\mathbf{AX}$. The assumptions about F and $\mathbf{X}$ imply all the assumptions of Theorem 2.3.1. Hence (6.5.6) follows from (3.2.6), (6.5.3) and (6.5.7) in an obvious fashion. □

Next, we state an analogous result for the L_2 - distances.

Theorem 6.5.2 *In addition to (1.1.1), H_s, (**NX**) and (6.5.3), assume that F in H_s and the integrating measure G satisfy (5.3.8), (5.5.44), (5.5.46) and (5.6.12). Then,*

$$K_1^+(\hat{\beta}) = \int \Big[W_1^+(y) + 2f(y)n^{1/2}\bar{\mathbf{x}}_n'(\hat{\beta} - \beta)\Big]^2 dG(y) + o_p(1),$$

$$K_2^+(\hat{\beta}) = \int \Big\|\mathbf{W}^+(y) + 2f(y)\mathbf{A}^{-1}(\hat{\beta} - \beta)\Big\|^2 dG(y) + o_p(1),$$

where $W_1^+(\cdot), \mathbf{W}^+(\cdot)$ now stand for $W_1^+(\cdot, \beta), \mathbf{W}^+(\cdot, \beta)$.

Proof. The proof follows from two applications of Theorem 5.5.2, once with $\mathbf{D} = n^{-1/2}[\mathbf{1}, \mathbf{0}, \cdots, \mathbf{0}]$ and once with $\mathbf{D} = \mathbf{XA}$. In both cases, take Y_{ni} and F_{ni} of that theorem to be equal to e_{ni} and $F, 1 \le i \le n$ respectively. The Claim 5.5.2 justifies the applicability of that theorem under the present assumptions. □

The next result is useful in obtaining the asymptotic null distribution's of the m.d. test statistics. Its proof uses Theorem 5.5.2 and 5.5.4 in a similar fashion as Theorems 5.5.1 and 5.5.3 are used in the proof of Theorem 6.3.1, and hence no details are given. Let

$$K_j^s := \inf\left\{K_j^+(\mathbf{t}); \mathbf{t} \in \mathbb{R}^p\right\}, \; j = 1, 2.$$

Theorem 6.5.3 *Assume that (1.1.1), H_s, (**NX**), (5.3.8), (5.5.44), (5.5.46) and (5.6.12) hold.*

(a) If, in addition, (5.6.9) and (5.6.10) hold, then

$$K_1^s = 2\int_0^\infty \Big\{W_1^+(y) - f(y)\int_0^\infty W_1^+ f dG\Big(\int_0^\infty f^2 dG\Big)^{-1}\Big\}^2 dG + o_p(1).$$

(b) Under no additional assumptions,

$$K_2^s = 2\int_0^\infty \Big\|\mathbf{W}^+(y) - f(y)\int_0^\infty \mathbf{W}^+ f dG\Big(\int_0^\infty f^2 dG\Big)^{-1}\Big\|^2 dG + o_p(1).$$

To obtain the asymptotic null distribution's of the given statistics from this theorem we now apply Lemma 6.3.1 to the approximating processes. The details will be given for K_2^s only as they are similar for K_1^s. Accordingly, let, for $n \ge 1, \; y \ge 0$,

$$\text{(6.5.8)} \qquad \mathbf{Z}_n(y) := \mathbf{W}^+(y) - f(y)\int_0^\infty \mathbf{W}^+ f dG\Big(\int_0^\infty f^2 dG\Big)^{-1}.$$

To determine the approximating r.v. for K_2^s we shall first obtain the covariance matrix function for this $\mathbf{Z}_n$, the computation of which is made easy by rewriting $\mathbf{Z}_n$ as follows. Recall the definition of ψ from (5.6.2) and define for $1 \le i \le n,\ y \in \mathbb{R}$,

$$\begin{aligned} \alpha_i(y) &:= I(e_i \le y) + I(e_i \le -y) - 1, \quad \overline{\alpha}_i := \int_0^\infty \alpha_i d\psi; \\ \boldsymbol{\alpha}' &:= (\alpha_1, \cdots, \alpha_n); \quad \overline{\boldsymbol{\alpha}}' := (\overline{\alpha}_1, \cdots, \overline{\alpha}_n); \quad a := \int_0^\infty f^2 dG. \end{aligned}$$

Then

$$\mathbf{Z}_n(y) = \mathbf{A}\mathbf{X}'[\boldsymbol{\alpha}(y) - f(y)\overline{\boldsymbol{\alpha}}a^{-1}], \quad y \ge 0. \tag{6.5.9}$$

Now observe that under H_s, $E\boldsymbol{\alpha} = \mathbf{0}$, $E\alpha_1(x)\alpha_1(y) = 2(1 - F(y))$, $0 \le x \le y$, and, because of the independence of the errors,

$$E\boldsymbol{\alpha}(x)\boldsymbol{\alpha}'(y) = 2(1 - F(y))\mathbf{I}_{p\times p}, \quad 0 \le x \le y. \tag{6.5.10}$$

Again, because of the symmetry and the continuity of F and Fubini, for $y \ge 0$,

$$\begin{aligned} E\alpha_1(y)\overline{\alpha}_1 &= \int_0^\infty E\Big([I(e_1 \le y) + I(e_1 \le -y) - 1] \\ &\qquad\qquad \times [I(e_1 \le x) + I(e_1 \le -x) - 1]\Big) d\psi(x) \\ &= \int_0^\infty [F(x \wedge y) + F(-x \wedge y) - F(y) + F(x \wedge -y) \\ &\qquad\qquad + F(-x \wedge -y) - F(-y)] d\psi(x) \\ &= 2(1 - F(y))\{\psi(y) - \psi(0)\} + \int_y^\infty 2(1 - F(x)) d\psi(x) \\ &= 2\int_y^\infty [\psi(x) - \psi(0)] dF(x) =: k(y), \quad \text{say.} \end{aligned}$$

The last equality is obtained by integrating the second expression in the previous one by parts. From this and the independence of the errors, we obtain

$$\begin{aligned} E\boldsymbol{\alpha}(y)\overline{\boldsymbol{\alpha}}' &= k(y)\mathbf{I}_{p\times p}, \quad y \ge 0, \\ E\overline{\boldsymbol{\alpha}}\,\overline{\boldsymbol{\alpha}}' &= \mathbf{I}_{p\times p} 4 \int_0^\infty \int_0^\infty (1 - F(y)) d\psi(x) d\psi(y) \\ &=: \mathbf{I}_{p\times p} r(F, G), \quad \text{say.} \end{aligned}$$

From these calculations one readily obtains that under H_s, for $0 \le x \le y$,

$$\begin{aligned}\mathbf{K}_n(x,y) &:= \mathbf{E}\mathbf{Z}_n(x)\mathbf{Z}_n'(y) \\ &= \Big[2(1-F(y)) - k(y)f(x)a^{-1} - k(x)f(y)a^{-1} + r(F,G)\Big]\mathbf{I}_{p\times p}.\end{aligned}$$

We also need the weak convergence of $\mathbf{W}^+$ to a continuous Gaussian process in uniform topology. One way to prove this is as follows. By (6.5.10),

$$E\mathbf{W}^+(x)\mathbf{W}^+(y)' = 2(1-F(y))\mathbf{I}_{p\times p}, \quad 0 \le x \le y, \tag{6.5.11}$$

From the definition (6.1.8) and the symmetry of F, we obtain

$$\begin{aligned}\mathbf{W}^+(y) &= \sum_{i=1}^n \mathbf{A}\mathbf{x}_{ni}\{I(e_{ni} \le y) - I(-e_{ni} < y)\} \qquad (6.5.12)\\ &= \sum_{i=1}^n \mathbf{A}\mathbf{x}_{ni}\{I(e_{ni} \le y) - F(y)\} \\ &\qquad - \sum_{i=1}^n \mathbf{A}\mathbf{x}_{ni}\{I(-e_{ni} \le y) - F(y)\} \\ &\qquad + \sum_{i=1}^n \mathbf{A}\mathbf{x}_{ni}\mathbf{I}(-e_{ni} = y) \\ &= \mathcal{W}_1(y) - \mathcal{W}_2(y) + \sum_{i=1}^n \mathbf{A}\mathbf{x}_{ni}I(-e_{ni} = y), \text{ say,} \qquad (6.5.13)\end{aligned}$$

for all $y \ge 0$.

Now, let $\mathbf{U}' := (\mathcal{U}_1, \mathcal{U}_2, \cdots, \mathcal{U}_p)$ be a vector of independent Wiener processes on [0, 1] such that $\mathbf{U}(0) = \mathbf{0}$, $E\mathbf{U} \equiv \mathbf{0}$, and $E\mathcal{U}_j(s)\mathcal{U}_j(t) = s \wedge t$, $1 \le j \le p$. Note that

$$E\mathbf{U}(2(1-F(x)))\mathbf{U}(2(1-F(y))) = 2(1-F(y))\mathbf{I}_{p\times p}, \quad 0 \le x \le y.$$

From (6.5.11) and (6.5.12), it hence follows, with the aid of the L-F CLT and the Cramér - Wold device, that under (**NX**), all finite dimensional distributions of $\mathbf{W}^+$ converge to those of $\mathbf{U}(2(1-F))$.

To prove the tightness in the uniform metric, proceed as follows. From (6.5.13) and the triangle inequality, because of (**NX**), it suffices to show that $\mathcal{W}_1$ and $\mathcal{W}_2$ are tight. But by the symmetry and the continuity of F,

$$\{\mathcal{W}_1(y),\ y \in \mathbb{R}\} =_d \{\mathcal{W}_2(y),\ y \in \mathbb{R}\} =_d \{\mathcal{W}_1(F^{-1}(t)), 0 \le t \le 1\}.$$

But, $\mathcal{W}_1(F^{-1})$ is obviously a p-vector of W.E.P.'s of the type W_d^* specified at (2.2.23). Thus the tightness follows from (2.2.25) of Corollary 2.2.1. We summarize this weak convergence result as

Lemma 6.5.1 *Let F be a continuous d.f. that is symmetric around* 0 *and* $\{e_{ni}, 1 \le i \le n\}$ *be i.i.d. F r.v.'s. Assume that* (**NX**) *holds. Then,*

$$\mathbf{W}^+(\cdot) \Rightarrow \mathbf{U}(2(1-F(\cdot))) \text{ in } (\mathcal{D}[0,\infty], \rho).$$

The above discussion suggests the approximating process for the $\mathbf{Z}_n$ of (6.5.10) to be

$$(6.5.14) \qquad \mathbf{Z}(y) := \mathbf{U}(2(1-F(y))) - \frac{f(y)\int_0^\infty \mathbf{U}(2(1-F))f\,dG}{\int_0^\infty f^2 dG}, \quad y \ge 0.$$

Straightforward calculations show that $\mathbf{K}_n(x,y) = E\mathbf{Z}(x)\mathbf{Z}'(y), 0 \le x \le y$, $n \ge 1$. This then verifies (i), (ii) and (iv) of Lemma 6.3.1 in the present case. Condition (iii) is verified as in the proof of Corollary 6.3.1(b) with the help of Lemma 6.5.1. To summarize, we have

Corollary 6.5.1 *(a) Under the conditions of Theorem 6.5.3(a),*

$$K_1^s \longrightarrow_d 2\int_0^\infty \Big[\mathcal{W}_1(2(1-F(y))) - \frac{f(y)\int_0^\infty \mathcal{W}_1(2(1-F))f\,dG}{\int_0^\infty f^2 dG}\Big]^2 dG(y).$$

(b) Under the conditions of Theorem 6.5.3(b),

$$K_2^s \longrightarrow_d 2\int_0^\infty \|\mathbf{Z}\|^2 dG(y), \quad \text{with } \mathbf{Z} \text{ given at (6.5.14).}$$

Remark 6.5.1 The distributions of the limiting r.v.'s appearing in (a) and (b) above have been studied by Martynov (1975, 1976) and Boos (1982) for some F and G. An interesting G in the present case is $G = \lambda$. But the corresponding tests are not ADF. Also because the F in H_s is unknown, one can not use $G = F$ or the Anderson - Darling integrating measures $dG = dF/\{F(1-F)\}$ in these test statistics.

One way to overcome this problem would be to use the signed rank analogues of the above tests which is equivalent to replacing the F in the integrating measure by an appropriate empirical of the residuals $\{Y_{nj} -$

$\mathbf{x}_{nj}'\mathbf{u}; 1 \le j \le n\}$. Let $R_{i\mathbf{u}}^+$ denote the rank of $|Y_{ni} - \mathbf{x}_{ni}'\mathbf{u}|$ among $\{|Y_{nj} - \mathbf{x}_{nj}'\mathbf{u}; 1 \le j \le n\}, 1 \le i \le n$, and define

$$\begin{aligned}\mathcal{Z}_1^+(t, \mathbf{u}) &:= n^{-1/2} \sum_{i=1}^n I(R_{i\mathbf{u}}^+ \le nt) sgn(Y_{ni} - \mathbf{x}_{ni}'\mathbf{u}),\\ \boldsymbol{\mathcal{Z}}_2^+(t, \mathbf{u}) &:= \mathbf{A} \sum_{i=1}^n \mathbf{x}_{ni} I(R_{i\mathbf{u}}^+ \le nt) sgn(Y_{ni} - \mathbf{x}_{ni}'\mathbf{u}),\end{aligned}$$

for $0 \le t \le 1$, $\mathbf{u} \in \mathbb{R}^p$. The signed rank analogues of K_1^s, K_2^s statistics, respectively, are

$$\mathcal{K}_1^s := \inf\{\mathcal{K}_1(\mathbf{u});\ \mathbf{u} \in \mathbb{R}^p\}, \quad \mathcal{K}_2^s := \inf\{\mathcal{K}_2(\mathbf{u});\ \mathbf{u} \in \mathbb{R}^p\},$$

where for $\mathbf{u} \in \mathbb{R}^p$,

$$\mathcal{K}_1(\mathbf{u}) := \int_0^1 [\mathcal{Z}_1^+(t, \mathbf{u})]^2 dL(t), \quad \mathcal{K}_2(\mathbf{u}) := \int_0^1 \|\boldsymbol{\mathcal{Z}}_2^+(t, \mathbf{u})\|^2 dL(t),$$

with $L \in \mathcal{DI}[0, 1]$. If $L(t) \equiv t$ then $\mathcal{K}_j^s, j = 1, 2$, are analogues of the Cramér - von Mises statistics. If L is specified by the relation $dL(t) = \{1/t(1-t)\}dt$, then the corresponding tests would be the Anderson - Darling type test of symmetry.

Note that if in (3.3.1) we put $d_{ni} \equiv n^{-1/2}, X_{ni} \equiv e_{ni}, F_{ni} \equiv F$, then $\mathbf{Z}_d^+$ of (3.3.1) reduces to $\mathcal{Z}_1^+$. Similarly, $\boldsymbol{\mathcal{Z}}_2^+$ corresponds to a p - vector of Z_d^+ - processes of (3.3.1) whose j^{th} component has $d_{ni} \equiv$ (j^{th} column of $\mathbf{A})'\mathbf{x}_{ni}$ and the rest of the entities the same as above. Consequently, from (3.3.12) and arguments like those used for Theorem 6.5.3, we can deduce the following

Theorem 6.5.4 *Assume that (1.1.1), H_s and (**NX**) hold; L is a d.f. on $[0, 1]$, and F of H_s satisfies (**F1**), (**F2**).*

(a) If, in addition, (5.6.9) and (5.6.10) hold, then

$$\mathcal{K}_1^s \longrightarrow_d \int_0^1 \Big[U_1(t) - \frac{q^+(t) \int_0^1 U_1 q^+ dL}{\int_0^1 (q^+)^2 dL}\Big]^2 dL(t).$$

(b) Under no additional assumptions,

$$\mathcal{K}_2^s \longrightarrow_d \int_0^1 \Big\|\mathbf{U}(t) - \frac{q^*(t) \int_0^1 \mathbf{U} q^* dL}{\int_0^1 (q^*)^2 dL}\Big\|^2 dL(t),$$

where $q^*(t) := 2[f(F^{-1}((t+1)/2) - f(0)]$, $0 \le t \le 1$. □

Clearly this theorem covers $L(t) \equiv t$ case but not the case where $dL(t) = \{1/t(1-t)\}dt$. The *problem of proving an analogue of the above theorem for a general L is unsolved* at the time of this writing. □

6.6 Regression Model Fitting

6.6.1 Introduction

In the previous sections of this chapter we assumed the model to be a linear regression model and then proceeded to develop tests for fitting an error distribution. In this section we shall consider the problem of fitting a given parametric model to a regression function. More precisely, let X, Y be r.v.'s with X being p-dimensional. In much of the existing literature the regression function is defined to be the conditional mean function $\mu(x) := E(Y|X = x)$, $x \in \mathbb{R}^p$, assuming it exists, and then one proceeds to fit a parametric model to this function, i.e., one assumes the existence of a parametric family

$$\mathcal{M} = \{m(x, \boldsymbol{\theta}) : x \in \mathbb{R}^p, \, \boldsymbol{\theta} \in \Theta\}$$

of functions and then proceeds to tests the hypothesis

$$\mu(x) = m(x, \boldsymbol{\theta}_0), \quad \text{for some } \boldsymbol{\theta}_0 \in \Theta \text{ and } \forall\, x \in \mathcal{I},$$

based on n i.i.d. observations $\{(X_i, Y_i);\ 1 \le i \le n\}$ on (X, Y). Here, and in the sequel, $\mathcal{I}$ is a compact subset of $\mathbb{R}^p$ and Θ is a proper subset of the q-dimensional Euclidean space $\mathbb{R}^q$.

The use of the conditional mean function as a regression function is justified partly for historical reasons and partly for convenience. In the presence of non-Gaussian errors it is desirable to look for other dispersions that would lead to different regression functions and which may well be equally appropriate to model the effect of the covariate vector X on the response variable Y. For example, if the errors are believed to form a white noise from a double exponential distribution then the conditional median function would be a proper regression function. We are thus motivated to propose a class of tests for testing goodness-of-fit hypotheses pertaining to a class of implicitly defined regression functions as follows.

Let ψ be a nondecreasing real valued function such that $E|\psi(Y - r)| < \infty$, for each $r \in \mathbb{R}$. Define the ψ-regression function m_ψ by the requirement that

$$(6.6.1) \qquad E[\psi(Y - m_\psi(X))|X] = 0, \quad a.s.$$

Observe that, if $\psi(x) \equiv x$, then $m_\psi = \mu$, and if

$$\psi(x) \equiv \psi_\alpha(x) := I(x > 0) - (1 - \alpha), \text{ for an } 0 < \alpha < 1,$$

then $m_\psi(x) \equiv m_\alpha(x)$, the αth quantile of the conditional distribution of Y, given $X = x$. In general, m_ψ will exist if ψ is skew symmetric and the conditional distribution of the error $Y - m_\psi(X)$, given X, is symmetric around zero.

The weighted empirical process that is of interest here is

$$S_{n,\psi}(x) := n^{-1/2} \sum_{i=1}^{n} \psi(Y_i - m_\psi(X_i))\, I(\ell(X_i) \le x),\ x \in [-\infty, \infty],$$

where ℓ is a known function from $\mathbb{R}^p$ to $\mathbb{R}$. Write m_I, $S_{n,I}$, for m_ψ, $S_{n,\psi}$ when $\psi(x) \equiv x$, respectively.

Tests of goodness-of-fit for fitting a model to m_ψ will be based on the process $S_{n,\psi}$. This process is an appropriate extension of the usual partial sum process useful in testing for the mean in the one sample problem to the current regression setup. To see this, recall the one sample mean testing problem where $Y_1, \cdots, Y_n$ is a random sample from a population with mean ν and the problem is to test $\nu = \nu_0$. The well celebrated cumulative sum test is based on the supremum over i of the process

$$\sum_{j=1}^{n} (Y_j - \nu_0) I(j \le i), \quad 1 \le i \le n.$$

Now consider the simple linear regression model $Y_i = (i/n)\beta + \varepsilon_i$, $1 \le i \le n$, $\beta \in \mathbb{R}$. The analogue of the above test for testing $\beta = \beta_0$ here is based on the maximum over i of the process

$$\sum_{j=1}^{n} (Y_j - (j/n)\beta_0) I(j \le i), \quad 1 \le i \le n.$$

In the regression model with the uniform non-random design points j/n in $[0, 1]$, where $Y_i = \mu(i/n) + \varepsilon_i$, the analogue of the above test for testing $\mu(x) = \mu_0(x)$, for $x \in [0, 1]$, and for some known function μ_0, is based on

$$\sup_{1 \le i \le n} \Big| \sum_{i=1}^{n} (Y_j - \mu_0(j/n)) I(j \le i) \Big| = \sup_{0 \le x \le 1} \Big| \sum_{i=1}^{n} (Y_j - \mu_0(j/n)) I((j/n) \le x) \Big|.$$

In the general regression model the role of j/n is played by X_j and one readily sees that $\mathcal{S}_{n,I}$ with $\ell(x) = x$ in the case $p = 1$, provides the right extension of the above partial sum process to the current regression setup. The function ℓ gives an extra flexibility in the choice of these processes.

In the sequel m_ψ is assumed to exist uniquely, the r.v. $\ell(X)$ is assumed to be continuous and

$$(6.6.2) \qquad E\psi^2(Y - m_\psi(X)) < \infty.$$

Under the assumptions (6.6.1) and (6.6.2), Theorem 6.6.1 below proves the weak convergence of $\mathcal{S}_{n,\psi}$ in $\mathcal{D}[-\infty, \infty]$, to a continuous Gaussian process $\mathcal{S}_\psi$ with the covariance function

$$K_\psi(x, y) = E\psi^2(Y - m_\psi(X))\, I(\ell(X) \le x \wedge y), \quad x,\, y \in \mathbb{R}.$$

Since the function

$$\tau_\psi^2(x) := K_\psi(x, x) = E\psi^2(Y - m_\psi(X))\, I(\ell(X) \le x)$$

is nondecreasing and nonnegative, $\mathcal{S}_\psi$ admits a representation

$$(6.6.3) \qquad \mathcal{S}_\psi(x) = B(\tau_\psi^2(x)), \quad \text{in distribution},$$

where B is a standard Brownian motion on the positive real line. Note that the continuity of $\ell(X)$ implies that of τ_ψ and hence that of $B(\tau_\psi^2)$. The representation (6.6.3), Theorem 6.6.1 and the continuous mapping theorem yield

$$\sup_{x \in \mathbb{R}} |\mathcal{S}_{n,\psi}(x)| \Longrightarrow \sup_{0 \le t \le \tau_\psi^2(\infty)} |B(t)| = \tau_\psi(\infty) \sup_{0 \le t \le 1} |B(t)|, \quad \text{in law.}$$

This result is useful for testing the simple hypothesis $\tilde{H}_0 : m_\psi = m_0$, where m_0 is a known function as follows. Estimate (under $m_\psi = m_0$) the variance $\tau_\psi^2(x)$ by

$$\tau_{n,\psi}^2(x) := n^{-1} \sum_{i=1}^{n} \psi^2(Y_i - m_0(X_i)) I(\ell(X_i) \le x), \quad x \in \mathbb{R},$$

and replace m_ψ by m_0 in the definition of $\mathcal{S}_{n,\psi}$. Write $s_{n,\psi}^2$ for $\tau_{n,\psi}^2(\infty)$ and let $\mathcal{I}_\ell := \ell(\mathcal{I})$. Then, for example, the Kolmogorov-Smirnov (K-S) test based on $\mathcal{S}_{n,\psi}$ of the given asymptotic level would reject the hypothesis $\tilde{H}_0$ if

$$\sup\{s_{n,\psi}^{-1} |\mathcal{S}_{n,\psi}(x)| : x \in \mathcal{I}_\ell\}$$

exceeds an appropriate critical value obtained from the boundary crossing probabilities of a Brownian motion on the unit interval which are readily available. More generally, the asymptotic level of any test based on a continuous function of $s_{n,\psi}^{-1}\mathcal{S}_{n,\psi}((\tau_{n,\psi}^2)^{-1})$ can be obtained from the distribution of the corresponding function of B on $[0,1]$, where $(\tau_{n,\psi}^2)^{-1}(t) := \inf\{x \in \mathbb{R} : \tau_{n,\psi}^2(x) \geq t\}$, $t \geq 0$. For example, the asymptotic level of the test based on the Cramér - von Mises statistic

$$s_{n,\psi}^{-2} \int_0^{s_{n,\psi}^2} \mathcal{S}_{n,\psi}^2((\tau_{n,\psi}^2)^{-1}(t))\, dH(t/s_{n,\psi}^2),$$

is obtained from the distribution of $\int_0^1 B^2\, dH$, where H is a d.f. on $[0,1]$.

Now consider the more realistic problem of testing the goodness-of-fit hypothesis

$$H_0 : m_\psi(x) = m(x, \boldsymbol{\theta}_0), \qquad \text{for some } \boldsymbol{\theta}_0 \in \Theta,\, x \in \mathcal{I}.$$

Let $\boldsymbol{\theta}_n$ be a consistent estimator of $\boldsymbol{\theta}_0$ under H_0 based on $\{(X_i, Y_i);\ 1 \leq i \leq n\}$. Let $m_n(x) \equiv m(x, \boldsymbol{\theta}_n)$ and define

$$\hat{\mathcal{S}}_{n,\psi}(x) = n^{-1/2} \sum_{i=1}^{n} \psi(Y_i - m_n(X_i)) I(\ell(X_i) \leq x), \quad x \in \mathbb{R}.$$

The process $\hat{\mathcal{S}}_{n,\psi}$ is a weighted empirical process, where the weights, at $\ell(X_i)$ are now given by the ψ-residuals $\psi(Y_i - m_n(X_i))$. Tests for H_0 can be based on an appropriately scaled discrepancy of this process. For example, an analogue of the K-S test would reject H_0 in favor of H_1 if $\sup\{\sigma_{n,\psi}^{-1}|\hat{\mathcal{S}}_{n,\psi}(x)| : x \in \mathcal{I}_\ell\}$ is too large, where $\sigma_{n,\psi}^2 := n^{-1}\sum_{i=1}^n \psi^2(Y_i - m_n(X_i))$. These tests, however, are not generally asymptotically distribution free (ADF).

In the next sub-section we shall describe a transform of the $\hat{\mathcal{S}}_{n,\psi}$ process whose limiting null distribution is known so that the tests based on it will be ADF. The description of this transformation needs some assumptions on the null model which are also stated in the same section. A computational formulas for computing some of these tests is given in the Sub-section 6.6.3. All proofs are given in the Sub-section 6.6.4. But before proceeding we discuss

Consistency. Here we shall briefly discuss some sufficient conditions that will imply the consistency of goodness-of-fit tests based on $\mathcal{S}_{n,\psi}$ for a simple hypothesis $m_\psi = m_0$ against the fixed alternative $m_\psi \neq m_0$, where m_0 is a known function.

By the statement $m_\psi \neq m_0$ it should be understood that the G-measure of the set $\{x \in \mathbb{R}^p;\, m_\psi(x) \neq m_0(x)\}$ is positive. Let $\lambda(x,z) := \mathbb{E}\{\psi(Y - m_\psi(X) + z)\,|X = x\}$, $x \in \mathbb{R}^p$, $z \in \mathbb{R}$. Note that ψ nondecreasing implies that $\lambda(x,z)$ is nondecreasing in z, for each x. Assume that for every $x \in \mathbb{R}^p$,

$$\lambda(x,z) = 0, \text{ if and only if } \ z = 0. \tag{6.6.4}$$

Let $d(x) := m_\psi(x) - m_0(x)$, $x \in \mathbb{R}^p$ and

$$D_n(x) := n^{-1/2} \sum_{i=1}^{n} \lambda(X_i, d(X_i))\, I(\ell(X_i) \le x), \quad x \in \mathbb{R}.$$

An adaptation of the Glivenko-Cantelli arguments (see (6.6.31) below) yields

$$\sup_{x\in\mathbb{R}} |n^{-1/2} D_n(x) - E\lambda(X, d(X)) I(\ell(X) \le x)| \to 0, \ a.s.,$$

where the expectation E is computed under the alternative m_ψ. Moreover, by Theorem 6.6.1 below, $\mathcal{S}_{n,\psi}$ converges weakly to a continuous Gaussian process. These facts together with the assumption (6.6.4) and a routine argument yield the consistency of the K-S and Cramér-von Mises tests based on $\mathcal{S}_{n,\psi}$.

Note that the condition (6.6.4) is trivially satisfied when $\psi(x) \equiv x$ while for $\psi = \psi_\alpha$, it is equivalent to requiring that zero be the unique α^{th} percentile of the conditional distribution of the error $Y - m_\psi(X)$, given X.

6.6.2 Transform T_n of $\mathcal{S}_{n,\psi}$

This section first discusses the asymptotic behavior of the processes introduced in the previous section under the simple and composite hypotheses. Then, in the special case of $p = 1$, $\ell(x) \equiv x$, a transformation T_n is given so that the process $T_n\hat{\mathcal{S}}_{n,\psi}$ has the weak limit with a known distribution. Consequently the tests based on the process $T_n\hat{\mathcal{S}}_{n,\psi}$ are ADF.

To begin with we give a general result of somewhat independent interest. For each n, let (Z_{ni}, X_i) be i.i.d. r.v.'s, $\{X_i, 1 \le i \le n\}$ independent of $\{Z_{ni}; 1 \le i \le n\}$,

$$\mathbb{E}Z_{n1} \equiv 0, \quad EZ_{n1}^2 < \infty, \quad \forall\, n \ge 1, \tag{6.6.5}$$

and define

$$\mathcal{Z}_n(x) = n^{-1/2} \sum_{i=1}^{n} Z_{ni} I(\ell(X_i) \le x), \qquad x \in \mathbb{R}. \tag{6.6.6}$$

The process $\mathcal{Z}_n$ takes its value in the Skorokhod space $D(-\infty, \infty)$. Extend it continuously to $\pm\infty$ by putting

$$\mathcal{Z}_n(-\infty) = 0, \quad \text{and} \quad \mathcal{Z}_n(\infty) = n^{-1/2}\sum_{i=1}^n Z_{ni}.$$

Then $\mathcal{Z}_n$ becomes a process in $\mathcal{D}[-\infty, \infty]$. Let $\sigma_n^2 := EZ_{n,1}^2$ and L denote the d.f. of $\ell(X)$. Note that under (6.6.5), the covariance function of $\mathcal{Z}_n$ is

$$K_n(x, y) := \sigma_n^2 L(x \wedge y), \quad x, y \in \mathbb{R},$$

The following lemma gives its weak convergence to an appropriate process.

Lemma 6.6.1 *In addition to the above setup and (6.6.5), assume that*

(6.6.7) $$\sigma_n^2 \longrightarrow \alpha^2, \quad \textit{for some } \alpha \geq 0.$$

Then,

(6.6.8) $$\mathcal{Z}_n \Longrightarrow B \circ \tau^2 \ \textit{in the space } \mathcal{D}[-\infty, \infty],$$

where $B \circ \tau^2$ is a continuous Brownian motion on $\mathbb{R}$ with respect to time τ^2, where $\tau^2(x) \equiv \alpha^2 L(x)$.

Proof. Apply the Lindeberg-Feller CLT and the Cramér - Wold device to show that all finite dimensional distributions converge weakly to the right limit, under the assumed conditions.

The argument for tightness is similar to that of the W_d process of Chapter 2.2. For convenience, write Z_i for Z_{ni} and fix $-\infty \leq t_1 < t_2 < t_3 \leq \infty$. Then we have

$$\begin{aligned}
&[\mathcal{Z}_n(t_3) - \mathcal{Z}_n(t_2)]^2[\mathcal{Z}_n(t_2) - \mathcal{Z}_n(t_1)]^2 \\
&= n^{-2}\Big[\sum_{i=1}^n Z_i I(t_2 < \ell(X_i) \leq t_3)\Big]^2\Big[\sum_{i=1}^n Z_i I(t_1 < \ell(X_i) \leq t_2)\Big]^2 \\
&= n^{-2}\sum_{i,j,k,l} \xi_i\xi_j\zeta_k\zeta_l,
\end{aligned}$$

with $\xi_i = Z_i I(t_2 < \ell(X_i) \leq t_3)$, $\zeta_i = Z_i I(t_1 < \ell(X_i) \leq t_2)$. Now, if the largest index among i, j, k, l is not matched by any other, then $\mathbb{E}\{\xi_i\xi_j\zeta_k\zeta_l\} = 0$. Also, since the two intervals $(t_2, t_3]$ and $(t_1, t_2]$ are disjoint, $\xi_i\zeta_i \equiv 0$. We thus obtain

(6.6.9) $$\begin{aligned}
&E\Big\{n^{-2}\sum_{i,j,k,l} \xi_i\xi_j\zeta_k\zeta_l\Big\} \\
&= n^{-2}\sum_{i,j<k} E\{\zeta_i\zeta_j\xi_k^2\} + n^{-2}\sum_{i,j<k} E\{\xi_i\xi_j\zeta_k^2\}.
\end{aligned}$$

But, by the assumed independence,

$$
\begin{aligned}
n^{-2} \sum_{i,j<k} E\{\zeta_i \zeta_j \xi_k^2\} &= \sum_{k=2}^{n} E\Big\{\Big(\sum_{i=1}^{k-1} \zeta_i\Big)^2 \xi_k^2\Big\} \\
&\leq \sigma_n^4 \, [L(t_2) - L(t_1)][L(t_3) - L(t_2)].
\end{aligned}
$$

One has a similar bound for the second term. This together with an argument like in Section 2.2 proves the tightness of the process $\mathcal{Z}_n$. □

Now we consider the asymptotic null behavior of $\mathcal{S}_{n,\psi}$ and $\hat{\mathcal{S}}_{n,\psi}$. For the sake of the clarity of the exposition, from now onwards we shall assume

(6.6.10) $$Y - m_\psi(X) \text{ is independent of } X.$$

Lemma 6.6.1 applied to $Z_{ni} \equiv \psi(Y_i - m_\psi(X_i))$ readily gives the following

Theorem 6.6.1 *Assume that (6.6.1), (6.6.2) and (6.6.10) hold. Then,*

(6.6.11) $$\mathcal{S}_{n,\psi} \Longrightarrow \mathcal{S}_\psi, \text{ in the space } \mathcal{D}[-\infty, \infty].$$

To study the asymptotic null behavior of $\hat{\mathcal{S}}_{n,\psi}$, the following additional regularity conditions on the underlying entities are needed. For technical reasons, the case of a smooth ψ (see (Ψ_1) below) and a non-smooth ψ (see (Ψ_2) below) are dealt with separately. All probability statements in these assumptions are understood to be made under H_0. We make the following assumptions.

About the estimator $\boldsymbol{\theta}_n$ assume

(6.6.12) $$\|n^{1/2}(\boldsymbol{\theta}_n - \boldsymbol{\theta}_0)\| = O_p(1).$$

About the model under H_0 assume the following: There exists a function $\dot{\mathbf{m}}$ from $\mathbb{R}^p \times \Theta$ to $\mathbb{R}^q$ such that $\dot{\mathbf{m}}(\cdot, \boldsymbol{\theta}_0)$ is measurable and satisfies the following: For all $k < \infty$,

(6.6.13) $$\sup n^{1/2} |m(X_i, \theta) - m(X_i, \boldsymbol{\theta}_0) - (\boldsymbol{\theta} - \boldsymbol{\theta}_0)' \dot{\mathbf{m}}(X_i, \boldsymbol{\theta}_0)| = o_p(1),$$

(6.6.14) $$E\|\dot{\mathbf{m}}(X, \boldsymbol{\theta}_0)\|^2 < \infty,$$

$$\boldsymbol{\Sigma}_0 := E\dot{\mathbf{m}}(X, \boldsymbol{\theta}_0)\dot{\mathbf{m}}'(X, \boldsymbol{\theta}_0) \quad \text{is positive definite,}$$

where in (6.6.13) the supremum is taken over $n^{1/2}\|\boldsymbol{\theta} - \boldsymbol{\theta}_0\| \leq k$ and $1 \leq i \leq n$.

(Ψ_1). (*Smooth* ψ). The function ψ is absolutely continuous with its almost everywhere derivative $\dot{\psi}$ such that the function $z \mapsto E|\dot{\psi}(\varepsilon - z) - \dot{\psi}(\varepsilon)|$ is continuous at 0.

(Ψ_2). (*Non-smooth* ψ). The function ψ is non-decreasing, right continuous, bounded and such that the function

$$z \mapsto \mathbb{E}\Big[\psi(\varepsilon + z) - \psi(\varepsilon)\Big]^2$$

is continuous at 0.

Recall the definition of $\sigma_\psi^2(x)$ from the previous section and note that under (6.6.10), it is a constant function, say, σ_ψ^2. Also, let

$$\begin{aligned} \gamma_\psi &:= E[\dot{\psi}(\varepsilon_1], && \text{for smooth } \psi, \\ &:= \int f(x)\,\psi(dx), && \text{for non-smooth } \psi, \end{aligned}$$

where f is the Lebesgue density of the error d.f. F. Let

$$\boldsymbol{\nu}(x) = E\dot{\mathbf{m}}(X, \boldsymbol{\theta}_0) I(\ell(X) \le x), \quad x \in \mathbb{R}.$$

Note that under (6.6.14) and (Ψ_1) or under (6.6.14), (Ψ_2) and (**F1**) these entities are well-defined.

We are now ready to formulate an asymptotic expansion of $\hat{\mathcal{S}}_{n,\psi}$, which is crucial for the subsequent results and the transformation T_n.

Theorem 6.6.2 *Assume that (6.6.1), (6.6.12), (6.6.13), (6.6.14), and H_0 hold. If, in addition, either*
*(**A**) (Ψ_1) holds, or*
*(**B**) (Ψ_2) and (**F1**) hold,*
then

$$|\hat{\mathcal{S}}_{n,\psi}(x) - \mathcal{S}_{n,\psi}(x) + \gamma_\psi\, \boldsymbol{\nu}'(x)\, n^{1/2}(\boldsymbol{\theta}_n - \boldsymbol{\theta}_0)| = u_p(1). \tag{6.6.15}$$

Remark 6.6.1 The assumption (Ψ_1) covers many interesting ψ's including the least square score $\psi(x) \equiv x$ and the Huber score $\psi(x) \equiv xI(|x| \le c) + c\,sign(x)I(|x| > c)$, where c is a real constant, while (Ψ_2) covers the α-quantile score $\psi(x) \equiv I(x > 0) - (1 - \alpha)$. □

Now, suppose additionally, $\gamma_\psi > 0$, and the estimator $\boldsymbol{\theta}_n$ satisfies

$$\begin{aligned} &n^{1/2}(\boldsymbol{\theta}_n - \boldsymbol{\theta}_0) \\ &\quad = (\gamma_\psi\, \boldsymbol{\Sigma}_0)^{-1} n^{-1/2} \sum_{i=1}^n \dot{\mathbf{m}}(X_i, \boldsymbol{\theta}_0)\psi(\varepsilon_i) + o_p(1), \end{aligned} \tag{6.6.16}$$

where $\varepsilon_i \equiv Y_i - m(X_i, \boldsymbol{\theta}_0)$. Then, the following corollary is an immediate consequence of the above theorem and Theorem 6.6.1. We shall state it for the smooth ψ- case only. The same holds in the non-smooth ψ.

Corollary 6.6.1 *Under the assumptions of Theorems 6.6.1 and 6.6.2(A),*

$$\hat{\mathcal{S}}_{n,\psi} \Longrightarrow \hat{\mathcal{S}}_\psi, \quad \textit{in the space } \mathcal{D}[-\infty, \infty],$$

where $\hat{\mathcal{S}}_\psi$ is a centered continuous Gaussian process with the covariance function

$$(6.6.17) \qquad K^1_\psi(x,y) = \sigma^2\,[L(x \wedge y) - \boldsymbol{\nu}'(x)\boldsymbol{\Sigma}_0^{-1}\boldsymbol{\nu}(y)],$$

Under (6.6.10), a class of M-estimators of $\boldsymbol{\theta}_0$ corresponding to a given ψ defined by the relation

$$\boldsymbol{\theta}_{n,\psi} := \operatorname{argmin}_{\mathbf{t}} \|n^{-1/2}\sum_{i=1}^n \dot{\mathbf{m}}(X_i, \mathbf{t})\psi(Y_i - m(X_i, \mathbf{t}))\|$$

generally satisfies (6.6.16). A set of sufficient conditions on the model $\mathcal{M}$ under which this holds includes (6.6.13), (6.6.14), and the following additional conditions. In these conditions $\mathbf{m}_i(\boldsymbol{\theta})$ stands for $\mathbf{m}(X_i, \boldsymbol{\theta})$, for the sake of brevity.

$$(6.6.18) \qquad n^{-1}\sum_{i=1}^n E\|\dot{\mathbf{m}}_i(\boldsymbol{\theta} + n^{-1/2}\mathbf{s}) - \dot{\mathbf{m}}_i(\boldsymbol{\theta})\|^2 = o(1), \quad \mathbf{s} \in \mathbb{R}^q.$$

$$(6.6.19) \qquad n^{-1/2}\sum_{i=1}^n \|\dot{\mathbf{m}}_i(\boldsymbol{\theta} + n^{-1/2}\mathbf{s}) - \dot{\mathbf{m}}_i(\boldsymbol{\theta})\| = O_p(1), \quad \mathbf{s} \in \mathbb{R}^q.$$

$$(6.6.20) \qquad \forall\, \epsilon > 0,\ \exists \text{ a } \delta > 0, \text{ and an } N < \infty, \ni\ \forall\, 0 < b < \infty,$$

$$\|\mathbf{s}\| \le b,\ n > N,$$

$$P\Big(\sup_{\|\mathbf{t}-\mathbf{s}\|\le\delta} n^{-1/2}\sum_{i=1}^n \|\dot{\mathbf{m}}_i(\boldsymbol{\theta} + n^{-1/2}\mathbf{t}) - \dot{\mathbf{m}}_i(\boldsymbol{\theta} + n^{-1/2}\mathbf{s})\| \le \epsilon\Big)$$

$$\ge 1 - \epsilon,$$

$$(6.6.21) \qquad \mathbf{e}'\mathbf{M}(\boldsymbol{\theta} + n^{-1/2}r\mathbf{e}) \text{ is monotonic in } r \in \mathbb{R},\ \forall\, \mathbf{e} \in \mathbb{R}^q,$$

$$\|\mathbf{e}\| = 1,\ n \ge 1, \quad \text{a.s.}$$

A proof of the above claim uses the methodology of Chapter 5.4. See also Section 8.2 below in connection with autoregressive models.

Unlike (6.6.3), the structure of K^1_ψ given at (6.6.17) does not allow for a simple representation of $\hat{\mathcal{S}}_\psi$ in terms of a process with a known distribution.

The situation is similar to the model checking for the underlying error distribution as in the previous sections of this chapter.

Now **focus on the case** $p = 1$, $\ell(x) \equiv x$. In this case it is possible to transform the process $\hat{S}_{n,\psi}$ so that it is still a statistic with a known limiting null distribution. To simplify the exposition further write $\dot{\mathbf{m}}(\cdot) = \dot{\mathbf{m}}(\cdot, \boldsymbol{\theta}_0)$. Set

$$\mathbf{A}(x) = \int \dot{\mathbf{m}}(y)\dot{\mathbf{m}}'(y)\, I(y \geq x)G(dy), \quad x \in \mathbb{R},$$

where G denotes the d.f. of X, assumed to be continuous. Assume that

(6.6.22) $\qquad \mathbf{A}(x_0)$ is nonsingular for some $x_0 < \infty$.

This and the nonnegative definiteness of $\mathbf{A}(x) - \mathbf{A}(x_0)$ implies that $\mathbf{A}(x)$ is non-singular for all $x \leq x_0$. Write $\mathbf{A}^{-1}(x)$ for $(\mathbf{A}(x))^{-1}$, and define, for $x \leq x_0$,

$$Tf(x) = f(x) - \int_{s \leq x} \dot{\mathbf{m}}'(s)\mathbf{A}^{-1}(s)\Big[\int \dot{\mathbf{m}}(z)\, I(z \geq s)\, f(dz)\Big]G(ds).$$

The transformation T will be applied to functions f which are either of bounded variation or Brownian motion. In the latter case the inner integral needs to be interpreted as a stochastic integral. Since T is a linear operator, $T(S_\psi)$ is a centered Gaussian process. Informally speaking, T maps $\hat{S}_{n,\psi}$ into the (approximate) martingale part of its Doob-Meyer decomposition. Moreover, we have the following fact.

Lemma 6.6.2 *Under the above setup and under (6.6.10),*

$$\text{Cov}[TS_\psi(x), TS_\psi(y)] = \sigma_\psi^2\, G(x \wedge y), \quad x,\, y \in \mathbb{R},$$

that is, TS_ψ/σ_ψ is a Brownian motion with respect to time G.

The proof uses the independence of increments of the Brownian motion S_ψ and properties of stochastic integrals. Details are left out for interesting readers.

To convince oneself about the validity of the above lemma, consider the empirical analog of this claim where S_ψ is replaced by $S_{n,\psi}$. Let

$$\begin{aligned} L_n(x) &:= n^{-1/2}\sum_{i=1}^{n} \dot{\mathbf{m}}(X_i)\, I(X_i \geq x)\,\psi(\varepsilon_i), \ \ s \in \mathbb{R}, \\ \mathcal{U}_n(x) &:= \int_{s \leq x} \dot{\mathbf{m}}'(s)\mathbf{A}^{-1}(s)L_n(s)G(ds), \ \ x \leq x_0. \end{aligned}$$

Notice that

$$TS_{n,\psi}(x) \equiv S_{n,\psi}(x) - \mathcal{U}_n(x).$$

Now, because of (6.6.1), (6.6.14), and the assumed i.i.d. setup,

$$\begin{aligned}
EL_n(s)L_n(t)' &= \sigma_\psi^2\, n^{-1} \sum_{i=1}^n E\dot{\mathbf{m}}(X_i)\,\dot{\mathbf{m}}'(X_i)\,I(X_i \ge s \vee t) \\
&= \sigma_\psi^2\, A(s \vee t), \\
EL_n(s)\mathcal{S}_{n,\psi}(y) &= \sigma_\psi^2\, E[\dot{\mathbf{m}}(X)\,I(X \ge s)\,I(X \le y)] \\
&= 0, \qquad\qquad\qquad \text{if } y < s, \\
&= \sigma_\psi^2\,[\boldsymbol{\nu}(y) - \boldsymbol{\nu}(s)], \qquad \text{if } y \ge s.
\end{aligned}$$

Use these facts to obtain, for $x \le y$,

$$\begin{aligned}
E\mathcal{S}_{n,\psi}(x)\mathcal{U}_n(y) &= \int_{s\le y} \dot{\mathbf{m}}'(s)\mathbf{A}^{-1}(s)E[L_n(s)\mathcal{S}_{n,\psi}(x)]G(ds) \\
&= \sigma_\psi^2 \int_{s\le x} \dot{\mathbf{m}}'(s)\mathbf{A}^{-1}(s)\,[\boldsymbol{\nu}(x) - \boldsymbol{\nu}(s)]G(ds) \\
E\mathcal{S}_{n,\psi}(y)\mathcal{U}_n(x) &= \int_{s\le x_0} \dot{\mathbf{m}}'(s)\mathbf{A}^{-1}(s)\,E[L_n(s)\mathcal{S}_{n,\psi}(y)]G(ds) \\
&= \sigma_\psi^2 \int_{s\le x} \dot{\mathbf{m}}'(s)\mathbf{A}^{-1}(s)\,[\boldsymbol{\nu}(y) - \boldsymbol{\nu}(s)]G(ds),
\end{aligned}$$

$$\begin{aligned}
&\sigma_\psi^{-2}\, E\mathcal{U}_n(x)\mathcal{U}_n(y) \\
&= E\Big[\int_{s\le x} \dot{\mathbf{m}}'(s)\mathbf{A}^{-1}(s)L_n(s)G(ds)\Big] \\
&\quad \times\Big[\Big(\int_{t\le x} + \int_{x<t\le y}\Big)\dot{\mathbf{m}}'(t)\mathbf{A}^{-1}(t)L_n(t)G(dt)\Big] \\
&= \int_{s\le x}\int_{t\le x} \dot{\mathbf{m}}'(s)\mathbf{A}^{-1}(s)E[L_n(s)L_n(t)']\mathbf{A}^{-1}(t)\dot{\mathbf{m}}(t)G(ds)G(dt) \\
&\quad + \int_{s\le x}\int_{x<t\le y} \dot{\mathbf{m}}'(s)\mathbf{A}^{-1}(s)E[L_n(s)L_n(t)']\mathbf{A}^{-1}(t)\dot{\mathbf{m}}(t) \\
&\qquad\qquad\qquad\qquad\qquad\qquad\qquad\qquad \times G(ds)G(dt) \\
&= \int_{s\le x}\int_{t\le x} \dot{\mathbf{m}}'(s)\mathbf{A}^{-1}(s)A(s\vee t)\mathbf{A}^{-1}(t)\dot{\mathbf{m}}(t)G(ds)G(dt) \\
&\quad + \int_{s\le x}\int_{x<t\le y} \dot{\mathbf{m}}'(s)\mathbf{A}^{-1}(s)\dot{\mathbf{m}}(t)G(ds)G(dt) \\
&= \sigma_\psi^2\Big[2\int_{s\le x} \dot{\mathbf{m}}'(s)\mathbf{A}^{-1}(s)\,[\boldsymbol{\nu}(x) - \boldsymbol{\nu}(s)]G(ds) \\
&\quad + \int_{s\le x} \dot{\mathbf{m}}'(s)\mathbf{A}^{-1}(s)\,[\boldsymbol{\nu}(y) - \boldsymbol{\nu}(x)]G(ds)\Big].
\end{aligned}$$

Using the above derivation one can now easily conclude that

$$ETS_{n,\psi}(x)TS_{n,\psi}(y) = \sigma_\psi^2\, G(x \wedge y).$$

The next result shows that the transformation T takes $\hat{S}_{n,\psi}$ in the limit to $TS_{n,\psi}$, and hence its limiting distribution is known.

Theorem 6.6.3 *(A). Assume, in addition to the assumptions of Theorem 6.6.2(A), that (6.6.16) and (6.6.22) hold. Then*

$$\sup_{x \le x_0} \left| T\hat{S}_{n,\psi}(x) - TS_{n,\psi}(x) \right| = o_p(1). \tag{6.6.23}$$

If in addition, (6.6.2) holds, then

$$TS_{n,\psi} \Longrightarrow TS_\psi \;\; \text{and} \;\; T\hat{S}_{n,\psi} \Longrightarrow TS_\psi \; \text{in} \; \mathcal{D}[-\infty, x_0]. \tag{6.6.24}$$

(B). The above claims continue to hold under the assumptions of Theorem 6.6.1(B), (6.6.16) and (6.6.22).

Note that $T\hat{S}_{n,\psi}$ is not a statistic as the transformation T depends on the parameter $\boldsymbol{\theta}_0$ and the covariate d.f. G. For statistical applications to the goodness-of-fit testing one needs to obtain an analogue of the above theorem where T is replaced by an estimator T_n. Let, for $x \in \mathbb{R}$,

$$\begin{aligned} G_n(x) &:= n^{-1} \sum_{i=1}^n I(X_i \le x), \\ \mathbf{A}_n(x) &:= \int \dot{\mathbf{m}}(y, \boldsymbol{\theta}_n) \dot{\mathbf{m}}'(y, \boldsymbol{\theta}_n)\, I(y \ge x)\, G_n(dy). \end{aligned}$$

An estimator of T is defined, for $x \le x_0$, to be

$$\begin{aligned} T_n f(x) = f(x) - \int_{-\infty}^x & \dot{\mathbf{m}}'(y, \boldsymbol{\theta}_n) \mathbf{A}_n^{-1}(y) \\ & \times \Big[\int \dot{\mathbf{m}}(z, \boldsymbol{\theta}_n)\, I(z \ge y)\, f(dz) \Big] G_n(dy). \end{aligned} \tag{6.6.25}$$

The next result is the most useful result of this section. It states the consistency of $T_n\hat{S}_{n,\psi}$ for $T\hat{S}_{n,\psi}$ under the following additional smoothness condition on $\dot{\mathbf{m}}$. For some $q \times q$ square matrix $\ddot{\mathbf{m}}(x, \boldsymbol{\theta}_0)$ and a non-negative function $K_1(x, \boldsymbol{\theta}_0)$, both measurable in the x-coordinate, the following holds:

$$\begin{aligned} & \mathbb{E}\|\dot{\mathbf{m}}(X, \boldsymbol{\theta}_0)\|^j K_1(X, \boldsymbol{\theta}_0) < \infty, \\ & \mathbb{E}\|\ddot{\mathbf{m}}(X, \boldsymbol{\theta}_0)\| \|\dot{\mathbf{m}}(X, \boldsymbol{\theta}_0)\|^j < \infty, \; j = 0, \; 1, \end{aligned} \tag{6.6.26}$$

and $\forall\, \epsilon > 0$, there exists a $\delta > 0$ such that $\|\boldsymbol{\theta} - \boldsymbol{\theta}_0\| < \delta$ implies

$$\|\dot{\mathbf{m}}(x, \boldsymbol{\theta}) - \dot{\mathbf{m}}(x, \boldsymbol{\theta}_0) - \ddot{\mathbf{m}}(x, \boldsymbol{\theta}_0)(\boldsymbol{\theta} - \boldsymbol{\theta}_0)\| \le \epsilon\, K_1(x, \boldsymbol{\theta}_0)\, \|\boldsymbol{\theta} - \boldsymbol{\theta}_0\|,$$

for G-almost all x.

We are ready to state

Theorem 6.6.4 *(A). Suppose, in addition to the assumptions of Theorem 6.6.3(A), (6.6.26) holds. Then,*

$$\sup_{x \le x_0} |T_n \hat{S}_{n,\psi}(x) - T S_{n,\psi}(x)| = o_p(1). \tag{6.6.27}$$

Consequently,

$$\sigma_{n,\psi}^{-1}\, T_n \hat{S}_{n,\psi}(\cdot) \Longrightarrow B \circ G, \quad in\ \mathcal{D}[-\infty, x_0]. \tag{6.6.28}$$

(B). The same continues to hold under (6.6.26) and the assumptions of Theorem 6.6.2(B).

Now, let $a < b$ be given real numbers and suppose one is interested in testing the hypothesis

$$H: \ m_\psi(x) = m(x, \boldsymbol{\theta}_0), \quad \text{for all } x \in [a, b] \text{ and for some } \boldsymbol{\theta}_0 \in \boldsymbol{\theta}.$$

Assume the support of G is $\mathbb{R}$, and $A(b)$ is positive definite. Then, $\mathbf{A}(x)$ is non-singular for all $x \le b$, continuous on $[a, b]$ and $\mathbf{A}^{-1}(x)$ is continuous on $[a, b]$ and

$$E\|\dot{\mathbf{m}}'(X)\mathbf{A}^{-1}(x)\| I(a < X \le b) < \infty.$$

Thus, under the conditions of the Theorem 6.6.4, $\sigma_{n,\psi}^{-1}\, T_n \hat{S}_{n,\psi}(\cdot) \Longrightarrow B \circ G$, in $\mathcal{D}[a, b]$ and we obtain

$$\sigma_{n,\psi}^{-1}\{T_n \hat{S}_{n,\psi}(\cdot) - T_n \hat{S}_{n,\psi}(a)\} \Longrightarrow B(G(\cdot)) - B(G(a)), \quad \text{in } D[a, b].$$

The stationarity of the increments of the Brownian motion then readily implies that

$$D_n := \sup_{a \le x \le b} \frac{|T_n \hat{S}_{n,\psi}(x) - T_n \hat{S}_{n,\psi}(a)|}{\{G_n(b) - G_n(a)\}^{1/2}\, \sigma_{n,\psi}} \Longrightarrow \sup_{0 \le u \le 1} |B(u)|.$$

Hence, the asymptotic null distribution of any test of H based on D_n is known.

6.6.3 Computation of $T_n\hat{S}_{n,\psi}$

In this sub-section we discuss some computational aspects of the proposed statistics. It is useful for the computational purpose to rewrite $T_n\hat{S}_{n,\psi}$ as follows: For convenience, let

$$
\begin{aligned}
(6.6.29)\qquad & m(x) := m(x, \boldsymbol{\theta}_0), & \dot{\mathbf{m}}(x) := \dot{\mathbf{m}}(x, \boldsymbol{\theta}_0), \\
& m_n(x) := m(x, \boldsymbol{\theta}_n), & \dot{\mathbf{m}}_n(x) := \dot{\mathbf{m}}(x, \boldsymbol{\theta}_n), \\
& \varepsilon_i := Y_i - m(X_i), & \varepsilon_{ni} := Y_i - m_n(X_i), \quad 1 \le i \le n.
\end{aligned}
$$

Let $X_{(i)}$ denote the ith order statistic among $\{X_j;\ 1 \le j \le n\}$, and $\tilde{\varepsilon}_{ni}$, denote the corresponding residuals ε_{ni}. Then,

$$
\begin{aligned}
(6.6.30)\quad & T_n\hat{S}_{n,\psi}(x) \\
&= n^{-1/2} \sum_{i=1}^{n-1} \Big[I(X_{(i)} \le x) \\
&\qquad - n^{-1} \sum_{j=1}^{n-1} \dot{\mathbf{m}}_n'(X_{(j)}) \mathbf{A}_n^{-1}(X_{(j)}) \dot{\mathbf{m}}_n(X_{(i)}) \\
&\qquad\qquad \times I(X_{(j)} \le X_{(i)} \wedge x) \Big] \psi(\tilde{\varepsilon}_{ni}).
\end{aligned}
$$

Let

$$
T_{li} := \sum_{k=1}^{l} \dot{\mathbf{m}}_n'(X_{(k)}) \mathbf{A}_n^{-1}(X_{(k)}) \dot{\mathbf{m}}_n(X_{(i)}), \quad 1 \le l, i \le n-1.
$$

If for some $1 \le j \le n-1$, $X_{(j)} \le x < X_{(j+1)}$, then one sees that $T_n\hat{S}_{n,\psi}(x) = \mathcal{S}_j$, where

$$
\mathcal{S}_j := \sum_{i=1}^{j} \Big[1 - T_{ii} \Big] \psi(\tilde{\varepsilon}_{ni}) - \sum_{i=j+1}^{n-1} T_{ji} \psi(\tilde{\varepsilon}_{ni}).
$$

This is useful in computing various tests based on the above transformation.

Now, let $g_1, \ldots, g_q$ be known real-valued G-square integrable functions on $\mathbb{R}$ and consider the class of models $\mathcal{M}$ with

$$
m(x, \boldsymbol{\theta}) = g_1(x)\theta_1 + \ldots + g_q(x)\theta_q.
$$

Then (6.6.13), (6.6.14) and (6.6.26) are trivially satisfied with

$$
\dot{\mathbf{m}}(x, \boldsymbol{\theta}) \equiv (g_1(x), \cdots, g_q(x))', \qquad \ddot{\mathbf{m}}(x, \boldsymbol{\theta}) \equiv \mathbf{0} \equiv K_1(x, \boldsymbol{\theta}).
$$

In particular this model includes the simple linear regression model. First consider the case when $q = 1$, $g_1(x) = x$ and assume $EX^2 < \infty$. In this case $\mathbf{A}(x) \equiv A(x) = \mathbb{E}X^2 I(X \geq x)$ is positive for all real x, uniformly continuous and decreasing on $\mathbb{R}$, and thus $A(x)$ is nonsingular for every $x \in \mathbb{R}$. A uniformly a.s. consistent estimator of A is

$$\mathbf{A}_n(x) \equiv n^{-1} \sum_{i=1}^{n} X_i^2 I(X_i \geq x).$$

Thus a test of the hypothesis that the regression mean function is simple linear through the origin on the interval $(-\infty, x_0]$ can be based on

$$\sup_{x \leq x_0} |T_n\hat{S}_{n,I}(x)| / \{\sigma_{n,I} G_n^{1/2}(x_0)\},$$

where

$$\begin{aligned} T_n\hat{S}_{n,I}(x) &= n^{-1/2} \sum_{i=1}^{n} (Y_i - X_i\theta_n) \Big[I(X_i \leq x) \\ &\quad - \sum_{j=1}^{n} \frac{X_j X_i I(X_j \leq X_i \wedge x)}{\sum_{k=1}^{n} X_k^2 I(X_k \geq X_j)} \Big], \\ \sigma_{n,I}^2 &= n^{-1} \sum_{i=1}^{n} (Y_i - X_i\theta_n)^2. \end{aligned}$$

Similarly, a test of the hypothesis that the regression *median* function is simple linear through the origin can be based on

$$\sup_{x \leq x_0} |T_n\hat{S}_{n,.5}(x)| / \{\sigma_{n,.5} G_n^{1/2}(x_0)\},$$

where

$$\begin{aligned} T_n\hat{S}_{n,.5}(x) &= n^{-1/2} \sum_{i=1}^{n} \{I(Y_i - X_i\theta_n > 0) - .5\} \Big[I(X_i \leq x_0) \\ &\quad - \sum_{j=1}^{n} \frac{X_j X_i I(X_j \leq X_i \wedge x)}{\sum_{k=1}^{n} X_k^2 I(X_k \geq X_j)} \Big], \end{aligned}$$

and

$$\sigma_{n,.5}^2 = n^{-1} \sum_{i=1}^{n} \{I(Y_i - X_i\theta_n > 0) - .5\}^2.$$

By Theorem 6.6.4, the asymptotic null distribution of both of these tests is free from the null model and other underlying parameters, as long as the estimator θ_n is the least square estimator in the former test and the least absolute deviation estimator in the latter.

In the case $q = 2$, $g_1(x) \equiv 1$, $g_2(x) \equiv x$, one obtains $\dot{\mathbf{m}}(x, \boldsymbol{\theta}_0) \equiv (1, x)'$ and $\mathbf{A}(x)$ is the 2×2 symmetric matrix

$$\mathbf{A}(x) = EI(X \geq x) \begin{pmatrix} 1 & X \\ X & X^2 \end{pmatrix}.$$

Clearly, $EX^2 < \infty$ implies $\mathbf{A}(x)$ is nonsingular for every real x and A^{-1} and A are continuous on $\mathbb{R}$. The matrix

$$\mathbf{A}_n(x) = n^{-1} \sum_{i=1}^{n} I(X_i \geq x) \begin{pmatrix} 1 & X_i \\ X_i & X_i^2 \end{pmatrix}.$$

provides a uniformly a.s. consistent estimator of $\mathbf{A}(x)$. Thus one may use $\sup_{x \leq x_0} |T_n \hat{S}_{n,I}(x)| / \{\sigma_{n,I} G_n(x_0)\}$ to test the hypothesis that the regression mean function is a simple linear model on the interval $(-\infty, x_0]$.

Similarly, one can use the test statistic

$$\sup_{x \leq x_0} |T_n \hat{S}_{n,.5}(x)| / \{\sigma_{n,.5} G_n^{1/2}(x_0)\}$$

to test the hypothesis that the regression median function is given by a simple linear function. In both cases A_n is as above and one should now use the general formula (6.6.30) to compute these statistics. Again, from Theorem 6.6.4 it readily follows that the asymptotic levels of both of these tests can be computed from the distribution of $\sup_{0 \leq u \leq 1} |B(u)|$, provided the estimator $\boldsymbol{\theta}_n$ is taken to be, respectively, the LS and the LAD.

Remark 6.6.2 Theorems 6.6.1 and 6.6.2 can be extended to the case where $\ell(X_i)$ is replaced by an r-vector of functions in the definitions of the $S_{n,\psi}$ and $\hat{S}_{n,\psi}$, for some positive integer r. In this case the time parameter of these processes is an r-dimensional vector. The difficulty in transforming such processes to obtain a limiting process that has a known limiting distribution is similar to that faced in transforming the multivariate empirical process in the i.i.d. setting. This, in turn, is related to the difficulty of having a proper definition of a multi-time parameter martingale. See Khmaladze (1988, 1993) for a discussion on the issues involved. For these reasons we restricted our attention here to the one dimensional case only. □

6.6.4 Proofs of some results of Section 6.6.2

Before proceeding further, we state two facts that will be used below repeatedly. Let $\{\xi_i\}$ be r.v.'s with finite first moment such that $\{(\xi_i, X_i)\}$ are i.i.d. and let ζ_i be i.i.d. square integrable r.v.'s. Then $\max_{1\le i\le n} n^{-1/2}|\zeta_i| = o_p(1)$ and

$$\text{(6.6.31)} \qquad \sup_{x\in\mathbb{R}} \Big| n^{-1}\sum_{i=1}^n \xi_i\, I(X_i \le x) - E\xi_1\, I(X \le x)\Big| \to 0, \ a.s.$$

The LLN's implies the pointwise convergence in (6.6.31). The uniformity is obtained with the aid of the triangle inequality and by decomposing each ξ_i into its negative and positive part and applying a Glivenko-Cantelli type argument to each part.

Remark 6.6.3 We are now ready to sketch an argument for the weak convergence of $\mathcal{S}_{n,\psi}((\tau^2_{n,\psi})^{-1})$ to B under the hypothesis $m_\psi = m_0$. For the sake of brevity, let $b_n := \tau^2_{n,\psi}(\infty)$, $b := \tau^2_\psi(\infty)$. First, note that

$$\begin{aligned}\sup_{0\le t\le b_n} |\tau^2_{n,\psi}((\tau^2_{n,\psi})^{-1}(t)) - t| &\le \max_{1\le i\le n} n^{-1}\psi^2(Y_i - m_0(X_i)) \\ &= o_p(1)\end{aligned}$$

by (6.6.2). Next, fix an $\epsilon > 0$ and let $\mathcal{A}_n := [|b_n - b| \le \epsilon]$ and $c_\epsilon := 1/[1-\frac{\epsilon}{b}]$. On $\mathcal{A}_n$,

$$1/[1+\frac{\epsilon}{b}] \le b/b_n \le 1/[1-\frac{\epsilon}{b}] = c_\epsilon$$

and

$$\begin{aligned}\sup_{0\le t\le b} |\tau^2_{n,\psi}((\tau^2_{n,\psi})^{-1}(t)) - t| &\le \sup_{0\le t\le b_n} |\tau^2_{n,\psi}((\tau^2_{n,\psi})^{-1}(t)) - t| \\ &\quad + \sup_{b_n < t\le b_n c_\epsilon} |\tau^2_{n,\psi}((\tau^2_{n,\psi})^{-1}(t)) - t|.\end{aligned}$$

The second term is further bounded from the above, on $\mathcal{A}_n$, by $\frac{b+\epsilon}{b-\epsilon}\,\epsilon$. But, by the ET, $P(\mathcal{A}_n) \to 1$. The arbitrariness of ϵ thus readily implies that

$$\sup_{0\le t\le \tau^2_\psi(\infty)} |\tau^2_{n,\psi}((\tau^2_{n,\psi})^{-1}(t)) - t| = o_p(1).$$

We thus obtain, in view of (6.6.31),

$$\begin{aligned}&\sup_{0\le t\le \tau^2_\psi(\infty)} |\tau^2_\psi((\tau^2_{n,\psi})^{-1}(t)) - t| \\ &\le \sup_{x\in\mathbb{R}} |\tau^2_\psi(x) - \tau^2_{n,\psi}(x)| + \sup_{0\le t\le \tau^2_\psi(\infty)} |\tau^2_{n,\psi}((\tau^2_{n,\psi})^{-1}(t)) - t| \\ &= o_p(1).\end{aligned}$$

These observations together with the continuity of the weak limit of $\mathcal{S}_{n,\psi}$ implies that

$$\sup_{0\le t\le \tau_\psi^2(\infty)\vee\tau_{n,\psi}^2(\infty)} |\mathcal{S}_{n,\psi}((\tau_{n,\psi}^2)^{-1}(t)) - \mathcal{S}_{n,\psi}((\tau_\psi^2)^{-1}(t))| = o_p(1).$$

Therefore, by Theorem 6.6.1, $s_{n,\psi}^{-1}\mathcal{S}_{n,\psi}((\tau_{n,\psi}^2)^{-1}) \Longrightarrow B$ and the limiting distribution of any continuous functional of $s_{n,\psi}^{-1}\mathcal{S}_{n,\psi}((\tau_{n,\psi}^2)^{-1})$ can be obtained from the distribution of the corresponding functional of B. In particular the asymptotic level of the test based on the Cramér - von Mises type statistic

$$s_{n,\psi}^{-2}\int_0^{s_{n,\psi}^2} \mathcal{S}_{n,\psi}^2((\tau_{n,\psi}^2)^{-1}(t))\,dH(t/s_{n,\psi}^2)$$

can be obtained from the distribution of $\int_0^1 B^2\,dH$, where H is a d.f. function on $[0,1]$. □

For our next lemma, recall the notation in (6.6.29) and Let

$$\mathcal{D}_n = n^{-1/2}\sum_{i=1}^n \Big|\psi(\varepsilon_{ni}) - \psi(\varepsilon_i) - (\varepsilon_{ni}-\varepsilon_i)\dot\psi(\varepsilon_i)\Big|\|r(X_i)\|.$$

where r is a measurable vector valued function such that

(6.6.32) $$\mathbb{E}\|r(X)\|^2 < \infty.$$

Let $\Delta_n = n^{1/2}(\boldsymbol{\theta}_n - \boldsymbol{\theta}_0)$.

Lemma 6.6.3 *Under the assumptions of Theorem 6.6.2(A) and (6.6.32),*

$$\mathcal{D}_n = o_p(1).$$

Proof. Fix an $\alpha > 0$. Let $h_i := \alpha + k\|\dot m(X_i)\|$ and

$$B_n = \Big\{\|\Delta_n\| \le k;\quad \max_i |m_n(X_i) - m(X_i) - \dot{\mathbf{m}}'(X_i)(\boldsymbol{\theta}_n - \boldsymbol{\theta}_0)| \le \frac{\alpha}{n^{1/2}}\Big\}.$$

Then by assumption (6.6.12) - (6.6.14) there exists a large $k < \infty$ and an integer N such that

$$P(B_n) > 1-\alpha, \qquad \forall n > N.$$

Now, on B_n, we obtain for $1 \leq i \leq n$, under H_0,

$$|m_n(X_i) - m(X_i)| \leq n^{-1/2} h(X_i).$$

Furthermore, by the absolute continuity of ψ, on B_n,

$$\mathcal{D}_n \leq n^{-1/2} \sum_{i=1}^{n} \|r(X_i)\| h(X_i) \int_{-n^{-1/2}}^{n^{-1/2}} |\dot{\psi}(\varepsilon_i - zh(X_i)) - \dot{\psi}(\varepsilon_i)| dz.$$

But, by (Ψ_1), the expected value of this upper bound equals

$$E(\|r(X)\| h(X))\, n^{1/2} \int_{-n^{-1/2}}^{n^{-1/2}} E|\dot{\psi}(\varepsilon - zh(X)) - \dot{\psi}(\varepsilon)| dz = o(1),$$

thereby completing the proof of Lemma 6.6.3. □

Proof of Theorem 6.6.2. Put, for an $x \in \mathbb{R}$,

$$R_n(x) := \hat{S}_{n,\psi}(x) - S_{n,\psi}(x) = n^{-1/2} \sum_{i=1}^{n} [\psi(\varepsilon_{ni}) - \psi(\varepsilon_i)] I(X_i \leq x).$$

Decompose R_n as

$$\begin{aligned} R_n(x) &= n^{-1/2} \sum_{i=1}^{n} [\psi(\varepsilon_{ni}) - \psi(\varepsilon_i) - (\varepsilon_{ni} - \varepsilon_i)\dot{\psi}(\varepsilon_i)] I(X_i \leq x) \\ &\quad - n^{-1/2} \sum_{i=1}^{n} [m_n(X_i) - m(X_i) \\ &\qquad\qquad - \dot{\mathbf{m}}'(X_i)\,(\boldsymbol{\theta}_n - \boldsymbol{\theta}_0)]\,\dot{\psi}(\varepsilon_i)\, I(X_i \leq x) \\ &\quad - n^{-1/2} \sum_{i=1}^{n} \dot{\mathbf{m}}'(X_i)\,\dot{\psi}(\varepsilon_i) I(X_i \leq x)\,(\boldsymbol{\theta}_n - \boldsymbol{\theta}_0) \\ &= R_{n1}(x) + R_{n2}(x) + R_{n3}(x)\,\Delta_n, \qquad \text{say}. \end{aligned}$$

The term $R_{n3}(x)$ is equal to

$$n^{-1} \sum_{i=1}^{n} \dot{\mathbf{m}}'(X_i)\,\dot{\psi}(\varepsilon_i) I(X_i \leq x).$$

By an application of (6.6.31) we readily obtain that

$$\sup_{x \in \mathbb{R}} \|R_{n3}(x) - \gamma_\psi\, \nu(x)\| = o_p(1).$$

Due to (6.6.12), it thus remains to show that R_{n1} and R_{n2} tend to zero, uniformly in x, in probability. The assertion for R_{n1} follows immediately from Lemma 6.6.3, because it is uniformly bounded by the $\mathcal{D}_n$ with $r \equiv 1$. As to R_{n2}, recall the event B_n from the proof of Lemma 6.6.3 and note that on B_n,

$$\sup_x |R_{n2}(x)| \le \alpha n^{-1} \sum_{i=1}^n |\dot\psi(\varepsilon_i)| = O(\alpha), \text{ a.s.},$$

by the LLN's. Since $\alpha > 0$ is arbitrarily chosen, this completes the proof of part (A).

As to the **proof of part (B)**, put, for $1 \le i \le n$, $t \in \mathbb{R}^q$, $a \in \mathbb{R}$,

$$\begin{aligned} d_{n,i}(t) &:= m(X_i, \boldsymbol{\theta}_0 + n^{-1/2}t) - m(X_i); \\ \gamma_{n,i} &:= n^{-1/2}(2\alpha + \delta\|\dot{\mathbf{m}}(X_i)\|), \ \alpha > 0, \delta > 0; \\ \mu_n(X_i, t, a) &:= E\psi(\varepsilon_i - d_{n,i}(t) + a\gamma_{n,i}). \end{aligned}$$

Define, for a, $x \in \mathbb{R}$ and $t \in \mathbb{R}^q$,

$$\begin{aligned} &D_n(x,t,a) \\ &:= n^{-1/2}\sum_{i=1}^n [\psi(\varepsilon_i - d_{n,i}(t) + a\gamma_{n,i}) - \mu_n(X_i,t,a) \\ &\qquad\qquad -\psi(\varepsilon_i)]\, I(X_i \le x). \end{aligned}$$

Write $D_n(x,t)$ and $\mu_n(X_i,t)$ for $D_n(x,t,0)$ and $\mu_n(X_i,t,0)$, respectively.

Note that by the i.i.d. assumption,

$$\begin{aligned} &Var(D_n(x,t,a)) \\ &\le \mathbb{E}[\psi(\varepsilon - d_{n,1}(t) + a\gamma_{n,1}) - \mu_n(X,t,a) - \psi(\varphi_1)]^2 \\ &\le \mathbb{E}[\psi(\varepsilon - d_{n,1}(t) + a\gamma_{n,1}) - \psi(\varepsilon)]^2 \to 0, \end{aligned}$$

by assumption (6.6.13) and (Ψ_2). Upon an application of Lemma 6.6.2 with $Z_{ni} = \psi(\varepsilon_i - d_{n,i}(t) + a\gamma_{n,i}) - \mu_n(X_i,t,a) - \psi(\varepsilon_i)$ we readily obtain that

$$\sup_{x\in\mathbb{R}} |D_n(x,t,a)| = o_p(1), \qquad \forall\, a \in \mathbb{R},\ t \in \mathbb{R}^q. \tag{6.6.33}$$

We need to prove that for every $b < \infty$,

$$\sup_{x\in\mathbb{R},\, \|t\|\le b} |D_n(x,t)| = o_p(1). \tag{6.6.34}$$

To that effect let

$$C_n := \Big\{ \sup_{\|t\|\le b} |d_{n,i}(t)| \le n^{-1/2}(\alpha + b\|\dot{\mathbf{m}}(X_i)\|),\ 1 \le i \le n \Big\},$$

and for an $\|s\| \le b$, let

$$A_n := \Big\{ \sup_{\|t\|\le b,\, \|t-s\|\le\delta} |d_{n,i}(t) - d_{n,i}(s)| \le \gamma_{n,i},\ 1 \le i \le n \Big\} \cap C_n.$$

By assumption (6.6.13), there is an $N < \infty$, depending only on α, such that $\forall\, b < \infty$ and $\forall\, \|s\| \le b$,

$$P(A_n) > 1 - \alpha, \qquad \forall n > N. \tag{6.6.35}$$

Now, by the monotonicity of ψ one obtains that on A_n, for each fixed $\|s\| \le b$ and $\forall\, \|t\| \le b$, with $\|t - s\| \le \delta$,

$$\begin{aligned}(6.6.36) \qquad & |D_n(x,t)| \\ \le\ & |D_n(x,s,1)| + |D_n(x,s,-1)| \\ & + \Big| n^{-1/2} \sum_{i=1}^n [\mu_n(X_i,s,1) - \mu_n(X_i,s,-1)]\, I(X_i \le x) \Big|.\end{aligned}$$

By (6.6.33), the first two terms converge to zero uniformly in x, in probability. Moreover, by the definition of $\gamma_{n,i}$, and by the LLN's

$$n^{-1/2} \sum_{i=1}^n \gamma_{n,i} = n^{-1} \sum_{i=1}^n (2\alpha + \delta \|\dot{\mathbf{m}}(X_i)\|) = O_p(\alpha + \delta).$$

In view of this and (**F1**), the last term in (6.6.34) is bounded above by

$$\begin{aligned} & n^{-1/2} \sum_{i=1}^n \int_{-\infty}^{\infty} |F(y + d_{n,i}(s) + \gamma_{n,i}) \\ & \qquad\qquad - F(y + d_{n,i}(s) - \gamma_{n,i})|\psi(dy) \\ \le\ & K\, n^{-1/2} \sum_{i=1}^n \gamma_{n,i} = O_p(\alpha + \delta), \end{aligned}$$

which can be made arbitrarily smaller than a positive constant multiple of α, by the choice of δ. This together with the compactness of the set $\{\|t\| \le b\}$ proves (6.6.34).

Next, by (6.6.1), (Ψ_2), (**F1**), Fubini's theorem, and the LLN's, we readily readily obtain

$$\sup_{x\in\mathbb{R},\|t\|\le b} \Big| n^{-1/2} \sum_{i=1}^n \mu_n(X_i,t)\, I(X_i \le x) + \boldsymbol{\nu}'(x)\gamma_\psi\, t\Big| = o_p(1).$$

This together with (6.6.34), (6.6.35) and the assumption (6.6.12) proves (6.6.15) and hence the part (B) of Theorem 6.6.2. □

Remark 6.6.4 By (6.6.22), $\lambda_1 := \inf\{a'\mathbf{A}(x_0)a;\, a \in \mathbb{R}^q,\, \|a\| = 1\} > 0$ and $\mathbf{A}(x)$ is positive definite for all $x \le x_0$, Hence,

$$\|\mathbf{A}^{-1/2}(x)\|^2 \le \lambda_1^{-1} < \infty, \quad \forall x \le x_0,$$

and (6.6.14) implies

$$E\|\dot{\mathbf{m}}'(X)\mathbf{A}^{-1}(x)\|I(X \le x_0) \le E\|\dot{\mathbf{m}}(X)\|\,\lambda_1^{-1} < \infty. \tag{6.6.37}$$

This fact is used in the following proofs repeatedly. □

Proof of Theorem 6.6.3. Details will be given for Part (A) only, they being similar for Part (B). We shall first prove (6.6.28).

From the definitions of T, we obtain that

$$\begin{aligned} T\hat{\mathcal{S}}_{n,\psi}(x) &= \hat{\mathcal{S}}_{n,\psi}(x) - \int_{-\infty}^{x} \dot{\mathbf{m}}'(y)\,\mathbf{A}^{-1}(y) \\ &\qquad \times\Big[\int_y^\infty \dot{\mathbf{m}}(t)\,\hat{\mathcal{S}}_{n,\psi}(dt)\Big]G(dy), \\ T\mathcal{S}_{n,\psi}(x) &= \mathcal{S}_{n,,\psi}(x) - \int_{-\infty}^{x} \dot{\mathbf{m}}'(y)\,\mathbf{A}^{-1}(y) \\ &\qquad \times\Big[\int_y^\infty \dot{\mathbf{m}}(t)\,\mathcal{S}_n(dt)\Big]G(dy). \end{aligned} \tag{6.6.38}$$

As before, set $\Delta_n := n^{1/2}(\boldsymbol{\theta}_n - \boldsymbol{\theta}_0)$. From (6.6.15) we obtain, uniformly in $x \in \mathbb{R}$,

$$\hat{\mathcal{S}}_{n,\psi}(x) = \mathcal{S}_{n,\psi}(x) - \gamma_\psi\,\nu'(x)\,\Delta_n + o_p(1). \tag{6.6.39}$$

The two integrals in (6.6.38) differ by

$$\int_\infty^x \dot{\mathbf{m}}'(y)\,\mathbf{A}^{-1}(y)\,\mathbf{D}_n(y)G(dy),$$

where

$$\mathbf{D}_n(y) := n^{-1/2}\sum_{i=1}^n \dot{\mathbf{m}}(X_i)\,[\psi(\varepsilon_i) - \psi(\varepsilon_{ni})]I(X_i \ge y).$$

This process is similar to the process R_n as studied in the proof of Theorem 6.6.2(A). Decompose D_n as

$$\begin{aligned}
\mathbf{D}_n(y) &= n^{-1/2}\sum_{i=1}^{n} \dot{\mathbf{m}}(X_i)\,[\psi(\varepsilon_i) - \psi(\varepsilon_{ni}) \\
&\qquad -(\varepsilon_i - \varepsilon_{ni})\dot{\psi}(\varepsilon_i)]I(X_i \geq y) \\
&\quad + n^{-1/2}\sum_{i=1}^{n} \dot{\mathbf{m}}(X_i)\,[m_n(X_i) - m(X_i) \\
&\qquad -\dot{\mathbf{m}}'(X_i)\,(\boldsymbol{\theta}_n - \boldsymbol{\theta}_0)]\,\dot{\psi}(\varepsilon_i)\,I(X_i \geq y) \\
&\quad + n^{-1/2}\sum_{i=1}^{n} \dot{\mathbf{m}}(X_i)\,\dot{\mathbf{m}}'(X_i)\,\dot{\psi}(\varepsilon_i)I(X_i \geq y) \\
&\qquad\qquad \times (\boldsymbol{\theta}_n - \boldsymbol{\theta}_0)
\end{aligned}$$

$$\text{(6.6.40)} \qquad = \mathbf{D}_{n1}(y) + \mathbf{D}_{n2}(y) + \mathbf{D}_{n3}(y)\,\Delta_n, \qquad \text{say.}$$

Lemma 6.6.3, with $r = \dot{m}$ and the triangle inequality readily imply

$$\sup_{y\in\mathbb{R}} \|\mathbf{D}_{n1}(y)\| = o_p(1).$$

This fact together with (6.6.37) yields

$$\text{(6.6.41)} \qquad \sup_{x\leq x_0} \Big|\int_{-\infty}^{x} \dot{\mathbf{m}}'(y)\,\mathbf{A}^{-1}(y)\,\mathbf{D}_{n1}(y)\,G(dy)\Big| = o_p(1).$$

Recall B_n from the proof of Lemma 6.6.3. Then, on B_n,

$$\begin{aligned}
\text{(6.6.42)} \qquad \sup_{y\in\mathbb{R}} \|\mathbf{D}_{n2}(y)\| &\leq \alpha\, n^{-1}\sum_{i=1}^{n} \|\dot{\mathbf{m}}(X_i)\|\,|\dot{\psi}(\varepsilon_i)| \\
&= O(\alpha), \ a.s.,
\end{aligned}$$

by the LLN's. Arbitrariness of α and (6.6.37) yield

$$\text{(6.6.43)} \qquad \sup_{x\leq x_0} \Big|\int_{-\infty}^{x} \dot{\mathbf{m}}'(y)\,\mathbf{A}^{-1}(y)\,\mathbf{D}_{n2}(y)\,G(dy)\Big| = o_p(1).$$

Now consider the third term. We have

$$\mathbf{D}_{n3}(y) = n^{-1}\sum_{i=1}^{n} \dot{\mathbf{m}}(X_i)\,\dot{\mathbf{m}}'(X_i)\,\dot{\psi}(\varepsilon_i)\,I(X_i \geq y).$$

An application of (6.6.31) together with (6.6.10) yield that

$$\sup_{y\in\mathbb{R}} \|\mathbf{D}_{n3}(y) - \gamma_\psi \mathbf{A}(y)\| \to 0, \ a.s.$$

This together with the fact $\|\Delta_n\| = o_p(1)$ entails that

$$\sup_{x \le x_0} \Big| \Big[\int_{-\infty}^{x} \dot{\mathbf{m}}'(y)\, \mathbf{A}^{-1}(y)\, \mathbf{D}_{n3}(y)\, G(dy) - \gamma_\psi\, \nu'(x) \Big] \Delta_n \Big| = o_p(1). \tag{6.6.44}$$

The proof of the claim (6.6.23) is completed upon combining (6.6.41) - (6.6.44) with (6.6.40).

Next, we turn to the proof of (6.6.24). In view of (6.6.23), it suffices to prove $T\mathcal{S}_{n,\psi} \Longrightarrow T\mathcal{S}_\psi$. To this effect, note that for each real x, $T\mathcal{S}_{n,\psi}(x)$ is a sum of centered finite variance i.i.d. r.v.'s. The convergence of the finite dimensional distributions thus follows from the classical CLT.

To verify the tightness, because $\mathcal{S}_{n,\psi}$ is tight and has a continuous limit by Theorem 6.6.1, it suffices to prove the same for the second term of $T\mathcal{S}_{n,\psi}$ in (6.6.38). To that effect, let

$$\phi(x) := \int_{-\infty}^{x} \|\dot{\mathbf{m}}'\mathbf{A}^{-1}\|\, dG, \qquad x \le x_0.$$

Note that ϕ is nondecreasing, continuous and by (6.6.37), $\phi(x_0) < \infty$. Now, rewrite the relevant term as

$$K_n(x) := n^{-1/2} \sum_{i=1}^{n} \psi(\varepsilon_i) \int_{-\infty}^{x} \dot{\mathbf{m}}'(y)\mathbf{A}^{-1}(y)\dot{\mathbf{m}}(X_i) I(X_i \ge y) \times G(dy).$$

Because the summands are martingale differences and because of (6.6.14) we obtain, with the help of Fubini's theorem, that for $x < y$,

$$\begin{aligned} &\mathbb{E}[K_n(y) - K_n(x)]^2 \\ &= \sigma_\psi^2 \int_x^y \int_x^y \dot{\mathbf{m}}'(s)\, \mathbf{A}^{-1}(s)\, \mathbf{A}(s \vee t)\, \mathbf{A}^{-1}(t)\, \dot{\mathbf{m}}(t)\, G(dt)\, G(ds). \end{aligned}$$

By (6.6.14), $\|\mathbf{A}\|_\infty := \sup_{x \in \mathbb{R}} \|\mathbf{A}(x)\| \le \int_{-\infty}^{\infty} \|\dot{\mathbf{m}}\|^2\, dG < \infty$. We thus obtain that

$$\begin{aligned} \mathbb{E}[K_n(y) - K_n(x)]^2 &\le \sigma_\psi^2 \|\mathbf{A}\|_\infty \Big[\int_x^y \|\dot{\mathbf{m}}'\, \mathbf{A}^{-1}\|\, dG\Big]^2 \\ &= \sigma_\psi^2 \|\mathbf{A}\|_\infty [\phi(y) - \phi(x)]^2. \end{aligned}$$

This then yields the tightness of the second term in (6.6.38) in a standard fashion and also completes the proof of the Theorem 6.6.3(A). □

For the proof of Theorem 6.6.4 the following lemma will be crucial.

Lemma 6.6.4 *Let $\mathcal{U}$ be a relatively compact subset of $\mathcal{D}[-\infty, x_0]$. Let L, L_n be a sequence of random distribution functions on $\mathbb{R}$ such that*

$$\sup_{t \le x_0} |L_n(t) - L(t)| \to 0, \ a.s.$$

Then

$$\sup_{t \le x_0,\, \alpha \in \mathcal{U}} \left| \int_{-\infty}^{t} \alpha(x)[L_n(dx) - L(dx)] \right| = o_p(1).$$

Its proof is similar to that of Lemma 3.1 of Chang (1990) and uses the fact that the uniform convergence over compact families of functions follows from the uniform convergence over intervals.

In the following proofs, the above lemma is used with $L_n \equiv G_n$ and $L \equiv G$ and more generally, with L_n and L given by the relations $dL_n \equiv hdG_n$, $dL \equiv hdG$, where h is an G-integrable function. As to the choice of $\mathcal{U}$, let $\{\alpha_n\}$ be a sequence of stochastic processes which are uniformly tight, i.e., for a given $\delta > 0$ there exists a compact set $\mathcal{U}$ such that $\alpha_n \in \mathcal{U}$ with probability at least $1-\delta$. Apply Lemma 6.6.4 with this $\mathcal{U}$ and observe that $\alpha_n \notin \mathcal{U}$ with small probability to finally obtain

$$\sup_{t} \left| \int_{-\infty}^{t} \alpha_n(x)[L_n(dx) - L(dx)] \right| = o_p(1).$$

As will be seen below, these types of integrals appear in the expansion of $T_n S^1_{n,\psi}$.

Proof of Theorem 6.6.4. Again, the details below are given for Part (A) only, and for convenience we do not exhibit $\boldsymbol{\theta}_0$ in K_1 and $\ddot{m}$. First, note that by (6.6.14) $n^{-1/2} max_i \|\dot{\mathbf{m}}(X_i)\| = o_p(1)$. Then by (6.6.26) and the LLN's, on an event with probability tending to one and for a given $\epsilon > 0$,

$$\begin{aligned} n^{-1} \sum_{i=1}^{n} &\|\dot{\mathbf{m}}(X_i)\| \|\dot{\mathbf{m}}_n(X_i) - \dot{\mathbf{m}}(X_i)\| \\ \le\ & n^{-1/2} max_i \|\dot{\mathbf{m}}(X_i)\| \, \|\Delta_n\| \Big\{ \epsilon\, n^{-1} \sum_{i=1}^{n} K_1(X_i) \\ & \qquad + n^{-1} \sum_{i=1}^{n} \|\ddot{m}(X_i)\| \Big\} = o_p(1). \end{aligned}$$

Similarly one obtains

$$n^{-1} \sum_{i=1}^{n} \|\dot{\mathbf{m}}_n(X_i) - \dot{\mathbf{m}}(X_i)\|^2 = o_p(1).$$

These bounds in turn together with (6.6.31) imply that

$$\begin{aligned}
&\sup_{y\in\mathbb{R}} \|\mathbf{A}_n(y) - \mathbf{A}(y)\| \\
&\le 2n^{-1}\sum_{i=1}^{n} \|\dot{\mathbf{m}}(X_i)\|\|\dot{\mathbf{m}}_n(X_i) - \dot{\mathbf{m}}(X_i)\| \\
&\quad + n^{-1}\sum_{i=1}^{n} \|\dot{\mathbf{m}}_n(X_i) - \dot{\mathbf{m}}(X_i)\|^2 \\
&\quad + \sup_{y\in\mathbb{R}} \|n^{-1}\sum_{i=1}^{n} \dot{\mathbf{m}}(X_i)\dot{\mathbf{m}}'(X_i)I(X_i \ge y) - A(y)\| \\
&= o_p(1).
\end{aligned}$$

Consequently we have

$$\sup_{y\le x_0} \|\mathbf{A}_n^{-1}(y) - \mathbf{A}^{-1}(y)\| = o_p(1). \tag{6.6.45}$$

Next, we shall prove (6.6.27). Let

$$\hat{\mathbf{U}}_n(y) := \int_y^\infty \dot{\mathbf{m}}_n(x)\,\hat{\mathcal{S}}_n(dx), \quad \mathbf{U}_n(y) := \int_y^\infty \dot{\mathbf{m}}(x)\,\mathcal{S}_n(dx).$$

Then we have

$$T_n\hat{\mathcal{S}}_n(x) = \hat{\mathcal{S}}_n(x) - \int_{-\infty}^{x} \dot{\mathbf{m}}_n'(y)\,\mathbf{A}_n^{-1}(y)\,\hat{\mathbf{U}}_n(y)\,G_n(dy),$$

so that from (6.6.38) we obtain, uniformly in $x \in \mathbb{R}$,

$$\begin{aligned}
&T_n\hat{\mathcal{S}}_n(x) - T\mathcal{S}_n(x) \\
&:= -\gamma\,\boldsymbol{\nu}'(x)\,\Delta_n + o_p(1) \\
&\quad + \int_{-\infty}^{x} \dot{\mathbf{m}}'(y)\,\mathbf{A}^{-1}(y)\,\mathbf{U}_n(y)\,G(dy) \\
&\quad - \int_{-\infty}^{x} \dot{\mathbf{m}}'(y,\boldsymbol{\theta}_n)\,\mathbf{A}_n^{-1}(y)\,\mathbf{U}_n^1(y)\,G_n(dy) \\
&= -\gamma\,\boldsymbol{\nu}'(x)\,\Delta_n + o_p(1) + B_{n1}(x) - B_{n2}(x), \quad \text{say.}
\end{aligned} \tag{6.6.46}$$

We shall shortly show that

$$\sup_{x\le x_0} \|\hat{\mathbf{U}}_n(x) - \mathbf{U}_n(x) + \gamma_\psi\,\mathbf{A}(x)\,\Delta_n\| = o_p(1). \tag{6.6.47}$$

Apply Lemma 6.6.2 k times, jth time with $Z_{ni} \equiv \dot{\mathbf{m}}_j(X_i)\psi(\varepsilon_i)$, where $\dot{\mathbf{m}}_j$ is the jth component of $\dot{\mathbf{m}}$, $1 \le j \le k$. Then under the assumed conditions it follows that $\mathbf{U}_n$ is tight. Using (6.6.45), (6.6.47), Lemma 6.6.4, and the assumption (6.6.26), we obtain

$$\begin{aligned} B_{n2}(x) &= \int_{-\infty}^{x} \dot{\mathbf{m}}'\, \mathbf{A}^{-1}\, \mathbf{U}_n\, dG_n - \gamma \int_{-\infty}^{x} \dot{\mathbf{m}}'\, dG\, \Delta_n + o_p(1), \\ &= \int_{-\infty}^{x} \dot{\mathbf{m}}'\mathbf{A}^{-1}\, \mathbf{U}_n\, dG - \gamma\, \boldsymbol{\nu}'(x)\, \Delta_n + o_p(1), \end{aligned}$$

uniformly in $x \le x_0$, which in turn together with (6.6.46) implies (6.6.27).

We shall now prove (6.6.47). Some of the arguments are similar to the proof of Theorem 6.6.3. Now, rewrite

$$\begin{aligned} \hat{\mathbf{U}}_n(y) &= n^{-1/2} \sum_{i=1}^{n} \dot{\mathbf{m}}_n(X_i)\, \psi(\varepsilon_{ni})\, I(X_i \ge y) \\ &= n^{-1/2} \sum_{i=1}^{n} \dot{\mathbf{m}}_n(X_i)\, [\psi(\varepsilon_{ni}) - \psi(\varepsilon_i) \\ &\qquad\qquad - (\varepsilon_{ni} - \varepsilon_i)\dot{\psi}(\varepsilon_i)] I(X_i \ge y) \\ &\quad + n^{-1/2} \sum_{i=1}^{n} \dot{\mathbf{m}}_n(X_i)\, [m(X_i) - m_n(X_i) \\ &\qquad\qquad - \dot{\mathbf{m}}_i'\, (\boldsymbol{\theta}_0 - \boldsymbol{\theta}_n)]\, \dot{\psi}(\varepsilon_i)\, I(X_i \ge y) \\ &\quad - n^{-1} \sum_{i=1}^{n} \dot{\mathbf{m}}_n(X_i)\, \dot{\mathbf{m}}_i'\, \dot{\psi}(\varepsilon_i) I(X_i \ge y)\, \Delta_n \\ &\quad + n^{-1/2} \sum_{i=1}^{n} \dot{\mathbf{m}}_n(X_i)\, \psi(\varepsilon_i)\, I(X_i \ge y) \\ &= -T_{n1}(y) - T_{n2}(y) - T_{n3}(y)\, \Delta_n + T_{n4}(y), \qquad \text{say.} \end{aligned}$$

Observe that T_{n1}, T_{n2} are, respectively, similar to D_{n1}, D_{n2} in the proof of Theorem 6.6.3 except the weights $\dot{\mathbf{m}}(X_i)$ are now replaced by $\dot{\mathbf{m}}_n(X_i)$. We shall first approximate T_{n1} by D_{n1}. We obtain, for a given $\epsilon > 0$,

$$\begin{aligned} &\sup_{y \in \mathbb{R}} \|T_{n1}(y) - D_{n1}(y)\| \\ &\quad \le n^{-1} \sum_{i=1}^{n} \Big[\|\ddot{m}(X_i)\| + \epsilon\, K_1(X_i) \Big]\, \|\Delta_n\| \\ &\qquad\qquad \times \int_{-h(X_i)/n^{1/2}}^{h(X_i)/n^{1/2}} |\dot{\psi}(\varepsilon_i - s) - \dot{\psi}(\varepsilon_i)|\, ds \\ &\quad = o_p(1). \end{aligned}$$

A similar, but simpler, argument using the assumption (6.6.13) shows that $\sup_{y \in \mathbb{R}} \|T_{n2}(y) - D_{n2}(y)\| = o_p(1)$. Since D_{n1} and D_{n2} tend to zero uniformly in y, we conclude that

$$\sup_{y\in\mathbb{R}}\{\|T_{n1}(y)\| + \|T_{n2}(y)\|\} = o_p(1).$$

Again, using (6.6.26) and (6.6.31) we obtain

$$\begin{aligned} T_{n3}(y) &= n^{-1}\sum_{i=1}^{n} \dot{\mathbf{m}}(X_i)\,\dot{\mathbf{m}}_i'\,\dot{\psi}(\varepsilon_i) I(X_i \geq y) + o_p(1) \\ &= \gamma \mathbf{A}(y) + o_p(1), \end{aligned}$$

uniformly in $y \in \mathbb{R}$. We now turn to T_{n4}. We shall prove

$$\text{(6.6.48)} \qquad \sup_{y\in\mathbb{R}} \|T_{n4}(y) - n^{-1/2}\sum_{i=1}^{n} \dot{\mathbf{m}}(X_i)\,\psi(\varepsilon_i)\, I(X_i \geq y)\| = o_p(1).$$

To that effect let $g_{ni} := \dot{\mathbf{m}}_n(X_i) - \dot{\mathbf{m}}(X_i)(\boldsymbol{\theta}_0) - \ddot{\mathbf{m}}(X_i)(\boldsymbol{\theta}_0)(\boldsymbol{\theta}_n - \boldsymbol{\theta}_0)$ and

$$\Gamma_n(y) := n^{-1/2}\sum_{i=1}^{n} g_{ni}\,\psi(\varepsilon_i)\, I(X_i \geq y).$$

Clearly, by (6.6.26), on a large set,

$$\sup_{y\in\mathbb{R}} \|\Gamma_n(y)\| \leq \epsilon\, k n^{-1}\sum_{i=1}^{n} K_1(X_i)\,|\psi(\varepsilon_i)| = O_P(\epsilon).$$

But, because of (6.6.1) and (6.6.31),

$$\sup_{y\in\mathbb{R}} \|n^{-1}\sum_{i=1}^{n} \ddot{\mathbf{m}}(X_i)(\boldsymbol{\theta}_0)\,\psi(\varepsilon_i)\, I(X_i \geq y)\| = o_p(1).$$

The claim (6.6.48) thus follows from these facts and the assumption that $\Delta_n = O_p(1)$, in a routine fashion. This also completes the proof of (6.6.47) and hence that of the theorem. □

Notes: We end this section with some historical remarks. An and Bing (1991) have proposed the K-S test based on $\hat{S}_{n,I}$ and a half sample splitting technique *a la* Rao (1972) and Durbin (1973) to make it ADF for fitting a simple linear regression model. They also discuss the problem of fitting a linear autoregressive model of order 1. See also section 7.6 of this monograph on this. Su and Wei (1991) proposed K-S test based on the $\hat{S}_{n,I}$-process to test for fitting a generalized linear regression model. Delgado (1993) constructed two sample type tests based on the $S_{n,I}$ for comparing two regression models. Diebolt (1995) has obtained the Hungarian-type strong approximation result for the analogue of $\mathcal{S}_{n,I}$ in a special regression setting. Stute (1997) investigated the large sample theory of the analogue of $\hat{\mathcal{S}}_{n,I}$ for model checking in a general regression setting. He also gave a nonparametric principal component analysis of the limiting process in a linear regression setup similar to the one given by Durbin *et al.* (1975) in the one sample setting.

The transformation T is based on the ideas of Khmaladze (1981). Stute, Thies and Zhou (1998), Koul and Stute (1999) discussed this in the regression and autoregressive settings, respectively. The above proofs are adapted from the latter two papers.

7

Autoregression

7.1 Introduction

The purpose of the Chapters 7 and 8 is to offer a unified functional approach to some aspects of robust estimation and goodness-of-fit testing problems in autoregressive (AR) and conditionally heteroscedastic autoregressive (ARCH) models. We shall first focus on the well celebrated p-th order linear AR models. For these models, the similarity of the functional approach developed in the previous chapters in connection with linear regression models is transparent. This chapter thus extends the domain of applications of the statistical methodology of the previous chapters to the one of the most applied models with dependent observations. Chapter 8 discusses the development of similar approach in some general non-linear AR and ARCH models.

As before, let F be a d.f. on $\mathbb{R}$, $p \geq 1$ be an integer, and $\mathbf{Y}_0 := (X_0, X_{-1}, \cdots, X_{1-p})'$ be an observable random vector. In a linear AR(p) model the observations $\{X_i\}$ are such that for some $\boldsymbol{\rho}' = (\rho_1, \rho_2, \cdots, \rho_p) \in \mathbb{R}^p$,

$$(7.1.1) \qquad \varepsilon_i \;=\; X_i - (\rho_1 X_{i-1} + \cdots + \rho_p X_{i-p}), \quad 1 \leq i \leq n,$$

are i.i.d. F r.v.'s, and independent of $\mathbf{Y}_0$. Processes that play a fundamental role in the robust estimation of $\boldsymbol{\rho}$ in this model are the *randomly weighted residual empirical* processes

$$(7.1.2) \qquad T_j(x, \mathbf{t}) \;:=\; n^{-1} \sum_{i=1}^{n} g_j(\mathbf{Y}_{i-1}) I(X_i - \mathbf{t}' Y_{i-1} \leq x),$$

for $x \in \mathbb{R}$, $\mathbf{t} \in \mathbb{R}^p$, $1 \le j \le p$, where $\boldsymbol{g} := (g_1, \cdots, g_p)$ is a p-vector of measurable functions from $\mathbb{R}^p$ to $\mathbb{R}$ and $\mathbf{Y}_{i-1} := (X_{i-1}, \cdots, X_{i-p})'$, $1 \le i \le n$. Let $\mathbf{T} := (T_1, \cdots, T_p)'$. Note that

$$\mathbf{T}(x, \mathbf{t}) = n^{-1} \sum_{i=1}^{n} \boldsymbol{g}(\mathbf{Y}_{i-1}) I(X_i - \mathbf{t}' Y_{i-1} \le x), \quad x \in \mathbb{R}^p, \mathbf{t} \in \mathbb{R}^p.$$

The generalized M - (GM) estimators of $\boldsymbol{\rho}$, are solution $\mathbf{t}$ of the p equations

$$\text{(7.1.3)} \qquad \mathcal{G}(\mathbf{t}) := \int \psi(x) \mathbf{T}(dx, \mathbf{t}) = \mathbf{0},$$

where ψ is a nondecreasing bounded measurable function from $\mathbb{R}$ to $\mathbb{R}$. These estimators are analogues of M-estimators of $\boldsymbol{\beta}$ in linear regression as discussed in Chapter 4. Note that taking

$$\begin{aligned} \psi(x) &= xI[|x| \le k] + kx|x|^{-1} I[|x| > k], && x \in \mathbb{R}, \\ \boldsymbol{g}(\mathbf{x}) &\equiv \mathbf{x} I[\|\mathbf{x}\| \le k] + k\mathbf{x}\|\mathbf{x}\|^{-1} I[\|\mathbf{x}\| > k], && \mathbf{x} \in \mathbb{R}^p, \end{aligned}$$

in (7.1.3) gives the Huber(k) estimators and taking $\boldsymbol{g}(\mathbf{x}) \equiv \mathbf{x}$, $\mathbf{x} \in \mathbb{R}^p$ and $\psi(x) \equiv x$, $x \in \mathbb{R}$, gives the famous least square estimator.

The minimum distance estimator $\boldsymbol{\rho}_{\boldsymbol{g}}^+$ that is an analogue of $\boldsymbol{\beta}_{\mathbf{D}}^+$ of (5.2.15) is defined as a minimizer, w.r.t. $\mathbf{t} \in \mathbb{R}^p$ of

$$\begin{aligned} \text{(7.1.4)} \qquad & K_{\boldsymbol{g}}(\mathbf{t}) \\ &= \sum_{j=1}^{p} \int \Big[n^{-1/2} \sum_{i=1}^{n} g_j(\mathbf{Y}_{i-1}) \Big\{ I(X_i \le x + \mathbf{t}'\mathbf{Y}_{i-1}) \\ &\qquad - I(-X_i < x - \mathbf{t}'\mathbf{Y}_{i-1} \Big\} \Big]^2 dG(x). \end{aligned}$$

Observe that $K_{\boldsymbol{g}}$ involves $\mathbf{T}$. In fact, $\forall$ $\mathbf{t} \in \mathbb{R}^p$,

$$\begin{aligned} K_{\boldsymbol{g}}(\mathbf{t}) &= \sum_{j=1}^{p} \int \Big[n^{1/2} \Big\{ T_j(x, \mathbf{t}) - \sum_{i=1}^{n} g_j(\mathbf{Y}_{i-1}) \\ &\qquad + T_j(-x, \mathbf{t}) \Big\} \Big]^2 dG(x). \end{aligned}$$

Three members of this class of estimators are of special interest. They correspond to the cases $\boldsymbol{g}(\mathbf{x}) \equiv \mathbf{x}$, $G(x) \equiv x$; $\boldsymbol{g}(\mathbf{x}) \equiv \mathbf{x}$, $G \equiv \delta_0$, the measure degenerate at 0; $\boldsymbol{g}(\mathbf{x}) \equiv \mathbf{x}$, $G \equiv F$ in the case F is known. The first gives an analogue of the *Hodges - Lehmann* (H-L) estimator of $\boldsymbol{\rho}$, the second gives the *least absolute deviation* (LAD) estimator, while the third gives an estimator that is more efficient at logistic (double exponential) errors than LAD (H-L) estimator.

Another important process in the model (7.1.1) is the ordinary residual empirical process

$$(7.1.5) \qquad F_n(x, \mathbf{t}) := n^{-1} \sum_{i=1}^{n} I(X_i - \mathbf{t}'\mathbf{Y}_{i-1} \le x), \quad x \in \mathbb{R}, \ \mathbf{t} \in \mathbb{R}^p.$$

An estimator of F or a test of goodness-of-fit pertaining to F are usually based on $F_n(x, \hat{\boldsymbol{\rho}})$, where $\hat{\boldsymbol{\rho}}$ is an estimator of $\boldsymbol{\rho}$.

Clearly F_n is a special case of (7.1.2). But, both F_n and T_j, $1 \le j \le p$, are special cases of

$$\begin{aligned}(7.1.6) \qquad & W_h(x, \mathbf{t}) \\ & := n^{-1} \sum_{i=1}^{n} h(\mathbf{Y}_{i-1}) I(X_i - \mathbf{t}'\mathbf{Y}_{i-1} \le x) \\ & = n^{-1} \sum_{i=1}^{n} h(\mathbf{Y}_{i-1}) I(\varepsilon_i \le x + (\mathbf{t} - \boldsymbol{\rho})'\mathbf{Y}_{i-1}),\end{aligned}$$

for $x \in \mathbb{R}$, $\mathbf{t} \in \mathbb{R}^p$, where h is a measurable function from $\mathbb{R}^p$ to $\mathbb{R}$. Choosing $h(\mathbf{Y}_{i-1}) \equiv g_j(\mathbf{Y}_{i-1})$ in W_h gives T_j, $1 \le j \le p$ and the choice of $h \equiv 1$ yields F_n.

From the above discussion it is apparent that the investigation of the large sample behavior of various inferential procedures pertaining to $\boldsymbol{\rho}$ and F, based on $\{\mathbf{T}\}$ and $F_n(\cdot, \hat{\boldsymbol{\rho}})$, is facilitated by the weak convergence properties of $\{W_h(x, \boldsymbol{\rho} + n^{-1/2}\mathbf{u}), \ x \in \mathbb{R}, \ \mathbf{u} \in \mathbb{R}^p\}$. This will be investigated in Section 7.2, with the aid of Theorem 2.2.4. In particular, this section contains an AUL result about $\{W_h(x, \boldsymbol{\rho} + n^{-1/2}\mathbf{u}), \ x \in R, \ \|\mathbf{u}\| \le b\}$ which in turn yields AUL results about $\{\mathbf{T}(x, \boldsymbol{\rho} + n^{-1/2}\mathbf{u}), \ x \in \mathbb{R}, \ \|\mathbf{u}\| \le b\}$ and $\{F_n(x, \boldsymbol{\rho} + n^{-1/2}\mathbf{u}), \ x \in \mathbb{R}, \ \|\mathbf{u}\| \le b\}$. These results are useful in studying GM - and R- estimators of $\boldsymbol{\rho}$, akin to Chapters 3 and 4 when dealing with linear regression models. They are also useful in studying the large sample behaviour of some tests of goodness-of-fit pertaining to F. Analogous results about the ordinary empirical of the residuals in autoregressive *moving average* models are briefly discussed in Remark 7.2.4.

Generalized M- estimators and analogues of Jaeckel's R- estimators are discussed in Section 7.3. In the same section we also discuss a class of generalized R- (GR-) estimators. In order to use *GR* or m.d. estimators to construct confidence intervals one often needs consistent estimators of the functional $Q(f)$ of the error density f. Appropriate analogues of estimators of $Q(f)$ of Section 4.5 are shown to be consistent under (**F1**) and (**F2**). This is also done in Section 7.3, with the help of the AUL property of $\{F_n(x, \boldsymbol{\rho} + n^{-1/2}\mathbf{u}),\ x \in \mathbb{R},\ \|\mathbf{u}\| \le b\}$. This result is also used to prove the AUL of serial rank correlations of the residuals in an AR(p) model. Such results should be useful in developing analogues of the method of moment estimators or Yule - Walker equations based on ranks in these models.

Section 7.4 investigates the behaviour of two classes of m.d. estimators of $\boldsymbol{\rho}$, including the class of estimators $\{\boldsymbol{\rho}_g^+\}$. A crucial result needed to obtain the asymptotic distributions of these estimators is the asymptotic uniform quadraticity of their defining dispersions. This result is also proved in Section 7.4.

The regression quantiles of Koenker and Basett (1978) are now accepted as an appropriate extension of the one sample quantiles in the one sample location models to the multiple linear regression model (1.1.1). Section 7.5 discusses an extension of these regression quantiles to the stationary ergodic linear autoregressive time series. The closely related autoregression rank scores are also discussed here.

Section 7.6 contains appropriate analogues of some of the goodness-of-fit tests of Chapter 6 pertaining to F, while Section 7.7 discusses at some length the goodness-of-fit tests for fitting an autoregressive model of order 1. The proposed tests are analogues of the tests of Section 6.6. These tests are based on certain martingale transforms of a partial sum process of the estimated innovations.

7.2 AUL of W_h and F_n

Recall the notation from (2.2.26), and the statement of Theorem 2.2.4. In (2.2.26), take

$$
\begin{aligned}
(7.2.1) \qquad \zeta_{ni} &\equiv \varepsilon_i, \quad h_{ni} \equiv h(\mathbf{Y}_{i-1}), \quad \delta_{ni} \equiv n^{-1/2}\mathbf{u}'\mathbf{Y}_{i-1},\\
\mathcal{A}_{n1} &= \sigma - \text{field } \{\mathbf{Y}_0\},\\
\mathcal{A}_{ni} &= \sigma - \text{field } \{\mathbf{Y}_0', \varepsilon_1, \cdots, \varepsilon_{i-1}\}, \quad 2 \le i \le n.
\end{aligned}
$$

Then one readily sees that the corresponding $V_h(x)$, $V_h^*(x)$ are, respectively, equal to $W_h(x, \boldsymbol{\rho} + n^{-1/2}\mathbf{u})$, $W_h(x, \boldsymbol{\rho})$ for each $\mathbf{u} \in \mathbb{R}^p$ and for all $x \in \mathbb{R}$.

Consequently, if we let

$$(7.2.2) \qquad \nu_h(x,\mathbf{t}) := n^{-1}\sum_{i=1}^{n} h(\mathbf{Y}_{i-1})F(x+(\mathbf{t}-\boldsymbol{\rho})'\mathbf{Y}_{i-1}),$$
$$\mathcal{W}_h(x,\mathbf{t}) := n^{1/2}[W_h(x,\mathbf{t}) - \nu_h(x,\mathbf{t})], \quad x\in\mathbb{R}, \quad \mathbf{t}\in\mathbb{R}^p,$$

then the corresponding $U_h^*(x)$, $U_h(x)$ are, respectively, equal to $\mathcal{W}_h(x,\boldsymbol{\rho})$, $\mathcal{W}_h(x,\boldsymbol{\rho}+n^{-1/2}\mathbf{u})$, for each $\mathbf{u}\in\mathbb{R}^p$ and for all $x\in\mathbb{R}$. Recall the conditions **(F1)** and **(F2)** from Corollary 2.3.1. We are now ready to state and prove the following

Theorem 7.2.1 *In addition to (7.1.1), assume that the following conditions hold:*

$$(7.2.3) \qquad \Big(n^{-1}\sum_{i=1}^{n} h^2(\mathbf{Y}_{i-1})\Big)^{1/2} = \alpha + o_p(1), \quad \alpha \text{ a positive r.v.}$$

$$(7.2.4) \qquad n^{-1/2}\max_{1\le i\le n}\|h(\mathbf{Y}_{i-1})\| = o_p(1).$$

$$(7.2.5) \qquad n^{-1/2}\max_{1\le i\le n}\|\mathbf{Y}_{i-1}\| = o_p(1).$$

$$(7.2.6) \qquad n^{-1}\sum_{i=1}^{n}\|h(\mathbf{Y}_{i-1})\mathbf{Y}_{i-1}\| = O_p(1).$$

(7.2.7) $\quad F$ *satisfies* **(F1)** *and* **(F2)**.

Then, for every $0<b<\infty$,

$$(7.2.8) \qquad \sup_{x\in\mathbb{R},\|\mathbf{u}\|\le b}|\mathcal{W}_h(x,\boldsymbol{\rho}+n^{-1/2}\mathbf{u}) - \mathcal{W}_h(x,\boldsymbol{\rho})| = o_p(1),$$

$$(7.2.9) \qquad n^{1/2}\Big[W_h(x,\boldsymbol{\rho}+n^{-1/2}\mathbf{u}) - W_h(x,\boldsymbol{\rho})\Big]$$
$$= -\mathbf{u}'n^{-1}\sum_{i=1}^{n} h(\mathbf{Y}_{i-1})\mathbf{Y}_{i-1}f(x) + u_p(1).$$

where $u_p(1)$ *is a sequence of stochastic processes that converges to zero, uniformly over the set* $x\in\mathbb{R}$, $\|\mathbf{u}\|\le b$, *in probability.*

Remark 7.2.1 If in the above theorem the condition (7.2.7) is weaken to

(7.2.10) $\qquad F$ has a continuous and positive density f on $\mathbb{R}$.

then, for every $k<\infty$, the analogues of (7.2.8) and (7.2.9) where the supremum over $x\in\mathbb{R}$ is changed to the supremum over the set $x\in[-k,k]$, continue to hold. This follows from adapting the following proof to the compact subsets. Note that f continuous on $\mathbb{R}$ implies that it is uniformly continuous on every bounded interval of $\mathbb{R}$. □

Proof of Theorem 7.2.1. In view of the discussion preceding the statement of the theorem it is clear that (2.2.32) of Theorem 2.2.4 applied to entities given in (7.2.1) above readily yields that

$$\sup_{x\in\mathbb{R}}\Big|\mathcal{W}_h(x,\boldsymbol{\rho}+n^{-1/2}\mathbf{u})-\mathcal{W}_h(x,\boldsymbol{\rho})\Big|=o_p(1),$$

for every fixed $\mathbf{u}\in\mathbb{R}^p$. It is the uniformity with respect to $\mathbf{u}$ that requires an extra argument and that also turns out to be a consequence of another application of (2.2.32) and a monotonic property inherent in these processes as we now show.

Since h is fixed, it will not be exhibited in the proof. Also, for convenience, write $\mathcal{W}(\cdot)$, $\mathcal{W}_{\mathbf{u}}(\cdot)$, $W_{\mathbf{u}}^{\pm}(\cdot)$, $\nu_{\mathbf{u}}^{\pm}(\cdot)$ etc. for $\mathcal{W}_h(\cdot,\boldsymbol{\rho})$, $\mathcal{W}_h(\cdot,\boldsymbol{\rho}+n^{-1/2}\mathbf{u})$, $W_h^{\pm}(\cdot,\boldsymbol{\rho}+n^{1/2}\mathbf{u})$, $\nu_h^{\pm}(\cdot,\boldsymbol{\rho}+n^{1/2}\mathbf{u})$ etc. with $\pm$ signifying the fact that $h^{\pm}$ now appears in the place of h in these processes where $h^+=0\vee h$, $h^-=h-h^+$. To conserve the space, write $\boldsymbol{\xi}_i$, h_i, $h_i^{\pm}$ for $\mathbf{Y}_{i-1}$, $h(\mathbf{Y}_{i-1})$, $h^{\pm}(\mathbf{Y}_{i-1})$, respectively, $i\ge 1$. Thus, e.g.,

$$(7.2.11)\qquad \mathcal{W}_{\mathbf{u}}^{\pm}(x)=n^{-1/2}\sum_{i=1}^{n}h_i^{\pm}\{I(\varepsilon_i\le x+n^{-1/2}\mathbf{u}'\boldsymbol{\xi}_i)-F(x+n^{-1/2}\mathbf{u}'\boldsymbol{\xi}_i)\}.$$

We also need the following processes:

$$(7.2.12)\qquad T^{\pm}(x;\mathbf{u},a):=n^{-1/2}\sum_{i=1}^{n}h_i^{\pm}I\left(\varepsilon_i\le x+n^{-1/2}\mathbf{u}'\boldsymbol{\xi}_i+n^{-1/2}a\|\boldsymbol{\xi}_i\|\right),$$

$$m^{\pm}(x;\mathbf{u},a):=n^{-1/2}\sum_{i=1}^{n}h_i^{\pm}F\left(x+n^{-1/2}\mathbf{u}'\boldsymbol{\xi}_i+n^{-1/2}a\|\boldsymbol{\xi}_i\|\right)$$

$$Z^{\pm}:=T^{\pm}-m^{\pm},\quad x\in\mathbb{R},\ \mathbf{u}\in\mathbb{R}^p,\quad a\in\mathbb{R}.$$

Observe that if in U_h of (2.2.26) we take $\zeta_{ni}\equiv\varepsilon_i$, $h_{ni}\equiv h^{\pm}(\boldsymbol{\xi}_i)$, $\delta_{ni}\equiv n^{-1/2}\{\mathbf{u}'\boldsymbol{\xi}_i+a\|\boldsymbol{\xi}_i\|\}$ and $\mathcal{A}_{ni}$, $1\le i\le n$, as in (7.2.1), we obtain

$$U_h(\cdot)=Z^{\pm}(\cdot;\mathbf{u},a),\quad\text{for every }\ \mathbf{u}\in\mathbb{R}^p,\ a\in\mathbb{R}.$$

Similarly, if we take $\delta_{ni}\equiv n^{-1/2}\mathbf{u}'\boldsymbol{\xi}_i$ and the rest of the quantities as above then

$$U_h(\cdot)=Z^{\pm}(\cdot;\mathbf{u},0)=\mathcal{W}_{\mathbf{u}}^{\pm}(\cdot),\quad\text{for every }\ \mathbf{u}\in\mathbb{R}^p.$$

It thus follows from two applications of (2.2.32) and the triangle inequality that for every $\mathbf{u} \in \mathbb{R}^p$, $a \in \mathbb{R}$,

$$\sup_{x \in \mathbb{R}} |Z^{\pm}(x; \mathbf{u}, a) - Z^{\pm}(x; \mathbf{u}, 0)| = o_p(1), \tag{7.2.13}$$

$$\sup_{x \in \mathbb{R}} |\mathcal{W}_{\mathbf{u}}^{\pm}(x) - \mathcal{W}^{\pm}(x)| = o_p(1). \tag{7.2.14}$$

Thus to prove (7.2.8), because of the compactness of the ball $\{\mathbf{u} \in \mathbb{R}^p; \|\mathbf{u}\| \le b\}$, it suffices to show that for every $\varepsilon > 0$ there is a $\delta > 0$ such that for every $\|\mathbf{u}\| \le b$,

$$\limsup_n P\left(\sup_{\|\mathbf{s}\| \le b, \|\mathbf{s}-\mathbf{u}\| \le \delta,\ x \in \mathbb{R}} |\mathcal{W}_{\mathbf{s}}(x) - \mathcal{W}_{\mathbf{u}}(x)| > 4\varepsilon \right) < \varepsilon. \tag{7.2.15}$$

By the definition of $\mathcal{W}^{\pm}$ and the triangle inequality, for $x \in \mathbb{R}$, $\mathbf{s}$, $\mathbf{u} \in \mathbb{R}^p$,

$$\begin{aligned} |\mathcal{W}_{\mathbf{s}}(x) - \mathcal{W}_{\mathbf{u}}(x)| &\le \left|\mathcal{W}_{\mathbf{s}}^{+}(x) - \mathcal{W}_{\mathbf{u}}^{+}(x))\right| \\ &\quad + \left|\mathcal{W}_{\mathbf{s}}^{-}(x) - \mathcal{W}_{\mathbf{u}}^{-}(x)\right|, \\ \left|\mathcal{W}_{\mathbf{s}}^{\pm}(x) - \mathcal{W}_{\mathbf{u}}^{\pm}(x)\right| &\le n^{1/2}\Big[\left|W_{\mathbf{s}}^{\pm}(x) - W_{\mathbf{u}}^{\pm}(x)\right| \\ &\quad + \left|\nu_{\mathbf{s}}^{\pm}(x) - \nu_{\mathbf{u}}^{\pm}(x)\right|\Big]. \end{aligned} \tag{7.2.16}$$

But $\|\mathbf{s}\| \le b$, $\|\mathbf{u}\| \le b$, $\|\mathbf{s} - \mathbf{u}\| \le \delta$ imply that for all $1 \le i \le n$,

$$\begin{aligned} & n^{-1/2}\mathbf{u}'\boldsymbol{\xi}_i - n^{-1/2}\delta\|\boldsymbol{\xi}_i\| \\ &\qquad \le n^{-1/2}\mathbf{s}'\boldsymbol{\xi}_i \\ &\qquad\qquad \le n^{-1/2}\mathbf{u}'\boldsymbol{\xi}_i + n^{-1/2}\delta\|\boldsymbol{\xi}_i\|. \end{aligned} \tag{7.2.17}$$

From (7.2.17), the monotonicity of the indicator function and the nonnegativity of $h^{\pm}$, we obtain

$$\begin{aligned} & T^{\pm}(x; \mathbf{u}, -\delta) - T^{\pm}(x; \mathbf{u}, 0) \\ &\qquad \le W_{\mathbf{s}}^{\pm}(x) - W_{\mathbf{u}}^{\pm}(x) \\ &\qquad\qquad \le T^{\pm}(x; \mathbf{u}, \delta) - T^{\pm}(x; \mathbf{u}, 0), \end{aligned}$$

for all $x \in \mathbb{R}$, $\|\mathbf{s}\| \le b$, $\|\mathbf{s} - \mathbf{u}\| \le \delta$. Now center $T^{\pm}$ appropriately to obtain

$$\begin{aligned} & n^{1/2}|W_{\mathbf{s}}^{\pm}(x) - W_{\mathbf{u}}^{\pm}(x)| \\ &\quad \le \left|Z^{\pm}(x; \mathbf{u}, \delta) - Z^{\pm}(x; \mathbf{u}, 0)\right| \\ &\quad\quad + \left|Z^{\pm}(x; \mathbf{u}, -\delta) - Z^{\pm}(x; \mathbf{u}, 0)\right| \\ &\quad\quad + \left|m^{\pm}(x; \mathbf{u}, \delta) - m^{\pm}(x; \mathbf{u}, 0)\right| \\ &\quad\quad + \left|m^{\pm}(x; \mathbf{u}, -\delta) - m^{\pm}(x; \mathbf{u}, 0)\right|, \end{aligned} \tag{7.2.18}$$

for all $x \in \mathbb{R}$, $\|\mathbf{s}\| \le b$, $\|\mathbf{s}-\mathbf{u}\| \le \delta$.

But, by **(F1)**, $\forall\ \|\mathbf{u}\| \le b$,

$$(7.2.19)\qquad \sup_{x\in\mathbb{R}} |m^{\pm}(x;\mathbf{u},\pm\delta) - m^{\pm}(x;\mathbf{u},0)| \le \delta \|f\|_\infty n^{-1}\sum_{i=1}^n \|h_i\boldsymbol{\xi}_i\|,$$

$$(7.2.20)\qquad \sup_{\|\mathbf{s}-\mathbf{u}\|\le\delta, x\in\mathbb{R}} n^{1/2}|\nu_{\mathbf{s}}^{\pm}(x) - \nu_{\mathbf{u}}^{\pm}(x)| \le \delta \|f\|_\infty n^{-1}\sum_{i=1}^n \|h_i\boldsymbol{\xi}_i\|.$$

From (7.2.19), (7.2.18), (7.2.13) applied with $a = \delta$ and $a = -\delta$ and the assumption (7.2.6) one concludes that for every $\epsilon > 0$ there is a $\delta > 0$ such that for each $\|\mathbf{u}\| \le b$,

$$\limsup_n P\left(\sup_{\|\mathbf{s}-\mathbf{u}\|\le\delta, x} n^{1/2}|W_{\mathbf{s}}^{\pm}(x) - W_{\mathbf{u}}^{\pm}(x)| > \epsilon\right) \le \epsilon/2.$$

From this, (7.2.20), (7.2.16), and (7.2.6) one now concludes (7.2.15) in a routine fashion. Finally, (7.2.9) follows from (7.2.8) and (7.2.7) by Taylor's expansion of F. □

An application of (7.2.9) with $h(\mathbf{Y}_{i-1}) = g_j(\mathbf{Y}_{i-1})$, $1 \le j \le p$, and the rest of the quantities as in (7.2.1) readily yields the AUL property of T_j - processes, $1 \le j \le p$ of (7.1.2). This together with integration by parts yields the following expansion of the M-scores $\mathcal{G}_j, 1 \le j \le p$ of (7.1.3).

Corollary 7.2.1 *In addition to (7.1.1), (7.2.5) and (7.2.7), assume that the following conditions hold.*

$$(7.2.21)(a)\quad \left(n^{-1}\sum_{i=1}^n g_j^2(\mathbf{Y}_{i-1})\right)^{1/2} = \alpha_j,\ \alpha_j,\ 1\le j\le p,\ \textit{positive r.v.'s},$$

$$(b)\quad n^{-1/2}\max_{1\le i\le n}\|\boldsymbol{g}(\mathbf{Y}_{i-1})\| = o_p(1).$$

$$(7.2.22)\ \psi \textit{ is nondecreasing, bounded and } \int \psi dF = 0.$$

$$(7.2.23)\ n^{-1}\sum_{i=1}^n \|g_j(\mathbf{Y}_{i-1})\mathbf{Y}_{i-1}\| = O_p(1),\ 1\le j\le p.$$

Then $\forall\ 0 < k, b < \infty$,

$$\sup\Big|n^{1/2}[\mathcal{G}_j(\boldsymbol{\rho} + n^{-1/2}\mathbf{u}) - \mathcal{G}_j(\boldsymbol{\rho})] - \mathbf{u}'n^{-1}\sum_{i=1}^n g_j(\mathbf{Y}_{i-1})\mathbf{Y}_{i-1}\int f d\psi\Big| = o_p(1)$$

where the supremum is taken over all ψ *with* $\|\psi\|_{tv} \le k < \infty, \|\mathbf{u}\| \le b, 1 \le j \le p$. □

We remark here that (7.2.21) is *a priori* satisfied whenever g is bounded.

Upon choosing $h \equiv 1$ in (7.2.9) one obtains an analogous result for the *ordinary residual empirical process* $F_n(x, \mathbf{t})$. Because of its importance and for an easy reference later on we state it as a separate result. Observe that in the following corollary the assumption (7.2.24) is nothing but the assumption (7.2.6) of Theorem 7.2.1 with $h \equiv 1$.

Corollary 7.2.2 *Suppose that (7.1.1), (7.2.5), and (7.2.7) hold. In addition, assume the following.*

(7.2.24)
$$n^{-1}\sum_{i=1}^{n}\|\mathbf{Y}_{i-1}\| = O_p(1).$$

Then, for every $0 < b < \infty$,

(7.2.25)
$$\sup\Big|n^{1/2}\{F_n(x, \boldsymbol{\rho} + n^{-1/2}\mathbf{u}) - F_n(x, \boldsymbol{\rho})\} \\ -\mathbf{u}'n^{-1}\sum_{i=1}^{n}\mathbf{Y}_{i-1}f(x)\Big| = o_p(1),$$

where the supremum is taken over $x \in \mathbb{R}, \|\mathbf{u}\| \le b$. □

Observe that none of the above results require that the process $\{X_i\}$ be stationary or any of its moments be finite. □

Remark 7.2.2 *On the assumptions (7.2.3) - (7.2.6).* If $\mathbf{Y}_0$ and $\{\varepsilon_i\}$ are so chosen as to make $\{X_i\}$ stationary, ergodic and if

$$E(h^2(\mathbf{Y}_0) + \|\mathbf{Y}_0\|^2 + \varepsilon_1^2) < \infty,$$

then all of these assumptions are *a priori* satisfied. See, e.g., Anderson (1971; p. 203). In particular, they hold for any bounded h, including the one corresponding to the Huber function $h(x) \equiv |x|I(|x| \le k) + sign(x)I(|x| > k), k > 0$.

Observe that (7.2.5) is *weaker than requiring the finiteness of the second moment.* To see this, consider, for example, an AR(1) model where X_0 and $\varepsilon_1, \varepsilon_2, \cdots$ are independent r.v.'s and for some $|\rho| < 1$,

$$X_i = \rho X_{i-1} + \varepsilon_i, \quad i \ge 1.$$

Then,

$$X_i = \rho^i X_0 + \sum_{j=1}^{i}\rho^{j-i}\varepsilon_j, \quad \mathbf{Y}_i = X_i, \quad i \ge 1.$$

Thus, here (7.2.5) is implied by

$$\max_{1 \le i \le n} n^{-1/2}|\varepsilon_i| = o_p(1).$$

But, this is equivalent to showing that $x^2 \ell n\{1 - P(|\varepsilon_1| > x)\} \to 0$ as $x \to \infty$, which, in turn is equivalent to requiring that $x^2 P(|\varepsilon_1| > x) \to 0$ as $x \to \infty$. This last condition is weaker than requiring that $E|\varepsilon_1|^2 < \infty$. For example, let the right tail of the distribution of $|\varepsilon_1|$ be given as follows:

$$\begin{aligned} P(|\varepsilon_1| > x) &= 1, \quad x < 2, \\ &= 1/(x^2 \ell n x), \quad x \ge 2. \end{aligned}$$

Then, $E|\varepsilon_1| < \infty, E\varepsilon_1^2 = \infty$, yet $x^2 P(|\varepsilon_1| > x) \to 0$ as $x \to \infty$.

A similar remark applies to (7.2.4) with respect to the square integrability of $h(\mathbf{Y}_0)$. □

Remark 7.2.3 An analogue of (7.2.25) was first proved by Boldin (1982) requiring $\{X_i\}$ to be stationary, $E\varepsilon_1 = 0$, $E(\varepsilon_1^2) < \infty$ and a uniformly bounded second derivative of F. The Corollary 7.2.2 is an improvement of Boldin's result in the sense that F needs to be smooth only up to the first derivative and the r.v.'s need not have finite second moment.

Again, if $\mathbf{Y}_0$ and $\{\varepsilon_i\}$ are so chosen that the Ergodic Theorem is applicable and $E(\mathbf{Y}_0) = \mathbf{0}$, then the coefficient $n^{-1}\sum_{i=1}^n \mathbf{Y}_{i-1}$ of the linear term in (7.2.25) will converge to $\mathbf{0}$, a.s.. Thus (7.2.25) becomes

$$\text{(7.2.26)} \qquad \sup_{\|\mathbf{u}\| \le b} |n^{1/2}\{F_n(x, \boldsymbol{\rho} + n^{-1/2}\mathbf{u}) - F_n(x, \boldsymbol{\rho})\}| = o_p(1).$$

In particular, this implies that if $\hat{\boldsymbol{\rho}}$ is an estimator of $\boldsymbol{\rho}$ such that

$$\|n^{1/2}(\hat{\boldsymbol{\rho}} - \boldsymbol{\rho})\| = O_p(1),$$

then

$$\|n^{1/2}\{F_n(\cdot, \hat{\boldsymbol{\rho}}) - F_n(\cdot, \boldsymbol{\rho})\}\|_\infty = o_p(1).$$

Consequently, *the estimation of $\boldsymbol{\rho}$ has asymptotically negligible effect on the estimation of the error d.f. F.* This is similar to the fact, observed in the previous chapter, that the estimation of the slope parameters in linear regression has asymptotically negligible effect on the estimation of the error d.f. as long as the design matrix is centered at the origin. □

Serial Rank Residual Correlations. An important application of (7.2.25) occurs when proving the AUL property of the serial rank correlations of the residuals as functions of $\mathbf{t}$. More precisely, let R_{it} denote the rank of $X_i - \mathbf{t}'\mathbf{Y}_{i-1}$ among $X_j - \mathbf{t}'\mathbf{Y}_{j-1}, 1 \le j \le n, 1 \le i \le n$. Define $R_{it} = 0$ for $i \le 0$. Residual rank correlations of lag j, for $1 \le j \le p$, $\mathbf{t} \in \mathbb{R}^p$, are defined as

$$
\begin{aligned}
(7.2.27) \qquad S_j(\mathbf{t}) & \\
:= & \frac{12}{n(n^2-1)} \sum_{i=j+1}^{n} \Big(R_{i-jt} - \frac{(n+1)}{2}\Big)\Big(R_{it} - \frac{(n+1)}{2}\Big), \\
\mathbf{S}' := & (S_1, \cdots, S_p).
\end{aligned}
$$

Simple algebra shows that

$$S_j(\mathbf{t}) = a_n[L_j(\mathbf{t}) - n(n+1)^2/4] + b_{nj}(\mathbf{t}), \quad 1 \le j \le p,$$

where a_n is a nonrandom sequence not depending on $\mathbf{t}$, $|a_n| = O(1)$, and

$$b_{nj}(\mathbf{t}) := \frac{6(n+1)}{\{n(n^2-1)\}} \left(\sum_{i=n-j+1}^{n} + \sum_{i=1}^{j} \right) R_{it},$$

$$L_j(\mathbf{t}) := n^{-3} \sum_{i=j+1}^{n} R_{i-jt} R_{it}, \quad 1 \le j \le p, \ \mathbf{t} \in \mathbb{R}^p.$$

Observe that $\sup\{|b_{nj}(\mathbf{t})|; \mathbf{t} \in \mathbb{R}^p\} \le 48p/n$, so that $n^{1/2} \sup\{|b_{nj}(\mathbf{t})|; \mathbf{t} \in \mathbb{R}^p\}$ tends to zero, a.s. It thus suffices to prove the AUL of $\{L_j\}$ only, $1 \le j \le p$. In order to state the AUL result we need to introduce

$$
\begin{aligned}
(7.2.28) \qquad Z_{ij} &:= f(\varepsilon_{i-j})F(\varepsilon_i) + f(\varepsilon_i)F(\varepsilon_{i-j}), & i > j, \\
&:= 0, & i \le j,
\end{aligned}
$$

$$
\begin{aligned}
\mathbf{U}_{ij} &:= \mathbf{Y}_{i-j-1}F(\varepsilon_i)f(\varepsilon_{i-j}) + \mathbf{Y}_{i-1}f(\varepsilon_i)F(\varepsilon_{i-j}), & i > j, \\
&:= 0, & i \le j,
\end{aligned}
$$

$$\overline{Z}_j := n^{-1} \sum_{i=j+1}^{n} Z_{ij}, \quad \overline{\mathbf{U}}_j := n^{-1} \sum_{i=j+1}^{n} \mathbf{U}_{ij}, \quad 1 \le j \le p.$$

$$\overline{\mathbf{Y}}_n := n^{-1} \sum_{i=1}^{n} \mathbf{Y}_{i-1}.$$

Observe that $\{Z_{ij}\}$ are bounded r.v.'s with $EZ_{ij} = \int f^2(x)dx$ for all i and j. Moreover, $\{\varepsilon_i\}$ i.i.d. F imply that $\{Z_{ij}, j < i \le n\}$ are stationary and ergodic. By the Ergodic Theorem,

$$\overline{Z}_j \to b(f) := \int f^2(x)dx, \quad \text{a.s.}, \quad j = 1, \cdots, p.$$

We are now ready to state and prove

Theorem 7.2.2 *Assume that (7.1.1), (7.2.5), (7.2.7) and (7.2.24) hold. Then for every $0 < b < \infty$ and for every $1 \le j \le p$,*

$$\sup_{\|\mathbf{u}\| \le b} \left| n^{1/2}[L_j(\boldsymbol{\rho} + n^{-1/2}\mathbf{u}) - \mathbf{L}_j(\boldsymbol{\rho})] - \mathbf{u}'[b(f)\overline{\mathbf{Y}}_n - \overline{\mathbf{U}}_j] \right| = o_p(1). \tag{7.2.29}$$

If (7.2.5) and (7.2.24) are strengthened to requiring $E(\|\mathbf{Y}_0\|^2 + \varepsilon_1^2) < \infty$ and $\{X_i\}$ stationary and ergodic then $\overline{\mathbf{Y}}_n$ and $\overline{\mathbf{U}}_j$ may be replaced by their respective expectations in (7.2.29).

Proof. Fix a j in $1 \le j \le p$. For the sake of simplicity of the exposition, write $L(\mathbf{u}), L(\mathbf{0})$ for $L_j(\boldsymbol{\rho} + n^{-1/2}\mathbf{u}), L_j(\boldsymbol{\rho})$, respectively. Apply similar convention to other functions of $\mathbf{u}$. Also write $\varepsilon_{i\mathbf{u}}$ for $\varepsilon_i - n^{-1/2}\mathbf{u}'\mathbf{Y}_{i-1}$ and $F_n(\cdot)$ for $F_n(\cdot, \boldsymbol{\rho})$. With these conventions $R_{i\mathbf{u}}$ is now the rank of $X_i - (\boldsymbol{\rho} + n^{-1/2}\mathbf{u})'\mathbf{Y}_{i-1} = \varepsilon_{i\mathbf{u}}$. In other words, $R_{i\mathbf{u}} \equiv nF_n(\varepsilon_{i\mathbf{u}}, \mathbf{u})$ and

$$L(\mathbf{u}) = n^{-1} \sum_{i=j+1}^{n} F_n(\varepsilon_{i-j\mathbf{u}}, \mathbf{u}) F_n(\varepsilon_{i\mathbf{u}}, \mathbf{u}), \quad \mathbf{u} \in \mathbb{R}^p.$$

The proof is based on the linearity properties of $F_n(\cdot, \mathbf{u})$ as given in (7.2.25) of Corollary 7.2.2 above. In fact if we let

$$B_n(x, \mathbf{u}) := F_n(x, \mathbf{u}) - F_n(x) - n^{-1/2}\mathbf{u}'\overline{\mathbf{Y}}_n f(x), \quad x \in \mathbb{R}$$

then (7.2.25) is equivalent to

$$\sup n^{1/2} |B_n(x, \mathbf{u})| = o_p(1).$$

All supremums, unless specified otherwise, in the proof are over $x \in \mathbb{R}, 1 \le i \le n$ and / or $\|\mathbf{u}\| \le b$. Rewrite

$$\begin{aligned}
& n^{1/2}(L(\mathbf{u}) - L(\mathbf{0})) \\
&= n^{-1/2} \sum_{i=j+1}^{n} \{F_n(\varepsilon_{i-j\mathbf{u}}, \mathbf{u}) F_n(\varepsilon_{i\mathbf{u}}, \mathbf{u}) - F_n(\varepsilon_{i-j}) F_n(\varepsilon_i)\} \\
&= n^{-1/2} \sum_{i=j+1}^{n} \Big[\{B_n(\varepsilon_{i-j\mathbf{u}}, \mathbf{u}) + F_n(\varepsilon_{i-j\mathbf{u}}) + n^{-1/2}\mathbf{u}'\overline{\mathbf{Y}}_n f(\varepsilon_{i-j\mathbf{u}})\} \\
&\qquad \cdot \{B_n(\varepsilon_{i\mathbf{u}}, \mathbf{u}) + F_n(\varepsilon_{i\mathbf{u}}) + n^{-1/2}\mathbf{u}'\overline{\mathbf{Y}}_n f(\varepsilon_{i\mathbf{u}})\} \\
&\qquad - F_n(\varepsilon_{i-j}) F_n(\varepsilon_i) \Big].
\end{aligned}$$

Hence, from (7.2.5), (7.2.20) and (7.2.24),

(7.2.30)
$$n^{1/2}(L(\mathbf{u}) - L(\mathbf{0})) = n^{-1/2}\sum_{i=j+1}^{n}[F_n(\varepsilon_{i-j\mathbf{u}})F_n(\varepsilon_{i\mathbf{u}}) - F_n(\varepsilon_i)F_n(\varepsilon_{i-j})] + n^{-1}\sum_{i=j+1}^{n}[F_n(\varepsilon_{i-j\mathbf{u}})f(\varepsilon_{i\mathbf{u}}) + F_n(\varepsilon_{i\mathbf{u}})f(\varepsilon_{i-j\mathbf{u}})](\mathbf{u}'\overline{\mathbf{Y}}_n) + u_p(1),$$

where, now, $u_p(1)$ is a sequence of stochastic processes converging to zero uniformly, in probability, over the set $\{\mathbf{u} \in \mathbb{R}^p;\ \|\mathbf{u}\| \le b)$.

Now recall that (7.2.7) and the asymptotic uniform continuity of the standard empirical process based on i.i.d. r.v.'s imply that

$$\sup_{|x-y|\le\delta} n^{1/2}|[F_n(x) - F(x)] - F_n(y) - F(y)]| = o_p(1)$$

when first $n \to \infty$ and then $\delta \to 0$. Hence from (7.2.5) and the fact that

$$\sup_{i,\mathbf{u}} |\varepsilon_{i\mathbf{u}} - \varepsilon_i| \le bn^{-1/2}\max_i \|\mathbf{Y}_{i-1}\|,$$

one readily obtains

$$\sup_{i,\mathbf{u}} n^{1/2}|[F_n(\varepsilon_{i\mathbf{u}}) - F(\varepsilon_{i\mathbf{u}})] - [F_n(\varepsilon_i) - F(\varepsilon_i)]| = o_p(1).$$

From this and (7.2.7) we obtain

(7.2.31)
$$\sup_{i,\mathbf{u}} n^{1/2}|F_n(\varepsilon_{i\mathbf{u}}) - F_n(\varepsilon_i) + n^{-1/2}\mathbf{u}'\mathbf{Y}_{i-1}f(\varepsilon_i)| = o_p(1).$$

From (7.2.30), (7.2.31), the uniform continuity of f and F, the Glivenko - Cantelli lemma, one obtains

(7.2.32)
$$n^{1/2}(L(\mathbf{u}) - L(\mathbf{0})) = n^{-1}\sum_{i=j+1}^{n}[F(\varepsilon_{i-j})f(\varepsilon_i) + F(\varepsilon_i)f(\varepsilon_{i-j})](\mathbf{u}'\overline{\mathbf{Y}}_n) - \mathbf{u}'n^{-1}\sum_{i=j+1}^{n}\{\mathbf{Y}_{i-j-1}f(\varepsilon_{i-j})F(\varepsilon_i) + \mathbf{Y}_{i-1}f(\varepsilon_i)F(\varepsilon_{i-j})\} + u_p(1).$$

In concluding (7.2.32) we also used the fact that by (7.2.5) and (7.2.24),

$$\sup_{\mathbf{u}} |n^{-3/2} \sum_{i=j+1}^{n} |\mathbf{u}'\mathbf{Y}_{i-j} \cdot \mathbf{u}'\mathbf{Y}_{i-1}|$$
$$\leq bn^{-1/2} \max_i \|\mathbf{Y}_{i-1}\| n^{-1} \sum_{i=j+1}^{n} \|\mathbf{Y}_{i-j}\| = o_p(1).$$

Now (7.2.29) readily follows from (7.2.32) and the notation introduced just before the statement of the theorem. The rest is obvious. □

Remark 7.2.4 *Autoregressive moving average models.* Boldin (1989) and Kreiss (1991) give an analogue of (7.2.26) for a moving average model of order q and an autoregressive moving average model of order (p, q) (ARMA (p, q)), respectively, when the error d.f. F has zero mean, finite second moment and bounded second derivative. Here we shall illustrate as to how Theorem 2.2.3 can be used to yield the same result under weaker conditions on F. For the sake of clarity, the details are carried out for an ARMA$(1, 1)$ model only.

Let $\varepsilon_0, \varepsilon_1, \varepsilon_2, \cdots$, be i.i.d. F r.v.'s and X_0 be a r.v. independent of $\{\varepsilon_i, i \geq 1\}$. Consider the process given by the relation

$$X_i = \rho X_{i-1} + \varepsilon_i + \beta\varepsilon_{i-1}, \quad i \geq 1, \tag{7.2.33}$$

where $|\rho| < 1, |\beta| < 1$. One can rewrite this model as

$$\begin{aligned} \varepsilon_i &= X_1 - (\rho X_0 + \beta\varepsilon_0), \quad i = 1, \\ &= X_i - \sum_{j=1}^{i-1} (-\beta)^j (\rho + \beta) X_{i-j-1} \\ &\quad + (-\beta)^{i-1}(\rho X_0 + \beta\varepsilon_0), \quad i \geq 2. \end{aligned} \tag{7.2.34}$$

Let $\boldsymbol{\theta} := (s, t)'$ denote a point in the open square $(-1, 1)^2$ and $\boldsymbol{\theta}_0 := (\rho, \beta)'$ denote the true parameter value. Assume that $\boldsymbol{\theta}$'s are restricted to the following sequence of neighbourhoods: For a $b \in (0, \infty)$,

$$n^{1/2}\{|s - \rho| + |t - \beta|\} \leq b. \tag{7.2.35}$$

Let $\{\bar{\varepsilon}_i, i \geq 1\}$ stand for the residuals $\{\varepsilon_i, i \geq 1\}$ of (7.2.34) after ρ and β are replaced by s and t, respectively, in (7.2.34). Let $F_n(\cdot, \boldsymbol{\theta})$ denote the empirical process of $\{\bar{\varepsilon}_i, 1 \leq i \leq n\}$. This empirical can be rewritten as

$$F_n(x, \boldsymbol{\theta}) = n^{-1} \sum_{i=1}^{n} I(\varepsilon_i \leq x + \delta_{ni}), \quad x \in \mathbb{R}, \tag{7.2.36}$$

where

$$(7.2.37)\quad \begin{aligned}\delta_{ni} &:= (s-\rho)X_0 + (t-b)\varepsilon_0, \quad i=1,\\ &= \sum_{j=1}^{i-2}\Big[(-t)^j(s+t) - (-\beta)^j(\rho+\beta)\Big]X_{i-j-1}\\ &\quad +(-t)^{i-1}(sX_0+t\varepsilon_0) - (-\beta)^{i-1}\Big](\rho X_0+\beta\varepsilon_0),\ i\ge 2.\\ &= \delta_{ni1}+\delta_{ni2},\ \text{say},\ i\ge 2.\end{aligned}$$

From (7.2.37), it follows that for every $\boldsymbol{\theta} \in (-1,1)^2$ satisfying (7.2.35),

$$\begin{aligned}|\delta_{n1}| &\le bn^{-1/2}(|X_0|+|\varepsilon_0|),\\ \max_{2\le i\le n}|\delta_{ni1}| &\le 2b\,n^{-1/2}\max_{1\le i\le n}|X_i|(1-bn^{-1/2}-\beta)^{-1}\\ &\qquad\qquad \times\{1+(1-|\beta|)^{-1}\},\\ \max_{2\le i\le n}|\delta_{ni2}| &\le 2bn^{-1/2}(1-bn^{-1/2}-\beta)^{-1}(|X_0|+|\varepsilon_0|).\end{aligned}$$

Consequently, if $n^{-1/2}\max_{1\le i\le n}|X_i| = o_p(1)$, then the $\{\delta_{ni}\}$ of (7.2.37) would satisfy (2.2.28) for every $\boldsymbol{\theta} \in (-1,1)^2$. But by (7.2.33),

$$(7.2.38)\quad \begin{aligned}X_i &= \rho X_0 + \beta\varepsilon_0 + \varepsilon_1,\ i=1,\\ &= \rho^{i-1}(\rho X_0+\beta\varepsilon_0) + \sum_{j=1}^{i-2}\rho^j(\rho+\beta)\varepsilon_{i-j-1} + \varepsilon_i,\ i\ge 2.\end{aligned}$$

Therefore, (2.2.28) will hold for the above $\{\delta_{ni}\}$ if

$$(7.2.39)\quad n^{-1/2}\max_{i\le i\le n}|\varepsilon_i| = o_p(1).$$

We now verify (2.2.30) for the above $\{\delta_{ni}\}$ and with $h_{ni}\equiv 1$. That is we must show that $n^{-1/2}\sum_{i=1}^n|\delta_{ni}| = O_p(1)$. We proceed as follows. Let $u = n^{1/2}(s-\rho)$, $v = n^{1/2}(t-\beta)$ and $Z_0 := |X_0|+|\varepsilon_0|$. By (7.2.35),

$|u| + |v| \le b$. From (7.2.37),

$$
\begin{aligned}
n^{-1/2} \sum_{i=1}^{n} |\delta_{ni}| & \\
\le\ & n^{-1} b Z_0 \\
& + n^{-1/2} \sum_{i=2}^{n} \Big| \sum_{j=0}^{i-2} [(-t)^j (s+t) - (-\beta)^j (\rho + \beta)] X_{i-j-1} \Big| \\
& + n^{-1/2} \sum_{i=2}^{n} |(-t)^{i-1}(sX_0 + t\varepsilon_0) - (-\beta)^{i-1} Z_0|, \\
=\ & A_{n1} + A_{n2} + A_{n3}, \quad \text{say.} \qquad (7.2.40)
\end{aligned}
$$

Clearly, $|A_{n1}| = o(1)$, a.s. Rewrite

$$
\begin{aligned}
A_{n2} =\ & n^{-1/2} \sum_{i=2}^{n} \Big| \sum_{j=0}^{i-2} \Big[(-t)^j \{ (u+v) n^{-1/2} + \rho + \beta \} \\
& \qquad - (-\beta)^j (\rho + \beta) \Big] X_{i-j-1} \Big| \\
\le\ & 2bn^{-1} \sum_{i=2}^{n} \sum_{j=0}^{i-2} |t|^j |X_{i-j-1}| \\
& + 2n^{-1/2} \sum_{i=2}^{n} \Big| \sum_{j=0}^{i-2} [(-t)^j - (-\beta)^j X_{i-j-1} \Big| \\
=\ & 2b\, A_{n21} + 2A_{n22}, \quad \text{say.} \qquad (7.2.41)
\end{aligned}
$$

By a change of variables and an interchange of summations one obtains

$$
A_{n21} \le n^{-1} \sum_{i=1}^{n} |X_i| (1 - |t|)^{-1}. \qquad (7.2.42)
$$

Next, use the expansion $a^j - c^j = (a - c) \sum_{k=0}^{j-1} a^{j-1-k} c^k$ for any real numbers a, c to obtain

$$
A_{n22} \le bn^{-1} \sum_{i=3}^{n} \sum_{j=1}^{i-2} \sum_{k=0}^{j-1} |t|^{j-1-k} |\beta|^k |X_{i-j-1}|.
$$

Again, use change of variables and interchange of summations repeatedly and the fact that $|\beta| \vee |t| < 1$, to conclude that this upper bound is bounded above by

$$
b(1 - |\beta|)^{-1} [(1 - |t|)^{-1} + 1] n^{-1} \sum_{i=1}^{n} |X_i|.
$$

This, (7.2.40), (7.2.41) and (7.2.42) together with (7.2.35) imply that

(7.2.43) $$A_{n2} \le 2bn^{-1}\sum_{i=1}^{n}|X_i|[(1-bn^{-1/2}-|\beta|)^{-1} \times\{1+(1-|\beta|)^{-1}\}+(1-|\beta|)^{-1}]$$

Finally, similar calculations show that

(7.2.44) $$A_{n3} = O_p(n^{-1/2}).$$

From (7.2.41), (7.2.44) and (7.2.45) it thus follows that if

$$n^{-1}\sum_{i=1}^{n}|X_i| = O_p(1),$$

then the $\{\delta_{ni}\}$ of (7.2.37) will satisfy (2.2.30) with $h_{ni} \equiv 1$. But in view of (7.2.38) and the assumption that $|\rho| \vee |\beta| < 1$, it readily follows that if

(7.2.45) $$n^{-1}\sum_{i=1}^{n}|\varepsilon_i| = O_p(1),$$

then (2.2.30) with $h_{ni} \equiv 1$ holds for the $\{\delta_{ni}\}$ of (7.2.37). We have thus proved the following:

If (7.2.33) holds with the error d.f. F satisfying (**F1**), (**F2**), (7.2.39) and (7.2.45), then $\forall\, \boldsymbol{\theta} \in (-1,1)^2$,

$$\sup_x |n^{-1/2}\sum_{i=1}^{n}\{I(\tilde{\varepsilon}_i \le x) - I(\varepsilon_i \le x) - F(x+\delta_{ni}) + F(x)\}| = o_p(1).$$

Now use an argument like the one used in the proof of Theorem 7.2.1 to conclude the following

Corollary 7.2.3 *In addition to (7.2.33), assume that the error d.f. F satisfies* (**F1**), (**F2**), *(7.2.39) and (7.2.45). Then, $\forall\, 0 < b < \infty$,*

$$\sup |n^{1/2}[F_n(x,\boldsymbol{\theta}) - F_n(x,\boldsymbol{\theta}_0)] - n^{-1/2}\sum_{i=1}^{n}\delta_{ni}f(x)| = o_p(1),$$

where the supremum is taken over $x \in \mathbb{R}$ and $\boldsymbol{\theta}$, $\boldsymbol{\theta}_0$ satisfying (7.2.35).

If (7.2.45) is strengthened to assuming that $E|\varepsilon| < \infty$, then

$$\sup \left| n^{-1/2}\sum_{i=1}^{n}\delta_{ni} - n^{1/2}\left[\frac{(s-\rho)}{(1-\rho)} + \frac{(t-\beta)}{(1+\beta)}\right]\mu\right| = o_p(1),$$

where the supremum is taken over s, t satisfying (7.2.35) and $\mu = E\varepsilon$. □

Consequently, if $E\varepsilon = 0$ and $(\hat{\rho}, \hat{\beta})$ is an estimator of (ρ, β) such that $\|n^{1/2}(\hat{\rho} - \rho, \hat{\beta} - \beta)\| = O_p(1)$, then an analogue of (7.2.26) holds in the present case also under weaker conditions than those given by Boldin or Kreiss.

The details for proving an analogue of Corollary 7.2.3 for a general ARMA (p, q) model are similar but some what more complicated than those given above.

Clearly, an analogue of Theorem 7.2.2 also remains trues here. Ferretti, Klemansky and Yohai (1991) used this result to study the asymptotic distribution of rank estimators obtained by solving Yule-Walker type equations using rank auto-covariances of the residuals in ARMA models. □

7.3 GM- and GR- Estimators

In this section we shall discuss the asymptotic distributions of GM- and R-estimators of $\boldsymbol{\rho}$. In addition, some consistent estimators of the functional $Q(f)$ will be also constructed. We begin with

7.3.1 GM-estimators

Here we shall state the asymptotic normality of the GM - estimators. Let $\boldsymbol{\rho}_M$ stand for a solution of (7.1.3) such that $\|n^{1/2}(\boldsymbol{\rho}_M - \boldsymbol{\rho})\| = O_p(1)$. That such an estimator $\boldsymbol{\rho}_M$ exists can be seen by an argument similar to the one given in Huber (1981) in connection with the linear regression model. To state the asymptotic normality of $\boldsymbol{\rho}_M$ we need to introduce some more notation. Let

$$(7.3.1) \qquad \boldsymbol{\mathcal{X}} := \begin{bmatrix} X_0 & X_{-1}, & \cdots, & X_{1-p} \\ X_1 & X_0, & \cdots, & X_{2-p} \\ \vdots & \vdots & & \vdots \\ X_{n-1} & X_{n-2} & \cdots, & X_{n-p} \end{bmatrix},$$

$$\mathbf{G} := \begin{bmatrix} g_1(\mathbf{Y}_0) & g_2(\mathbf{Y}_{-1}), & \cdots, & g_p(\mathbf{Y}_{1-p}) \\ g_1(\mathbf{Y}_1) & g_2(\mathbf{Y}_0), & \cdots, & g_p(\mathbf{Y}_{2-p}) \\ \vdots & \vdots & & \vdots \\ g_1(\mathbf{Y}_{n-1}) & g_2(\mathbf{Y}_{n-2}), & \cdots, & g_p(\mathbf{Y}_{n-p}) \end{bmatrix},$$

$$\mathcal{B}_n \;:=\; g'\mathcal{X} = \sum_{i=1}^{n}(g_1(\mathbf{Y}_{i-1})\mathbf{Y}'_{i-1}\cdots, g_p(\mathbf{Y}_{i-p})\mathbf{Y}'_{i-1})'.$$

Proposition 7.3.1 *In addition to (7.1.1), (7.2.5), (7.2.7), (7.2.21), (7.2.22) and (7.2.23) assume that*

(7.3.2) $\quad n^{-1}\mathcal{B}_n = \mathcal{B} + o_p(1),$ *for some $p\times p$ non-random positive definition matrix $\mathcal{B}$.*

Then

$$n^{1/2}(\boldsymbol{\rho}_M - \boldsymbol{\rho}) = -(\mathcal{B}a)^{-1}n^{1/2}\mathcal{G}(\boldsymbol{\rho}) + o_p(1).$$

If, in addition, we assumes that

(7.3.3) $\quad n^{-1}\mathbf{G}'\mathbf{G} = \mathbf{G}^* + o_p(1),\;\; \mathbf{G}^*$ *a $p\times p$ non-random positive definite matrix,*

then

(7.3.4)
$$n^{1/2}(\boldsymbol{\rho}_M - \boldsymbol{\rho}) \longrightarrow_d \mathcal{N}(0, v(\psi, F)\mathbf{J}),$$
$$\mathbf{J} := \mathcal{B}^{-1}\mathbf{G}^*\mathcal{B}^{-1}, \quad v(\psi, F) := (\int f d\psi)^{-2}(\int \psi^2 dF).$$

Proof. Follows from Corollary 7.2.1, the Cramér - Wold device and Lemma 9.1.3 in the Appendix applied to $n^{1/2}\mathcal{G}(\boldsymbol{\rho})$. □

Again, if $\mathbf{Y}_0$ and $\{\varepsilon_i\}$ are so chosen as to make $\{X_i\}$ stationary, ergodic and $E(\|\mathbf{Y}_0\|^2 + \varepsilon_1^2) < \infty$ then (7.3.2) and (7.3.3) are *a priori* satisfied. See, e.g., Anderson (1971; p. 203).

Note : For a more general class of GM-estimators see Bustos (1982) where a result analogous to the above corollary for smooth score functions ψ is obtained.

7.3.2 GR-estimators

This section will discuss some extensions of the analogues of Jaeckel's (1972) R- estimators and generalize R- estimators of $\boldsymbol{\rho}$ and their asymptotic distributions.

Let $\xi_i(\mathbf{t}) := X_i - \mathbf{t}'\mathbf{Y}_{i-1},\ 1 \le i \le n$. Recall that $R_{i\mathbf{t}}$ is the rank of $\xi_i(\mathbf{t})$ among $\{\xi_k(\mathbf{t}),\ 1 \le k \le n\}$, for $1 \le i \le n$. Also, $R_{i\mathbf{t}} \equiv 0$ for $i \le 0$. Let φ be a nondecreasing score function from $[0, 1]$ to the real line such that

(7.3.5)
$$\sum_{i=1}^{n}\varphi(i/(n+1)) = 0.$$

For example, if $\varphi(t) = -\varphi(1-t)$ for all $t \in [0,1]$, i.e., if φ is skew symmetric, then it satisfies (7.3.5). Let $\boldsymbol{g} := (g_1, \cdots, g_2)'$ be a p-vector of measurable functions from $\mathbb{R}^p$ to $\mathbb{R}$ and define

$$\begin{aligned}
\mathbf{S}_{\boldsymbol{g}}(\mathbf{t}) &:= n^{-1}\sum_{i=1}^{n} \boldsymbol{g}(\mathbf{Y}_{i-1})\varphi(R_{i\mathbf{t}}/(n+1)),\\
\mathbf{Z}_{\boldsymbol{g}}(u,\mathbf{t}) &:= n^{-1}\sum_{i=1}^{n} \boldsymbol{g}(\mathbf{Y}_{i-1})I(R_{i\mathbf{t}} \le nu),\\
\mathcal{Z}_{\boldsymbol{g}}(u,\mathbf{t}) &:= \mathbf{Z}(u,\mathbf{t}) - \bar{\boldsymbol{g}}\,u, \quad \mathbf{t} \in \mathbb{R}^p,\, 0 \le u \le 1.
\end{aligned}$$

Write $\mathbf{S}$, $\mathbf{Z}$ for $\mathbf{S}_{\boldsymbol{g}}$, $\mathbf{Z}_{\boldsymbol{g}}$ whenever $\boldsymbol{g}(\mathbf{y}) \equiv \mathbf{y}$. The class of rank statistics $\mathbf{S}_{\boldsymbol{g}}$, one for each φ, is an analogue of the class of rank statistics T_d of (3.1.2) discussed in Chapter 3 in connection with linear regression models. A test of the hypothesis $\boldsymbol{\rho} = \boldsymbol{\rho}_0$ may be based on a suitably standardized $\mathbf{S}(\boldsymbol{\rho}_0)$, the large values of the statistic being significant.

We now give an analogue of the Jaeckel estimator for the AR(p) model. Accordingly, for a $\mathbf{t} \in \mathbb{R}^p$, let $\xi_{(i)}(\mathbf{t})$ denote the the ith largest residual among $\{\xi_k(\mathbf{t}),\, 1 \le k \le n\}$, $1 \le i \le n$, and set

$$\mathcal{J}(\mathbf{t}) \;:=\; \sum_{i=1}^{n} \varphi(i/(n+1))\xi_{(i)}(\mathbf{t}) \equiv \sum_{i=1}^{n} \varphi(R_{i\mathbf{t}}/(n+1))\xi_i(\mathbf{t}).$$

Then Jaeckel's estimator $\tilde{\boldsymbol{\rho}}_J$ is defined by the relation

$$\tilde{\boldsymbol{\rho}}_J = \operatorname{argmin}\{\mathcal{J}(\mathbf{t});\, \mathbf{t} \in \mathbb{R}^p\}.$$

Jaeckel's argument about the existence of an analogue of $\tilde{\boldsymbol{\rho}}_J$ in the context of linear regression model can be adapted to the present situation. This follows from the following *two* lemmas, the first of which is of a general interest.

Lemma 7.3.1 *Let $d_1, d_2, \cdots, d_n, v_1, v_2, \cdots, v_n$ be real numbers such that not all $\{d_i\}$ are the same and no two $\{v_i\}$ are the same. Let r_{iu} denote the rank of $v_i - ud_i$ among $\{v_j - ud_j; 1 \le j \le n\}, u \in \mathbb{R}$. Let $\{b_n(i); 1 \le i \le n\}$ be a set of real numbers that are nondecreasing in i. Let*

$$T(u) := \sum_{i=1}^{n} d_i b_n(r_{iu}), \quad u \in \mathbb{R}.$$

Then, $T(u)$ is a nondecreasing step function in all those $u \in \mathbb{R}$ for which there are no ties among $\{v_j - ud_j; 1 \le j \le n\}$.

Proof. See Theorem II.7E, p. 35 of Hájek (1969). □

Lemma 7.3.2 *Assume that the model (7.1.1) holds with* $\mathbf{Y}_0', X_1, X_2, \cdots, X_n$ *having a continuous joint distribution. Then the following hold.*

(a) For each realization $(\mathbf{Y}_0', X_1, \cdots, X_n)$*, the assumption (7.3.5) implies that* $\mathcal{J}(\mathbf{t})$ *is nonnegative, continuous and convex function of* $\mathbf{t}$ *with its a.e. derivative equal to* $-n\mathbf{S}(\mathbf{t})$.

(b) If the realization $(\mathbf{Y}_0', X_1, X_2, \cdots, X_n)$ *is such that the rank of* $\mathcal{X}_c$ *is* p *then, for every* $0 < b < \infty$*, the set* $\{\mathbf{t} \in \mathbb{R}^p; \mathcal{J}(\mathbf{t}) \le b\}$ *is bounded, where* $\mathcal{X}_c$ *is the* $\mathcal{X}$ *of (7.3.1), centered at the origin.*

Proof. (a) For any $\mathbf{x}' = (x_1, x_2, \cdots, x_n) \in \mathbb{R}^n$, let $x(1) \le x(2) \le \cdots \le x(n)$ denote the ordered $x_1, x_2, \cdots, x_n$. Let $\mathbf{\Pi} := \{\boldsymbol{\pi} = (\pi_1, \pi_2, \cdots, \pi_n)'; \boldsymbol{\pi}$ a permutation of the integers $1, 2, \cdots, n\}$, $b_n(i) := \varphi(i/(n+1)), 1 \le i \le n$, and define

$$\begin{aligned} D(\mathbf{x}) &:= \sum_{i=1}^n b_n(i) x(i), \quad D_{\boldsymbol{\pi}}(\mathbf{x}) := \sum_{i=1}^n b_n(i) x_{\pi_i}, \quad \mathbf{x} \in \mathbb{R}^n, \\ k &:= min\{1 \le j \le n;\, b_n(j) > 0\}. \end{aligned}$$

Observe that $\mathcal{J}(\mathbf{t}) = D(\boldsymbol{\xi}(\mathbf{t}))$, where $\boldsymbol{\xi}(\mathbf{t}) := (\xi_1(\mathbf{t}), \cdots, \xi_n(\mathbf{t}))'$.

Now, (7.3.5) and φ nondecreasing implies that

$$\begin{aligned} D(\mathbf{x}) &= \sum_{i=1}^n b_n(i)(x(i) - x(k)) \\ &= \sum_{i=1}^{k-1} b_n(i)(x(i) - x(k)) + \sum_{i=k}^n b_n(i)(x(i) - x(k)) \\ &\ge 0, \quad \forall\ \mathbf{x} \in \mathbb{R}^n, \end{aligned}$$

because each summand is nonnegative. This proves that $\mathcal{J}(\mathbf{t}) \ge 0$, $\mathbf{t} \in \mathbb{R}$.

By Theorem 368 of hardy, Littlewood and Polya (1952),

$$D(\mathbf{x}) = \max_{\pi \in \Pi} D_\pi(\mathbf{x}), \quad \forall\ \mathbf{x} \in \mathbb{R}^n.$$

Therefore, $\forall\ \mathbf{t} \in \mathbb{R}^p$,

$$(7.3.6) \qquad \begin{aligned} \mathcal{J}(\mathbf{t}) &= D(\boldsymbol{\xi}(\mathbf{t})) = \max_{\boldsymbol{\pi} \in \mathbf{\Pi}} D_{\boldsymbol{\pi}}(\boldsymbol{\xi}(\mathbf{t})) \\ &= \max_{\boldsymbol{\pi} \in \mathbf{\Pi}} \sum_{i=1}^n b_n(i)(X_{\pi_i} - t'\mathbf{Y}_{\pi_{i-1}}). \end{aligned}$$

This shows that $\mathcal{J}(\mathbf{t})$ is a maximal element of a finite number of continuous and convex functions, which itself is continuous and convex. The statement about a.e. differential being $-nS(\mathbf{t})$ is obvious. This completes the proof of (a).

(b) Without the loss of generality assume $b > \mathcal{J}(\mathbf{0})$. Write a $\mathbf{t} \in \mathbb{R}^p$ as $\mathbf{t} = s\mathbf{e}, s \in \mathbb{R}, \mathbf{e} \in \mathbb{R}^p$, $\|\mathbf{e}\| = 1$. Let $d_i \equiv \mathbf{e}'\mathbf{Y}_{i-1}$. The assumptions about $\mathcal{J}$ imply that not all $\{d_i\}$ are equal. Rewrite

$$\begin{aligned}\mathcal{J}(\mathbf{t}) &= \mathcal{J}(s\mathbf{e}) = \sum_{i=1}^{n} b_n(i)(X - sd)(i) \\ &= \sum_{i=1}^{n} b_n(r_{is})(X_i - sd_i)\end{aligned}$$

where now r_{is} is the rank of $X_i - sd_i$ among $\{X_j - sd_j; 1 \le j \le n\}$. From (7.3.6) it follows that $\mathcal{J}(s\mathbf{e})$ is linear and convex in s, for every $\mathbf{e} \in \mathbb{R}^p, \|\mathbf{e}\| = 1$. Its a.e. derivative w.r.t. s is $-\sum_{i=1}^{n} d_i b_n(r_{is})$, which by Lemma 7.3.1 and because of the assumed continuity, is nondecreasing in u and eventually positive. Hence $\mathcal{J}(s\mathbf{e})$ will eventually exceed b, for every $\mathbf{e} \in \mathbb{R}^p$, $\|\mathbf{e}\| = 1$.

Thus, there exists a $s_{\mathbf{e}}$ such that $\mathcal{J}(s_{\mathbf{e}}\mathbf{e}) > b$. Since $\mathcal{J}$ is continuous, there is an open set $O_{\mathbf{e}}$ of unit vectors $\boldsymbol{\nu}$, containing $\mathbf{e}$ such that $\mathcal{J}(s_{\mathbf{e}}\boldsymbol{\nu}) > b$. Since $b > \mathcal{J}(0)$, and $\mathcal{J}$ is convex, $J(s\boldsymbol{\nu}) > b, \forall s \ge s_{\mathbf{e}}$ and $\forall\ \boldsymbol{\nu} \in O_{\mathbf{e}}$. Now, for each unit vector $\mathbf{e}$, there is an open set $O_{\mathbf{e}}$ covering it. Since the unit sphere is compact, a finite number of these sets covers it. Let m be the maximum of the corresponding finite set of $s_{\mathbf{e}}$. Then for all $s \ge m$, for all unit vectors $\boldsymbol{\nu}, \mathcal{J}(s\boldsymbol{\nu}) > b$. This proves the claim (b), hence the lemma. □

Note : Lemma 7.3.2 *and its proof is an adaptation of Theorems 1 and 2 of Jaeckel* (1972) *to the present case.* □

From the above lemma it follows that if the r.v.'s $\mathbf{Y}_0, X_1, X_2, \cdots, X_n$ are continuous and the matrix $n^{-1}\sum_{i=1}^{n}(\mathbf{Y}_{i-1} - \overline{\mathbf{Y}})(\mathbf{Y}_{i-1} - \overline{\mathbf{Y}})'$ is a.s. positive definite, then the rank of $\boldsymbol{\mathcal{X}}_c$ is a.s. p and the set $\{\mathbf{t} \in \mathbb{R}^p; \mathcal{J}(\mathbf{t}) \le b\}$ is a.s. bounded for every $0 \le b < \infty$. Thus a minimizer $\tilde{\boldsymbol{\rho}}_J$ of $\mathcal{J}$ exists, a.s., and has the property that it makes $\|\mathbf{S}\|$ small. As is shown in Jaeckel (1972) in connection with the linear regression model, it follows from the AUL result given in Theorem 7.3.1 below that $\tilde{\boldsymbol{\rho}}_J$ and $\tilde{\boldsymbol{\rho}}_R$ are asymptotically equivalent. Note that the score function φ need not satisfy (7.3.5) in this theorem.

Unlike in the regression model (1.1.1), these estimators are not robust against outliers in the errors because the weights in the scores $\mathbf{S}$ are now unbounded functions of the errors. Akin to GM- estimators, we thus define GR- estimators as

$$\rho_{Rg} := \text{argmin}_{\mathbf{t}} \|\mathbf{S}_{\boldsymbol{g}}(\mathbf{t})\|^2. \tag{7.3.7}$$

Strictly speaking these estimators are not determined only by the residual ranks, as here the weights in $\mathbf{S}_{\boldsymbol{g}}$ involve the observations also. But we borrow this terminology from linear regression setup.

For the convenience of the statement of the assumptions and results, from now onwards we shall assume that the observed time series comes from the following model.

$$X_i = \rho_1 X_{i-1} + \rho_2 X_{i-2} + \cdots + \rho_p X_{i-p} + \varepsilon_i, \qquad i = 0, \pm 1, \pm 2, \cdots, \boldsymbol{\rho} \in \mathbb{R}^p, \tag{7.3.8}$$

with all roots of the equation

$$x^p - \rho_1 x^{p-1} - \rho_2 x^{p-2} - \cdots - \rho_p = 0 \tag{7.3.9}$$

inside the interval $(-1, 1)$, where $\{\varepsilon_i, i = 0, \pm 1, \pm 2, \cdots\}$ are i.i.d. F r.v.'s, with

$$E\varepsilon = 0, \quad E\varepsilon^2 < \infty. \tag{7.3.10}$$

It is well known that such a time series admits the representation

$$X_i = \sum_{k \le i} \theta_{i-k}\varepsilon_k, \quad i = 0, \pm 1, \pm 2, \cdots, \quad \text{in } L_2 \text{ and a.s.,} \tag{7.3.11}$$

where the constants $\{\theta_j, j \ge 0\}$ are such that $\theta_0 = 1$, $\sum_{j \ge 0} |\theta_j| < \infty$, and where the unspecified lower limit on the index of summation is $-\infty$. See, e.g., Anderson (1971) and Brockwell and Davis (1987, pp. 76-86). Thus $\{X_i\}$ is stationary, ergodic and $E\|\mathbf{Y}_0\|^2 < \infty$. Hence (7.2.3) implies (7.2.6). Moreover, the stationarity of $\{\mathbf{Y}_{i-1}\}$ and $E\|\mathbf{Y}_0\|^2 < \infty$ imply that $\forall\, \eta > 0$,

$$\begin{aligned} P(\max_{1 \le i \le n} \|\mathbf{Y}_{i-1}\| \ge \eta n^{1/2}) &\le \{\eta n^{1/2}\}^{-2} \sum_{i=1}^{n} E\|\mathbf{Y}_{i-1}\|^2 I(\|\mathbf{Y}_{i-1}\| \ge \eta n^{1/2}) \\ &= \eta^{-2} E\|\mathbf{Y}\|^2 I(\|\mathbf{Y}_0\| \ge \eta n^{1/2}) = o(1). \end{aligned} \tag{7.3.12}$$

Thus (7.2.5) holds. By the same reason, the square integrability of $\boldsymbol{g}(\mathbf{Y}_0)$ will imply

$$\text{(7.3.13)} \qquad \max_i n^{-1/2}\|\boldsymbol{g}(\mathbf{Y}_{i-1})\| = o_p(1),$$

$$n^{-1}\sum_{i=1}^n |\boldsymbol{g}'(\mathbf{Y}_{i-1})\mathbf{Y}_{i-1}| = O_p(1),$$

$$\boldsymbol{\Gamma}_{\boldsymbol{g}} := \text{plim}_n n^{-1}\sum_{i=1}^n (\boldsymbol{g}(\mathbf{Y}_{i-1}) - \bar{\boldsymbol{g}})(\boldsymbol{g}(\mathbf{Y}_{i-1}) - \bar{\boldsymbol{g}})' \text{ exists},$$

$$\boldsymbol{\Sigma}_{\boldsymbol{g}} := \text{plim}_n n^{-1}\sum_{i=1}^n (\boldsymbol{g}(\mathbf{Y}_{i-1}) - \bar{\boldsymbol{g}})(\mathbf{Y}_{i-1} - \bar{\mathbf{Y}})' \text{ exists}.$$

These observations are frequently used in the proof of Theorem 7.3.1 below, without mentioning. Let

$$\widehat{\boldsymbol{\mathcal{Z}}}_{\boldsymbol{g}}(u) := n^{-1}\sum_{i=1}^n (\boldsymbol{g}(\mathbf{Y}_{i-1}) - \bar{\boldsymbol{g}})[I(F(\varepsilon_i) \le u) - u], \quad 0 \le u \le 1,$$

$$\widehat{\mathbf{S}}_{\boldsymbol{g}} := n^{-1}\sum_{i=1}^n (\boldsymbol{g}(\mathbf{Y}_{i-1}) - \bar{\boldsymbol{g}})\,[\varphi(F(\varepsilon_i)) - \bar{\varphi}], \quad \bar{\varphi} = \int_0^1 \varphi(u)du,$$

$$q(u) := f(F^{-1}(u)), \ 0 \le u \le 1, \quad Q := \int f d\varphi.$$

With this preliminary background, we now state

Theorem 7.3.1 *(AUL of R-statistics). Assume that (7.3.8), (7.3.9) and (7.3.10) above hold. In addition, assume that F satisfies* (**F1**)*,* (**F2**)*, $E\|\boldsymbol{g}(\mathbf{Y}_0)\|^2 < \infty$, and that $\boldsymbol{\Sigma}_{\boldsymbol{g}}$, $\boldsymbol{\Gamma}_{\boldsymbol{g}}$ of (7.3.13) are positive definite. Then, for every $0 < b < \infty$,*

$$\text{(7.3.14)} \qquad \sup_{0\le u\le 1, \|\mathbf{s}\|\le b} \|n^{1/2}[\boldsymbol{\mathcal{Z}}_{\boldsymbol{g}}(u, \boldsymbol{\rho} + n^{-1/2}\mathbf{s}) - \widehat{\boldsymbol{\mathcal{Z}}}_{\boldsymbol{g}}(u)] - \boldsymbol{\Sigma}_{\boldsymbol{g}}\mathbf{s}\, q(u)\| = o_p(1).$$

$$\text{(7.3.15)} \qquad \sup_{\varphi\in\mathcal{C}, \|\mathbf{s}\|\le b} \|n^{1/2}\{\mathbf{S}(\boldsymbol{\rho} + n^{-1/2}\mathbf{s}) - \widehat{\mathbf{S}}_{\boldsymbol{g}}\} + \boldsymbol{\Sigma}_{\boldsymbol{g}}'\mathbf{s}\, Q\| = o_p(1).$$

Proof. Integration by parts shows that

$$\mathbf{S}_{\boldsymbol{g}}(\mathbf{t}) = \bar{\boldsymbol{g}}\bar{\varphi} - \int_0^1 \boldsymbol{\mathcal{Z}}_{\boldsymbol{g}}(u(n+1)/n, \mathbf{t})\varphi(du).$$

This and (7.3.14) implies (7.3.15) trivially. The proof of the claim (7.3.14) is similar to that of Theorem 3.2.2. One uses Theorem 7.2.1 wherever (2.3.25) is used in there. We give details in brief. Drop the suffix $\boldsymbol{g}$ in these details.

Recall the definition of T_j, $1 \le j \le p$, and $\mathbf{T}$ from (7.1.2), and that of F_n from (7.1.5). Let, for $0 \le u \le 1$, $\mathbf{s}, \mathbf{t} \in \mathbb{R}^p$,

$$\begin{aligned}
\tilde{\mathbf{T}}(u,\mathbf{t}) &:= \mathbf{T}(F^{-1}(u),\mathbf{t}) = n^{-1}\sum_{i=1}^{n} \boldsymbol{g}(\mathbf{Y}_{i-1})I(\xi_i(\mathbf{t}) \le F^{-1}(u)) \\
&= n^{-1}\sum_{i=1}^{n} \boldsymbol{g}(\mathbf{Y}_{i-1})I(\varepsilon_i \le F^{-1}(u) + (\mathbf{t}-\boldsymbol{\rho})'\mathbf{Y}_{i-1}) \\
F_{n\mathbf{s}}^{-1}(u) &:= \inf\{x;\ F_n(x,\ \boldsymbol{\rho}+n^{-1/2}\mathbf{s}) \ge u\}, \\
\tilde{\mathcal{Z}}(u,\mathbf{s}) &:= n^{-1}\sum_{i=1}^{n} \boldsymbol{g}(\mathbf{Y}_{i-1})I(\varepsilon_i \le F_{n\mathbf{s}}^{-1}(u) + n^{-1/2}\mathbf{s}'\mathbf{Y}_{i-1}) \\
&= \tilde{\mathbf{T}}(FF_{n\mathbf{s}}^{-1}(u), \boldsymbol{\rho}+n^{-1/2}\mathbf{s}).
\end{aligned}$$

Note also that

$$\begin{aligned}
&\sup_{0\le u\le 1, \|\mathbf{s}\|\le b} \|\mathcal{Z}_{\boldsymbol{g}}(u,\boldsymbol{\rho}+n^{-1/2}\mathbf{s}) - \tilde{\mathcal{Z}}(u,\mathbf{s})\| \\
&\le\ 2\max_i n^{-1/2}\|\boldsymbol{g}(\mathbf{Y}_{i-1})\| = o_p(1).
\end{aligned}$$

Thus it suffices to prove (7.3.14) with $\mathcal{Z}$ replaced by $\tilde{\mathcal{Z}}$. But this follows from the fact $\tilde{\mathcal{Z}}(u,\mathbf{s}) \equiv \tilde{\mathbf{T}}(FF_{n\mathbf{s}}^{-1}(u), \boldsymbol{\rho}+n^{-1/2}\mathbf{s})$ and Theorem 7.2.1 applied p times, j^{th} time to $h(\mathbf{Y}_{i-1}) = g_j(\mathbf{Y}_{i-1})$ and from Corollary 7.2.2, in a fashion similar to that in the proof of Theorem 3.2.1. □

The next result gives the asymptotic normality of the estimator $\hat{\boldsymbol{\rho}}_{R\boldsymbol{g}}$.

Theorem 7.3.2 *In addition to the assumptions of Theorem 7.3.1, suppose the following hold:* $\forall\, \mathbf{e} \in \mathbb{R}^p$, $\|\mathbf{e}\| = 1$, $1 \le i \le n$,

$$\begin{aligned}
(7.3.16)\qquad &\textit{Either}\ \ \mathbf{e}'(\boldsymbol{g}(\mathbf{Y}_{i-1}) - \bar{\boldsymbol{g}})(\mathbf{Y}_{i-1} - \bar{\mathbf{Y}})'\mathbf{e} \ \ge\ 0, \\
&\textit{Or}\ \ \mathbf{e}'(\boldsymbol{g}(\mathbf{Y}_{i-1}) - \bar{\boldsymbol{g}})(\mathbf{Y}_{i-1} - \bar{\mathbf{Y}})'\mathbf{e} \ \le\ 0.
\end{aligned}$$

Then,

$$n^{1/2}(\hat{\boldsymbol{\rho}}_{R\boldsymbol{g}} - \boldsymbol{\rho}) \Longrightarrow \mathcal{N}\Big(0, \frac{\sigma_\varphi^2}{Q^2}\,\boldsymbol{\Sigma}_{\boldsymbol{g}}^{-1}\boldsymbol{\Gamma}_{\boldsymbol{g}}\boldsymbol{\Sigma}_{\boldsymbol{g}}^{-1}\Big).$$

Proof. From the methods in Sections 5.4 and 5.5, Theorem 7.3.1 and (7.3.16) imply that $n^{1/2}\|(\hat{\boldsymbol{\rho}}_{\boldsymbol{g}} - \boldsymbol{\rho})\| = O_p(1)$ and that

$$n^{1/2}(\hat{\boldsymbol{\rho}}_{R\boldsymbol{g}} - \boldsymbol{\rho}) = (Q\boldsymbol{\Sigma}_{\boldsymbol{g}})^{-1}n^{1/2}\hat{\mathbf{S}}_{\boldsymbol{g}}(\boldsymbol{\rho}) + o_p(1).$$

Observe that $n^{1/2}\hat{\mathbf{S}}_{\boldsymbol{g}}(\boldsymbol{\rho})$ is a vector of square integrable mean zero martingale arrays with $nE\hat{\mathbf{S}}_{\boldsymbol{g}}\hat{\mathbf{S}}_{\boldsymbol{g}}' = \sigma_\varphi^2\boldsymbol{\Gamma}_{\boldsymbol{g}}$, $\sigma_\varphi^2 := \int_0^1[\varphi(u) - \bar{\varphi}]^2du$. Thus, by the routine Cramér-Wold device and by Lemma 9.1.3 in the Appendix, one readily obtains the claim of the asymptotic normality. □

Remark 7.3.1 Write $\hat{\rho}_R$ for $\hat{\rho}_{Rg}$ when $g(\mathbf{y}) \equiv \mathbf{y}$. Argue either as in the Section 3.4 or as in Jaeckel (1972) to conclude that $\|n^{1/2}(\hat{\rho}_R - \hat{\rho}_J)\| = o + p(1)$. Consequently by Theorem 7.3.2,

$$(7.3.17) \quad n^{1/2}(\hat{\rho}_R - \rho) = n^{1/2}(\tilde{\rho}_j - \rho) + o_p(1) = Q^{-1}\Sigma^{-1}\hat{\mathbf{S}} + o_p(1). \quad \square$$

Remark 7.3.2 Recently Mukherjee and Bai (2001) have extended the AUL result (7.3.15) to any nondecreasing square integrable φ for the case $g(\mathbf{y}) \equiv \mathbf{y}$, provided the error d.f. F has the finite Fisher information for location, the condition (a) of Theorem 3.2.3. Their proof uses the contiguity argument, approximating the square integrable φ by bounded scores and the causality property of the linear model (7.3.8)-(7.3.11). □

7.3.3 Estimation of $Q(f) := \int f d\varphi(F)$

As is evident from Theorem 7.3.1, the rank analysis of an AR(p) model via the above GR-estimators will need a consistent estimator of the functional Q. In this subsection we give two classes of consistent estimators of this functional in the AR(p) model (7.3.8) and (7.3.9). One class of estimators is obtained by replacing f and F in Q by a kernel density estimator and the empirical d.f. based on the estimated residuals, respectively. This is analogous to the class of estimators discussed in Theorem 4.5.3. The other class is an analogue of the class of estimators discussed in Theorem 4.5.1 in connection with the linear regression setup.

Accordingly, let $\tilde{\rho}$ be an estimator of ρ, K be a probability density on $\mathbb{R}$, h_n be a sequence of positive numbers, $h_n \to 0$ and define, for $x \in \mathbb{R}$,

$$\begin{aligned} \tilde{\varepsilon}_i &:= X_i - \tilde{\rho}'\mathbf{Y}_{i-1}, 1 \le i \le n; \quad \tilde{F}_n(x) := n^{-1}\sum_{i=1}^n I(\tilde{\varepsilon}_i \le x), \\ \tilde{f}_n(x) &:= (nh_n)^{-1}\sum_{i=1}^n K(\frac{x - \tilde{\varepsilon}_i}{h_n}), \; f_n(x) := (nh_n)^{-1}\sum_{i=1}^n K(\frac{x - \varepsilon_i}{h_n}). \end{aligned}$$

Finally, let

$$\tilde{Q}_n := \int \tilde{f}_n d\varphi(\tilde{F}_n).$$

Recall the definition of $\mathcal{C}$ from (3.2.1).

Theorem 7.3.3 *In addition to (7.3.8)-(7.3.10), assume that* (**F1**), (**F2**) *and the following conditions hold.*

(7.3.18) $h_n > 0;\ h_n \to 0, n^{1/2}h_n \to \infty.$

(7.3.19) *K is absolutely continuous with its a.e. derivative* $\dot{K}$ *satisfying* $\int |\dot{K}| < \infty$.

(7.3.20) $\|n^{1/2}(\tilde{\rho} - \rho)\| = O_p(1).$

Then,

$$\sup_{\varphi \in \mathcal{C}} |\tilde{Q}_n - Q(f)| = o_p(1). \tag{7.3.21}$$

Proof. The proof is similar to that of Theorem 4.5.3, so we shall be brief, indicating only one major difference. Unlike in the linear regression setup, i.e., unlike (4.5.9), here we have from Remark 7.2.3,

$$\sup_x n^{1/2}|\tilde{F}_n(x) - F_n(x)| = o_p(1). \tag{7.3.22}$$

where $F_n(x) \equiv F_n(x, \rho)$. In other words the linearity term involving $n^{1/2}(\tilde{\rho} - \rho)$ is not present in the approximation of $\tilde{F}_n$. Proceeding as in the proof of Theorem 4.5.3, (7.3.22) will yield

$$\begin{aligned}\|\tilde{f}_n - f_n\|_\infty &\le (n^{1/2}h_n)^{-1} \cdot \|n^{1/2}[\tilde{F}_n - F_n]\|_\infty \cdot \int |\dot{K}| \\ &= o_p\left((n^{1/2}h_n)^{-1}\right) = o_p(1).\end{aligned}$$

Compare this with (4.5.16) where the analogous term is of the order $O_p((n^{1/2}h_n)^{-1})$ instead of $o_p((n^{1/2}h_n)^{-1})$. The rest of the proof is exactly the same as there with the proviso that one uses (7.3.22) instead of (4.5.9), whenever needed. □

The reader may wish to modify the above proof to see that $\tilde{Q}_n$ continues to be consistent for Q even when $E\varepsilon \neq 0$, so that the term that is linear in $n^{1/2}(\tilde{\rho} - \rho)$ is now present in the expansion of $\tilde{F}_n$.

We shall now describe an analogue of $\hat{Q}_n^\alpha$ of (4.5.5). The motivation is the same as in Section 4.5, so we shall be brief on that also. Accordingly, let

$$\tilde{p}(y) := \int [\tilde{F}_n(y + x) - \tilde{F}_n(-y + x)] d\varphi(\tilde{F}_n(x)), \quad y \ge 0.$$

Observe that $\tilde{p}$ is an estimator of the d.f. of the absolute difference $|\varepsilon - \eta|$, where ε and η are independent r.v.'s with respective d.f.'s F and $\varphi(F)$. As in

Section 4.5, one can use the following representation for the computational purposes. For $y \geq 0$,

$$\tilde{p}(y) = n^{-1} \sum_{j=1}^{n} [\varphi(j/n) - \varphi(j-1)/n] \sum_{i=1}^{n} I(|\tilde{\varepsilon}_{(i)} - \tilde{\varepsilon}_{(j)}| \leq y),$$

where $\{\tilde{\varepsilon}_{(i)}\}$ are the ordered residuals $\{\tilde{\varepsilon}_i\}$ from the smallest to the largest.

Now let $\tilde{t}_n^{\alpha}$ denote an α-th percentile of the d.f. $\tilde{p}(y)$ and define

$$\tilde{Q}_n^{\alpha} = n^{1/2} \tilde{p}(n^{-1/2} \tilde{t}_n^{\alpha})/2\tilde{t}_n^{\alpha}, \quad 0 < \alpha < 1.$$

The consistency of these estimators may be proved using the method of the proof of Theorem 4.5.1 and the results given in Corollary 7.2.1. The discussion about the choice of α etc. that appears in Remark 4.5.1 is also pertinent here.

Another class of estimators is obtained by modifying $\tilde{Q}_n$ by replacing $\hat{F}_n$ by the estimator $\overline{F}_n(x) = \int \tilde{f}_n(y) I(-\infty < y \leq x) dy$. The consistency of these estimators can be also proved by the help of Corollary 7.2.1. □

7.4 Minimum Distance Estimation

In this section we shall discuss two classes of m.d. estimators. They are the analogues of the classes of estimators defined in the linear regressive setup at (5.2.8) and (5.2.15). To be precise, consider the autoregressive model (7.1.1) and define, for a $G \in \mathcal{DI}(\mathbb{R})$, and a $\mathbf{t} \in \mathbb{R}^p$,

$$\text{(7.4.1)} \qquad K_{\boldsymbol{g}}(\mathbf{t}) = \sum_{j=1}^{p} \int \Big[n^{-1} \sum_{i=1}^{n} g_j(\mathbf{Y}_{i-1}) \{ X_i \leq x + \mathbf{t}'\mathbf{Y}_{i-1}) - F(x) \} \Big]^2 dG(x),$$

$$K_{\boldsymbol{g}}^{+}(\mathbf{t}) = \sum_{j=1}^{p} \int \Big[n^{-1/2} \sum_{i=1}^{n} g_j(\mathbf{Y}_{i-1}) \{ I(X_i \leq x + \mathbf{t}'\mathbf{Y}_{i-1}) - I(-X_i < x - \mathbf{t}'\mathbf{Y}_{i-1}) \} \Big]^2 dG(x).$$

In the case the error d.f. F is *known*, define a class of m.d. estimators of $\boldsymbol{\rho}$ to be

$$\text{(7.4.2)} \qquad \hat{\boldsymbol{\rho}}_{\boldsymbol{g}} := \text{argmin} \ \{ K_{\boldsymbol{g}}(\mathbf{t}); \mathbf{t} \in \mathbb{R}^p \}.$$

In the case F is *unknown but symmetric around* 0, define a class of m.d. estimators of $\boldsymbol{\rho}$ to be

$$(7.4.3) \qquad \boldsymbol{\rho}_{\boldsymbol{g}}^{+} := \text{argmin } \{K_{\boldsymbol{g}}^{+}(\mathbf{t}); \mathbf{t} \in \mathbb{R}^p\}.$$

Note that the role played by the vectors

$$\{n^{-1/2}[g_1(\mathbf{Y}_{i-1}), g_2(\mathbf{Y}_{i-1}), \cdots, g_p(\mathbf{Y}_{i-1})];\ 1 \le i \le n\}$$

is similar to that of the vectors $\{\mathbf{d}_{ni}; 1 \le i \le n\}$ of Chapter 5. To put it in matrices, the precise analogue of $\mathbf{D}$ is the matrix $n^{-1/2}\boldsymbol{g}$, where $\boldsymbol{g}$ is as in (7.3.1).

The existence of these estimators has been discussed in Dhar (1991a) for $p = 1$ and in Dhar (1991c) for $p \ge 1$. For $p = 1$, these results are relatively easy to state and prove. We give an existence result for the estimator defined at (7.4.3) in the case $p = 1$.

Lemma 7.4.1 *In addition to (7.1.1) with* $p = 1$, *assume that*

$$(7.4.4) \qquad \textit{Either} \quad xg(x) \ \ge\ 0, \qquad \forall\, x \in \mathbb{R},$$

$$(7.4.5) \qquad \textit{Or} \quad xg(x) \ \le\ 0, \qquad \forall\, x \in \mathbb{R}.$$

Then, a minimizer of $K_{\boldsymbol{g}}^{+}$ *exists if either* $G(\mathbb{R}) = \infty$ *or* $G(\mathbb{R}) < \infty$ *and* $\boldsymbol{g}(0) = \mathbf{0}$.

The proof of this lemma is precisely similar to that of Lemma 5.3.1. The discussion about the computation of their analogues that appears in Section 5.3 is also relevant here with appropriate modifications. Thus, for example, if G is continuous and symmetric around 0, i.e., satisfies (5.3.10), then, analogous to (5.3.12),

$$\begin{aligned} K_{\boldsymbol{g}}^{+}(\mathbf{t}) \ &= \ \sum_{j=1}^{p}\sum_{i=1}^{n}\sum_{k=1}^{n} g_j(\mathbf{Y}_{i-1})g_j(\mathbf{Y}_{k-1}) \\ &\qquad \times\Big\{|G(X_i - \mathbf{t}'\mathbf{Y}_{i-1}) - G(-X_k + \mathbf{t}'\mathbf{Y}_{k-1})| \\ &\qquad\qquad -|G(X_i - \mathbf{t}'\mathbf{Y}_{i-1}) - G(X_k - \mathbf{t}'\mathbf{Y}_{k-1})|\Big\}. \end{aligned}$$

If G is degenerate at 0 then one obtains, assuming the continuity of the errors, that

$$(7.4.6) \qquad K_{\boldsymbol{g}}^{+}(\mathbf{t}) = \sum_{j=1}^{p}\Big[\sum_{i=1}^{n} g_j(\mathbf{Y}_{i-1})\, sign(X_i - \mathbf{t}'\mathbf{Y}_{i-1})\Big]^2, \quad \text{w.p. } 1.$$

One has similar expressions for a general G. See (5.3.7) and (5.3.11).

If $g(\mathbf{x}) \equiv x$, $G(x) \equiv x$, $\hat{\rho}_g$ is m.l.e. of ρ if F is logistic, while ρ_g^+ is an analogue of the Hodges - Lehmann estimator. Similarly, if $g(x) \equiv x$ and G is degenerate at 0, then ρ_g^+ is the LAD estimator.

We shall now focus on proving their asymptotic normality. The approach is the same as that of Section 5.4 and 5.5, i.e., we shall prove that these dispersions satisfy (5.4.A1) - (5.4.A5) by using the techniques that are similar to those used in Section 5.5. Only the tools are somewhat different because of the dependence structure.

To begin with we state the additional assumptions needed under which an asymptotic uniform quadraticity result for a general dispersion of the above type holds. Because here the weights are random, we have to be somewhat careful if we do not wish to impose more than necessary moment conditions on the underlying entities. For the same reason, unlike the linear regression setup where the asymptotic uniform quadraticity of the underlying dispersions was obtained in L_1, we shall obtain these results in probability only. This is also reflected in the formulation of the following assumptions.

$$\text{(7.4.7)} \qquad \text{(a)} \quad Eh^2(\mathbf{Y}_0) < \infty. \qquad \text{(b)} \quad 0 < E\varepsilon^2 < \infty.$$

$$\text{(7.4.8)} \qquad \forall \ \|\mathbf{u}\| \le b, a \in \mathbb{R},$$
$$\int Eh^2(\mathbf{Y}_0)|F(x + n^{-1/2}(\mathbf{u}'\mathbf{Y}_0 + a\|\mathbf{Y}_0\|)) - F(x)|dG(x)$$
$$= o(1)$$

There exists a constant $0 < k < \infty$, $\ni \ \forall \delta > 0$, $\forall \, \|\mathbf{u}\| \le b$,

$$\text{(7.4.9)} \quad \liminf_n P\Big(\int n^{-1}\Big[\sum_{i=1}^n h^{\pm}(\mathbf{Y}_{i-1})\Big\{F(x + n^{-1/2}\mathbf{u}'\mathbf{Y}_{i-1} + \delta_{ni})$$
$$- F(x + n^{-1/2}\mathbf{u}'\mathbf{Y}_{i-1} - \delta_{ni})\Big\}\Big]^2 dG(x) \le k\delta^2\Big) = 1,$$

where $\delta_{ni} := n^{-1/2}\delta\|\mathbf{Y}_{i-1}\}\|$ and $h^{\pm}$ is as in the proof of Theorem 7.2.1. For every $\|\mathbf{u}\| \le b$,

$$\text{(7.4.10)} \qquad \int n^{-1}\Big[\sum_{i=1}^n h(\mathbf{Y}_{i-1})\Big\{F(x + n^{-1/2}\mathbf{u}'\mathbf{Y}_{i-1}) - F(x)$$
$$- n^{-1/2}\mathbf{u}'\mathbf{Y}_{i-1}f(x)\Big\}\Big]^2 dG(x) = o_p(1),$$

and (5.5.44b) holds.

Now, recall the definitions of $W_h, \nu_h, \mathcal{W}_h, \mathcal{W}^{\pm}, T^{\pm}, W^{\pm}, Z^{\pm}, m^{\pm}$ from (7.1.6), (7.2.2), (7.2.11) and (7.2.12). Let $|\cdot|_G$ denote the L_2 - norm w.r.t. the measure G. In the *proofs below*, we have adopted the notation and conventions used in the proof of Theorem 7.2.1. Thus, e.g., $\xi_i \equiv \mathbf{Y}_{i-1}$; $\mathcal{W}_{\mathbf{u}}(\cdot)$, $\nu_{\mathbf{u}}(\cdot)$ stand for $\mathcal{W}_h(\cdot, \boldsymbol{\rho}+n^{-1/2}\mathbf{u})$, $\nu_h(\cdot, \boldsymbol{\rho}+n^{-1/2}\mathbf{u})$, etc.

Lemma 7.4.2 *Suppose that the autoregression model (7.3.8) and (7.3.9) holds. Then the following hold.*

(7.4.11) *Assumption (7.4.8) implies that* $\forall\ 0 < b < \infty$,

$$E\int [Z^{\pm}(x;\mathbf{u},a) - Z^{\pm}(x;\mathbf{u},0)]^2 dG(x) = O(1),\ \|\mathbf{u}\| \le b,\ a \in \mathbb{R}.$$

(7.4.12) *Assumption (7.4.9) implies that* $\forall\ 0 < b < \infty$,

$$\liminf_n P\Big(\sup_{\|\mathbf{v}-\mathbf{u}\|\le\delta} n^{1/2} |\nu_h^{\pm}(x, \boldsymbol{\rho} + n^{-1/2}\mathbf{v}) - \nu_h^{\pm}(x, \boldsymbol{\rho} + n^{-1/2}\mathbf{u})|_G^2 \le k\delta^2\Big) = 1,\ \ \forall \|\mathbf{u}\| \le b.$$

where k and δ are as in (7.4.9).

(7.4.13) *Assumptions (7.4.7), (7.4.9) and (7.4.11) imply that* $\forall\ 0 < b < \infty,\ \|\mathbf{u}\| \le b,$

$$\sup_{\|\mathbf{u}\|\le b} \int \Big[n^{1/2}\{\nu_h(x, \boldsymbol{\rho} + n^{-1/2}\mathbf{u}) - \nu_h(x, \boldsymbol{\rho})\} - \mathbf{u}'n^{-1}\sum_{i=1}^n h(\mathbf{Y}_{i-1} f(x)\Big]^2 dG(x) = o_p(1).$$

Proof. Let, for $x, a \in \mathbb{R}$; $\mathbf{u}, \mathbf{y} \in \mathbb{R}^p$,

(7.4.14)
$$p(x, \mathbf{u}, a; \mathbf{y}) := \Big|F(x + n^{-1/2}(\mathbf{u}'\mathbf{y} + a\|\mathbf{y}\|) - F(x + n^{-1/2}\mathbf{u}'\mathbf{y})\Big|$$

Now, observe that $n^{1/2}[Z^{\pm}(x;\mathbf{u},a) - Z^{\pm}(x;\mathbf{u},0)]$ is a sum of n r.v.'s whose i^{th} summand is conditionally centered, given $\mathcal{F}_{i-1}$, and whose conditional variance, given $\mathcal{F}_{i-1}$, is

$$E[\{h^{\pm}(\xi_i)\}^2 p(x, \mathbf{u}, a; \xi_i)\{1 - p(x, \mathbf{u}, a; \xi_i)\}],\ 1 \le i \le n.$$

Hence, by Fubini, the stationarity of $\{\xi_i\}$ and the fact that $(h^{\pm})^2 \le h^2$, $\forall\ \|\mathbf{u}\| \le b$,

$$\text{L.H.S. } (7.4.11) \le \int Eh^2(\mathbf{Y}_0) p(x, \mathbf{u}, a; \mathbf{Y}_0) dG(x) = o(1),$$

by (7.4.8) applied with the given a and with a = 0 and the triangle inequality.

To prove (7.4.12), use the nonnegativity of $h^{\pm}$, the monotonicity of F and (7.2.17), to obtain that $\|\mathbf{v}\| \leq b$, $\|\mathbf{v}-\mathbf{u}\| \leq \delta$ imply that $\forall\ \|\mathbf{u}\| \leq b$,

$$\text{(7.4.15)} \qquad \begin{aligned} & n^{1/2}|\nu_{\mathbf{v}}^{\pm}(x) - \nu_{\mathbf{u}}^{\pm}(x)| \\ & \leq |m^{\pm}(x;\mathbf{u},\delta) - m^{\pm}(x;\mathbf{u},\delta) - m^{\pm}(x;\mathbf{u},-\delta)|, \\ & \forall\ x \in \mathbb{R}. \end{aligned}$$

This and (7.4.9) readily imply(7.4.12) as the r.v. in the L.H.S. of (7.4.9) is precisely the $|\cdot|_G^2$ of the R.H.S. of (7.4.15) for each $n \geq 1$.

The proof of (7.4.13) is obtained from (7.4.7), (7.4.9) and (7.4.10) in the same way as that of (5.5.11) from (5.5.g), (5.5.h) and (5.5.i), hence no details are given. □

Lemma 7.4.3 *Suppose that the autoregressive model (7.3.8) and (7.3.9) holds. In addition, assume that (7.4.8) and (7.4.9) hold.*

Then, $\forall\ 0 < b < \infty$,

$$\text{(7.4.16)} \qquad \sup_{\|\mathbf{u}\| \leq b} \int [\mathcal{W}_h^{\pm}(x, \boldsymbol{\rho} + n^{-1/2}\mathbf{u}) - \mathcal{W}_h^{\pm}(x, \boldsymbol{\rho})]^2 dG(x) = o_p(1).$$

$$\text{(7.4.17)} \qquad \sup_{\|\mathbf{u}\| \leq b} \int [\mathcal{W}_h^{\pm}(x, \boldsymbol{\rho} + n^{-1/2}\mathbf{u}) - \mathcal{W}_h(x, \boldsymbol{\rho})]^2 dG(x) = o_p(1).$$

Proof. Let $q(x, \mathbf{u}; \mathbf{y}) := |F(x + n^{-1/2}\mathbf{u}'\mathbf{y}) - F(x)|, x \in \mathbb{R}; \mathbf{u}, \mathbf{y} \in \mathbb{R}^p$. The r.v. $n^{1/2}[\mathcal{W}_{\mathbf{u}}^{\pm}(\cdot) - \mathcal{W}^{\pm}(\cdot)]$ is a sum of n r.v.'s whose i^{th} summand is conditionally centered, given $\mathcal{F}_{i-1}$, and whose conditional variance, given $\mathcal{F}_{i-1}$, is $E[\{h^{\pm}(\xi_i)\}^2 q(\cdot, \mathbf{u}; \xi_i)\{1 - q(\cdot, \mathbf{u}; \xi_i)\}], 1 \leq i \leq n$. Hence, by Fubini, the stationarity of $\{\xi_i\}$ and the fact that $(h^{\pm})^2 \leq h^2$, $\forall\ \|\mathbf{u}\| \leq b$,

$$\begin{aligned} & E|\mathcal{W}_{\mathbf{u}}^{\pm} - \mathcal{W}^{\pm}|_G^2 \\ & \leq \int n^{-1} \sum_{i=1}^{n} Eh^2(\mathbf{Y}_{i-1}|F(x + n^{-1/2}\mathbf{u}'\mathbf{Y}_{i-1}) - F(x)|dG(x) \\ & \leq \int Eh^2(\mathbf{Y}_0)|F(x + n^{-1/2}\mathbf{u}'\mathbf{Y}_0) - F(x)|dG(x). \end{aligned}$$

Therefore, by (7.4.8) with $a = 0$ and the Markov inequality,

$$\text{(7.4.18)} \qquad |\mathcal{W}_{\mathbf{u}}^{\pm} - \mathcal{W}^{\pm}|_G^2 = o_p(1), \quad \forall\ \|\mathbf{u}\| \leq b.$$

Thus, to prove (7.4.16), because of the compactness of the ball $\{\mathbf{u} \in \mathbb{R}^p;\ \|\mathbf{u}\| \leq b\}$, it suffices to show that for every $\eta > 0$ there is a $\delta > 0$ such that for every $\|\mathbf{u}\| \le b$,

$$\liminf_n P\Big(\sup_{\|\mathbf{v}-\mathbf{u}\|\le\delta} |\mathcal{L}_{\mathbf{v}} - \mathcal{L}_{\mathbf{u}}| < \eta \Big) = 1, \tag{7.4.19}$$

where $\mathcal{L}_{\mathbf{u}} := |\mathcal{W}^{\pm}_{\mathbf{u}} - \mathcal{W}^{\pm}|^2_G$, $\mathbf{u} \in \mathbb{R}^p$.

Expand the quadratic, apply the C-S inequality to the cross product terms, to obtain

$$|\mathcal{L}_{\mathbf{u}} - \mathcal{L}_{\mathbf{v}}| \le |\mathcal{W}^{\pm}_{\mathbf{u}} - \mathcal{W}^{\pm}_{\mathbf{v}}|^2_G + 2|\mathcal{W}^{\pm}_{\mathbf{u}} - \mathcal{W}^{\pm}_{\mathbf{v}}|_G |\mathcal{W}^{\pm}_{\mathbf{v}} - \mathcal{W}^{\pm}|_G. \tag{7.4.20}$$

Observe that $h^{\pm} \ge 0$, F nondecreasing and (7.2.17) imply that

$$0 \le |m^{\pm}(x;\mathbf{u},\pm\delta) - m^{\pm}(x;\mathbf{u},0)| \le m^{\pm}(x;\mathbf{u},\delta) - m^{\pm}(x;\mathbf{u},-\delta),$$

for all $x \in \mathbb{R}$, $\|\mathbf{s}\| \le b$, $\|\mathbf{s}-\mathbf{u}\| \le \delta$. Use this, the second inequality in (7.2.16), (7.2.17), (7.2.18), and the fact that $(a+b)^2 \le 2(a^2+b^2)$, $a \in \mathbb{R}$, to obtain

$$\begin{aligned} &|\mathcal{W}^{\pm}_{\mathbf{v}} - \mathcal{W}^{\pm}_{\mathbf{u}}|^2_G \\ &\le 16\Big\{ \int \Big[Z^{\pm}(x;\mathbf{u},\delta) - Z^{\pm}(x;\mathbf{u},0)\Big]^2 dG(x) \\ &\quad + \int \Big[Z^{\pm}(x;\mathbf{u},-\delta) - Z^{\pm}(x;\mathbf{u},0)\Big]^2 dG(x) \\ &\quad + \int [m^{\pm}(x;\mathbf{u},\delta) - m^{\pm}(x;\mathbf{u},-\delta)]^2 dG(x) + |n^{1/2}(\nu^{\pm}_{\mathbf{v}} - \nu^{\pm}_{\mathbf{u}})|^2_G \Big\}, \end{aligned}$$

for all $\|\mathbf{v}\| \le b$, $\|\mathbf{v}-\mathbf{u}\| \le \delta$. This together with (7.4.9), (7.4.12), (7.4.13), (7.4.18), (7.4.20) and the C-S inequality proves (7.4.19) and hence (7.4.16).

The proof of (7.4.17) follows from (7.4.16) and the first inequality in (7.2.15). □

Now define, for $\mathbf{t} \in \mathbb{R}^p$,

$$\begin{aligned} &K_h(\mathbf{t}) \\ &:= \int \Big[n^{-1/2}\sum_{i=1}^n h(\mathbf{Y}_{i-1})\{I(X_i \le x + \mathbf{t}'\mathbf{Y}_{i-1}) - F(y)\}\Big]^2 dG(x), \\ &\hat{K}_h(\mathbf{t}) \\ &:= \int \Big[\mathcal{W}_h(x,\boldsymbol{\rho}) + n^{1/2}(\mathbf{t}-\boldsymbol{\rho})' n^{-1}\sum_{i=1}^n h(\mathbf{Y}_{i-1})\mathbf{Y}_{i-1} f(x)\Big]^2 dG(x). \end{aligned}$$

Theorem 7.4.1 *Suppose that the autoregressive model (7.3.8) - (7.3.10) holds and that (5.5.45), (7.4.7) - (7.4.10) hold. Then, $\forall\ 0 < b < \infty$,*

$$\sup_{\|\mathbf{u}\|\le b} \Big|K_h(\boldsymbol{\rho} + n^{-1/2}\mathbf{u}) - \hat{K}_h(\boldsymbol{\rho} + n^{-1/2}\mathbf{u})\Big| = o_p(1). \tag{7.4.21}$$

Proof. Observe that, by (5.5.45), (7.4.7),

$$E\int \mathcal{W}_h^2(x,\boldsymbol{\rho})dG(x) = Eh^2(\mathbf{Y}_0)\int F(1-F)dG < \infty. \tag{7.4.22}$$

The rest of the proof of (7.4.21) follows from Lemmas 7.4.2 and 7.4.3 in a similar way as that of (5.5.9) from Lemmas 5.5.1, 5.5.2 and the result (5.5.11). □

Now we shall apply this result to obtain the required quadraticity of the dispersion $K_{\boldsymbol{g}}$ and $K_{\boldsymbol{g}}^{+}$. For that purpose recall the matrices $\boldsymbol{\mathcal{X}}, \mathbf{G}$ and $\boldsymbol{\mathcal{B}}_n$ from (7.3.1). Note that X_{i-j}, $g_j(\mathbf{Y}_{i-1})$ are the $(i,j)^{th}$ entries of $\boldsymbol{\mathcal{X}}, \mathbf{G}$ respectively, $1 \le i \le n$, $1 \le j \le p$. Also observe that the

$$j^{th} \text{ row of } \boldsymbol{\mathcal{B}}_n \text{ is } \sum_{i=1}^{n} g_j(\mathbf{Y}_{i-1})\mathbf{Y}_{i-1}', \quad 1 \le j \le p. \tag{7.4.23}$$

To obtain the desired result about $K_{\boldsymbol{g}}$, we need to apply the above theorem p times, j^{th} time with

$$h(\mathbf{Y}_{i-1}) \equiv g_j(\mathbf{Y}_{i-1}), \quad j = 1, \cdots, p. \tag{7.4.24}$$

Now write $\mathcal{W}_j$ for $\mathcal{W}_h$ when h is as in (7.4.24) and $\mathcal{W}_j(\cdot)$ for $\mathcal{W}_j(\cdot,\boldsymbol{\rho})$, $1 \le j \le p$. Note that for $1 \le j \le p$, $x \in \mathbb{R}$,

$$\mathcal{W}_j(x) \quad := \quad n^{-1/2}\sum_{i=1}^{n} g_j(\mathbf{Y}_{i-1})\{I(\varepsilon_i \le x) - F(x)\}.$$

We also need to define the approximating quadratic forms: For $\mathbf{t} \in \mathbb{R}^p$, let

$$\hat{K}_{\boldsymbol{g}}(\mathbf{t}) := \sum_{j=1}^{p}\int \Big[\mathcal{W}_j(x) + n^{1/2}(\mathbf{t}-\boldsymbol{\rho})' \times n^{-1}\sum_{i=1}^{n} g_j(\mathbf{Y}_{i-1})\mathbf{Y}_{i-1}f(x)\Big]^2 dG(x) \tag{7.4.25}$$

$$\hat{K}_{\boldsymbol{g}}^{+}(\mathbf{t}) := \sum_{j=1}^{p}\int \Big[\mathcal{W}_j^{+}(x) + 2n^{1/2}(\mathbf{t}-\boldsymbol{\rho})' \times n^{-1}\sum_{i=1}^{n} g_j(\mathbf{Y}_{i-1})\mathbf{Y}_{i-1}f(x)\Big]^2 dG(x), \tag{7.4.26}$$

where, for $1 \le j \le p, x \in \mathbb{R}$,

$$\mathcal{W}_j^{+}(x) := n^{-1/2}\sum_{i=1}^{n} g_j(\mathbf{Y}_{i-1})\{I(\varepsilon_i \le x) - I(-\varepsilon_i < x)\}.$$

The following theorem readily follows from (7.4.21).

Theorem 7.4.2 *Suppose that the autoregressive model (7.3.8) and (7.3.9) holds and that (5.5.44(a)), (5.5.45), (7.4.7) - (7.4.11) hold for the p functions h given at (7.4.24). Then,* $\forall\ 0 < b < \infty$,

$$\text{(7.4.27)} \qquad \sup_{\|\mathbf{u}\| \le b} |K_{\boldsymbol{g}}(\boldsymbol{\rho} + n^{-1/2}\mathbf{u}) - \hat{K}_{\boldsymbol{g}}(\boldsymbol{\rho} + n^{-1/2}\mathbf{u})| = o_p(1). \qquad \square$$

Lemmas 7.4.2 and 7.4.3 can be directly used to obtain the following

Theorem 7.4.3 *In addition to the assumptions of Theorem 7.4.2, except (5.5.45), assume that F is symmetric around* 0, *G satisfies (5.3.8) and that (5.6.12) holds. Then,* $\forall\ 0 < b < \infty$,

$$\text{(7.4.28)} \qquad \sup_{\|\mathbf{u}\| \le b} |K^+_{\boldsymbol{g}}(\boldsymbol{\rho} + n^{-1/2}\mathbf{u}) - \hat{K}^+_{\boldsymbol{g}}(\boldsymbol{\rho} + n^{-1/2}\mathbf{u}))| = o_p(1). \qquad \square$$

Upon expanding the quadratic and using an appropriate analogue of (7.4.22) obtained when h is as in (7.4.24), one can rewrite

$$\begin{aligned}\hat{K}_{\boldsymbol{g}}(\mathbf{t}) &= \hat{K}_{\boldsymbol{g}}(\boldsymbol{\rho}) + 2(\mathbf{t} - \boldsymbol{\rho})' n^{-1/2} \mathcal{B}_n' \int \mathcal{W}(x) f(x) dG(x) \\ &\quad + (\mathbf{t} - \boldsymbol{\rho})' n^{-1} \mathcal{B}_n' \mathcal{B}_n (\mathbf{t} - \boldsymbol{\rho}) |f|_G^2, \quad \mathbf{t} \in \mathbb{R}^p,\end{aligned}$$

where $\mathcal{W} := \{\mathcal{W}_1, \cdots, \mathcal{W}_p)'$. Now consider the r.v.'s in the second term. Recalling the definition of ψ from (5.6.2), one can rewrite

$$\mathcal{S}_n := \int \mathcal{W}(x) f(x) dG(x) = -n^{-1/2} \sum_{i=1}^n \boldsymbol{g}_i [\psi(\varepsilon_i) - E\psi(\varepsilon)],$$

where

$$\boldsymbol{g}_i' := (g_1(\mathbf{Y}_{i-1}), g_2(\mathbf{Y}_{i-1}), \cdots, g_p(\mathbf{Y}_{i-1})), \ 1 \le i \le n.$$

Since $\boldsymbol{g}_i$ is a function of $\mathbf{Y}_{i-1}$, it is $\mathcal{F}_{i-1}$ - measurable. Therefore, in view of (7.4.7(a)) applied to h at (7.4.24), and by (5.5.44(a)), $\{(\mathcal{S}_n, \mathcal{F}_{n-1}), n \ge 1\}$ is a mean zero square integrable martingale array. Hence, it follows from Lemma 9.1.3 in the Appendix that

$$\mathcal{S}_n \to_d \mathcal{N}(\mathbf{0}, \mathbf{G}^* \tau^2 \mathbf{I}_{p\times p}), \ \mathbf{G}^* = E\boldsymbol{g}_1 \boldsymbol{g}_1', \ \tau^2 = \frac{Var\,\psi(\varepsilon)}{(\int f^2 dG)^2}.$$

By the stationarity and the Ergodic Theorem, we also obtain

$$\text{(7.4.29)} \qquad n^{-1}\mathcal{B}_n \to \mathcal{B}, \ \text{a.s.}, \ \mathcal{B} := En^{-1}\mathcal{B}_n = E\boldsymbol{g}_1 \mathbf{Y}_0'.$$

Consequently it follows that the dispersion K_g satisfies (5.4.A1) to (5.4.A3) with

$$\theta_0 = \rho, \quad \delta_n \equiv n^{1/2}, \quad \mathbf{S}_n \equiv n^{-1/2}\mathcal{B}_n'\mathcal{S}_n,$$
$$\mathbf{W}_n \equiv n^{-1}\mathcal{B}_n'\mathcal{B}_n, \quad \mathbf{W} = \mathcal{B}, \quad \Sigma = \mathcal{B}'\mathbf{G}^*\mathcal{B}\tau^2,$$

and hence it is an U.L.A.N.Q. dispersion.

In view of (7.4.22) applied to h as in (7.4.24), the condition (5.4.A4) is trivially implied by (7.4.7(a)) and (5.5.45).

Recall, from Section 5.5, that in the linear regression setup the condition (5.4.A5) was shown to be implied by (5.5.k) and (5.5.l). In the present situation, the role of $\mathbf{\Gamma}_n, \overline{\mathbf{\Gamma}}_n$ of (5.5.k) is being played by $n^{-1}\mathcal{B}_n f$, $n^{-1}\mathcal{B}_n \int f dG$, respectively. Thus, in view of (7.4.29) and (5.5.44(a)), an analogue of (5.5.k) would hold in the present case if we additionally assumed that $\mathcal{B}$ is positive definite. An exact analogue of (5.5.l) in the present case is

$$\text{(7.4.30)} \qquad \begin{aligned} &\text{Either } \mathbf{e}'\boldsymbol{g}_i\mathbf{Y}_{i-1}'\mathbf{e} \geq 0, \forall 1 \leq i \leq n, \forall\, \mathbf{e} \in \mathbb{R}^p, \ \|\mathbf{e}\| = 1, \text{ a.s.,} \\ &\text{Or } \mathbf{e}'\boldsymbol{g}_i\mathbf{Y}_{i-1}'\mathbf{e} \leq 0, \forall 1 \leq i \leq n, \ \forall \mathbf{e} \in \mathbb{R}^p, \ \|\mathbf{e}\| = 1, \text{ a.s..} \end{aligned}$$

We are now ready to state the following

Theorem 7.4.4 *In addition to the assumptions of Theorem 7.4.2, assume that the $\mathcal{B}$ of (7.4.29) is positive definite and that (7.4.30) holds. Then,*

$$\text{(7.4.31)} \qquad n^{1/2}(\hat{\rho}_{\boldsymbol{g}} - \rho) = -\Big\{n^{-1}\mathcal{B}_n \int f^2 dG\Big\}^{-1} \mathcal{S}_n + o_p(1).$$

Consequently,

$$\text{(7.4.32)} \qquad n^{1/2}(\hat{\rho}_{\boldsymbol{g}} - \rho) \longrightarrow_d \mathcal{N}(\mathbf{0}, (\mathcal{B})^{-1}\mathbf{G}^*(\mathcal{B}')^{-1}\tau^2). \qquad \square$$

Let $\hat{\rho}_{\boldsymbol{x}}$ denote the estimator $\hat{\rho}_{\boldsymbol{g}}$ when $\boldsymbol{g}(\mathbf{x}) \equiv \mathbf{x}$. Observe that in this case $\mathbf{G}^* = \mathcal{B} = En^{-1}\mathcal{X}'\mathcal{X} = E\mathbf{Y}_0\mathbf{Y}_0$. Moreover, the assumption (7.4.30) is *a priori* satisfied and (7.3.8), (7.3.9) and (7.4.7(b)) imply that $E\mathbf{Y}_0\mathbf{Y}_0'$ is positive definite. Consequently, we have obtained

Corollary 7.4.1 *Suppose that the autoregressive model (7.3.8), (7.3.9) holds and that (5.5.44), (5.5.45), (7.4.7(b)), (7.4.9) - (7.4.11) with $h(\mathbf{x}) \equiv \mathbf{x}$ hold. Then,*

$$\text{(7.4.33)} \qquad n^{1/2}(\hat{\rho}_{\boldsymbol{x}} - \rho) \rightarrow_d \mathcal{N}(\mathbf{0}, (E\mathbf{Y}_0\mathbf{Y}_0')^{-1}\tau^2). \qquad \square$$

Remark 7.4.1 *Asymptotic Optimality of $\hat{\rho}_{\boldsymbol{\mathcal{X}}}$.* Because $E\mathbf{Y}_0\mathbf{Y}_0'$ and $\boldsymbol{\mathcal{B}}$ are positive definite, and because of $n^{-1}\boldsymbol{\mathcal{X}}'\boldsymbol{\mathcal{X}} \to E\mathbf{Y}_0\mathbf{Y}_0'$, a.s., and because of (7.4.29), there exists an N_0 such that $n^{-1}\boldsymbol{\mathcal{X}}'\boldsymbol{\mathcal{X}}$ and $n^{-1}\boldsymbol{\mathcal{B}}_n$ are positive definite, for all $n \geq N_0$, a.s.

Recall the inequality (5.6.8). Take $\mathbf{J} = n^{-1/2}\boldsymbol{g}', \mathbf{L} = n^{-1/2}\boldsymbol{\mathcal{X}}'$ in that inequality to obtain

$$n^{-1}\mathbf{G}'\mathbf{G} \geq n^{-1}\mathbf{G}'\boldsymbol{\mathcal{X}}(n^{-1}\boldsymbol{\mathcal{X}}'\boldsymbol{\mathcal{X}})^{-1}n^{-1}\boldsymbol{\mathcal{X}}'\mathbf{G}, \ \forall\, n \geq N_0, \ \text{ a.s.},$$

with equality holding if, and only if $\boldsymbol{\mathcal{X}} \propto \boldsymbol{G}$. Letting n tend to infinity in this inequality yields

$$(\boldsymbol{\mathcal{B}})^{-1}\mathbf{G}^*(\boldsymbol{\mathcal{B}}')^{-1} \geq (E\mathbf{Y}_0\mathbf{Y}_0')^{-1}.$$

We thus have proved the following:

(7.4.34) *Among all estimators $\{\hat{\rho}_{\boldsymbol{g}}\}$, where the components of $\boldsymbol{g}$ satisfy* (7.4.7(a)), (7.4.8) – (7.4.10), *for the given* (F, G) *that satisfy (7.4.7(b)), (5.5.44), (5.5.45), the one that minimizes the asymptotic variance is* $\hat{\rho}_{\boldsymbol{\mathcal{X}}}$!

We shall now state analogous results for $\rho_{\boldsymbol{g}}^+$. Arguments for their proofs are similar to those appearing above and, hence, will not be given.

Theorem 7.4.5 *In addition to the assumptions of Theorem 7.4.4, except (5.5.45), assume that F is symmetric around* 0, *G satisfies (5.3.8) and that (5.6.12) holds. Then,*

$$n^{1/2}(\rho_{\boldsymbol{g}}^+ - \rho) = -\{n^{-1}\boldsymbol{\mathcal{B}}_n \int f^2 dG\}^{-1}\boldsymbol{\mathcal{S}}_n^+ + o_p(1), \tag{7.4.35}$$

where

$$\boldsymbol{\mathcal{S}}_n^+ := \int \boldsymbol{\mathcal{W}}^+(x)f(x)dG(x) = n^{-1/2}\sum_{i=1}^{n} \boldsymbol{g}_i[\psi(-\varepsilon_i) - \psi(\varepsilon_i)].$$

Consequently,

$$n^{1/2}\rho_{\boldsymbol{g}}^+ - \rho) \to_d \mathcal{N}(\mathbf{0}, (\boldsymbol{\mathcal{B}})^{-1}\mathbf{G}^*(\boldsymbol{\mathcal{B}}')^{-1}\tau^2),$$
$$n^{1/2}(\rho_{\boldsymbol{\mathcal{X}}}^+ - \rho) \to_d \mathcal{N}(\mathbf{0}, (E\mathbf{Y}_0\mathbf{Y}_0')^{-1}\tau^2). \qquad \square$$

Obviously the optimality property like (7.4.34) holds here also.

Remark 7.4.2 *On assumptions for the asymptotic normality of* $\hat{\rho}_{\boldsymbol{x}}$, $\rho^+_{\boldsymbol{x}}$. If G is a finite measure and F has uniformly continuous density then it is not hard to see that (7.4.8) - (7.4.11), with $h(\mathbf{Y}_0)$ equivalent to the components of $\mathbf{Y}_0$, are all implied by (7.4.7(b)).

Consider the following assumptions for a general G:

(7.4.36) $$E|\varepsilon|^3 < \infty, \quad E\varepsilon^2 > 0.$$

For each $1 \le j \le p$, as a function of $s \in \mathbb{R}$,

(7.4.37) $$\int E|X_{1-j}|^2 \|\mathbf{Y}_0\| f(x + s\|\mathbf{Y}_0\|) dG(x) \text{ is continuous at } 0.$$

(7.4.38) $$\int_{-1}^{1} \int E\{\|\mathbf{Y}_0\| [f(x + n^{-1/2}(\mathbf{u}'\mathbf{Y}_0 + t\delta\|\mathbf{Y}_0\|)) - f(x + n^{-1/2}\mathbf{u}'\mathbf{Y}_0)]\}^2 dG(x)dt = o(1), \ \forall\, \delta > 0, \, \mathbf{u} \in \mathbb{R}^p.$$

(7.4.39) For every $\mathbf{u} \in \mathbb{R}^p$,

$$\int [n^{-1} \sum_{i=1}^{n} X^{\pm}_{1-j} \|\mathbf{Y}_{i-1}\| f(x + n^{-1/2}\mathbf{u}'\mathbf{Y}_{i-1})]^2 dG(x) = O_p(1), \quad 1 \le j \le p.$$

An argument similar to the one used in verifying the Claim 5.5.1 shows that (5.5.44(a)), (7.4.36) and (7.4.37) imply (7.4.8) while (5.5.44(b)), (7.4.38) and (7.4.39) imply (7.4.9) and (7.4.11) with h as in (7.4.24).

In particular if $G(x) \equiv x$, then (5.5.44), (7.4.36) and f continuous imply all of the above conditions, (5.5.45) and (5.6.12). This is seen with the help of a version of Lemma 9.1.6. □

Remark 7.4.3 *Asymptotic relative efficiency of* $\hat{\rho}_{\boldsymbol{x}}$ *and* $\rho^+_{\boldsymbol{x}}$. Since their asymptotic variances are the same, we shall carry out the discussion in terms of $\hat{\rho}_{\boldsymbol{x}}$ only, as the same applies to $\rho^+_{\boldsymbol{x}}$ under the additional assumption of the symmetry of F and G.

Consider the case $p = 1$. Let $\sigma^2 = Var(\varepsilon)$ and $\hat{\rho}_{ls}$ denote the least square estimator of ρ_1. Then it is well known that under (7.4.7b), $n^{1/2}(\hat{\rho}_{ls} - \rho_1) \rightarrow_d \mathcal{N}(0, 1 - \rho_1^2)$. See, e.g., Anderson (1971). Also note that in this case $(EY_0Y_0')^{-1} = (1 - \rho_1^2)/\sigma^2$. Hence the asymptotic relative efficiency e of $\hat{\rho}_x$, relative to $\hat{\rho}_{ls}$, obtained by taking the ratio of the inverses of their asymptotic variances, is

(7.4.40) $$e = e(\hat{\rho}_x, \hat{\rho}_{ls}) = \sigma^2/\tau^2.$$

Note that $e > 1$ means $\hat{\rho}_x$ is asymptotically more efficient than $\hat{\rho}_{ls}$. It follows that $\hat{\rho}_x$ is to be preferred to $\hat{\rho}_{ls}$ for the heavy tailed error d.f.'s F. Also note that if $G(x) \equiv x$ then $\tau^2 = 1/12[\int f^2(x)dx]^2$ and $e = 12\sigma^2[\int f^2(x)dx]^2$. If G is degenerate at 0, then $\tau^2 = 1/4[f^2(0)]$ and $e = 4\sigma^2 f^2(0)$. These expressions are well known in connection with the Wilcoxon and median rank estimators of the slope parameters in linear regression models. For example if F is $\mathcal{N}(0,1)$ then the first expression is $3/\pi$ while the second is $2/\pi$. See Lehmann (1975) for some bounds on these expressions. Similar conclusions remain valid for $p > 1$. □

Remark 7.4.4 *Least Absolute Deviation Estimator.* As mentioned earlier, if we choose $\boldsymbol{g}(\mathbf{x}) \equiv \mathbf{x}$ and G to be degenerate at 0 then $\boldsymbol{\rho}_{\boldsymbol{x}}^{+}$ is the LAD estimator, v.i.z.,

$$\boldsymbol{\rho}_{\ell ad}^{+} := argmin_{\mathbf{t}} \sum_{j=1}^{p} \Big[\sum_{i=1}^{n} X_{i-j}\, \text{sign}(X_i - \mathbf{t}'\mathbf{Y}_{i-1})\Big]^2. \tag{7.4.41}$$

See also (7.4.6). Because of its importance we shall now summarize sufficient conditions under which it is asymptotically normally distributed. Of course we could use the stronger conditions (7.4.36) - (7.4.39) but they do not use the given information about G.

Clearly, (7.4.7(b)) implies (7.4.7(a)) in the present case. Moreover, in this case the L.H.S. of (7.4.8) becomes

$$EX_{1-j}^2 |F(n^{-1/2}(\mathbf{u}'\mathbf{Y}_0 + a\|\mathbf{Y}_0\|)) - F(n^{-1/2}\mathbf{u}'\mathbf{Y}_0)|$$

which tends to 0 by the D.C.T., (7.4.7(b)) and the continuity of F, $1 \le j \le p$.

Now consider (7.4.9). Assume the following

(7.4.42) *F has a density f, continuous and positive at* 0.

Recall from (7.3.12) that under (7.3.8), (7.3.9) and (7.4.7(b)),

$$n^{-1/2} \max\{\|\mathbf{Y}_{i-1}\|; 1 \le i \le n\} = o_p(1). \tag{7.4.43}$$

The r.v.'s involved in the L.H.S. of (7.4.9) in the present case are

$$n^{-1}\Big[\sum_{i=1}^{n} X_{i-j}^{\pm}\Big\{F(n^{-1/2}\mathbf{u}'\mathbf{Y}_{i-1} + n^{-1/2}\delta\|\mathbf{Y}_{i-1}\|)$$
$$-F(n^{-1/2}\mathbf{u}'\mathbf{Y}_{i-1} - n^{-1/2}\delta\|\mathbf{Y}_{i-1}\|)\Big\}\Big]^2$$

which, in view of (7.4.42), can be bounded above by

$$4\delta^2[n^{-1}\sum_{i=1}^{n} X_{i-j}^{\pm}\|\mathbf{Y}_{i-1}\| f(\eta_{ni})]^2, \tag{7.4.44}$$

where $\{\eta_{ni}\}$ are r.v.'s, with

$$\eta_{ni} \in n^{-1/2}\Big[\mathbf{u}'\mathbf{Y}_{i-1} - \delta\|\mathbf{Y}_{i-1}\|, \mathbf{u}'\mathbf{Y}_{i-1} + \delta\|\mathbf{Y}_{i-1}\|\Big], \quad 1 \le i \le n.$$

Hence, by the stationarity and the ergodicity of the process $\{X_i\}$, (7.4.7(b)), (7.4.42) and (7.4.43) imply that the r.v.'s in (7.4.44) converge to

$$4\delta^2[EX_{1-j}^{\pm}\|\mathbf{Y}_0\| f(0)]^2, \ 1 \le j \le p, \ a.s.,$$

This verifies (7.4.9) in the present case. The condition (7.4.10) is verified similarly.

Also note that here (5.5.44) is implied by (7.4.42) and (5.5.45) is trivially satisfied as $\int F(1-F)dG \le 1/4$ in the present case. We summarize the above discussion in

Corollary 7.4.2 *Assume that the autoregressive model (7.3.8) - (7.3.10) holds. In addition, assume that the error d.f. F has finite second moment, $F(0) = 1/2$ and satisfies (7.4.42). Then,*

$$n^{1/2}(\boldsymbol{\rho}_{\ell ad}^{+} - \boldsymbol{\rho}) \rightarrow_d \mathcal{N}(\mathbf{0}, (E\mathbf{Y}_0\mathbf{Y}_0')^{-1}/4f^2(0)),$$

where $\boldsymbol{\rho}_{\ell ad}^{+}$ is defined at (7.4.41).

7.5 Autoregression Quantiles and Rank Scores

The regression quantiles of Koenker and Basett (1978) (KB) are now accepted as an appropriate extension of the one sample quantiles in the one sample location models to the multiple linear regression model (1.1.1). In this section we shall discuss an extension of these regression quantiles to the stationary ergodic linear autoregressive time series of order p as specified at (7.3.8) - (7.3.10). We shall also assume that the error d.f. F is continuous.

Let $\mathbf{Z}_i' := (1, \mathbf{Y}_i')$, and for an $0 \le \alpha \le 1$, $\mathbf{t} \in \mathbb{R}^{p+1}$,

$$\psi_\alpha(u) := \alpha\, u\, I(u > 0) - (1-\alpha)\, u\, I(u \le 0),$$

$$Q_\alpha(\mathbf{t}) := \sum_{i=1}^{n} \psi_\alpha(X_i - \mathbf{Z}_{i-1}'\mathbf{t}),$$

$$\mathbf{S}_\alpha(\mathbf{t}) := n^{-1} \sum_{i=1}^{n} \mathbf{Y}_{i-1}\{I(X_i - \mathbf{Z}_{i-1}'\mathbf{t} \le 0) - \alpha\}.$$

The extension of the one sample order statistics to the linear AR(p) model (7.3.8) - (7.3.10) is given by the autoregression quantiles defined as a minimizer

$$\hat{\rho}(\alpha) := \operatorname{argmin}_{\mathbf{t}} Q_\alpha(\mathbf{t}). \tag{7.5.1}$$

We also need to define

$$\hat{\rho}_{md}(\alpha) := \operatorname{argmin}_{\mathbf{t}} \|\mathbf{S}_\alpha(\mathbf{t})\|^2, \tag{7.5.2}$$

Note that $\hat{\rho}(.5)$ and $\hat{\rho}_{md}(.5)$ are both equla to the LAD (least absolute deviation) estimator, which provides the extension of the one sample median to the above model.

Let $\mathbf{1}_n' := (1, \cdots, 1)_{1\times n}$ be an n-dimensional vector of 1's, π_n be a subset of size $p+1$ of the set of integers $\{1, 2, \cdots, n\}$, $\mathbf{X}_n' := (X_1, \cdots, X_n)$, $\mathbf{X}_{\pi_n}$ be the vector of X_i, $i \in \pi_n$, $\mathbf{H}_n$ be the $n \times (p+1)$ matrix with rows $\mathbf{Z}_{i-1}'; i = 1, \cdots, n$, and $\mathbf{H}_{\pi_n}$ be the $(p+1) \times (p+1)$ matrix with rows $\mathbf{Z}_{i-1}'$; $i \in \pi_n$.

Now recall that the above model is casual and invertible satisfying a relation like (7.3.11). This and the continuity of F implies that the rows of $\mathbf{H}_n$ are linearly independent as are its columns, w.p.1. Hence, the various inverses below exist w.p.1.

Now, let

$$\mathcal{B}_n(\alpha) := \left\{\mathbf{t} \in \mathbb{R}^{p+1}; Q_\alpha(\mathbf{t}) = \text{minimum}\right\},$$

and consider the following linear programming problem.

$$\begin{aligned} &\text{minimize } \alpha \mathbf{1}_n'\mathbf{r}^+ + (1-\alpha)\mathbf{1}_n'\mathbf{r}^-, \text{ w.r.t. } (\mathbf{t}, \mathbf{r}^+, \mathbf{r}^-), \\ &\text{subject to } \mathbf{X}_n - \mathbf{H}_n\mathbf{t} = \mathbf{r}^+ - \mathbf{r}^-, \\ &\text{over all } \ (\mathbf{t}, \mathbf{r}^+, \mathbf{r}^-) \in \mathbb{R}^{p+1} \times (0,\infty)^n \times (0,\infty)^n. \end{aligned} \tag{7.5.3}$$

Note that $\hat{\rho}(\alpha) \in \mathcal{B}_n(\alpha)$. Moreover, the set $\mathcal{B}_n(\alpha)$ is the convex hull of one or more basic solutions of the form

$$\mathbf{b}_{\pi_n} = \mathbf{H}_{\pi_n}^{-1}\mathbf{X}_{\pi_n}, \quad \pi_n \subset \{1, 2, \cdots, n\}. \tag{7.5.4}$$

This is proved in the same fashion as in KB.

A closely related entity is the so called *autoregression rank scores* defined as follows. Consider the following dual of the above linear programming problem.

$$\begin{aligned} &\text{Maximize } \mathbf{X}_n'\mathbf{a}, \text{ w.r.t. } \mathbf{a}, \text{ subject to} \\ &\mathbf{X}_n'\mathbf{a} = (1-\alpha)\mathbf{X}_n'\mathbf{1}_n, \quad \mathbf{a} \in [0,1]^n. \end{aligned} \tag{7.5.5}$$

By the linear programming theory the optimal solution $\hat{\mathbf{a}}_n(\alpha) = (a_{n1}(\alpha), \cdots, a_{nn}(\alpha))'$ of this problem can be computed in terms of $\hat{\rho}(\alpha)$ as follows: If $\hat{\rho}(\alpha) = \mathbf{H}_{\pi(\alpha)}^{-1}\mathbf{X}_{\pi(\alpha)}$, for some $(p+1)$-dimensional subset $\pi(\alpha)$ of $\{1, \cdots, n\}$, then, for $i \notin \pi(\alpha)$,

$$\begin{aligned} \hat{a}_{ni}(\alpha) &= 1, \quad X_i > \mathbf{Z}_{i-1}'\hat{\rho}(\alpha), \\ &= 0, \quad X_i < \mathbf{Z}_{i-1}'\hat{\rho}(\alpha), \end{aligned} \tag{7.5.6}$$

and, for $i \in \pi(\alpha)$, $\hat{a}_{ni}(\alpha)$ is the solution of the $p+1$ linear equations

$$\begin{aligned} &\sum_{j \in \pi_n(\alpha)} \mathbf{Z}_{j-1}\hat{a}_{nj}(\alpha) \\ &= (1-\alpha)\sum_{j=1}^{n} \mathbf{Z}_{j-1} - \sum_{j=1}^{n} \mathbf{Z}_{j-1} I\big(X_j > \mathbf{Z}_{j-1}'\hat{\rho}(\alpha)\big). \end{aligned} \tag{7.5.7}$$

The continuity of F implies that the autoregression rank scores $\hat{\mathbf{a}}_n(\alpha)$ are unique for all $0 < \alpha < 1$, w.p.1. The process $\hat{\mathbf{a}}_n \in [0,1]^n$ has piecewise linear paths in $[\mathcal{C}(0,1)]^n$ and $\hat{\mathbf{a}}_n(0) = \mathbf{1}_n = \mathbf{1}_n - \hat{\mathbf{a}}_n(1)$. It is invariant in the sense that $\hat{\mathbf{a}}_n(\alpha)$ based on the vector $\mathbf{X}_n + \mathbf{H}_n\mathbf{t}$ is the same as the $\hat{\mathbf{a}}_n(\alpha)$ based on $\mathbf{X}_n$, for all $\mathbf{t} \in \mathbb{R}^{p+1}$, $0 < \alpha < 1$. One can use the computational algorithm of Koenker and d'Odrey (1987, 1993) to compute these entities.

In the next two subsections we shall discuss the asymptotic distributions of $\hat{\rho}_n(\alpha)$ and $\hat{\mathbf{a}}_n(\alpha)$.

7.5.1 Autoregression quantiles

In this sub-section we shall show how the results of section 7.2 can be used to obtain the limiting distribution of $\hat{\rho}_n(\alpha)$. All the needed results are given in the following lemma. Its statement needs the additional notation:

$$\rho(\alpha) := \rho + F^{-1}(\alpha)\mathbf{e}_1, \quad \mathbf{e}_1 := (1, 0, \cdots, 0)',$$
$$q(\alpha) := f(F^{-1}(\alpha)), \quad 0 < \alpha < 1;$$
$$\mathbf{\Sigma}_n := n^{-1}\mathbf{H}_n'\mathbf{H}_n = n^{-1}\sum_{i=1}^{n}\mathbf{Z}_{i-1}\mathbf{Z}_{i-1}', \ \mathbf{\Sigma} = \text{plim}_n \mathbf{\Sigma}_n.$$

By the Ergodic Theorem, Σ exists and is positive definite. In this subsection, for any process $\mathcal{Z}_n(\mathbf{s}, \alpha)$, the statement $\mathcal{Z}_n(\mathbf{s}, \alpha) = o_p^*(1)$ means that for every $0 < a \le 1/2$, $0 < b < \infty$, $\sup\{|\mathcal{Z}_n(\alpha)|; \|\mathbf{s}\| \le b, a \le \alpha \le 1 - a\} = o_p(1)$.

Lemma 7.5.1 *Suppose the assumptions made at (7.3.8) - (7.3.10) hold. If, in addition (7.2.10) holds, then, for every* $0 < a \le 1/2$, $0 < b < \infty$,

$$\|\mathbf{S}_\alpha(\rho(\alpha) + n^{-1/2}\mathbf{s}) - \mathbf{S}_\alpha(\rho(\alpha)) - \mathbf{\Sigma}_n \mathbf{s} q(\alpha)\| = o_p^*(1). \tag{7.5.8}$$

Moreover,

$$\begin{aligned} n^{1/2}(\hat{\rho}_{md}(\alpha) - \rho(\alpha)) \\ = -\{q(\alpha)\mathbf{\Sigma}_n\}^{-1} n^{1/2}\mathbf{S}_\alpha(\rho(\alpha)) + o_p^*(1), \end{aligned} \tag{7.5.9}$$

$$n^{1/2}(\hat{\rho}_{md}(\alpha) - \hat{\rho}(\alpha)) = o_p^*(1). \tag{7.5.10}$$

If, (7.2.10) is strengthened to (**F1**) *and* (**F2**)*, then, for every* $0 < b < \infty$,

$$\sup \|n^{1/2}[\mathbf{S}_\alpha(\rho(\alpha) + n^{-1/2}\mathbf{s}) - \mathbf{S}_\alpha(\rho(\alpha))] - \mathbf{\Sigma}_n \mathbf{s} q(\alpha)\| = o_p(1), \tag{7.5.11}$$

where the supremum is taken over $(\alpha, \mathbf{s}) \in [0, 1] \times \{\mathbf{s} \in \mathbb{R}^{p+1}; \|\mathbf{s}\| \le b\}$.

A sketch of the proof. The claims (7.5.8) and (7.5.11) follow from Theorem 7.2.1 and Remark 7.2.1 in an obvious fashion: apply these results once with $h \equiv 1$ and p times, j^{th} time with $h(\mathbf{Y}_{i-1}) \equiv X_{i-j}$. In view of the Ergodic Theorem all conditions of Theorem 7.2.1 are a priori satisfied, in view of the current assumptions.

The proof of (7.5.9) is similar to that of Theorem 5.5.3. It amounts to first showing that

$$\sup_{a \le \alpha \le 1-a} \|n^{1/2}(\hat{\rho}_{md}(\alpha) - \rho(\alpha))\| = O_p(1), \tag{7.5.12}$$

and then using the result (7.5.8) to conclude the claim. But the proof of (7.5.12) is similar to that of Lemma 5.5.4, and hence no details are given.

To prove (7.5.10), we shall first show that for every $0 < a \le 1/2$,

$$n^{1/2}\|\mathbf{S}_\alpha(\hat{\boldsymbol{\rho}}(\alpha))\| = o_p^*(1). \tag{7.5.13}$$

To that effect, let

$$\begin{aligned}\mathbf{w}_n(\alpha) \quad :=& \quad \sum_{i \notin \pi_n(\alpha)} \mathbf{Z}'_{i-1}\{I(X_i - \mathbf{Z}'_{i-1}\hat{\boldsymbol{\rho}}(\alpha) \le 0) - \alpha\}\,\mathbf{H}^{-1}_{\pi_n(\alpha)} \\ & + \sum_{i \notin \pi_n(\alpha)} \mathbf{Z}'_{i-1} I\Big(X_i - \mathbf{Z}'_{i-1}\hat{\boldsymbol{\rho}}(\alpha) = 0\Big)\,\mathbf{H}^{-1}_{\pi_n(\alpha)},\end{aligned}$$

where $\pi_n(\alpha)$ is as in (7.5.4). Using $\text{sgn}(x) = 1 - 2I(x \le 0) + I(x = 0)$ we have the following inequalities w.p.1. For all $0 < \alpha < 1$,

$$(\alpha - 1)\mathbf{1}_p < \mathbf{w}_n(\alpha) < \alpha\mathbf{1_p}.$$

Note that from (7.5.4) we have $I((X_i - \mathbf{Z}'_{i-1}\hat{\boldsymbol{\rho}}(\alpha) = 0) = 0$, for all $i \notin \pi_n(\alpha)$. Thus we obtain

$$\begin{aligned}&\Big[\sum_{i=1}^{n} \mathbf{Z}'_{i-1}\{I(X_i - \mathbf{Z}'_{i-1}\hat{\boldsymbol{\rho}}(\alpha) \le 0) - \alpha\} \\ &\qquad - \sum_{i \in \pi_n(\alpha)} \mathbf{Z}'_{i-1}\{I(X_i - \mathbf{Z}'_{i-1}\hat{\boldsymbol{\rho}}(\alpha) \le 0) - \alpha\}\Big]\mathbf{H}^{-1}_{\pi_n(\alpha)} \\ =& \quad \mathbf{w}'_n(\alpha).\end{aligned}$$

Again, by (7.5.4), $I(X_i - \mathbf{Z}'_{i-1}\hat{\boldsymbol{\rho}}(\alpha) \le 0) = 1$, $i \in \pi_n(\alpha)$, $0 < \alpha < 1$,, w.p.1. Hence, w.p.1., $\forall\, 0 < \alpha < 1$,

$$n^{1/2}\mathbf{S}_\alpha(\hat{\boldsymbol{\rho}}(\alpha)) = n^{-1/2} \sum_{i \in \pi_n(\alpha)} \mathbf{Z}'_{i-1}(1 - \alpha) + n^{-1/2}\mathbf{H}'_{\pi_n(\alpha)}\mathbf{w}_n(\alpha),$$

so that

$$\sup_{a \le \alpha \le 1-a} \|n^{1/2}\mathbf{S}_\alpha(\hat{\boldsymbol{\rho}}(\alpha))\| \le 2(p+1) \max_{1 \le i \le n} n^{-1/2}\|\mathbf{Z}_{i-1}\| = o_p(1),$$

in view of the square integrability of X_0 and the stationarity of the process. This completes the proof of (7.5.13). Hence we obtain

$$\sup_{a \le \alpha \le 1-a} \inf_{\mathbf{s}} \|n^{1/2}\mathbf{S}_\alpha(\mathbf{s})\| = o_p(1). \tag{7.5.14}$$

This and (7.5.13) essentially then show that

$$\sup_{a \le \alpha \le 1-a} n^{1/2}\|\mathbf{S}_\alpha(\hat{\boldsymbol{\rho}}(\alpha)) - \mathbf{S}_\alpha(\hat{\boldsymbol{\rho}}_{md})(\alpha))\| = o_p(1),$$

which together with (7.5.8) proves the claim (7.5.10). □

The following corollary is immediate.

Corollary 7.5.1 *Under the assumptions (7.2.10) and (7.3.8) - (7.3.10),*

(7.5.15) $$n^{1/2}(\hat{\boldsymbol{\rho}}(\alpha) - \boldsymbol{\rho}(\alpha)) = -\{q(\alpha)\boldsymbol{\Sigma}_n\}^{-1} n^{1/2} \mathbf{S}_\alpha(\boldsymbol{\rho}(\alpha)) + o_p^*(1).$$

Moreover, for every $0 < \alpha_1 < \cdots < \alpha_k < 1$, the asymptotic joint distribution of the vector $n^{1/2}[(\hat{\boldsymbol{\rho}}(\alpha_1) - \boldsymbol{\rho}(\alpha_1)), \cdots, (\hat{\boldsymbol{\rho}}(\alpha_k) - \boldsymbol{\rho}(\alpha_k))]$ is $(p+1) \times k$ normal distribution with the mean matrix bf0 and the covariance matrix $\mathbf{A} \oplus \boldsymbol{\Sigma}^{-1}$, where

$$\mathbf{A} = \Big((\alpha_i \wedge \alpha_j - \alpha_i \alpha_j)/q(\alpha_i)q(\alpha_j) \Big)_{1 \le i,j \le k},$$

and where $\oplus$ denotes the Kronecker matrix product.

7.5.2 Autoregression rank scores

Now we shall discuss the asymptotic behaviour of the autoregression rank scores defined at (7.5.6). To that effect we need to introduce some more notation. Let q be a positive integer, $\{k_{nij};\ 1 \le j \le q\}$ be $\mathcal{F}_{i-1} := \sigma - \{\mathbf{Y}_0, \varepsilon_0, \varepsilon_1, \cdots, \varepsilon_{i-1}\}$ measurable and independent of ε_i, $1 \le i \le n$. Let $\mathbf{k}_{ni} := (k_{ni1}, \cdots, k_{niq})'$ and $\mathbf{K}$ denote the matrix whose i^{th} row is $\mathbf{k}_{ni}$, $1 \le i \le n$. Define the processes

$$\hat{\mathbf{U}}_{\mathbf{k}}(\alpha) := n^{-1} \sum_{i=1}^{n} \mathbf{k}_{ni} \{\hat{a}_{ni}(\alpha) - (1-\alpha)\},$$

$$\mathbf{U}_{\mathbf{k}}(\alpha) := n^{-1} \sum_{i=1}^{n} \mathbf{k}_{ni} \{I(\varepsilon_i > F^{-1}(\alpha) - (1-\alpha)\}, \quad 0 \le \alpha \le 1.$$

Let

$$\mathcal{K}_n := n^{-1} \mathbf{K}' \mathbf{H}_n, \quad \hat{\boldsymbol{\Delta}}_n(\alpha) := n^{1/2}(\hat{\boldsymbol{\rho}}(\alpha) - \boldsymbol{\rho}(\alpha)).$$

We are now ready to state

Lemma 7.5.2 *In addition to the model assumptions (7.3.8) - (7.3.10), suppose the following two conditions hold. For some positive definite matrix $\boldsymbol{\Gamma}_{q \times q}$,*

(7.5.16) $$n^{-1} \mathbf{K}' \mathbf{K} = \boldsymbol{\Gamma} + o_p(1).$$

(7.5.17) $$n^{-1/2} \max_{1 \le i \le n} \|\mathbf{k}_{ni}\| = o_p(1).$$

Then, for every $0 < a \le 1/2$,

(7.5.18) $$n^{1/2}[\hat{\mathbf{U}}_{\mathbf{k}}(\alpha) - \mathbf{U}_{\mathbf{k}}(\alpha)] = -\mathcal{K}_n \cdot \hat{\boldsymbol{\Delta}}_n(\alpha) q(\alpha) + o_p^*(1).$$

Consequently,

$$(7.5.19)\qquad n^{1/2}\hat{\mathbf{U}}_{\mathbf{k}}(\alpha) = n^{1/2}[\mathbf{U}_{\mathbf{k}}(\alpha) - \mathcal{K}_n \cdot \mathbf{\Sigma}_n^{-1}\mathbf{S}_\alpha(\boldsymbol{\rho}(\alpha))] + o_p^*(1).$$

Proof. From (7.5.6), we obtain that $\forall\, 1 \le i \le n,\ 0 < \alpha < 1$,

$$\begin{aligned}\hat{a}_{ni}(\alpha) &= I(\varepsilon_i > F^{-1}(\alpha) + n^{-1/2}\mathbf{Z}_{i-1}'\hat{\mathbf{\Delta}}_n(\alpha)) \\ &\qquad\qquad + \hat{a}_{ni}(\alpha) I(X_i = \mathbf{Z}_{i-1}'\hat{\boldsymbol{\rho}}(\alpha)),\end{aligned}$$

which in turn yields the following identity:

$$\begin{aligned}&\hat{a}_{ni}(\alpha) - (1-\alpha) \\ &= I(\varepsilon_i > F^{-1}(\alpha) - (1-\alpha) \\ &\quad -\{I(\varepsilon_i \le F^{-1}(\alpha) + n^{-1/2}\mathbf{Z}_{i-1}'\hat{\mathbf{\Delta}}_n(\alpha)) - I(\varepsilon_i \le F^{-1}(\alpha))\} \\ &\quad + \hat{a}_{ni}(\alpha) I(X_i = \mathbf{Z}_{i-1}'\hat{\boldsymbol{\rho}}(\alpha)),\end{aligned}$$

for all $1 \le i \le n$, $0 < \alpha < 1$, w.p.1. This and (7.5.4) yield

$$\begin{aligned}&n^{1/2}\hat{\mathbf{U}}_{\mathbf{k}}(\alpha) \\ &= n^{1/2}\mathbf{U}_{\mathbf{k}}(\alpha) - \mathcal{K}_n\hat{\mathbf{\Delta}}_n(\alpha)q(\alpha) \\ &\quad -\Big[n^{-1/2}\sum_{i=1}^n \mathbf{k}_{ni}\Big\{I(\varepsilon_i \le F^{-1}(\alpha) + n^{-1/2}\mathbf{Z}_{i-1}'\hat{\mathbf{\Delta}}_n(\alpha)) \\ &\qquad\qquad\qquad -I(\varepsilon_i \le F^{-1}(\alpha))\Big\} - \mathcal{K}_n\hat{\mathbf{\Delta}}_n(\alpha)q(\alpha)\Big] \\ &\quad + n^{-1/2}\sum_{i\in\pi_n(\alpha)} \mathbf{k}_{ni}\hat{a}_{ni}(\alpha) I(X_i = \mathbf{Z}_{i-1}'\hat{\boldsymbol{\rho}}(\alpha)) \\ &= n^{1/2}\mathbf{U}_{\mathbf{k}}(\alpha) - \mathcal{K}_n\hat{\mathbf{\Delta}}_n(\alpha)q(\alpha) - R_1(\alpha) + R_2(\alpha), \quad \textit{say}.\end{aligned}$$

Now, by the C-S inequality and by (7.5.16), $\|\mathcal{K}_n\| = O_p(1)$. Apply Remark 7.2.1 to $\gamma_{ni} \equiv k_{nij}$ and other entities as in the previous section to conclude that $\sup\{\|R_1(\alpha)\|; 0 \le \alpha \le 1\} = o_p(1)$. Also, note that from the results of the previous section we have $\|\sup_{a\le\alpha\le 1-a}\|\hat{\mathbf{\Delta}}_n(\alpha)\| = O_p(1)$. Use this and (7.5.17) to obtain $\sup\{\|R_1(\alpha)\|; a \le \alpha \le 1-a\} = o_p(1)$, thereby completing the proof of (7.5.18). The rest is obvious. □.

Corollary 7.5.2 *Under the assumptions of Lemma 7.5.2, the autoregression quantile and autoregression rank score processes are asymptotically independent. Moreover, for every $k \ge 1$, and for every $0 < \alpha_1 < \cdots < \alpha_k$,*

$$\begin{aligned}&n^{1/2}(\hat{\mathbf{U}}_{\mathbf{k}}(\alpha_1), \cdots, \hat{\mathbf{U}}_{\mathbf{k}}(\alpha_1)) \Longrightarrow \mathcal{N}(\mathbf{0}, \mathcal{B}), \\ &\mathcal{B} := \mathbf{B} \oplus \mathit{plim}_n n^{-1}[\mathbf{K}_n' - \mathcal{K}_n\mathbf{\Sigma}_n\mathbf{H}_n'][\mathbf{K}_n' - \mathcal{K}_n\mathbf{\Sigma}_n\mathbf{H}_n']',\end{aligned}$$

where $\mathbf{B} := ((\alpha_i \wedge \alpha_j - \alpha_i\alpha_j))_{1\le i,j\le k}$.

Proof. Let $s_i(\alpha) := I(\varepsilon_i > F^{-1}(\alpha)) - (1-\alpha)$, $1 \le i \le n$, $\mathbf{s}(\alpha) := (s_1(\alpha), \cdots, s_n(\alpha))'$. The leading r.v.'s in the right hand sides of (7.5.15) and (7.5.19) are equal to

$$-\mathbf{\Sigma}_n^{-1} n^{-1/2}\mathbf{H}_n'\mathbf{s}(\alpha)/q(\alpha), \quad n^{-1/2}[\mathbf{K}_n' - \mathcal{K}_n\mathbf{\Sigma}_n\mathbf{H}_n']\mathbf{s}(\alpha),$$

respectively. By the stationarity, ergodicity of the underlying process, Lemma 9.1.3 in the Appendix, and by the Cramér-Wold device, it follows that for each $\alpha \in (0,1)$, the asymptotic joint distribution of $\hat{\mathbf{\Delta}}_n(\alpha)$ and $n^{1/2}\hat{\mathbf{U}}_{\mathbf{k}}(\alpha)$ is $(p+1+q)$-dimensional normal with the mean vector $\mathbf{0}$ and the covariance matrix

$$\mathcal{D} = \begin{bmatrix} \mathcal{D}_{11} & \mathcal{D}_{12} \\ \mathcal{D}_{12} & \mathcal{D}_{22} \end{bmatrix},$$

where

$$\mathcal{D}_{11} := [\alpha(1-\alpha)/q^2(\alpha)]\Sigma^{-1}$$
$$\mathcal{D}_{22} := \mathrm{plim}_n n^{-1}[\mathbf{K}_n' - \mathcal{K}_n\mathbf{\Sigma}_n\mathbf{H}_n']'[\mathbf{K}_n' - \mathcal{K}_n\mathbf{\Sigma}_n\mathbf{H}_n']$$
$$\mathcal{D}_{12} := [\alpha(1-\alpha)/q(\alpha)]\,\mathrm{plim}_n n^{-1}\mathbf{\Sigma}_n^{-1}\mathbf{H}_n'[\mathbf{K}_n' - \mathcal{K}_n\mathbf{\Sigma}_n\mathbf{H}_n']'.$$

But, by definition, , w.p.1, $\forall n \ge 1$,

$$n^{-1}\mathbf{\Sigma}_n^{-1}\mathbf{H}_n'[\mathbf{K}_n' - \mathcal{K}_n\mathbf{\Sigma}_n\mathbf{H}_n']' = \mathbf{\Sigma}_n^{-1}\mathcal{K}_n' - \mathbf{\Sigma}_n^{-1}\mathcal{K}_n' = 0$$

This proves the claim of independence for each α. The result is proved similarly for any finite dimensional joint distribution. □

Note: The above results were first obtained in Koul and Saleh (1995), using numerous facts available in linear regression from the works of Koenker and Bassett (1978) and Grutenbrunner and Jurečková (1992), and of course the AUL result given in Theorem 7.2.1.

7.6 Goodness-of-fit Testing for F

Once again consider the AR(p) model given by (7.3.8), (7.3.9) and let F_0 be a known d.f.. Consider the problem of testing $H_0 : F = F_0$. One of the common tests of H_0 is based on the Kolmogorov - Smirnov statistic

$$D_n := n^{1/2}\sup_x |F_n(x, \hat{\rho}) - F_0(x)|.$$

From Corollary 7.2.1 one readily has the following:

If F_0 has finite second moment and a uniformly continuous density f_0, $f_0 > 0$ a.e.; $\hat{\rho}$ satisfies (7.3.20) under F_0, then, under H_0,

$$D_n = \sup\left|B(F_0(x)) + n^{1/2}(\hat{\rho} - \rho)' n^{-1} \sum_{i=1}^{n} \mathbf{Y}_{i-1} f_0(x)\right| + o_p(1).$$

In addition, if $E\mathbf{Y}_0 = 0 = E\varepsilon_1$, then $D_n \to_d \sup\{|B(t)|, 0 \leq t \leq 1\}$, thereby rendering D_n asymptotically distribution free.

Next, consider, $H_{01} : F = \mathcal{N}(\mu, \sigma^2)$, $\mu \in \mathbb{R}$, $\sigma^2 > 0$. In other words, H_{01} states that the AR(p) process is generated by some normal errors. Let $\hat{\mu}_n$, $\hat{\sigma}_n$ and $\hat{\rho}_n$ be estimators of μ, σ, and ρ, respectively. Define

$$\hat{F}_n(x) := n^{-1} \sum_{i=1}^{n} I(X_i \leq x\hat{\sigma}_n + \hat{\mu}_n + \hat{\rho}_n' \mathbf{Y}_{i-1}), \quad x \in \mathbb{R},$$

$$\hat{D}_n := n^{1/2} \sup_x |\hat{F}_n(x) - \Phi(x)|, \quad \Phi = \mathcal{N}(0,1)\ d.f..$$

Corollary 7.2.1 can be readily modified in a routine fashion to yield that if

$$n^{1/2}|(\hat{\mu}_n - \mu) + (\hat{\sigma}_n - \sigma)|\sigma^{-1} + n^{1/2}\|\hat{\rho}_n - \rho\| = O_p(1)$$

then

$$\hat{D}_n := \sup_x |B(\Phi(x)) + n^{1/2}\{(\hat{\mu}_n - \mu) + (\hat{\sigma}_n - \sigma)\}\sigma^{-1}\phi(x)| + o_p(1),$$

where ϕ is the density of Φ. Thus the asymptotic null distribution of $\hat{D}_n$ is similar to its analogue in the one sample location-scale model: *the estimation of ρ has no effect on the large sample null distribution of $\hat{D}_n$.*

Clearly, similar conclusions can be applied to other goodness-of-fit tests. In particular we leave it as *an exercise* for an interested reader to investigate the large sample behaviour of the goodness - of - fit tests based on L_2 - distances, analogous to the results obtained in Section 6.3. Lemma 6.3.1 and the results of the previous section are found useful here. □

7.7 Autoregressive Model Fitting

7.7.1 Introduction

In this section we shall consider the problem of fitting a given parametric autoregressive model of order 1 to a real valued stationary ergodic Markovian time series X_i, $i = 0, \pm 1, \pm 2, \cdots$. Much of the development here is parallel to that of Section 6.6 above. We shall thus be brief on motivation and details here.

Let ψ be a nondecreasing real valued function such that $E|\psi(X_1 - r)| < \infty$, for each $r \in \mathbb{R}$. Define the ψ-autoregressive function m_ψ by the requirement that

$$E[\psi(X_1 - m_\psi(X_0))|X_0] = 0, \quad a.s. \tag{7.7.1}$$

Observe that, if $\psi(x) \equiv x$, then $m_\psi = \mu$, and if

$$\psi(x) \equiv \psi_\alpha(x) := I(x > 0) - (1 - \alpha), \text{ for an } 0 < \alpha < 1,$$

then $m_\psi(x) \equiv m_\alpha(x)$, the αth quantile of the conditional distribution of X_1, given $X_0 = x$. The choice of ψ is up to the practitioner. If the desire is to have a goodness-of-fit procedure that is less sensitive to outliers in the innovations $X_i - m_\psi(X_{i-1})$, then one may choose a bounded ψ. In the sequel m_ψ is assumed to exist uniquely.

The process of interest here is

$$M_{n,\psi}(x) := n^{-1/2} \sum_{i=1}^{n} \psi(X_i - m_\psi(X_{i-1}))\, I(X_{i-1} \le x), \; x \in [-\infty, \infty].$$

Note that $M_{n,\psi}$ is an analogue of $S_{n,\psi}$ with $p = 1$, $\ell(x) \equiv x$, of Section 6.6. Write μ, $M_{n,I}$, for m_ψ, $M_{n,\psi}$ when $\psi(x) \equiv x$, respectively. Tests of goodness-of-fit for fitting a model to m_ψ will be based on the process $M_{n,\psi}$.

We shall also assume throughout that the d.f. G of X_0 is continuous and

$$E\psi^2(X_1 - m_\psi(X_0)) < \infty. \tag{7.7.2}$$

Under (7.7.1), (7.7.2), and under some additional assumptions that involve some moments and conditional innovation density, Theorem 7.7.1 below, which in turn follows from the Theorem 2.2.6, gives the weak convergence of $M_{n,\psi}$ to a continuous mean zero Gaussian process M_ψ with the covariance function

$$K_\psi(x, y) = E\psi^2(X_1 - m_\psi(X_0))\, I(X_0 \le x \wedge y), \quad x, y \in \mathbb{R}.$$

Arguing as for (6.6.3), M_ψ admits a representation

$$M_\psi(x) = B(\tau_\psi^2(x)), \quad \text{in distribution}, \tag{7.7.3}$$

where B is a standard Brownian motion on the positive real line. Note that the continuity of the d.f. G implies that of τ_ψ and hence that of $B(\tau_\psi^2)$. The representation (7.7.3), Theorem 7.7.1 and the continuous mapping theorem yield

$$\sup_{x\in\mathbb{R}} |M_{n,\psi}(x)| \Longrightarrow \sup_{0\le t\le \tau_\psi^2(\infty)} |B(t)| = \tau_\psi(\infty) \sup_{0\le t\le 1} |B(t)|, \quad \text{in law.}$$

Thus, to test the simple hypothesis $\tilde{H}_0 : m_\psi = m_0$, where m_0 is a known function proceed as follows. Estimate (under $m_\psi = m_0$) the variance $\tau_\psi^2(x)$ by

$$\tau_{n,\psi}^2(x) := n^{-1} \sum_{i=1}^{n} \psi^2(X_i - m_0(X_{i-1}))I(X_{i-1} \le x), \quad x \in \mathbb{R},$$

and replace m_ψ by m_0 in the definition of $M_{n,\psi}$. Write $s_{n,\psi}^2$ for $\tau_{n,\psi}^2(\infty)$. Then, for example, the Kolmogorov-Smirnov (K-S) test based on $M_{n,\psi}$ of the given asymptotic level would reject the hypothesis $\tilde{H}_0$ if

$$\sup\{s_{n,\psi}^{-1}|M_{n,\psi}(x)| : x \in \mathbb{R}\}$$

exceeds an appropriate critical value obtained from the boundary crossing probabilities of a Brownian motion on the unit interval which are readily available. More generally, the asymptotic level of any test based on a continuous function of $s_{n,\psi}^{-1} M_{n,\psi}((\tau_{n,\psi}^2)^{-1})$ can be obtained from the distribution of the corresponding function of B on $[0,1]$, where $(\tau_{n,\psi}^2)^{-1}(t) := \inf\{x \in \mathbb{R} : \tau_{n,\psi}^2(x) \ge t\}$, $t \ge 0$. For example, the asymptotic level of the test based on the Cramér - von Mises statistic

$$s_{n,\psi}^{-2} \int_0^{s_{n,\psi}^2} M_{n,\psi}^2((\tau_{n,\psi}^2)^{-1}(t))\, dH(t/s_{n,\psi}^2),$$

is obtained from the distribution of $\int_0^1 B^2\, dH$, where H is a d.f. on $[0,1]$.

Now, let $\mathcal{M}$ be as in Section 6.6 and consider the problem of testing the goodness-of-fit hypothesis

$$H_0 : m_\psi(x) = m(x, \boldsymbol{\theta}_0), \quad \text{for some } \boldsymbol{\theta}_0 \in \Theta,\, x \in \mathcal{I},$$

where $\mathcal{I}$ is now a compact subset of $\mathbb{R}$. Let $\boldsymbol{\theta}_n$ be an consistent estimator of $\boldsymbol{\theta}_0$ under H_0 based on $\{X_i,\, 0 \le i \le n\}$. Define, for an $-\infty \le x \le \infty$,

$$\hat{M}_{n,\psi}(x) = n^{-1/2} \sum_{i=1}^{n} \psi(X_i - m(X_{i-1}, \boldsymbol{\theta}_n))I(X_{i-1} \le x).$$

The process $\hat{M}_{n,\psi}$ is a weighted empirical process, where the weights, at X_{i-1} are now given by the ψ-residuals $\psi(X_i - m(X_{i-1}, \hat{\theta}_n))$. Tests for H_0 can be based on an appropriately scaled discrepancy of this process. For example, an analogue of the K-S test would reject H_0 in favor of H_1 if $\sup\{\sigma_{n,\psi}^{-1}|\hat{M}_{n,\psi}(x)| : x \in \mathbb{R}\}$ is too large, where $\sigma_{n,\psi}^2 := n^{-1}\sum_{i=1}^n \psi^2(X_i - m(X_{i-1}, \boldsymbol{\theta}_n))$. These tests, however, are not generally asymptotically distribution free.

In the next sub-section we shall show that under the same conditions on $\mathcal{M}$ as in Section 6.6.2, under H_0, the weak limit of $\tilde{T}_n\hat{M}_{n,\psi}$ is $B(G)$, so that the asymptotic null distribution of various tests based on it will be known. Here $\tilde{T}_n$ is an analogue of the transformation T_n of (6.6.25). Computational formulas for computing some of these tests is also given in the same section. Tests based on various discrepancies of $\hat{M}_{n,\psi}$ are consistent here also under the condition (6.6.4) with λ properly defined: Just change X, Y to X_0, X_1, respectively.

7.7.2 Transform T_n of $M_{n,\psi}$

This section first discusses the asymptotic behavior of the processes introduced in the previous section under the simple and composite hypotheses. Then a transformation T and its estimate $\tilde{T}_n$ are given so that the processes $T\hat{M}_{n,\psi}$ and $\tilde{T}_n\hat{M}_{n,\psi}$ have the same weak limit with a known distribution.

Let

$$\begin{aligned} F_y(x) &:= P(X_1 - m_\psi(X_0) \le x\,|X_0 = y), \qquad x,\, y \in \mathbb{R}, \\ \varepsilon_i &:= X_i - m_\psi(X_{i-1}),\ i = 0, \pm 1, \pm 2, \cdots . \end{aligned}$$

We are ready to state our first result.

Theorem 7.7.1 *Assume that (7.7.1) and (7.7.2) holds. Then all finite dimensional distributions of $M_{n,\psi}$ converge weakly to those of a centered continuous Gaussian process M_ψ with the covariance function K_ψ.*

(I). Suppose, in addition, that for some $\eta > 0$, $\delta > 0$,

(7.7.4)
$$\begin{aligned} &(a)\ \ E\,\psi^4(\varepsilon_1) < \infty, \quad (b)\ \ E\,\psi^4(\varepsilon_1)|X_0|^{1+\eta} < \infty, \\ &(c)\ \ E\{\psi^2(\varepsilon_2)\psi^2(\varepsilon_1)|X_1|\}^{1+\delta} < \infty, \end{aligned}$$

and that the family of d.f.'s $\{F_y,\, y \in \mathbb{R}\}$ have Lebesgue densities $\{f_y,\, y \in \mathbb{R}\}$ that are uniformly bounded:

(7.7.5)
$$\sup_{x,y} f_y(x) < \infty.$$

Then

$$M_{n,\psi} \Longrightarrow \mathcal{M}_\psi, \text{ in the space } D[-\infty,\infty]. \tag{7.7.6}$$

(II). Instead of (7.7.4) and (7.7.5), suppose that ψ is bounded and the family of d.f.'s $\{F_y,\, y \in \mathbb{R}\}$ have Lebesgue densities $\{f_y,\, y \in \mathbb{R}\}$ satisfying

$$\int \Big[\mathbb{E}\Big\{ f_{X_0}^{1+\delta}(x - m_\psi(X_0))\Big\}\Big]^{\frac{1}{1+\delta}} dx < \infty, \tag{7.7.7}$$

for some $\delta > 0$. Then also (7.7.6) holds.

Proof . Part (I) follows from Theorem 2.2.4 while part (II) follows from Theorem 2.2.7 upon choosing $\varphi = m_\psi$, $Z_{n,i} \equiv \varepsilon(X_i - m_\psi(X_{i-1})$, $L_y \equiv F_y$, $l_y \equiv f_y$ in there. □

Remark 7.7.1 Conditions (7.7.4) and (7.7.5) are needed to ensure the uniform tightness in the space $D[-\infty,\infty]$ for a general ψ while (7.7.7) suffices for a bounded ψ. Condition (7.7.4) is satisfied when, as is assumed in most standard time series models, the innovations are independent of the past, and when for some $\delta > 0$, $\mathbb{E}\psi^{4(1+\delta)}(\varepsilon_1)$ and $\mathbb{E}X_1^{2(1+\delta)}$ are finite. Moreover, in this situation the conditional distributions do not depend on y, so that (7.7.5) amounts to assuming that the density of ε_1 is bounded. In the case of bounded ψ, $\mathbb{E}|X_1|^{1+\delta} < \infty$, for some $\delta > 0$, implies (7.7.4).

Now consider the assumption (7.7.7). Note that the stationary distribution G has Lebesgue density $g(x) \equiv \mathbb{E}f_{X_0}(x - m_\psi(X_0))$. This fact together with (7.7.5) implies that the left hand side of (7.7.7) is bounded from the above by a constant $C := [\sup_{x,y} f_y(x)]^{\frac{\delta}{1+\delta}}$ times

$$\int \big[\mathbb{E}f_{X_0}(x - m_\psi(X_0))\big]^{\frac{1}{1+\delta}} dx = \int g^{\frac{1}{1+\delta}}(x)dx.$$

Thus, (7.7.7) is implied by assuming

$$\int g^{\frac{1}{1+\delta}}(x)dx < \infty.$$

Alternately, suppose m_ψ is bounded, and that $f_y(x) \le f(x)$, for all $x, y \in \mathbb{R}$, where f is a bounded and unimodal Lebesgue density on $\mathbb{R}$. Then also the left hand side of (7.7.7) is finite. One thus sees that in the particular case of the i.i.d. homoscedastic errors, (7.7.7) is satisfied for either all bounded error densities and for all stationary densities that have an exponential tail or for all bounded unimodal error densities in the case of bounded m_ψ. Summarizing, we see that (7.7.4), (7.7.5) and (7.7.7) are fulfilled in many models under standard assumptions on the relevant densities and moments.

Perhaps the differences between Theorem 6.6.1 and the above theorem are worth pointing out. In the former no additional moment conditions, beyond the finite second moment of the ψ-innovation, were needed nor did it require the the error density to be bounded or to satisfy any thing like (7.7.7).

Next, we need to study the asymptotic null behaviour of $\hat{M}_{n,\psi}$. To that effect, the following additional regularity conditions on the underlying entities will be needed. To begin with, the regularity conditions for asymptotic expansion of $\hat{M}_{n,\psi}$ are stated without assuming X_{i-1} to be independent of ε_i, $i \geq 1$. All probability statements in these assumptions are understood to be made under H_0. Unlike in the regression setup, the d.f. of X_0 here in general depends on $\boldsymbol{\theta}_0$ but this dependence is not exhibited for the sake of convenience. We make the following assumptions.

The estimator $\boldsymbol{\theta}_n$ satisfies

$$n^{1/2}(\boldsymbol{\theta}_n - \boldsymbol{\theta}_0) = n^{-1/2} \sum_{i=1}^{n} \phi(X_{i-1}, X_i, \boldsymbol{\theta}_0) + o_p(1) \qquad (7.7.8)$$

for some q-vector valued function ϕ such that $\mathbb{E}\{\phi(X_0, X_1, \boldsymbol{\theta}_0)|X_0\} = 0$ and $\phi(\boldsymbol{\theta}_0) := E\phi(X_0, X_1, \boldsymbol{\theta}_0)\phi'(X_0, X_1, \boldsymbol{\theta}_0)$ exists and is positive definite.

(F). The family of d.f.'s $\{F_y,\ y \in \mathbb{R}\}$ has Lebesgue densities $\{f_y,\ y \in \mathbb{R}\}$ that are equicontinuous: For every $\alpha > 0$ there exists a $\delta > 0$ such that

$$\sup_{y \in \mathbb{R}, |x-z|<\delta} |f_y(x) - f_y(z)| \leq \alpha.$$

Let, for $1 \leq j \leq q$, $x \in \mathbb{R}$,

$$\begin{aligned}
\dot{\mathbf{m}}_j(x, \boldsymbol{\theta}_0) &= \mathbb{E}\dot{\mathbf{m}}_j(X_0, \boldsymbol{\theta}_0)\dot{\psi}(X_1 - m(X_0, \boldsymbol{\theta}_0))I(X_0 \leq x), \\
\dot{\Gamma}_j(x, \boldsymbol{\theta}_0) &= \mathbb{E}\dot{\mathbf{m}}_j(X_0, \boldsymbol{\theta}_0) \int f_{X_0}\, d\psi\, I(X_0 \leq x), \\
\dot{\mathbf{M}}(x, \boldsymbol{\theta}_0) &= (\dot{\mathbf{m}}_1(x, \boldsymbol{\theta}_0), \ldots, \dot{\mathbf{m}}_q(x, \boldsymbol{\theta}_0))', \\
\dot{\boldsymbol{\Gamma}}(x, \boldsymbol{\theta}_0) &= (\dot{\Gamma}_1(x, \boldsymbol{\theta}_0), \ldots, \dot{\Gamma}_q(x, \boldsymbol{\theta}_0))'.
\end{aligned}$$

Note that by (6.6.14) and (Ψ_1) or by (6.6.14), (Ψ_2) and (F) these entities are well-defined.

We are now ready to formulate an asymptotic expansion of $\hat{M}_{n,\psi}$, which is crucial for the subsequent results and the transformation T_n. Recall the conditions (Ψ_1) and (Ψ_2) from Section 6.6.2.

Theorem 7.7.2 *Assume that (7.7.1), (7.7.8) and H_0 hold. About the model $\mathcal{M}$, assume that (6.6.13) and (6.6.14) hold with X_i, X replaced by X_{i-1}, X_0, respectively.*

(A). *If, in addition (Ψ_1) holds, then*

$$\Big|\hat{M}_{n,\psi}(x) - M_{n,\psi}(x) + \dot{\mathbf{M}}'(x,\boldsymbol{\theta}_0)\, n^{-1/2}\sum_{i=1}^{n}\phi(X_{i-1},X_i,\boldsymbol{\theta}_0)\Big| = u_p(1). \tag{7.7.9}$$

(B). *Assume, in addition, that (Ψ_2) and (F) hold, and that either $\mathbb{E}|X_0|^{1+\delta} < \infty$, for some $\delta > 0$ and (7.7.5) holds or (7.7.7) holds. Then the conclusion (7.7.9) with $\dot{\mathbf{M}}$ replaced by $\dot{\mathbf{\Gamma}}$ continues to hold.*

We note that the Remark 6.6.1 applies to the autoregressive setup also.

The following corollary is an immediate consequence of the above theorem and Theorems 7.7.1. We shall state it for the smooth ψ- case only. The same holds in the non-smooth ψ. Note that under H_0, $\varepsilon_i \equiv X_i - m(X_{i-1},\boldsymbol{\theta}_0)$.

Corollary 7.7.1 *Under the assumptions of Theorems 7.7.1 and 7.7.2(A),*

$$\hat{M}_{n,\psi} \Longrightarrow \hat{M}_{\psi}, \quad \textit{in the space } D[-\infty,\infty],$$

where $\hat{M}_{\psi}$ is a centered continuous Gaussian process with the covariance function

$$\begin{aligned} &K_{\psi}^{1}(x,y)\\ &\quad= K_{\psi}(x,y) + \dot{\mathbf{M}}'(x,\boldsymbol{\theta}_0)\phi(\boldsymbol{\theta}_0)\mathbf{M}(y,\boldsymbol{\theta}_0)\\ &\qquad -\dot{\mathbf{M}}'(x,\boldsymbol{\theta}_0)\mathbb{E}\Big\{I(X_0 \le y)\psi(\varepsilon_1)\phi(X_0,X_1,\boldsymbol{\theta}_0)\Big\}\\ &\qquad -\dot{\mathbf{M}}'(y,\boldsymbol{\theta}_0)\mathbb{E}\Big\{I(X_0 \le x)\psi(\varepsilon_1)\phi(X_0,X_1,\boldsymbol{\theta}_0)\Big\}. \end{aligned}$$

The above complicated looking covariance function can be further simplified if we choose $\boldsymbol{\theta}_n$ to be related to the function ψ in the following fashion. Recall from the previous section that $\sigma^2_\psi(x) = \mathbb{E}[\psi^2(\varepsilon_1)|X_0 = x]$ and let, for $x \in \mathbb{R}^p$,

$$\gamma_\psi(x) := E[\dot\psi(\varepsilon_1)|X_0 = x], \qquad \text{for smooth } \psi,$$
$$:= \int f_X(x)\,\psi(dx), \qquad \text{for non-smooth } \psi.$$

From now onwards we shall assume that

(7.7.10) The errors ε_i are i.i.d. F, ε_i independent of X_{i-1}, for each $i = 0, \pm1, \cdots$, and F satisfies **F1** and **F2**.

Then it readily follows that

$$\sigma^2_\psi(x) = \sigma^2_\psi, \textit{ a positive constant in } x, \textit{ a. s.},$$
$$\gamma_\psi(x) = \gamma_\psi, \textit{ a positive constant in } x, \textit{ a. s.},$$

and that $\boldsymbol{\theta}_n$ satisfies (7.7.8) with

$$\phi(x, y, \boldsymbol{\theta}_0) = (\gamma_\psi\,\Sigma_0)^{-1}\dot{\mathbf{m}}(x, \boldsymbol{\theta}_0)\psi(y - m(x, \boldsymbol{\theta}_0)), \tag{7.7.11}$$

for $x, y \in \mathbb{R}$, where $\boldsymbol{\Sigma}_0 := E\dot{\mathbf{m}}(X_0, \boldsymbol{\theta}_0)\dot{\mathbf{m}}'(X_0, \boldsymbol{\theta}_0)$, so that $\phi(\boldsymbol{\theta}_0) = \tau\,\boldsymbol{\Sigma}_0^{-1}$, with $\tau := \sigma^2_\psi/\gamma^2_\psi$. Then direct calculations show that the above covariance function simplifies to

$$K^1_\psi(x, y) = E\psi^2(\varepsilon_1)\,[G(x \wedge y) - \nu'(x)\boldsymbol{\Sigma}_0^{-1}\nu(y)],$$
$$\boldsymbol{\nu}(x) = E\dot{\mathbf{m}}(X_0, \boldsymbol{\theta}_0)\,I(X_0 \le x), \; x, \, y \in \mathbb{R}.$$

Under (7.7.10), a set of sufficient conditions on the model $\mathcal{M}$ is given in Section 8.2 below under which a class of M-estimators of $\boldsymbol{\theta}_0$ corresponding to a given ψ defined by the relation

$$\theta_{n,\psi} := \text{argmin}_t \|n^{-1/2}\sum_{i=1}^n \dot{\mathbf{m}}(X_{i-1}, t)\psi(X_i - m(X_{i-1}, t))\|$$

satisfies (7.7.11). See, Theorem 8.2.1 and Corollary 8.2.1 below.

Throughout the rest of the section we shall assume that (7.7.10) holds. To simplify the exposition further write $\dot{\mathbf{m}}(\cdot) = \dot{\mathbf{m}}(\cdot, \boldsymbol{\theta}_0)$. Set

$$\mathbf{A}(x) = \int \dot{\mathbf{m}}(y)\dot{\mathbf{m}}'(y)\; I(y \ge x)G(dy), \quad x \in \mathbb{R}.$$

Assume that

(7.7.12) $\mathbf{A}(x)$ is nonsingular for some $x_0 < \infty$.

This and the nonnegative definiteness of $\mathbf{A}(x) - \mathbf{A}(x_0)$ implies that $\mathbf{A}(x)$ is non-singular for all $x \le x_0$. Write $\mathbf{A}^{-1}(x)$ for $(\mathbf{A}(x))^{-1}$, and define, for $x \le x_0$,

$$Tf(x) = f(x) - \int_{s \le x} \dot{\mathbf{m}}'(s)A^{-1}(s)\Big[\int \dot{\mathbf{m}}(z)\; I(z \ge s)\; f(dz)\Big]G(ds).$$

It is clear that an analogue of the Lemma 6.6.2 holds here also as do the derivations following this lemma with obvious modifications.

The next result is analogous to Theorem 6.6.3.

Theorem 7.7.3 *(A). Assume, in addition to the assumptions of Theorem 7.7.2(A), that (7.7.10) and (7.7.12) hold. Then*

$$\sup_{x \le x_0} \Big|T\hat{M}_{n,\psi}(x) - TM_{n,\psi}(x)\Big| = o_p(1).$$

If in addition, (7.7.1), (7.7.4) and (7.7.5) hold, then

$$TM_{n,\psi} \Longrightarrow TM_\psi \text{ and } T\hat{M}_{n,\psi} \Longrightarrow TM_\psi \text{ in } D[-\infty, x_0].$$

(B). The above claims continue to hold under the assumptions of Theorem 7.7.1(B), (7.7.10) and (7.7.12).

We now describe the analog of T_n of Section 6.6. Let, for $x \in \mathbb{R}$,

$$\begin{aligned} G_n(x) &:= n^{-1}\sum_{i=1}^{n} I(X_{i-1} \le x) \\ \mathbf{A}_n(x) &:= \int \dot{\mathbf{m}}(y, \boldsymbol{\theta}_n)\dot{\mathbf{m}}'(y, \boldsymbol{\theta}_n)\; I(y \ge x)\; G_n(dy). \end{aligned}$$

An estimator of T is defined, for $x \le x_0$, to be

$$\begin{aligned} \tilde{T}_n f(x) &= f(x) - \int_{-\infty}^{x} \dot{\mathbf{m}}'(y, \boldsymbol{\theta}_n)\mathbf{A}_n^{-1}(y) \\ &\qquad \times \Big[\int \dot{\mathbf{m}}(z, \boldsymbol{\theta}_n)\; I(z \ge y)\; f(dz)\Big]G_n(dy). \end{aligned}$$

The next result is the most useful result of this section. It proves the consistency of $\tilde{T}_n\hat{M}_{n,\psi}$ for $T\hat{M}_{n,\psi}$ under the same additional smoothness condition on $\dot{\mathbf{m}}$ as in Section 6.6.

Theorem 7.7.4 *(A). Suppose, in addition to the assumptions of Theorem 7.7.3(A), (6.6.26) holds and that (7.7.4) with $\psi(\varepsilon_1)$, $\psi(\varepsilon_2)$ replaced by $\|\dot{\mathbf{m}}(X_0, \boldsymbol{\theta}_0)\|\,\psi(\varepsilon_1)$, $\|\dot{\mathbf{m}}(X_1, \boldsymbol{\theta}_0)\|\;\psi(\varepsilon_2)$, respectively, holds. Then,*

$$\sup_{x \le x_0} |\tilde{T}_n M^1_{n,\psi}(x) - TM_{n,\psi}(x)| = o_p(1), \tag{7.7.13}$$

and consequently,

$$\sigma_{n,\psi}^{-1} \tilde{T}_n \hat{M}_{n,\psi}(\cdot) \Longrightarrow B \circ G, \quad \text{in } D[-\infty, x_0]. \tag{7.7.14}$$

(B). The same continues to hold under (6.6.26) and the assumptions of Theorem 7.7.2(B).

Remark 7.7.2 By (7.7.12), $\lambda_1 := \inf\{a'\mathbf{A}(x_0)a;\ a \in \mathbb{R}^q,\ \|a\| = 1\} > 0$ and $\mathbf{A}(x)$ is positive definite for all $x \leq x_0$, Hence,

$$\|\mathbf{A}^{-1/2}(x)\|^2 \leq \lambda_1^{-1} < \infty, \quad \forall x \leq x_0,$$

and (6.6.14) implies

$$E\|\dot{\mathbf{m}}'(X_0)\mathbf{A}^{-1}(X_0)\| I(X_0 \leq x_0) \leq \mathbb{E}\|\dot{\mathbf{m}}(X_0)\|\, \lambda_1^{-1} < \infty. \tag{7.7.15}$$

This fact is used in the proofs repeatedly.

Now, let $a < b$ be given real numbers and suppose one is interested in testing the hypothesis

$$H:\ m_\psi(x) = m(x, \boldsymbol{\theta}_0), \quad \text{for all } x \in [a, b] \text{ and for some } \boldsymbol{\theta}_0 \in \boldsymbol{\theta}.$$

Assume the support of G is $\mathbb{R}$, and $\mathbf{A}(b)$ is positive definite. Then, $\mathbf{A}(x)$ is non-singular for all $x \leq b$, continuous on $[a, b]$ and $\mathbf{A}^{-1}(x)$ is continuous on $[a, b]$ and

$$E\|\dot{\mathbf{m}}'(X_0)\mathbf{A}^{-1}(X_0)\| I(a < X_0 \leq b) < \infty.$$

Thus, under the conditions of the Theorem 7.7.4, $\sigma_{n,\psi}^{-1} T_n \hat{M}_{n,\psi}(\cdot) \Longrightarrow B \circ G$, in $D[a, b]$ and we obtain

$$\sigma_{n,\psi}^{-1}\{T_n \hat{M}_{n,\psi}(\cdot) - T_n \hat{M}_{n,\psi}(a)\} \Longrightarrow B(G(\cdot)) - B(G(a)), \quad \text{in } D[a, b].$$

The stationarity of the increments of the Brownian motion then readily implies that

$$D_n := \sup_{a \leq x \leq b} \frac{|T_n \hat{M}_{n,\psi}(x) - T_n \hat{M}_{n,\psi}(a)|}{\{G_n(b) - G_n(a)\}^{1/2}\, \sigma_{n,\psi}} \Longrightarrow \sup_{0 \leq u \leq 1} |B(u)|.$$

Hence, any test of H based on D_n is ADF. Proofs of the last three theorems is given at the end of this section.

7.7.3 Some examples

In this sub-sections we discuss some examples of non-linear time series to which the above results may be applied. This section is some what different from its analogous Section 6.6.3 primarily because one can have non-linearity in autoregressive modeling from the way the lag variables appear in the model.

Again, it is useful for computational purposes to rewrite $\tilde{T}_n \hat{M}_{n,\psi}$ as follows: For convenience, let

$$m_n(x) := m(x, \boldsymbol{\theta}_n), \quad \varepsilon_{ni} := X_i - m_n(X_{i-1}), \quad 1 \le i \le n.$$

Then for for all $x \le x_0$,

$$\begin{aligned} (7.7.16) \qquad & \tilde{T}_n \hat{M}_{n,\psi}(x) \\ = \; & n^{-1/2} \sum_{i=1}^{n} \Big[I(X_{i-1} \le x) \\ & \qquad - n^{-1} \sum_{j=1}^{n} \dot{\mathbf{m}}_n'(X_{j-1}) A_n^{-1}(X_{j-1}) \dot{\mathbf{m}}_n(X_{i-1}) \\ & \qquad\qquad \times I(X_{j-1} \le X_{i-1} \wedge x) \Big] \psi(\varepsilon_{ni}). \end{aligned}$$

Now, let $g_1, \ldots, g_q$ be known real-valued G-square integrable functions on $\mathbb{R}$ and consider the class of models $\mathcal{M}$ with

$$m(x, \boldsymbol{\theta}) = g_1(x)\theta_1 + \ldots + g_q(x)\theta_q.$$

Then (6.6.13) and (6.6.26) are trivially satisfied with $\dot{\mathbf{m}}(x, \theta) \equiv (g_1(x), \cdots, g_q(x))'$ and $\ddot{m}(x, \theta) \equiv 0 \equiv K_1(x, \theta)$.

A major difference between the regression setup and the autoregressive setup is that this model includes a large class of the first order autoregressive models. Besides including the first order linear autoregressive (AR(1)) model where $q = 1$, $g_1(x) \equiv x$, this class also includes some nonlinear autoregressive models. For example the choice of $q = 2$, $g_1(x) = x$, $g_2(x) = xe^{-x^2}$ gives an exponential-amplitude dependent AR(1) (EXPAR(1)) model of Ozaki and Oda (1978) or the choice of $p = 1$, $q = 4$, $g_1(x) = I(x \le 0)$, $g_2(x) = xI(x \le 0)$, $g_3(x) = I(x > 0)$, $g_4(x) = xI(x > 0)$ gives the self exciting threshold AR(1) model

$$m(x, \boldsymbol{\theta}) = (\theta_1 + \theta_2 \, x) I(x \le 0) + (\theta_3 + \theta_4 \, x) I(x > 0).$$

For more on these and several other non-linear AR(1) models see Tong (1990). In the following discussion the assumption (7.7.10) is in action.

In the linear AR(1) model $\dot{m}(x,\theta) \equiv x$ and $\mathbf{A}(x) \equiv A(x) \equiv \mathbb{E}X_0^2 I(X_0 \geq x)$ is positive for all real x, uniformly continuous and decreasing on $\mathbb{R}$, and thus trivially satisfies (7.7.12). A uniformly a.s. consistent estimator of A is

$$A_n(x) \equiv n^{-1} \sum_{k=1}^{n} X_{k-1}^2 \, I(X_{k-1} \geq x).$$

Thus a test of the hypothesis that the first order autoregressive mean function is linear AR(1) on the interval $(-\infty, x_0]$ can be based on

$$\sup_{x \leq x_0} |\tilde{T}_n \hat{M}_{n,I}(x)| / \{\sigma_{n,I} G_n^{1/2}(x_0)\},$$

where

$$\begin{aligned} &\tilde{T}_n \hat{M}_{n,I}(x) \\ &= n^{-1/2} \sum_{i=1}^{n} (X_i - X_{i-1}\theta_n) \Big[I(X_{i-1} \leq x) \\ &\qquad - n^{-1} \sum_{j=1}^{n} \frac{X_{j-1} X_{i-1} I(X_{j-1} \leq X_{i-1} \wedge x)}{n^{-1} \sum_{k=1}^{n} X_{k-1}^2 \, I(X_{k-1} \geq X_{j-1})} \Big] \\ \sigma_{n,I}^2 &= n^{-1} \sum_{i=1}^{n} (X_i - X_{i-1}\theta_n)^2. \end{aligned}$$

Similarly, a test of the hypothesis that the first order autoregressive *median* function is linear AR(1) can be based on

$$\sup_{x \leq x_0} |\tilde{T}_n \hat{M}_{n,.5}(x)| / \{\sigma_{n,.5} G_n^{1/2}(x_0)\},$$

where

$$\begin{aligned} &\tilde{T}_n \hat{M}_{n,.5}(x) \\ &= n^{-1/2} \sum_{i=1}^{n} \{I(X_i - X_{i-1}\theta_n > 0) - .5\} \Big[I(X_{i-1} \leq x) \\ &\qquad - n^{-1} \sum_{j=1}^{n} \frac{X_{j-1} X_{i-1} I(X_{j-1} \leq X_{i-1} \wedge x)}{n^{-1} \sum_{k=1}^{n} X_{k-1}^2 \, I(X_{k-1} \geq X_{j-1})} \Big] \end{aligned}$$

and

$$\sigma_{n,.5}^2 = n^{-1} \sum_{i=1}^{n} \{I(X_i - X_{i-1}\theta_n > 0) - .5\}^2.$$

By Theorem 7.7.4, both of these tests are ADF as long as the estimator $\boldsymbol{\theta}_n$ is the least square (LS) estimator in the former test and the least absolute deviation (LAD) estimator in the latter. For the former test we additionally require $\mathbb{E}\varepsilon_1^{4(1+\delta)} < \infty$, for some $\delta > 0$, while for the latter test $\mathbb{E}\varepsilon_1^2 < \infty$ and f being uniformly continuous and positive suffice.

In the EXPAR(1) model, $\dot{\mathbf{m}}(x, \boldsymbol{\theta}_0) \equiv (x,\, xe^{-x^2})'$ and $\mathbf{A}(x)$ is the 2×2 symmetric matrix

$$\mathbf{A}(x) = \mathbb{E}I(X_0 \geq x) \begin{pmatrix} X_0^2 & X_0^2 e^{-X_0^2} \\ X_0^2 e^{-X_0^2} & X_0^2 e^{-2X_0^2} \end{pmatrix}$$

From Theorem 4.3 of Tong (1990: p 128), if $\mathbb{E}\varepsilon_1^4 < \infty$, f is absolutely continuous and positive on $\mathbb{R}$ then the above EXPAR(1) process is stationary, ergodic, the corresponding stationary d.f. G is strictly increasing on $\mathbb{R}$, and $\mathbb{E}X_0^4 < \infty$. Moreover, one can directly verify that $\mathbb{E}X_0^2 < \infty$ implies $\mathbf{A}(x)$ is nonsingular for every real x and $\mathbf{A}^{-1}$ and $\mathbf{A}$ are continuous on $\mathbb{R}$. The matrix

$$\mathbf{A}_n(x) = n^{-1} \sum_{i=1}^{n} I(X_{i-1} \geq x) \begin{pmatrix} X_{i-1}^2 & X_{i-1}^2\, e^{-X_{i-1}^2} \\ X_{i-1}^2\, e^{-X_{i-1}^2} & X_{i-1}^2\, e^{-2X_{i-1}^2} \end{pmatrix}$$

provides a uniformly a.s. consistent estimator of $\mathbf{A}(x)$. Thus one may use $\sup_{x \leq x_0} |T_n \hat{M}_{n,I}(x)| / \{\sigma_{n,I} G_n(s_0)\}$ to test the hypothesis that the autoregressive mean function is given by an EXPAR(1) function on an interval $(-\infty, x_0]$. Similarly, one can use the test statistic

$$\sup_{x \leq x_0} |T_n\ M_{n,.5}(x)| / \{\sigma_{n,.5} G_n^{1/2}(x_0)\}$$

to test the hypothesis that the autoregressive median function is given by an EXPAR(1) function. In both cases $\mathbf{A}_n$ is as above and one should now use the general formula (7.7.16) to compute these statistics. Again, from Theorem 7.7.4 it readily follows that the asymptotic levels of both of these tests can be computed from the distribution of $\sup_{0 \leq u \leq 1} |B(u)|$, provided the estimator $\boldsymbol{\theta}_n$ is taken to be, respectively, the LS and the LAD. Again one needs the $(4+\delta)$th moment assumption for the former test and the uniform continuity of f for the latter test. The relevant asymptotics of the LS estimator and a class of M-estimators with bounded ψ in a class of non-linear time series models is given in Tjøstheim (1986) and Koul (1996), respectively. In particular these papers include the above EXPAR(1) model.

7.7.4 Proofs of some results of Section 7.7.2

Many proofs are similar to those of Section 6.6.4. For example the analogue of (6.6.31) holds here also with X_i replaced by X_{i-1} and with i.i.d. replaced by assuming that the r.v.'s $\{(\xi_i, X_i)\}$ are stationary and ergodic. In many arguments just replace the LLN's by the Ergodic Theorem and the classical CLT by the central limit theorem for martingales as given in Lemma A3 of the Appendix. So many details are not given or are shortened.

The Remark 6.6.3 applies here also without any change. The proof of part (A) of Theorem 7.7.2 is exactly similar to that of part (A) of Theorem 6.6.2 while that of part (B) is some what different. We give the details for this part only.

Proof part (B) of Theorem 7.7.2. Put, for $1 \le i \le n$, $t \in \mathbb{R}^q$,

$$\begin{aligned} d_{n,i}(t) &:= m(X_{i-1}, \boldsymbol{\theta}_0 + n^{-1/2}t) - m(X_{i-1}, \boldsymbol{\theta}_0); \\ \gamma_{n,i} &:= n^{-1/2}(2\alpha + \delta \| \dot{\mathbf{m}}(X_{i-1}, \boldsymbol{\theta}_0)\|), \ \alpha > 0, \ \delta > 0; \\ \mu_n(X_{i-1}, t, a) &:= \mathbb{E}[\psi(\varepsilon_i - d_{n,i}(t) + a\gamma_{n,i}) \,|\, X_{i-1}]. \end{aligned}$$

Define, for a, $x \in \mathbb{R}$ and $t \in \mathbb{R}^q$,

$$\begin{aligned} & D_n(x, t, a) \\ & := \ n^{-1/2} \sum_{i=1}^n [\psi(\varepsilon_i - d_{n,i}(t) + a\gamma_{n,i}) - \mu_n(X_{i-1}, t, a) \\ & \qquad\qquad -\psi(\varepsilon_i)]\, I(X_{i-1} \le x). \end{aligned}$$

Write $D_n(x,t)$ and $\mu_n(X_{i-1}, t)$ for $D_n(x,t,0)$ and $\mu_n(X_{i-1}, t, 0)$, respectively.

Note that the summands in $D_n(x,t,a)$ form mean zero bounded martingale differences, for each x, t and a. Thus

$$\begin{aligned} & Var(D_n(x,t,a)) \\ & \le \ \mathbb{E}[\psi(\varepsilon_1 - d_{n,1}(t) + a\gamma_{n,1}) - \mu_n(X_0, t, a) - \psi(\varepsilon_1)]^2 \\ & \le \ \mathbb{E}[\psi(\varepsilon_1 - d_{n,1}(t) + a\gamma_{n,1}) - \psi(\varepsilon_1)]^2 \to 0, \end{aligned}$$

by assumption (6.6.13) and (Ψ_2). Upon an application of Theorem 2.2.5 with $Z_{n,i} = \psi(\varepsilon_i - d_{n,i}(t) + a\gamma_{n,i}) - \mu_n(X_{i-1}, t, a) - \psi(\varepsilon_i)$ we readily obtain that

$$(7.7.17) \qquad \sup_{x \in \mathbb{R}} |D_n(x,t,a)| = o_p(1), \qquad \forall\, a \in \mathbb{R},\ t \in \mathbb{R}^q.$$

The assumption (C) of Theorem 2.2.5 with these $\{Z_{n,i}\}$ and $\tau^2 \equiv 0$ is implied by (Ψ_2) while (7.7.7) implies (2.2.74) here.

We need to prove that for every $b < \infty$,

$$\sup_{x\in\mathbb{R},\,\|t\|\le b} |D_n(x,t)| = o_p(1). \tag{7.7.18}$$

To that effect let

$$C_n := \Big\{ \sup_{\|t\|\le b} |d_{n,i}(t)| \le n^{-1/2}(\alpha + b\|\dot{\mathbf{m}}(X_{i-1})\|),\ 1 \le i \le n \Big\},$$

and for an $\|s\| \le b$, let

$$A_n := \Big\{ \sup_{\|t\|\le b,\,\|t-s\|\le\delta} |d_{n,i}(t) - d_{n,i}(s)| \le \gamma_{n,i},\ 1 \le i \le n \Big\} \cap C_n.$$

By assumption (6.6.13), there is an $N < \infty$, depending only on α, such that $\forall\, b < \infty$ and $\forall\, \|s\| \le b$,

$$P(A_n) > 1 - \alpha, \qquad \forall n > N. \tag{7.7.19}$$

Now, by the monotonicity of ψ one obtains that on A_n, for each fixed $\|s\| \le b$ and $\forall\, \|t\| \le b$ with $\|t - s\| \le \delta$,

$$\begin{aligned}
&|D_n(x,t)| \\
&\le |D_n(x,s,1)| + |D_n(x,s,-1)| \\
&\quad + \Big|n^{-1/2}\sum_{i=1}^n [\mu_n(X_{i-1},s,1) - \mu_n(X_{i-1},s,-1)]\, I(X_{i-1} \le x)\Big|.
\end{aligned}$$

By (7.7.17), the first two terms converge to zero uniformly in x, in probability, while the last term is bounded above by

$$\begin{aligned}
n^{-1/2}\sum_{i=1}^n \int_{-\infty}^{\infty} &|F_{X_{i-1}}(y + d_{n,i}(s) + \gamma_{n,i}) \\
&- F_{X_{i-1}}(y + d_{n,i}(s) - \gamma_{n,i})|\psi(dy).
\end{aligned}$$

Observe that for every $\|s\| \le b$, on A_n, $|d_{n,i}(s)| + \gamma_{n,i} \le a_n$, for all $1 \le i \le n$, where $a_n := max_{1\le i\le n}\, n^{-1/2}[3\alpha + (b+\delta)\|\dot{\mathbf{m}}(X_{i-1})\|]$. By (Ψ_2), the above bound in turn is bounded from above by

$$\begin{aligned}
n^{-1/2}\sum_{i=1}^n \gamma_{n,i}\Big[&\sup_{y\in\mathbb{R},\,|x-z|\le a_n} |f_y(x) - f_y(z)| \\
&+ 2\int_{-\infty}^{\infty} f_{X_{i-1}}(y)\,\psi(dy)\Big].
\end{aligned} \tag{7.7.20}$$

Now, by the ET, (6.6.14) and (F),

$$n^{-1/2}\sum_{i=1}^{n}\gamma_{n,i} = n^{-1}\sum_{i=1}^{n}(2\alpha + \delta\|\dot{\mathbf{m}}(X_{i-1},\boldsymbol{\theta}_0)\|) = o_p(1),$$

and

$$n^{-1/2}\sum_{i=1}^{n}\gamma_{n,i}\int_{-\infty}^{\infty} f_{X_{i-1}}(y)\,\psi(dy)$$
$$\to 2\alpha\int_{-\infty}^{\infty}\mathbb{E}f_{X_0}(y)\,\psi(dy) \,+\, \delta\int_{-\infty}^{\infty}\mathbb{E}\|\dot{\mathbf{m}}(X_0)\|f_{X_0}(y)\,\psi(dy).$$

Observe that, by (6.6.14), the functions

$$q_j(y) := \mathbb{E}\|\dot{\mathbf{m}}(X_0)\|^j f_{X_0}(y),\ j = 0,1,\ y \in \mathbb{R},$$

are Lebesgue integrable on $\mathbb{R}$. By (F), they are also uniformly continuous and hence bounded on $\mathbb{R}$ so that $\int_{-\infty}^{\infty} q_j(y)\,\psi(dy) < \infty$ for $j = 0, 1$. We thus obtain that the bound in (7.7.20) converges in probability to

$$4\,\alpha\int_{-\infty}^{\infty} q_0(y)\,\psi(dy) \,+\, 2\,\delta\int_{-\infty}^{\infty} q_1(y)\,\psi(dy)$$

which can be made less than δ by the choice of α. This together with (7.7.17) applied with $a = 0$ and the compactness of $\mathcal{N}_b$ proves (7.7.18).

Next, by (7.7.1) and Fubini's theorem, we have

$$
\begin{aligned}
& n^{-1/2}\sum_{i=1}^{n}\mu_n(X_{i-1},t)\,I(X_{i-1}\le x) \\
&= n^{-1/2}\sum_{i=1}^{n}\Big[\mu_n(X_{i-1},t) - \mu_n(X_{i-1},0)\Big]\,I(X_{i-1}\le x) \\
&= -n^{-1/2}\sum_{i=1}^{n} I(X_{i-1}\le x) \\
&\quad\times\int_{-\infty}^{\infty}\Big[F_{X_{i-1}}(y + d_{n,i}(t)) - F_{X_{i-1}}(y)\Big]\,\psi(dy) \\
&= -n^{-1}\sum_{i=1}^{n}\dot{\mathbf{m}}'(X_{i-1},\boldsymbol{\theta}_0)\,t\,I(X_{i-1}\le x) \\
&\quad\times\int_{-\infty}^{\infty} f_{X_{i-1}}(y)\,\psi(dy) + o_p(1) \\
&= -\mathbb{E}\dot{\mathbf{m}}'(X_0,\boldsymbol{\theta}_0)\,I(X_0\le x)\int_{-\infty}^{\infty} f_{X_0}(y)\,\psi(dy)\,t \\
&\qquad + o_p(1),
\end{aligned}
$$

uniformly in $x \in \mathbb{R}$ and $\|t\| \le b$. In the above, the last equality follows from (6.6.31) while the one before that follows from the assumptions (6.6.13), (Ψ_2) and (F). This together with (7.7.18), (7.7.19) and the assumption (7.7.8) proves (7.7.9) and hence the part (B) of Theorem 7.7.2. □

The details of the proofs of the remaining two theorems are similar to their analogues in the regression setup and will not be given here. They also appear in Koul and Stute (1999).

8

Nonlinear Autoregression

8.1 Introduction

The decade of 1990's has seen an exponential growth in the applications of nonlinear autoregressive models (AR) to economics, finance and other sciences. Tong (1990) illustrates the usefulness of homoscedastic AR models in a large class of applied examples from physical sciences while Gouriéroux (1997) contains several examples from economics and finance where the ARCH (autoregressive conditional heteroscedastic) models of Engle (1982) and its various generalizations are found useful. Most of the existing literature has focused on developing classical inference procedures in these models. The theoretical development of the analogues of the estimators discussed in the previous sections that are known to be robust against outliers in the innovations in linear AR models has relatively lagged behind.

In this chapter we shall discuss the asymptotic distributions of the analogues of some of the procedures of the previous chapters for a class of nonlinear AR and ARCH time series models in two main sections. The main point of the chapter is to exhibit the significance of the weak convergence approach in providing a unified methodology for establishing these results for non-smooth underlying score functions ψ, φ and L in these dynamic models. The main technical tool used are Theorems 2.2.4 and 2.2.5. We shall first focus on nonlinear AR models. The main reason for doing this is that the motivation for various estimators is simple and the details of the proofs are relatively transparent.

8.2 AR Models

Let p, q, be positive integers, $\mathbf{Y}_0 := (X_0, X_{-1}, \cdots, X_{1-p})'$ be an observable random vector, Ω be an open subset of $\mathbb{R}^q$, and μ be real valued function from $\mathbb{R}^p \times \Omega$. In a nonlinear AR($p$) model of interest here, the observations $\{X_i\}$ are such that for some $\boldsymbol{\theta}' = (\theta_1, \theta_2, \cdots, \theta_q) \in \Omega$,

$$(8.2.1) \qquad \varepsilon_i = X_i - \mu(\mathbf{Y}_{i-1}, \boldsymbol{\theta}), \quad i = 0, \pm 1, \cdots,$$

are i.i.d. r.v.'s, and independent of $\mathbf{Y}_0$. We do not need to have stationary and ergodic time series for the results of this chapter to hold.

To proceed further we need to make the following additional model smoothness assumption about the functions μ. There exists a function $\dot{\boldsymbol{\mu}}$ from $\mathbb{R}^p \times \Omega$ to $\mathbb{R}^q$, measurable in the first p coordinates, such that for every $\epsilon > 0$, $k < \infty$, $\mathbf{s} \in \Omega$,

$$(8.2.2) \qquad \sup n^{1/2} |\mu(\mathbf{Y}_{i-1}, \mathbf{t}) - \mu(\mathbf{Y}_{i-1}, \mathbf{s}) - (\mathbf{t} - \mathbf{s})' \dot{\boldsymbol{\mu}}_i(\mathbf{Y}_{i-1}, \mathbf{s})| = o_p(1),$$

where the supremum is taken over $1 \le i \le n$, $n^{1/2}\|\mathbf{t} - \mathbf{s}\| \le k$.

Note that the differentiability of μ in $\boldsymbol{\theta}$ alone need not imply this assumption. To see this consider the simple case $q = 1 = p$, $\mu(y, \theta) = \theta^2 y$. Then the left hand side of (8.2.2) is bounded above by a constant multiple of $n^{-1/2} \max_i |Y_{i-1}|$, which may not tend to zero in probability unless some additional conditions are satisfied.

Now we are ready to define analogues of M- and R- and m.d. estimators for $\boldsymbol{\theta}$. For the sake of brevity we shall often write $\mu_i(\mathbf{t})$, μ_i for $\mu(\mathbf{Y}_{i-1}, \mathbf{t})$, $\mu(\mathbf{Y}_{i-1}, \boldsymbol{\theta})$ and $\dot{\boldsymbol{\mu}}_i(\mathbf{s})$, $\dot{\boldsymbol{\mu}}_i$ for $\dot{\boldsymbol{\mu}}(Y_{i-1}, \mathbf{s})$, $\dot{\boldsymbol{\mu}}(Y_{i-1}, \boldsymbol{\theta})$, respectively. The basic scores needed are as follows. Let $\varepsilon_i(\mathbf{t}) := X_i - \mu(\mathbf{Y}_{i-1}, \mathbf{t})$, and $R_{i\mathbf{t}}$ denote the rank of $\varepsilon_i(\mathbf{t})$ among $\{\varepsilon_j(\mathbf{t});\ 1 \le j \le n\}$, $1 \le i \le n$, $\mathbf{t} \in \boldsymbol{\Omega}$. Also let ψ, φ be as in Section 7.3, L be a d.f. on $[0, 1]$ and define

$$(8.2.3) \qquad \begin{aligned}
\mathbf{M}(\mathbf{t}) &:= n^{-1/2} \sum_{i=1}^n \dot{\boldsymbol{\mu}}_i(\mathbf{t}) \psi(\varepsilon_i(\mathbf{t})), \\
\mathbf{S}(\mathbf{t}) &:= n^{-1/2} \sum_{i=1}^n \dot{\boldsymbol{\mu}}_i(\mathbf{t}) \varphi(R_{i\mathbf{t}}/(n+1)), \\
\mathbf{Z}(u, \mathbf{t}) &:= n^{-1/2} \sum_{i=1}^n \dot{\boldsymbol{\mu}}_i(\mathbf{t}) I(R_{i\mathbf{t}} \le nu), \\
\mathbb{Z}(u, \mathbf{t}) &:= \mathbf{Z}(u, \mathbf{t}) - n^{-1/2} \sum_{i=1}^n \dot{\boldsymbol{\mu}}_i(\mathbf{t})\, u, \quad 0 \le u \le 1,
\end{aligned}$$

$$\mathcal{K}(\mathbf{t}) := \int_0^1 \|\mathbb{Z}(u,\mathbf{t})\|^2 L(du), \qquad \mathbf{t} \in \boldsymbol{\Omega}.$$

Also, define

$$\text{(8.2.4)} \qquad \hat{\boldsymbol{\theta}}_M := \text{argmin}_{\mathbf{t}} \|\mathbf{M}(\mathbf{t})\|^2, \quad \hat{\boldsymbol{\theta}}_R := \text{argmin}_{\mathbf{t}} \|\mathbf{S}(\mathbf{t})\|^2,$$
$$\hat{\boldsymbol{\theta}}_{md} := \text{argmin}_{\mathbf{t}} \|\mathbb{Z}(\mathbf{t})\|^2.$$

Note that $\hat{\boldsymbol{\theta}}_M$, $\hat{\boldsymbol{\theta}}_R$ are the extensions of M- and R- estimators of Section 7.3 and $\hat{\boldsymbol{\theta}}_{md}$ is an analogue of the m.d. estimator (5.2.18) appropriate for the model here. Thus, for example, the choice of $\psi(x) \equiv sign(x)$, $\varphi(u) \equiv u$ give an extension of the LAD and the Hodges-Lehmann estimators to nonlinear autoregressive models. As seen in Section 5.6.1, in linear regression models, the estimator $\hat{\boldsymbol{\theta}}_{md}$ with $L(u) \equiv u$ is asymptotically more efficient than the LAD (Hodges-Lehmann) estimator at the logistic (double exponential) errors. It is thus of interest to investigate sufficient conditions on the model (8.2.1) and (8.2.2) for the asymptotic normality of these estimators. Such results are useful in showing, among other things, that the various asymptotic relative efficiency facts that hold for linear models about these estimators continue to hold for a large class of nonlinear time series models.

The next subsection gives the additional assumptions and the main results. Section 8.2.3 contains some examples of homoscedastic nonlinear AR models.

8.2.1 Main Results in AR models

The basic weighted empirical process that plays the role of W_h of (7.1.6) here is as follows:

$$\text{(8.2.5)} \qquad d_i(\mathbf{t}) := \mu_i(\mathbf{t}) - \mu_i(\boldsymbol{\theta}), \quad 1 \le i \le n,$$
$$\mathbf{W}(y,\mathbf{t}) := n^{-1} \sum_{i=1}^n \dot{\boldsymbol{\mu}}_i(\mathbf{t}) I(X_i - \mu_i(\mathbf{t}) \le y)$$
$$= n^{-1} \sum_{i=1}^n \dot{\boldsymbol{\mu}}_i(\mathbf{t}) I(\varepsilon_i \le y + d_i(\mathbf{t})),$$
$$\boldsymbol{\nu}(y,\mathbf{t}) := n^{-1} \sum_{i=1}^n \dot{\boldsymbol{\mu}}_i(\mathbf{t})) F(y + d_i(\mathbf{t})),$$
$$\mathcal{W}(y,\mathbf{t}) := n^{1/2}[\mathbf{W}(y, \boldsymbol{\theta} + n^{-1/2}\mathbf{t}) - \boldsymbol{\nu}(y, \boldsymbol{\theta} + n^{-1/2}\mathbf{t})],$$

for $y \in \mathbb{R}$, $\mathbf{t} \in \mathbb{R}^q$. The components of the vector of processes $\mathbf{W}$ and $\boldsymbol{\mathcal{W}}$ are different from W_h and $\mathcal{W}_h$ of (7.2.1) or any other previous process because now the weights also involve the parameter. In linear models, the weights $\dot{\boldsymbol{\mu}}_i(\mathbf{t}) \equiv \mathbf{Y}_{i-1}$, thereby making them free of $\mathbf{t}$. The dependence of weights on the parameter creates additional technical difficulty. Some of the assumptions stated below are needed to overcome this new difficulty with minimal constraints on the model μ. In particular we avoid requiring the second order differentiability assumption on this function. These assumptions are as follows.

(8.2.6) There exists a positive definite matrix, $\boldsymbol{\Sigma}$ possibly depending on $\boldsymbol{\theta}$, such that $n^{-1}\sum_{i=1}^{n} \dot{\boldsymbol{\mu}}_i \dot{\boldsymbol{\mu}}_i' = \boldsymbol{\Sigma} + o_p(1)$.

(8.2.7) $\max_{1 \le i \le n} n^{-1/2}\|\dot{\boldsymbol{\mu}}_i\| = o_p(1)$.

(8.2.8) $n^{-1}\sum_{i=1}^{n} E\|\dot{\boldsymbol{\mu}}_i(\boldsymbol{\theta} + n^{-1/2}\mathbf{s}) - \dot{\boldsymbol{\mu}}_i(\boldsymbol{\theta})\|^2 = o(1), \quad \mathbf{s} \in \mathbb{R}^q$.

(8.2.9) $n^{-1/2}\sum_{i=1}^{n} \|\dot{\boldsymbol{\mu}}_i(\boldsymbol{\theta} + n^{-1/2}\mathbf{s}) - \dot{\boldsymbol{\mu}}_i(\boldsymbol{\theta})\| = O_p(1), \quad \mathbf{s} \in \mathbb{R}^q$.

(8.2.10) $\forall \epsilon > 0, \exists$ a $\delta > 0$, and an $N < \infty, \ni \forall\, 0 < b < \infty$, $\|\mathbf{s}\| \le b$, $n > N$,

$$P\Big(\sup_{\|\mathbf{t}-\mathbf{s}\|<\delta} n^{-1/2}\sum_{i=1}^{n} \|\dot{\boldsymbol{\mu}}_i(\boldsymbol{\theta} + n^{-1/2}\mathbf{t}) - \dot{\boldsymbol{\mu}}_i(\boldsymbol{\theta} + n^{-1/2}\mathbf{s})\| \le \epsilon\Big) \ge 1 - \epsilon.$$

The first two assumptions are similar to (7.2.3)-(7.2.5) of Theorem 7.2.1, while the last three assumptions are needed to achieve the required equicontinuity of the $\boldsymbol{\mathcal{W}}$ processes with respect to the parameter in the weights. Clearly in the case of linear models, the last three assumptions are vacuously satisfied. The analogue of Theorem 7.2.1 and Remark 7.2.1 is given by the following lemma.

Lemma 8.2.1 *Suppose the assumptions (8.2.1), (8.2.2), (8.2.6) - (8.2.10) hold. Then, for every continuity point y of F, and for every $0 < b < \infty$,*

(8.2.11) $$\sup_{\|\mathbf{t}\| \le b} |\boldsymbol{\mathcal{W}}(y, \mathbf{t}) - \boldsymbol{\mathcal{W}}(y, \mathbf{0})| = o_p(1).$$

If, in addition, the error d.f. satisfies (7.2.10), then for every $0 <$

k, $b < \infty$,

$$\sup_{|y| \le k, \|\mathbf{t}\| \le b} |\mathcal{W}(y, \mathbf{t}) - \mathcal{W}(y, \mathbf{0})| = o_p(1). \tag{8.2.12}$$

Finally, if, in addition, the error d.f. satisfies (**F1**) *and* (**F2**), *then for every* $0 < b < \infty$,

$$|\mathcal{W}(y, \mathbf{t}) - \mathcal{W}(y, \mathbf{0})| = u_p(1), \tag{8.2.13}$$

$$\begin{aligned} & n^{1/2}\,[\mathbf{W}(y, \mathbf{t}) - \mathbf{W}(y, \mathbf{0})] - \mathbf{t}'\mathbf{\Sigma} f(y) \\ & \quad = n^{-1/2} \sum_{i=1}^{n} [\dot{\boldsymbol{\mu}}_i(\boldsymbol{\theta} + n^{-1/2}\mathbf{t}) - \dot{\boldsymbol{\mu}}_i(\boldsymbol{\theta})] F(y) + u_p(1), \end{aligned} \tag{8.2.14}$$

where $u_p(1)$ *is a sequence of stochastic processes that converges to zero, uniformly over the set* $y \in \mathbb{R}$, $\|\mathbf{t}\| \le b$, *in probability.*

The proof of this lemma will be given at the end of this subsection. We now proceed to give several results about the estimators defined in the previous section. As will be seen the above lemma plays the same role in their proofs as did Theorem 7.2.1 in the linear AR model.

Recall the definition of the class of score functions Ψ from (4.2.3). Let

$$\Psi_F := \Big\{\psi \in \Psi;\ \int \psi dF = 0\Big\}, \quad \lambda := \int f d\psi.$$

The first result of this section gives the AUL of the M-scores.

Theorem 8.2.1 *In addition to the assumptions (8.2.1), (8.2.2), (8.2.6) - (8.2.10), suppose that the error d.f. F satisfies* (**F1**) *and* (**F2**).

Then, for every $0 < b < \infty$,

$$\sup_{\psi \in \Psi_F, \|\mathbf{t}\| \le b} \Big\| \mathbf{M}(\boldsymbol{\theta} + n^{-1/2}\mathbf{t}) - \mathbf{M}(\boldsymbol{\theta}) - \mathbf{\Sigma}\mathbf{t}\,\lambda \Big\| = o_p(1).$$

This follows from (8.2.14), the fact that $\int F d\psi = \psi(\infty)$, for all $\psi \in \Psi$, and integration by parts that shows that

$$\begin{aligned} \mathbf{M}(\mathbf{t}) - \mathbf{M}(\mathbf{s}) &= n^{-1/2} \sum_{i=1}^{n} [\dot{\boldsymbol{\mu}}_i(\mathbf{t}) - \dot{\boldsymbol{\mu}}_i(\mathbf{s})]\psi(\infty) \\ &\quad - \int n^{1/2}[\mathbf{W}(y, \mathbf{t}) - \mathbf{W}(y, \mathbf{s})]\psi(dy), \quad \mathbf{s}, \mathbf{t} \in \mathbb{R}^q. \end{aligned}$$

This theorem is useful in obtaining the limiting distribution of $\hat{\boldsymbol{\theta}}_M$ provided $n^{1/2}\|\hat{\boldsymbol{\theta}}_M - \boldsymbol{\theta}\| = O_p(1)$. Using the methodology of sections 5.4

and 5.5, one such sufficient condition that guarantees this is the following:

$$(8.2.15)\qquad \boldsymbol{e}'\mathbf{M}(\boldsymbol{\theta}+n^{-1/2}r\boldsymbol{e}) \text{ is monotonic in } r\in\mathbb{R},\ \forall\,\boldsymbol{e}\in\mathbb{R}^q,\quad \|\boldsymbol{e}\|=1,\ n\ge 1,\ \text{ a.s.}$$

Corollary 8.2.1 *In addition to the assumptions of Theorem 8.2.1 and (8.2.15), assume that $\lambda > 0$. Then, for each $\psi \in \Psi_F$,*

$$(8.2.16)\qquad \begin{aligned} n^{1/2}(\hat{\boldsymbol{\theta}}_M - \boldsymbol{\theta}) &= \{\lambda\,\boldsymbol{\Sigma}\}^{-1}\mathbf{M}(\boldsymbol{\theta}) + o_p(1), \\ n^{1/2}(\hat{\boldsymbol{\theta}}_M - \boldsymbol{\theta}) &\longrightarrow_d \mathcal{N}\left(\mathbf{0}, v(\psi, F)\,\boldsymbol{\Sigma}^{-1}\right), \end{aligned}$$

where $v(\psi, F)$ is as in (7.3.4).

The following corollary gives the analogue of the above corollary for the LAD estimator under weaker conditions on the error d.f. Its proof uses (8.2.12).

Corollary 8.2.2 *Assume (8.2.1) and (8.2.2) - (8.2.10) hold. In addition, if (8.2.15) with $\psi(x) \equiv sgn(x)$ holds and if the error d.f. F has positive and continuous density in an open neighborhood of 0, then $n^{1/2}(\hat{\boldsymbol{\theta}}_{lad} - \boldsymbol{\theta}) \Rightarrow \mathcal{N}(\mathbf{0}, \boldsymbol{\Sigma}^{-1}/4f^2(0))$.*

To state analogous results about the other two classes of estimators we need to introduce some more notation. Let

$$\begin{aligned} \overline{\varphi} &:= \int_0^1 \varphi(u)du, \quad \overline{\dot{\boldsymbol{\mu}}} := n^{-1}\sum_{i=1}^n \dot{\boldsymbol{\mu}}_i, \quad q(u) := f(F^{-1}(u)), \\ \hat{\mathbb{Z}}(u) &:= n^{-1/2}\sum_{i=1}^n (\dot{\boldsymbol{\mu}}_i - \overline{\dot{\boldsymbol{\mu}}})[I(F(\varepsilon_i) \le u) - u], \quad 0 \le u \le 1, \\ \hat{\mathbf{S}} &:= n^{-1/2}\sum_{i=1}^n (\dot{\boldsymbol{\mu}}_i - \overline{\dot{\boldsymbol{\mu}}})[\varphi(F(\varepsilon_i)) - \overline{\varphi}], \\ \hat{\mathcal{K}}(\mathbf{t}) &:= \int_0^1 \|\hat{\mathbb{Z}}(u) + \boldsymbol{\Gamma}\,\mathbf{t}\,q(u)\|^2 L(du), \qquad \mathbf{t} \in \mathbf{R}^q. \end{aligned}$$

Theorem 8.2.2 *Suppose the assumptions (8.2.1), (8.2.2), (8.2.7) (8.2.10), (**F1**) and (**F2**) hold. In addition, suppose that for some positive definite $q \times q$ matrix $\boldsymbol{\Gamma}$, possibly depending on $\boldsymbol{\theta}$,*

$$(8.2.17)\qquad n^{-1}\sum_{i=1}^n (\dot{\boldsymbol{\mu}}_i - \overline{\dot{\boldsymbol{\mu}}})(\dot{\boldsymbol{\mu}}_i - \overline{\dot{\boldsymbol{\mu}}})' = \boldsymbol{\Gamma} + o_p(1).$$

Then, for every $0 < b < \infty$,

$$\text{(8.2.18)} \qquad \sup \left\| \mathbb{Z}(u, \boldsymbol{\theta} + n^{-1/2}\mathbf{t}) - \hat{\mathbb{Z}}(u) - \boldsymbol{\Gamma}\,\mathbf{t}\,q(u) \right\| = o_p(1),$$

$$\text{(8.2.19)} \qquad \sup \left\| \mathbf{S}(\boldsymbol{\theta} + n^{-1/2}\mathbf{t}) - n^{1/2}\overline{\mu}\overline{\varphi} - \hat{\mathbf{S}} - \boldsymbol{\Gamma}\,\mathbf{t} \int q\,d\varphi \right\| = o_p(1),$$

$$\text{(8.2.20)} \qquad \sup_{L \in \Phi, \|\mathbf{t}\| \le b} \left| \mathcal{K}(\boldsymbol{\theta} + n^{-1/2}\mathbf{t}) - \hat{\mathcal{K}}(\mathbf{t}) \right| = o_p(1).$$

where the supremum in (8.2.18) is taken over all $0 \le u \le 1$, $\|\mathbf{t}\| \le b$ *and in (8.2.19) over* $\varphi \in \Phi$, $\|\mathbf{t}\| \le b$.

The claim (8.2.20) follows from (8.2.18) trivially while (8.2.19) follows from (8.2.18) and the integration by parts operation that gives

$$\mathbf{S}(\mathbf{t}) \equiv \overline{\mu}\overline{\varphi} - \int \mathbb{Z}\Big(\frac{u(n+1)}{n}, \mathbf{t}\Big)\varphi(du), \quad \forall\, \mathbf{t} \in \mathbb{R}^q, \ w.p.1.$$

The proof of (8.2.18) will be indicated at the end of this section.

The next two corollaries give the asymptotic distribution of the R- and m.d- estimators.

Corollary 8.2.3 *In addition to the assumptions of Theorem 8.2.2, assume that*

$$\text{(8.2.21)} \qquad \mathbf{e}'\mathbf{S}(\boldsymbol{\theta} + n^{-1/2} r\mathbf{e}) \text{ is monotonic in } r \in \mathbb{R},\ \forall\, \mathbf{e} \in \mathbb{R}^q, \quad \|\mathbf{e}\| = 1,\ n \ge 1, \quad a.s.$$

$$\text{(8.2.22)} \qquad \overline{\varphi} = 0.$$

Then, for every $\varphi \in \Phi$,

$$\text{(8.2.23)} \qquad n^{1/2}(\hat{\boldsymbol{\theta}}_R - \boldsymbol{\theta}) = \{\gamma\boldsymbol{\Gamma}\}^{-1}\hat{\mathbf{S}} + o_p(1), \quad \gamma := \int_0^1 q\,d\varphi$$

$$n^{1/2}(\hat{\boldsymbol{\theta}}_R - \boldsymbol{\theta}) \longrightarrow_d \mathcal{N}\left(\mathbf{0}, \tau(\varphi, F)\,\boldsymbol{\Gamma}^{-1}\right),$$

where $\tau(\varphi, F) = \int \varphi^2(u)du/\gamma^2$.

Corollary 8.2.4 *In addition to the assumptions of Theorem 8.2.2, assume that for some* $0 < \ell \in L_2([0,1], L)$,

$$\text{(8.2.24)} \qquad \int_0^1 \mathbf{e}'\mathbb{Z}(u, \boldsymbol{\theta} + n^{-1/2} r\mathbf{e})\ell(u)L(du) \text{ is monotonic in } r \in \mathbb{R}, \quad \forall\, \mathbf{e} \in \mathbb{R}^q,\ \|\mathbf{e}\| = 1,\ n \ge 1, \quad a.s.$$

Then, for every $\varphi \in \Phi$,

$$n^{1/2}(\hat{\boldsymbol{\theta}}_{md} - \boldsymbol{\theta}) = \{\gamma_1 \boldsymbol{\Gamma}\}^{-1} \int_0^1 \hat{\mathbb{Z}}(u) q(u) L(du) + o_p(1), \tag{8.2.25}$$
$$n^{1/2}(\hat{\boldsymbol{\theta}}_R - \boldsymbol{\theta}) \longrightarrow_d \mathcal{N}\left(\mathbf{0}, \sigma_0^2 \boldsymbol{\Gamma}^{-1}\right),$$

where σ_0^2 is as in (5.6.21) and (5.6.15).

Remark 8.2.1 *On the assumptions (8.2.6) - (8.2.9).* These assumptions are stated for a general time series satisfying (8.2.1) and (8.2.2). If the underlying time series and μ is such that the series is stationary and ergodic, then the following hold: Conditions (8.2.6) and (8.2.7) are implied by

$$E\|\dot{\boldsymbol{\mu}}_1\|^2 < \infty, \tag{8.2.26}$$

while (8.2.8) is equivalent to

$$E\|\dot{\boldsymbol{\mu}}_1(\boldsymbol{\theta} + n^{-1/2}\mathbf{t}) - \dot{\boldsymbol{\mu}}_1(\boldsymbol{\theta})\|^2 = o(1), \quad \mathbf{t} \in \mathbb{R}^{\mathbf{q}}, \tag{8.2.27}$$

and the assumption (8.2.9) is implied by

$$n^{1/2} E\|\dot{\boldsymbol{\mu}}_1(\boldsymbol{\theta} + n^{-1/2}\mathbf{t}) - \dot{\boldsymbol{\mu}}_1(\boldsymbol{\theta})\| = O(1), \; \mathbf{t} \in \mathbb{R}^{\mathbf{q}}, \tag{8.2.28}$$

where E denotes the expectation under the stationary distribution of $\mathbf{Y}_0$, which typically will depend on the true parameter $\boldsymbol{\theta}$.

Remark 8.2.2 *On the conditions (8.2.15), (8.2.21) and (8.2.24).* These conditions may be replaced by any other conditions that will guarantee the tightness of the respective standardized estimators. However, these assumptions are satisfied in all those time series models where the parameter vector $\boldsymbol{\theta}$ in the function μ appears in a linear fashion. Consider the model

$$\mu(\mathbf{y}, \mathbf{t}) = \mathbf{t}'\mathbf{g}(\mathbf{y}), \tag{8.2.29}$$

where $\mathbf{g}$ is a q-vector of functions $g_1, \cdots, g_q$, each from $\mathbb{R}^p$ to $\mathbb{R}$. In this case $\dot{\boldsymbol{\mu}}(\mathbf{y}, \mathbf{t}) \equiv \mathbf{g}(\mathbf{y})$ and

$$\mathbf{e}'\mathbf{M}(\boldsymbol{\theta} + n^{-1/2} r\mathbf{e}) = n^{-1/2} \sum_{i=1}^{n} \mathbf{e}'\mathbf{g}(\mathbf{Y}_{i-1}) \psi(X_i - r\mathbf{e}'\mathbf{g}(\mathbf{Y}_{i-1}))$$

which, in view of ψ being nondecreasing, is clearly seen to be monotonic decreasing in r, thereby verifying (8.2.15).

Similarly, one obtains

$$\mathbf{e}'\mathbf{S}(\boldsymbol{\theta}+n^{-1/2}r\mathbf{e}) = n^{-1/2}\sum_{i=1}^{n}\mathbf{e}'\mathbf{g}(\mathbf{Y}_{i-1})\varphi(R_{ir\mathbf{e}}/(n+1)),$$

$$\int_0^1 \mathbf{e}'\mathbb{Z}(u,\boldsymbol{\theta}+n^{-1/2}r\mathbf{e})\ell(u)\,L(du)$$

$$= -n^{-1/2}\sum_{i=1}^{n}\mathbf{e}'(\mathbf{g}(\mathbf{Y}_{i-1})-\bar{\mathbf{g}})\varphi_0(R_{ir\mathbf{e}}/n),$$

where $\varphi_0(u) \equiv \int_0^u \ell\,dL$. Now use Lemma 7.3.1 of Section 7.3.2 above and the nondecreasing nature of φ, φ_0 to verify (8.2.21) and (8.2.24) in the present case.

Note that for the stationary and ergodic models of the type (8.2.29), the assumptions (8.2.2), (8.2.6) and (8.2.7) are implied by the square integrability of $\|\mathbf{g}(\mathbf{Y}_0)\|$ while (8.2.8) - (8.2.10) are vacuously satisfied. In the next section we give some examples of time series satisfying (8.2.29). These series are linear in parameter and nonlinear in the lag variables. Tong (1990) gives numerous examples of such models and their practical importance. The above results are thus applicable to such models. □

Remark 8.2.3 *Efficiency.* Observe that the scale factors $v(\psi, F)$, $\tau(\varphi, F)$ and σ_0 that appear in the asymptotic variances above also appear in the asymptotic variances of their analogues in linear regression and autoregressive models. Thus the asymptotic relative efficiency statements that hold for these linear models continue to hold in the present nonlinear time series setup. □

Remark 8.2.4 *M.D. Estimation.* In the case the errors in (8.2.1) are symmetrically distributed, the analogues of the m.d. estimators (7.4.3) in the current setup are defined as follows: Let $\dot{\mu}_{i,j}(\mathbf{t})$ denote the jth component of the vector $\dot{\boldsymbol{\mu}}_i(\mathbf{t})$, $1 \le i \le n$, $1 \le j \le p$, and define

$$\begin{aligned} K^+(\mathbf{t}) &:= \sum_{j=1}^{p}\int\Big[n^{-1/2}\sum_{i=1}^{n}\dot{\mu}_{i,j}(\mathbf{t})\{I(X_i \le x+\mu_i(\mathbf{t})) \\ &\qquad\qquad -I(-X_i < x-\mu_i(\mathbf{t}))\}\Big]^2 dG(x), \\ \boldsymbol{\theta}^+ &:= \text{argmin}\{K^+(\mathbf{t});\ \mathbf{t}\in\mathbb{R}^p\}, \end{aligned}$$

where $G \in \mathcal{DI}(\mathbb{R})$. Assuming the symmetry of the error density and of G, and using the techniques of Chapters 5 and 7, it is possible to give a set of minimal sufficient conditions under which $n^{1/2}(\boldsymbol{\theta}^+ - \boldsymbol{\theta})$ will converge weakly to $\mathcal{N}(\mathbf{0}, \Sigma^{-1}v(F,G))$, where $v(F,G) ;= \mathrm{Var}\{\int_{-\infty}^{\varepsilon} f dG\}/(\int f^2 dG)^2$. Details are left out as an exercise for an interested reader. □

Remark 8.2.5 The above estimators will typically be not robust against outliers in the errors whenever $\{\dot{\boldsymbol{\mu}}_i\}$ are unbounded functions of $\{\mathbf{Y}_{i-1}\}$. One way to robustify these estimators is to replace the weights $\dot{\boldsymbol{\mu}}_i(\mathbf{t})$ in their defining scores by a vector of bounded functions of $\dot{\boldsymbol{\mu}}_i(\mathbf{t})$, just as was done in the previous chapter when defining GM- and GR- estimators. The asymptotic theory of such procedures can also be established using the methodology developed here. □

Next, we give some results about sequential weighted residual empirical processes. These results are found useful for testing some change point hypotheses pertaining to the error distribution in these models. Accordingly, let $\{h_{ni}\}$ be an array of r.v.'s, h_{ni} being past measurable and independent of ε_i, for each i. Define, for $y \in \mathbb{R}$, $\mathbf{t} \in \mathbb{R}^q$, $u \in [0,1]$,

$$T_h(y, \mathbf{t}, u) := n^{-1/2} \sum_{i=1}^{[nu]} h_{ni} I(\varepsilon_i \le y + d_i(\mathbf{t})),$$

$$\tilde{T}_h(y, \mathbf{t}, u) := n^{-1/2} \sum_{i=1}^{[nu]} h_{ni} \Big[I(\varepsilon_i \le y + d_i(\mathbf{t})) - F(y + d_i(\mathbf{t})) \Big].$$

Theorem 8.2.3 *In addition to the assumptions (8.2.1), (8.2.2), (**F1**), (**F2**), (2.2.52), (8.2.7), assume that $\{h_{ni}\}$ satisfy the following.*

$$(8.2.30) \qquad n^{-1} \sum_{i=1}^{n} E h_{ni}^4 \|\dot{\boldsymbol{\mu}}_i(\boldsymbol{\theta})\|^2 = O(1),$$

$$(8.2.31) \qquad \sum_{i=1}^{n} E h_{ni}^4 [\mu_i(\boldsymbol{\theta} + n^{-1/2}\mathbf{t}) - \mu_i(\boldsymbol{\theta})]^2 = O(1), \quad \mathbf{t} \in \mathbb{R}^{\mathrm{q}}.$$

Then, for every $0 < b < \infty$,

$$(8.2.32) \qquad \sup_{y,\mathbf{t},u} \Big| \tilde{T}_h(y, \boldsymbol{\theta} + n^{-1/2}\mathbf{t}, u) - \tilde{T}_h(y, \boldsymbol{\theta}, u) \Big| = o_p(1),$$

$$(8.2.33) \qquad \sup_{y,\mathbf{t},u} \Big| T_h(y, \boldsymbol{\theta} + n^{-1/2}\mathbf{t}, u) - T_h(y, \boldsymbol{\theta}, u) - n^{-1/2} \sum_{i=1}^{[nu]} h_{ni} \dot{\boldsymbol{\mu}}_i' \mathbf{t} f(y) \Big| = o_p(1),$$

where the supremum is taken over $y \in \mathbb{R}$, $\|\mathbf{t}\| \le b$, $0 \le u \le 1$.

Now consider the residual sequential empirical process

$$F_n(y, \mathbf{t}, u) := n^{-1} \sum_{i=1}^{[nu]} I(X_i - \mu_i(\mathbf{t}) \le y),$$

$$\nu(y, \mathbf{t}, u) := n^{-1} \sum_{i=1}^{[nu]} F(y + d_i(\mathbf{t})),$$

$$\tilde{T}_1(y, \mathbf{t}, u) := n^{1/2}[F_n(y, \mathbf{t}, u) - \nu(y, \mathbf{t}, u)].$$

Note that F_n is the T_h with $h_{ni} \equiv 1$ and $\tilde{T}_1$ is equal to $\tilde{T}_h$ with $h_{ni} \equiv 1$. Write $F_n(y, \mathbf{t})$, $\tilde{T}_1(y, \mathbf{t})$ etc. for $F_n(y, \mathbf{t}, 1), \tilde{T}_1(y, \mathbf{t}, 1)$. The following two corollaries trivially follow from Lemma 8.2.1 and Theorem 8.2.3, respectively.

Corollary 8.2.5 *Under (8.2.1), (8.2.2),* (**F1**)*,* (**F2**)*, (8.2.7), we obtain*

$$\sup_{y,\mathbf{t}} |\tilde{T}_1(y, \boldsymbol{\theta} + n^{-1/2}\mathbf{t}) - \tilde{T}_1(y, \boldsymbol{\theta})| = o_p(1). \tag{8.2.34}$$

If, in addition,

$$n^{-1} \sum_{i=1}^{n} \|\dot{\boldsymbol{\mu}}_i\| = O_p(1), \tag{8.2.35}$$

then, for every $0 < b < \infty$,

$$\sup_{y,\mathbf{t}} \Big| n^{1/2}[F_n(y, \boldsymbol{\theta} + n^{-1/2}\mathbf{t}) - F_n(y, \boldsymbol{\theta})] - \mathbf{t}'\overline{\dot{\boldsymbol{\mu}}}\, f(y) \Big| = o_p(1), \tag{8.2.36}$$

where the supremum is taken over $y \in \mathbb{R}$, $\|\mathbf{t}\| \le b$.

Corollary 8.2.6 *Assume (8.2.1), (8.2.2),* (**F1**)*, and (8.2.7), hold. In addition suppose*

$$n^{-1} \sum_{i=1}^{n} E\|\dot{\boldsymbol{\mu}}_i\|^2 = O(1), \tag{8.2.37}$$

$$\sum_{i=1}^{n} E[\mu_i(\boldsymbol{\theta} + n^{-1/2}\mathbf{t}) - \mu_i(\boldsymbol{\theta})]^2 = O(1). \tag{8.2.38}$$

Then, for every $0 < b < \infty$,

$$\sup_{y,\mathbf{t},u} |\tilde{T}_1(y, \boldsymbol{\theta} + n^{-1/2}\mathbf{t}, u) - \tilde{T}_1(y, \boldsymbol{\theta}, u)| = o_p(1). \tag{8.2.39}$$

$$\sup_{y,\mathbf{t},u} \Big| n^{1/2}[F_n(y, \boldsymbol{\theta} + n^{-1/2}\mathbf{t}, u) - F_n(y, \boldsymbol{\theta}, u)] - n^{-1} \sum_{i=1}^{[nu]} \dot{\boldsymbol{\mu}}_i'\mathbf{t}\, f(y) \Big| = o_p(1), \tag{8.2.40}$$

where the supremum is taken over $y \in \mathbb{R}$, $\|\mathbf{t}\| \le b$, $0 \le u \le 1$.

Proof of Lemma 8.2.1. We shall first prove (8.2.13). Let

$$\dot{\boldsymbol{\mu}}_{ni}(\mathbf{t}) \equiv \dot{\boldsymbol{\mu}}_i(\boldsymbol{\theta} + n^{-1/2}\mathbf{t}), \quad d_{ni}(\mathbf{t}) := \mu_i(\boldsymbol{\theta} + n^{-1/2}\mathbf{t}) - \mu_i(\boldsymbol{\theta}),$$
$$\mathcal{W}^*(y, \mathbf{t}) := n^{-1/2} \sum_{i=1}^n \dot{\boldsymbol{\mu}}_{ni}(\mathbf{t})[I(\varepsilon_i \le y) - F(y)],$$
$$\mathbf{U}(y, \mathbf{t}) := n^{-1/2} \sum_{i=1}^n [\dot{\boldsymbol{\mu}}_{ni}(\mathbf{t}) - \dot{\boldsymbol{\mu}}_{ni}(\mathbf{0})]\,[I(\varepsilon_i \le y) - F(y)].$$

Then one can rewrite

$$\mathcal{W}(y, \mathbf{t}) - \mathcal{W}(y, \mathbf{0}) \equiv [\mathcal{W}(y, \mathbf{t}) - \mathcal{W}^*(y, \mathbf{t})] + \mathbf{U}(y, \mathbf{t}).$$

Thus, it suffices to prove that for every $0 < b < \infty$,

$$\text{(8.2.41)} \qquad \sup_{y \in \mathbb{R}, \|\mathbf{t}\| \le b} \|\mathcal{W}(y, \mathbf{t}) - \mathcal{W}^*(y, \mathbf{t})\| = o_p(1),$$

$$\text{(8.2.42)} \qquad \sup_{y \in \mathbb{R}, \|\mathbf{t}\| \le b} \|\mathbf{U}(y, \mathbf{t})\| = o_p(1).$$

We proceed to **prove (8.2.41)**. This will follow from Theorem 2.2.4 in a similar fashion as in (7.2.8), once we verify the conditions of that theorem, which we shall do now.

Fix a $0 < b < \infty$. In the sequel, the indices in the $\sup_{y,i,\mathbf{t}}$ vary from $y \in \mathbb{R}$, $1 \le i \le n$, $\|\mathbf{t}\| \le b$. Let $\mathcal{F}_{ni} := \sigma-$field$\{\mathbf{Y}_0, \varepsilon_1, \cdots, \varepsilon_{i-1}\}$, $1 \le i \le n$. Let μ_{nij}, W_j, $\mathcal{W}_j$, etc. denote the jth coordinates of $\dot{\boldsymbol{\mu}}_{ni}$, $\mathbf{W}$, $\mathcal{W}$, etc. Note that if in (2.2.26) we take

$$\text{(8.2.43)} \qquad h_{ni} \equiv \mu_{nij}(\mathbf{s}), \quad \zeta_{ni} \equiv \varepsilon_i, \quad \delta_{ni} \equiv d_{ni}(\mathbf{s}), \quad \mathcal{A}_{ni} \equiv \mathcal{F}_{ni},$$

then $V_h(y)$, $V_h^*(y)$, respectively, are equal to $\mathcal{W}_j(y, \mathbf{s})$, $W_j(y, \mathbf{0})$, for all $y \in \mathbb{R}$, $\mathbf{s} \in \mathbb{R}^q$.

Now, by (8.2.2), $\forall \alpha > 0$, $\exists n_1$, $\ni$ $\forall n > n_1$,

$$\text{(8.2.44)} \qquad P\Big(\sup_{i,\mathbf{t}} |\mu_i(\boldsymbol{\theta} + n^{-1/2}\mathbf{t}) - \mu_i(\boldsymbol{\theta}) - n^{-1/2}\mathbf{t}\dot{\boldsymbol{\mu}}_i| \le b\alpha n^{-1/2}\Big) \ge 1 - \alpha.$$

Hence from (8.2.7), we readily obtain

$$\text{(8.2.45)} \qquad \sup_{i,\mathbf{t}} |d_{ni}(\mathbf{t})| = o_p(1).$$

This verifies (2.2.28) for the δ_{ni} of (8.2.43).

Next, by (8.2.6) and (8.2.8),

$$\Big(n^{-1}\sum_{i=1}^{n}\dot{\mu}_{nij}^{2}(\mathbf{s})\Big)^{1/2} = \sigma_{jj} + o_p(1),$$

where σ_{jj} is the jth diagonal element of the matrix $\mathbf{\Sigma}$, so that (2.2.34) is verified for the h_{ni} of (8.2.43) for every $\mathbf{s} \in \mathbb{R}^q, 1 \le j \le q$. Finally, because for every $\mathbf{s} \in \mathbb{R}^q, 1 \le j \le q$,

$$\begin{aligned} n^{-1/2}\max_i |\dot{\mu}_{nij}(\mathbf{s})| &\le n^{-1/2}\max_i |\dot{\mu}_{nij}(\mathbf{s}) - \dot{\mu}_{ij}| \\ &\quad + n^{-1/2}\max_i |\dot{\mu}_{ij}|, \\ n^{-1/2}\max_i |\dot{\mu}_{nij}(\mathbf{s}) - \dot{\mu}_{ij}| &\le \Big(n^{-1}\sum_{i=1}^{n}\{\dot{\mu}_{nij}(\mathbf{s}) - \dot{\mu}_{ij}\}^2\Big)^{1/2}, \end{aligned}$$

(2.2.27) is implied by (8.2.7) and (8.2.8) for the h_{ni} of (8.2.43). Hence, (2.2.32) readily implies that

$$(8.2.46) \qquad \sup_{y\in\mathbb{R}} \|\mathcal{W}(y,\mathbf{s}) - \mathcal{W}^*(y,\mathbf{s})\| = o_p(1), \quad \|\mathbf{s}\| \le b.$$

To complete the proof of (8.2.41), it suffices to show that $\forall\, \alpha > 0$, $\exists\, \delta > 0$ and n_0, $\ni \forall n > n_0$, and for each $\|\mathbf{s}\| \le b$,

$$(8.2.47) \qquad P\Big(\sup_{y\in\mathbb{R}, \|\mathbf{t}-\mathbf{s}\|<\delta} \|\mathbf{D}(y,\mathbf{t}) - \mathbf{D}(y,\mathbf{s})\| > \alpha\Big) \le \alpha,$$

where $\mathbf{D}(y,\mathbf{t}) = \mathcal{W}(y,\mathbf{t}) - \mathcal{W}^*(y,\mathbf{t})$.

For the sake of brevity, let $\alpha_i(y,\mathbf{t}) := I(\varepsilon_i \le y + d_{ni}(\mathbf{t})) - F(y + d_{ni}(\mathbf{t}))$ and write $\alpha_i(y)$ for $\alpha_i(y,\mathbf{0})$. Then one can rewrite

$$\begin{aligned} \mathbf{D}(y,\mathbf{t}) - \mathbf{D}(y,\mathbf{s}) &= n^{-1/2}\sum_{i=1}^{n}[\dot{\boldsymbol{\mu}}_{ni}(\mathbf{t}) - \dot{\boldsymbol{\mu}}_{ni}(\mathbf{s})]\,[\alpha_i(y,\mathbf{t}) - \alpha_i(y)] \\ &\quad + n^{-1/2}\sum_{i=1}^{n}\dot{\boldsymbol{\mu}}_{ni}(\mathbf{s})[\alpha_i(y,\mathbf{t}) - \alpha_i(y,\mathbf{s})] \\ &= \mathbf{D}_1(y,\mathbf{s},\mathbf{t}) + \mathbf{D}_2(y,\mathbf{s},\mathbf{t}), \quad \text{say} \end{aligned}$$

To prove (8.2.46) it thus suffices to prove analogous result for $\mathbf{D}_1$ and $\mathbf{D}_2$. But, (8.2.10) readily implies this for $\mathbf{D}_1$ as $|\alpha_i(y,\mathbf{t}) - \alpha_i(y)| \le 1$, uniformly in i, y, $\mathbf{t}$.

We proceed to prove it for $\mathbf{D}_2$. Write $\mu_{nij}(\mathbf{s}) = \mu^+_{nij}(\mathbf{s}) - \mu^-_{nij}(\mathbf{s})$. Let $D^{\pm}_{2j}$ denote D_{2j} where $\mu_{nij}(\mathbf{s})$ is replaced by $\mu^{\pm}_{nij}(\mathbf{s})$. It suffices to prove the analog of (8.2.46) for $D^{\pm}_{2j}$, $1 \le j \le q$.

Now, fix an $1 \le j \le q$, $\alpha > 0$, $\|\mathbf{s}\| \le b$, and a $\delta > 0$. Let $\Delta_{ni} := n^{-1/2}(\delta\|\dot{\boldsymbol{\mu}}_i\| + 2b\alpha)$ and

$$A_n := \Big\{ \sup_{\|\mathbf{t}\| \le b, \|\mathbf{t}-\mathbf{s}\| < \delta} |d_{ni}(\mathbf{t}) - d_{ni}(\mathbf{s})| \le \Delta_{ni},\ 1 \le i \le n \Big\}.$$

From (8.2.44), it follows that

$$\text{(8.2.48)} \qquad P(A_n) > 1 - \alpha, \quad \forall n > n_1.$$

Next, let, for y, $a \in \mathbb{R}$,

$$\begin{aligned} \mathcal{D}^{\pm}_j(y, \mathbf{s}, a) \quad := \quad & n^{-1/2} \sum_{i=1}^n \dot{\mu}^{\pm}_{nij}(\mathbf{s}))[I(\varepsilon_i \le y + d_{ni}(\mathbf{s}) + a\Delta_{ni}) \\ & \qquad - F(y + d_{ni}(\mathbf{s}) + a\Delta_{ni})], \end{aligned}$$

Let

$$\text{(8.2.49)} \qquad \delta_{ni} \equiv d_{ni}(\mathbf{s}) + a\Delta_{ni}.$$

By definition, δ_{ni} are $\mathcal{F}_{ni}$- measurable, for every $a \in \mathbb{R}$, $1 \le i \le n$. Moreover, by (8.2.7) and (8.2.45),

$$\begin{aligned} \text{(8.2.50)} \qquad & \max_i |\delta_{ni}| \\ & \le \max_i |d_{ni}(\mathbf{s})| + \max_i n^{-1/2}(\delta\|\dot{\boldsymbol{\mu}}_i\| + 2b\alpha) = o_p(1). \end{aligned}$$

The rest of the argument being the same as for (8.2.46), one more application of (2.2.34) with these δ_{ni} and the other entities as in (8.2.43), yields that

$$\text{(8.2.51)} \qquad \sup_y |\mathcal{D}^{\pm}_j(y, \mathbf{s}, a) - \mathcal{D}^{\pm}_j(y, \mathbf{s}, 0)| = o_p(1), \qquad \forall\, a \in \mathbb{R}.$$

Now, use the monotonicity of the indicator function and the d.f. F to obtain that, on A_n, for $\|\mathbf{t}\| \le b$, $\|\mathbf{t} - \mathbf{s}\| < \delta$,

$$\begin{aligned} & D^{\pm}_{2j} \\ \le\ & |\mathcal{D}^{\pm}_j(y, \mathbf{s}, 1) - \mathcal{D}^{\pm}_j(y, \mathbf{s}, 0)| + |\mathcal{D}^{\pm}_j(y, \mathbf{s}, -1) - \mathcal{D}^{\pm}_j(y, \mathbf{s}, 0)| \\ & + \Big| n^{-1/2} \sum_{i=1}^n \dot{\mu}^{\pm}_{nij}(\mathbf{s}))[F(y + d_{ni}(\mathbf{s}) + a\Delta_{ni}) \\ & \qquad\qquad - F(y + d_{ni}(\mathbf{s}) + a\Delta_{ni})]\Big|. \end{aligned}$$

By (**F1**), the last term in this bound is no larger than

$$2\|f\|_\infty n^{-1}\sum_{i=1}^{n}|\dot{\mu}^{\pm}_{nij}(\mathbf{s})|(\delta\|\dot{\mu}_i\|+2b\alpha)$$

which, in view of (8.2.6) and (8.2.8), can be made to be smaller than a constant multiple of α, by the choice of δ, with arbitrarily large probability, for all sufficiently large n. This together with (8.2.48) and (8.2.51) completes the proof of an analogue of (8.2.46) for $\mathbf{D}_2$, and hence of the claim (8.2.41).
Proof of (8.2.42). Fix a $1 \le j \le q$ and let $h_{ni} \equiv \dot{\mu}_{nij}(\mathbf{t}) - \dot{\mu}_{nij}(\mathbf{0})$. Write $h_{ni} = h^+_{ni} - h^-_{ni}$ so that the jth component of $\mathbf{U}$ is written as $U_j = U^+_j - U^-_j$, with

$$U^{\pm}_j(y,\mathbf{t}) := n^{-1/2}\sum_{i=1}^{n} h^{\pm}_{ni}\alpha_i(y).$$

From (8.2.8), we obtain,

$$\begin{aligned}
(8.2.52)\qquad & Var\left(U^{\pm}_j(y,\mathbf{t})\right)\\
&= n^{-1}\sum_{i=1}^{n} E(h^{\pm}_{ni})^2 F(y)(1-F(y))\\
&\le n^{-1}\sum_{i=1}^{n} E(h^{\pm}_{ni})^2 = o(1), \quad \forall\, y\in\mathbb{R}.
\end{aligned}$$

Next, fix an $\alpha > 0$ and let $-\infty = y_0 < y_1 < \cdots < y_r = \infty$ be a partition of $\mathbb{R}$ such that $F(y_k) - F(y_{k-1}) \le \alpha$, $k = 0, 1, \cdots, r$. Then, once again using the monotonicity of the indicator function and F, we obtain,

$$\sup_y |U^{\pm}_j(y,\mathbf{t})| \;\le\; \max_{0\le k\le r} |U^{\pm}_j(y_k,\mathbf{t})| + \alpha n^{-1/2}\sum_{i=1}^{n}|h_{ni}|.$$

This, (8.2.52), (8.2.9), and the arbitrariness of α implies

$$\sup_y \|\mathbf{U}(y,\mathbf{t})\| = o_p(1), \quad \forall\, \|\mathbf{t}\| \le b.$$

To finish the proof of (8.2.42), we need to prove an analogue of (8.2.47) for the $\mathbf{U}(y,\mathbf{t})$-process. But this is implied by (8.2.10), because $|\alpha_i(y)| \le 1$, and because

$$\mathbf{U}(y,\mathbf{t}) - \mathbf{U}(y,\mathbf{s}) = n^{-1/2}\sum_{i=1}^{n}[\dot{\boldsymbol{\mu}}_{ni}(\mathbf{t}) - \dot{\boldsymbol{\mu}}_{ni}(\mathbf{s})]\,\alpha_i(y).$$

This completes the proof of (8.2.42), and hence of (8.2.13), while that of (8.2.14) follows from (8.2.13) and the Taylor expansion of F. The proof of (8.2.12) is similar and simpler while that of (8.2.11) follows from the Chebbychev inequality. The proof of Lemma 8.2.1 is terminated. □

Proof of Theorem 8.2.3. The proof of this theorem is facilitated by the following two lemmas. Recall the definitions of δ_{ni} from (8.2.49).

Lemma 8.2.2 *Under (F1) and (8.2.2) - (8.2.8), for some $K < \infty$ and for every $\alpha > 0$,*

$$\limsup_n P\Big(\sup_{x,y} n^{-1/2}\sum_{i=1}^n \Big|h_{ni}[F(y+\delta_{ni}) - F(x+\delta_{ni})]\Big| \le K\alpha\Big) = 1,$$

for every $a \in \mathbb{R}$, $\|\mathbf{s}\| \le b$, where the supremum is taken over the set $\{x, y \in \mathbb{R};\ |F(y) - F(x)| \le n^{-1/2}\alpha\}$.

Proof. Let $u_n := \max_i |\delta_{ni}|$, $\tau_n := \{|f(y) - f(x)|;\ |F(y) - F(x)| \le n^{-1/2}\alpha\}$, and $\omega_n := \{|f(z) - f(v)|;\ |z - v| \le u_n\}$. From (**F1**), (**F2**) and (8.2.50), $\tau_n = o(1)$, $\omega_n = o_p(1)$. Also, from (8.2.2) - (8.2.8), we obtain $n^{-1/2}\sum_{i=1}^n |\delta_{ni}| = O_p(1)$. The lemma follows from these facts and the inequality

$$\begin{aligned} n^{-1/2}&\sum_{i=1}^n \Big|h_{ni}\,[F(y+\delta_{ni}) - F(x+\delta_{ni})]\Big| \\ &\le n^{-1/2}\sum_{i=1}^n |h_{ni}|\,|F(y) - F(x)| + n^{-1/2}\sum_{i=1}^n |h_{ni}\delta_{ni}|\{\tau_n + 2\omega_n\}. \quad \square \end{aligned}$$

Lemma 8.2.3 *Let F be a continuous strictly increasing d.f. on $\mathbb{R}$,, ε_i be i.i.d. F, $\alpha > 0$, $n \ge 1$, $N := [n^{1/2}/\alpha]$ and $\{y_j\}$ be the partition of $\mathbb{R}$ such that $F(y_j) = j/N$, $1 \le j \le N$, $y_0 = -\infty$, $y_{N+1} = \infty$. Then, under (2.2.52),*

$$\text{(8.2.53)}\quad \sup_{u,j}\Big|n^{-1/2}\sum_{i=1}^{[nu]} h_{ni}^{\pm}\Big\{I(\varepsilon_i \le y_{j+1}) - I(\varepsilon_i \le y_j) - 1/N\Big\}\Big| = o_p(1),$$

where the supremum is taken over $0 \le u \le 1$, $0 \le j \le N+1$.

Proof. Let

$$V_{i,j} := h_{ni}^{\pm}\Big\{I(\varepsilon_i \le y_{j+1}) - I(\varepsilon_i \le y_j) - 1/N\Big\}, \quad S_{i,j} := \sum_{k=1}^i V_{k,j}.$$

Clearly, for each $0 \le j \le N+1$, $\{S_{i,j}, \mathcal{F}_{ni}, 1 \le i \ge n,\}$ is a mean zero martingale array. By the inequality (9.1.4), for some $C < \infty$,

$$P(\max_{1\le i\le n} |S_{i,j}| > \alpha) \le \alpha^{-4} ES^4_{n,j},$$
$$ES^4_{n,j} \le C\Big\{E\Big[\sum_{i=1}^n E\Big(V^2_{i,j}\Big|\mathcal{F}_{n,i-1}\Big)\Big]^2 + \sum_{i=1}^n EV^4_{i,j}\Big\}.$$

But, because $h^{\pm}_{ni} \le |h_{ni}|$, for all i,

$$\sum_{i=1}^n EV^4_{i,j} \le \sum_{i=1}^n Eh^4_{ni},$$
$$E\Big(V^2_{i,j}\Big|\mathcal{F}_{n,i-1}\Big) \le h^2_{ni}\, N^{-1} \le n^{-1/2}\, h^2_{ni}\, \frac{\alpha}{1-\alpha}, \qquad 1 \le i \le n,$$
$$E\Big[\sum_{i=1}^n E(V^2_{i,j}|\mathcal{F}_{n,i-1})\Big]^2 \le \Big(\frac{\alpha}{1-\alpha}\Big)^2 \sum_{i=1}^n Eh^4_{ni}, \quad \forall 0 \le j \le N+1.$$

Hence, for some constant C_α, depending only on C and α,

$$P(L.H.S.(8.2.53) > \alpha) \le C_\alpha\, n^{-2}\, N \sum_{i=1}^n Eh^4_{ni} = O(n^{-1/2}) = o(1),$$

Now we proceed to **prove (8.2.32)**. Write $h_{ni} = h^+_{ni} - h^-_{ni}$. By the triangle inequality, it suffices to prove (8.2.32) with h_{ni} replaced by $h^{\pm}_{ni}$. Let $V^{\pm}(y, \mathbf{t}, u)$ denote the difference inside the absolute value on the L.H.S. of (8.2.32) with h_{ni} replaced by $h^{\pm}_{ni}$. Now let $\delta_{ni}(\mathbf{t}) \equiv d_{ni}(\mathbf{t}) + a\Delta_{ni}$, and as before we will continue writing δ_{ni} for $\delta_{ni}(\mathbf{s})$. Define

$$U^{\pm}(a, y, \mathbf{t}, u) \;:=\; n^{-1/2} \sum_{i=1}^{[nu]} h^{\pm}_{ni}\Big[I(\varepsilon_i \le y + \delta_{ni}(\mathbf{t})) - I(\varepsilon_i \le y)$$
$$-F(y + \delta_{ni}(\mathbf{t})) + F(y)\Big]$$

We have, by (**F**), (2.2.52) , and the DCT, for every fixed $y \in \mathbb{R}$, $\mathbf{t} \in \mathbb{R}^q$, and $0 \le u \le 1$,

$$\mathrm{Var}\Big(V^{\pm}(y, \mathbf{t}, u)\Big) \le n^{-1} \sum_{i=1}^n Eh^2_{ni}|F(y + d_{ni}(\mathbf{t})) - F(y)| = o(1).$$

Now, fix an $\alpha > 0$, $\|\mathbf{s}\| \le b$ and a $\delta > 0$. Let A_n be as in the proof of (8.2.41). Arguing as in there, we obtain that on the set A_n, $\forall\, \|\mathbf{t}\| \le$

b, $\|\mathbf{t} - \mathbf{s}\| < \delta$, and for all $y \in \mathbb{R}$, $0 \le u \le 1$,

$$
\begin{aligned}
&V^{\pm}(y, \mathbf{t}, u) \\
&\le \sup_{y,u} |U^{\pm}(1, y, \mathbf{s}, u)| + \sup_{y,u} |U^{\pm}(-1, y, \mathbf{s}, u)| \\
&\quad + \sup_{y,u} n^{-1/2} \sum_{i=1}^{[nu]} h_{ni}^{\pm}[F(y + d_{ni}(\mathbf{s}) + \Delta_{ni}) - F(y + d_{ni}(\mathbf{s}) - \Delta_{ni})].
\end{aligned}
$$

But, by (**F1**), (2.2.52) , and (8.2.30), the last term in this bound is bounded above by

$$
C\Big(\delta n^{-1} \sum_{i=1}^{n} \|h_{ni}\boldsymbol{\mu}_i\| + 2bn^{-1} \sum_{i=1}^{n} |h_{ni}|\alpha\Big) = O_p(\alpha),
$$

by the choice of δ. Thus to complete the proof of (8.2.32), it suffices to show that

$$
\sup_{y,u} |U^{\pm}(a, y, \mathbf{s}, u)| = o_p(1), \quad a \in \mathbb{R}, \ \|\mathbf{s}\| \le b. \tag{8.2.54}
$$

Let N and $\{y_j\}$ be as in the proof of Lemma 8.2.3. Then we obtain

$$
\begin{aligned}
&\sup_{y,u} |U^{\pm}(a, y, \mathbf{s}, u)| \\
&\le 2 \sup_{j,u} |U^{\pm}(a, y_j, \mathbf{s}, u)| + n^{-1/2} \sum_{i=1}^{n} |h_{ni}| \max_j [F(y_{j+1}) - F(y_j)] \\
&\quad + \max_j n^{-1/2} \sum_{i=1}^{n} |h_{ni}| \, |F(y_{j+1} + \delta_{ni}) - F(y_j + \delta_{ni})| \\
&\quad + \sup_{j,u} \Big| n^{-1/2} \sum_{i=1}^{[nu]} h_{ni}^{\pm} \Big\{ I(\varepsilon_i \le y_{j+1}) - I(\varepsilon_i \le y_j) - 1/N \Big\} \Big|
\end{aligned}
$$

The second term is $O_p(\alpha)$ by the definition of y_j's and (2.2.52) , while the last two terms are $o_p(1)$ by Lemmas 8.2.2 and 8.2.3.

Using the fact that for each a, y, $\mathbf{s}$, $n^{1/2}U^{\pm}(a, y, \mathbf{s}, i/n)$ is a martingale in i, and arguing as in the proof of Lemma 8.2.3, we obtain

$$
\begin{aligned}
&P\Big(\sup_{j,u} |U^{\pm}(a, y_j, \mathbf{s}, u)| > \alpha\Big) \\
&\le N \max_j P\Big(\sup_{1 \le i \le n} |U^{\pm}(a, y_j, \mathbf{s}, i/n)| > \alpha\Big) \\
&\le N n^{-2} \max_j E\Big\{n^{1/2} U^{\pm}(a, y_j, \mathbf{s}, 1)\Big\}^4 = O(n^{-1/2}).
\end{aligned}
$$

This and the arbitrariness of α completes the proof of (8.2.39).

The claim (8.2.33) follows from (8.2.32) in a routine fashion, thereby ending the proof of Theorem 8.2.3. □

8.2.2 Examples of AR models

We shall first discuss the problem of testing for a change in the error distribution of the model (8.2.1), (8.2.2).

Example 8.2.1 *Testing for a change in the error d.f.*
Let F_1, F_2 be two different distribution functions, not necessarily known, and $F_1 \neq F_2$. Consider the problem of testing the change point hypothesis

H_0 : the errors $\varepsilon_1, \cdots \varepsilon_n$ in (8.2.1) are i.i.d., against

$H_1 : \varepsilon_1, \cdots, \varepsilon_j$ are i.i.d. F_1, $\varepsilon_{j+1}, \cdots, \varepsilon_n$ are i.i.d. F_2, for some $1 \leq j < n$.

That is, we are interested in testing the hypothesis that the time series (8.2.1) is generated by i.i.d. errors, versus the alternatives that for some $1 \leq j < n$, the first j and the last $n - j$ observations are generated from possibly two different error distributions.

To describe a test for this problem, let $\hat{\boldsymbol{\theta}}$ be estimators of $\boldsymbol{\theta}$ based on X_i, $1 - p \leq i \leq n$. Let $\mathbf{d}_n := n^{1/2}(\hat{\boldsymbol{\theta}} - \boldsymbol{\theta})$. Assume that

$$\mathbf{d}_n = O_p(1), \qquad \text{under } H_0. \tag{8.2.55}$$

Also, let $\hat{F}_{nu}$, $\hat{F}_{n(1-u)}$, denote residual empirical processes based on the first $[nu]$ residuals $X_i - \mu_i(\hat{\boldsymbol{\theta}})$; $1 \leq i \leq [un]$, and the last $n - [nu]$ residuals $X_i - \mu_i(\hat{\boldsymbol{\theta}})$; $[nu] + 1 \leq i \leq n$, $u \in [0, 1]$, where $[x]$ denotes the greatest integer less than or equal to the real number x. The Kolmogorov-Smirnov type test of this hypothesis is based on the process

$$\Delta_n(y, u) := \frac{[nu]}{n}\left(1 - \frac{[nu]}{n}\right) n^{1/2} \left\{ \hat{F}_{nu}(y) - \hat{F}_{n(1-u)}(y) \right\},$$

where $y \in \mathbb{R}$, $u \in [0, 1]$.

For the sake of brevity write here $W(y, u)$ for $W(y, \boldsymbol{\theta}, u)$ of Corollary 8.2.5. Also, let the common error d.f. be denoted by F and its density by f. All the needed assumptions for the validity of (8.2.39) and (8.2.40) are assumed to be in action.

From (8.2.40) we obtain that under H_0,

$$\begin{aligned}\Delta_n(y,u) &= (1-u)\Big[W(y,u) + n^{-1}\sum_{i=1}^{[nu]} \dot{\mu}_i' \mathbf{d}_n f(y)\Big] \\ &\quad - u\Big[W(y,1) - W(y,u) + n^{-1}\sum_{i=1+[nu]}^{n} \dot{\mu}_i' \mathbf{d}_n f(y)\Big] \\ &\quad + u_p(1).\end{aligned}$$

Now, suppose additionally that for some random vector $\mathbf{m}$,

$$n^{-1}\sum_{i=1}^{n} \dot{\mu}_i = \mathbf{m} + o_p(1). \tag{8.2.56}$$

Then one readily obtains

$$\sup_{0\le u\le 1}\Big|n^{-1}\sum_{i=1}^{[nu]} \dot{\mu}_i - u\,\mathbf{m}\Big| = o_p(1).$$

Hence, under H_0, and under the above appropriate regularity conditions,

$$\Delta_n(y,u) = \Big[W(y,u) - uW(y,1)\Big] + u_p(1).$$

Thus, it follows, say from Bickel and Wichura (1971), that under H_0,

$$\sup_{y\in\mathbb{R},0\le u\le 1} |\Delta_n(y,u)| \longrightarrow_d \sup_{0\le t,u\le 1} |\Delta(t,u)|, \tag{8.2.57}$$

where Δ is a zero mean continuous Gaussian process on $[0,1]^2$ with

$$E\{\Delta(s,u),\Delta(t,v)\} = [s\wedge t - st]\,[u\wedge v - uv].$$

Consequently, the test based on $\sup\{|\Delta_n(y,u)|;\ y\in\mathbb{R}, 0\le u\le 1\}$ is asymptotically distribution free for testing H_0 versus H_1. We end this example by noting that the condition (8.2.56) is typically satisfied if the process is stationary and ergodic and the summands involved here have finite expectations as will be typically the case in the following few examples.

Example 8.2.2 *SETAR(2;1,1) model.* If in (8.2.1) we take

$$q=2,\ p=1,\ \mu(y,\boldsymbol{\theta}) = \theta_1 y I(y>0) + \theta_2 y I(y\le 0), \tag{8.2.58}$$

then it becomes the SETAR(2;1,1) [self-exciting threshold] model of Tong (1990; p130). Note that here $\mathbf{Y}_{i-1} \equiv X_{i-1}$. Let

$$y^+ = \max\{0,y\},\ y^- = \min\{0,y\},\ \mathbf{W}_i \equiv (X_{i-1}^+, X_{i-1}^-)'.$$

Tong (1990) contains some sufficient conditions for the stationarity and ergodicity of the SETAR(2;1,1) process. For example this holds if the error density f is positive everywhere, $E|\varepsilon| < \infty$, and $\theta_1 < 1$, $\theta_2 < 1$, $\theta_1\theta_2 < 1$. Moreover, if additionally $E\varepsilon^2 < \infty$, then $EX_0^2 < \infty$. Hence, by the Ergodic Theorem, in this model the assumptions (8.2.2), (8.2.6) to (8.2.10) are satisfied with $\dot{\boldsymbol{\mu}}_i(\mathbf{t}) \equiv \mathbf{W}_i$,

$$\text{(8.2.59)} \qquad \boldsymbol{\Sigma} = \begin{bmatrix} E(X_0^+)^2 & 0 \\ 0 & E(X_0^-)^2 \end{bmatrix},$$

$$\boldsymbol{\Gamma} = \begin{bmatrix} \mathrm{Var}(X_0^+) & -\mu_0^+\mu_0^- \\ -\mu_0^+\mu_0^- & \mathrm{Var}(X_0^-) \end{bmatrix}, \quad \mu_0^{\pm} := E(X_0^{\pm}).$$

We emphasize the fact that all expectations here depend on the parameter $\boldsymbol{\theta}$.

Note this model is also an example of the sub-model (8.2.29) with $\mathbf{g}(y) \equiv (y^+, y^-)'$. From the discussion in Remark 8.2.2, it follows that the conditions (8.2.15), (8.2.21) and (8.2.24) are also satisfied here. We thus obtain the following

Corollary 8.2.7 *In addition to (8.2.1) and (8.2.58), assume the following.*

(8.2.60) *The error d.f. F has a uniformly continuous everywhere positive density f and $E\varepsilon^2 < \infty$, $E\varepsilon = 0$.*

Then

$$n^{1/2}(\hat{\boldsymbol{\theta}}_M - \boldsymbol{\theta}) \longrightarrow_d \mathcal{N}(\mathbf{0}, \boldsymbol{\Sigma}^{-1}v(\psi, F)), \quad \forall \psi \in \Psi,$$
$$n^{1/2}\hat{(}\boldsymbol{\theta}_R - \boldsymbol{\theta}) \longrightarrow_d \mathcal{N}(\mathbf{0}, \boldsymbol{\Gamma}^{-1}\tau(\varphi, F)), \quad \forall \varphi \in \Phi,$$
$$n^{1/2}(\hat{\boldsymbol{\theta}}_{md} - \boldsymbol{\theta}) \longrightarrow_d \mathcal{N}(\mathbf{0}, \boldsymbol{\Gamma}^{-1}\sigma_0^2), \quad \forall L \in \Phi,$$

where $\boldsymbol{\Sigma}$ and $\boldsymbol{\Gamma}$ are as in (8.2.59).

Now consider the problem of *testing the goodness-of-fit hypothesis H_0 : $F = F_0$*, against the alternative $F \neq F_0$, where F_0 is a known d.f. having a uniformly continuous everywhere positive density f. Let $\hat{\boldsymbol{\theta}}$ be any estimator satisfying

$$\text{(8.2.61)} \qquad n^{1/2}\|\hat{\boldsymbol{\theta}} - \boldsymbol{\theta}\| = O_p(1), \qquad \text{under } H_0.$$

Let $D_n := n^{1/2}\sup_y |F_n(y, \hat{\boldsymbol{\theta}}) - F_0(y)|$. From (8.2.36) we readily obtain

$$D_n = \sup_y |W(y, \boldsymbol{\theta}) + n^{1/2}(\hat{\boldsymbol{\theta}} - \boldsymbol{\theta})'\mathbf{m}\, f_0(y)| + o_p(1),$$

where $W(y, \boldsymbol{\theta})$ is as in (8.2.34) - the standardized empirical of the i.i.d. r.v.'s $\{\varepsilon_i\}$- and $\mathbf{m} := (\mu_0^+, \mu_0^-)'$.

Compare this finding with that in Remark 7.2.3 pertaining to the linear AR model. In linear AR(p) models with zero mean errors, the analogous D_n statistic satisfies $D_n = \sup_y |W(y, \boldsymbol{\theta})| + o_p(1)$, thereby rendering the tests based on D_n ADF. But in the current case, even if $E\varepsilon = 0$, the vector $\mathbf{m} \neq \mathbf{0}$ and hence the tests based on D_n are not ADF. Note that SETAR models are piecewise linear, a very simple departure from the usual linearity, yet the above mentioned property fails.

Next, consider the testing problem of Example 8.2.1 for this model. Under the conditions of Corollary 8.2.7 on the common error d.f. under H_0 and under (8.2.55), all needed conditions for the validity of (8.2.57) are trivially satisfied, and hence any test based on Δ_n is asymptotically d.f. for testing the hypothesis of no change in the error d.f.

Example 8.2.3 *EXPAR model.* Let $\Omega := (-1, 1) \times \mathbb{R} \times (0, \infty)$ and let

$$q = 3,\ p = 1, \quad \mu(y, \boldsymbol{\theta}) = \{\theta_1 + \theta_2 exp(-\theta_3 y^2)\}y, \quad \boldsymbol{\theta} \in \Omega.$$

Then (8.2.1) becomes an example of an amplitude-dependent exponential autoregressive model of order 1 (EXPAR(1)). From Tong (1990), one obtains that under (8.2.60), this times series is stationary and ergodic, and $EX_0^2 < \infty$. Because $x^k exp(-\alpha x^2)$ is a smooth function of α with all derivatives bounded in x, for all $k \geq 0$, (8.2.2) - (8.2.10), (8.2.10), (8.2.37) and (8.2.37) are readily seen to hold with

$$\dot{\boldsymbol{\mu}}_i(\boldsymbol{\theta}) := \begin{pmatrix} X_{i-1} \\ X_{i-1} exp(-\theta_3 X_{i-1}^2) \\ -\theta_2 X_{i-1}^3 exp(-\theta_3 X_{i-1}^2) \end{pmatrix}$$

$$n^{-1} \sum_{i=1}^{n} \dot{\boldsymbol{\mu}}_i(\boldsymbol{\theta}) = E\dot{\boldsymbol{\mu}}_1(\boldsymbol{\theta}) + o_p(1).$$

The analogue of the D_n statistic asymptotically behaves here like

$$D_n = \sup_{y \in \mathbb{R}} |W(y, \boldsymbol{\theta}) + \mathbf{d}_n' E\dot{\boldsymbol{\mu}}_1(\boldsymbol{\theta})\, f_0(y)| + o_p(1).$$

Now, if F_0 is such that the stationary distribution is symmetric around zero, then $E\dot{\boldsymbol{\mu}}_1(\boldsymbol{\theta}) = \mathbf{0}$ and here also, like in the linear AR(1) model with zero mean errors, the test based on D_n is asymptotically distribution free.

Similarly the conclusions of Example 8.2.1 are also valid here in connections with the change point testing problem, assuming of course among other things, that an estimator $\hat{\boldsymbol{\theta}}$ satisfying (8.2.55) exists here.

But note that if $\theta_2 = 0$, then θ_3 is not identifiable. However, in many applications one takes θ_3 to be a known number. In that case we again have a model of the type (8.2.29), and hence all the limit results about M-, R-, and m.d. estimators are valid under the assumption that the error d.f. satisfy (8.2.60) with $\boldsymbol{\Sigma} = E\mathbf{H}\mathbf{H}'$, $\boldsymbol{\Gamma} = \boldsymbol{\Sigma} - \boldsymbol{\nu}_1\boldsymbol{\nu}_1'$, where

$$\mathbf{H} = \begin{bmatrix} X_0 & X_0^2 exp(-\theta_3 X_0^2) \\ X_0^2 exp(-\theta_3 X_0^2) & X_0^2 exp(-2\theta_3 X_0^2) \end{bmatrix}$$

$$\boldsymbol{\nu}_1 = E\begin{bmatrix} X_0 \\ X_0 exp(-\theta_3 X_0^2) \end{bmatrix}$$

We end this example by mentioning that the above theory is seen easily to hold for the general EXPAR(m) model given by

$$\mu(\mathbf{y}, \boldsymbol{\theta}) := \sum_{j=1}^{p} [\alpha_j + \beta_j exp(-\delta X_{i-j}^2)] X_{i-j},$$

where now $\boldsymbol{\theta} = (\alpha_1, \cdots, \alpha_p, \beta_1, \cdots, \beta_p, \delta)' \in (-1, 1)^p \times R^p \times (0, \infty)$.

Remark 8.2.6 *An Extension.* Analogues of the most of the above results are valid in more general AR models with possibly a nonlinear covariate effects present. Let $\mathbf{Z}_{ni}$, $1 \le i \le n$, be another set of $r \times 1$ random vectors denoting a covariate vector and ℓ be a known function from $\mathbb{R}^p \times \mathbb{R}^r \times \Theta$ to the real line and consider the model

$$X_{ni} = \ell(\mathbf{Y}_{n,i-1}, \mathbf{Z}_{ni}, \boldsymbol{\theta}) + \varepsilon_{ni}, \quad 1 \le i \le n,$$

where, starting with $\mathbf{Y}_0$, $\mathbf{Y}_{n,i-1} := (X_{n,i-1}, \cdots, X_{n,i-p})'$, $1 \le i \le n$. Moreover, $\mathbf{Y}_{n,i-1}$, $\mathbf{Z}_{ni}$ and ε_{ni}, are assumed to be mutually independent for each $1 \le i \le n$. Koul (1996) developed a general asymptotic theory analogous to the above discussion in these models. In fact the above discussion is an adaptation of the results in this paper to AR models without trend. □

8.3 ARCH Models

In this section we shall discuss analogues of some of the results of the previous sections for some autoregressive conditionally heteroscedastic (ARCH) models.

8.3.1 ARCH Models and Some Definitions

As before, let $\{X_i, i \geq 1-p\}$ be an observable time series, and $\mathbf{Y}_{i-1} := (X_{i-1}, X_{i-2}, \cdots, X_{i-p})'$, $i = 1, 2, \cdots$. Let Ω_j, $j = 1, 2$, be open subsets of $\mathbb{R}^q$, $\mathbb{R}^r$, respectively, where p, q, r, are known positive integer. Set $\Omega := \Omega_1 \times \Omega_2$, $m = q + r$. Let μ and σ be known functions, respectively, from $\mathbb{R}^p \times \Omega_1$ to $\mathbb{R}$, and $\mathbb{R}^p \times \Omega_2$ to $\mathbb{R}^+ := (0, \infty)$, both measurable in the first p coordinates. In the ARCH models of interest one observes a process $\{X_i, i \geq 1-p\}$ such that for some $\boldsymbol{\alpha} \in \Omega_1$, $\boldsymbol{\beta} \in \Omega_2$,

$$X_i = \mu(\mathbf{Y}_{i-1}, \boldsymbol{\alpha}) + \sigma(\mathbf{Y}_{i-1}, \boldsymbol{\beta})\, \varepsilon_i, \qquad i \geq 1, \tag{8.3.1}$$

where the errors $\{\varepsilon_i, i \geq 1\}$ are independent of $\mathbf{Y}_0$, and standard i.i.d. F r.v.'s.

Just as in the previous section, the focus of this chapter is to show how the weak convergence results of certain basic randomly weighted empirical processes can be used to obtain the asymptotic distributions of various estimators of $\boldsymbol{\alpha}$ in a unified fashion. To proceed further we need to make the following basic model smoothness assumptions about the functions μ and σ, analogous to (8.2.2). There exist functions $\dot{\boldsymbol{\mu}}$ and $\dot{\boldsymbol{\sigma}}$, respectively, from $\mathbb{R}^p \times \Omega_1$ to $\mathbb{R}^q$ and $\mathbb{R}^p \times \Omega_2$ to $\mathbb{R}^r$, both measurable in the first p coordinates, such that for every $k < \infty$, ,

$$\sup \frac{n^{1/2}|\mu(\mathbf{Y}_{i-1}, \mathbf{t}) - \mu(\mathbf{Y}_{i-1}, \mathbf{s}) - (\mathbf{t}-\mathbf{s})'\dot{\boldsymbol{\mu}}(\mathbf{Y}_{i-1}, \mathbf{s})|}{\sigma(\mathbf{Y}_{i-1}, \boldsymbol{\beta})} = o_p(1), \qquad \mathbf{s} \in \Omega_1, \tag{8.3.2}$$

$$\sup \frac{n^{1/2}|\sigma(\mathbf{Y}_{i-1}, \mathbf{t}) - \sigma(\mathbf{Y}_{i-1}, \mathbf{s}) - (\mathbf{t}-\mathbf{s})'\dot{\boldsymbol{\sigma}}(\mathbf{Y}_{i-1}, \mathbf{s})|}{\sigma(\mathbf{Y}_{i-1}, \boldsymbol{\beta})}, = o_p(1), \qquad \mathbf{s} \in \Omega_2. \tag{8.3.3}$$

where the supremum is taken over $1 \leq i \leq n$, $n^{1/2}\|\mathbf{t}-\mathbf{s}\| \leq k$.

To define analogues of M- and m.d.- estimators of $\boldsymbol{\alpha}$, we need to introduce the following scores. Write $\mathbf{t} := (\mathbf{t}_1', \mathbf{t}_2')' \in \Omega := \Omega_1 \times \Omega_2$. Let

$$\varepsilon_i(\mathbf{t}) := \frac{X_i - \mu(\mathbf{Y}_{i-1}, \mathbf{t}_1)}{\sigma(\mathbf{Y}_{i-1}, \mathbf{t}_2)},$$

and let $R_{i\mathbf{t}}$ denote the rank of $\varepsilon_i(\mathbf{t})$ among $\{\varepsilon_j(\mathbf{t});\ 1 \leq j \leq n\}$, $1 \leq i \leq n$.

Define, for $\mathbf{t} \in \Omega_1 \times \Omega_2$,

$$(8.3.4) \qquad \begin{aligned} \boldsymbol{M}(\mathbf{t}) &:= n^{-1/2} \sum_{i=1}^{n} \frac{\dot{\boldsymbol{\mu}}(\mathbf{Y}_{i-1}, \mathbf{t}_1)}{\sigma(\mathbf{Y}_{i-1}, \mathbf{t}_2)} \psi(\varepsilon_i(\mathbf{t})), \\ \boldsymbol{Z}(u; \mathbf{t}) &:= n^{-1/2} \sum_{i=1}^{n} \frac{\dot{\boldsymbol{\mu}}(\mathbf{Y}_{i-1}, \mathbf{t}_1)}{\sigma(\mathbf{Y}_{i-1}, \mathbf{t}_2)} I(R_{it} \le nu), \\ \mathbb{Z}(u; \mathbf{t}) &:= \mathbf{Z}(u; \mathbf{t}) - n^{-1/2} \sum_{i=1}^{n} \frac{\dot{\boldsymbol{\mu}}(\mathbf{Y}_{i-1}, \mathbf{t}_1)}{\sigma(\mathbf{Y}_{i-1}, \mathbf{t}_2)} u, \ 0 \le u \le 1, \\ \mathcal{K}(\mathbf{t}) &:= \int_0^1 \|\mathbb{Z}(u; \mathbf{t}_1, \mathbf{t}_2)\|^2 L(du), \end{aligned}$$

where ψ and L are as in (8.2.3). Some times we shall write $\boldsymbol{M}(\mathbf{t}_1, \mathbf{t}_2)$ etc. for $\boldsymbol{M}(\mathbf{t})$, etc. Note that the above scores are the analogues of the scores defined at (8.2.3).

Now let $\hat{\boldsymbol{\beta}}$ be a preliminary $n^{1/2}$ - consistent estimator of $\boldsymbol{\beta}$. Based on $\hat{\boldsymbol{\beta}}$, analogues of M- and m.d.- estimators of $\boldsymbol{\alpha}$ are defined, akin to (8.2.4), by the relation

$$(8.3.5) \quad \hat{\boldsymbol{\alpha}} := \operatorname{argmin}_{\mathbf{t}_1 \in \Omega_1} \|\boldsymbol{M}(\mathbf{t}_1, \hat{\boldsymbol{\beta}})\|, \ \ \hat{\boldsymbol{\alpha}}_{md} := \operatorname{argmin}_{\mathbf{t}_1 \in \Omega_1} \mathcal{K}(\mathbf{t}_1, \hat{\boldsymbol{\beta}}).$$

This is motivated by noting that (8.3.1) is equivalent to

$$X_i/\sigma(\mathbf{Y}_{i-1}, \boldsymbol{\beta}) = \mu(\mathbf{Y}_{i-1}, \boldsymbol{\alpha})/\sigma(\mathbf{Y}_{i-1}, \boldsymbol{\beta}) + \varepsilon_i,$$

which in turn can be approximated by

$$X_i/\sigma(\mathbf{Y}_{i-1}, \hat{\boldsymbol{\beta}}) \approx \mu(\mathbf{Y}_{i-1}, \boldsymbol{\alpha})/\sigma(\mathbf{Y}_{i-1}, \hat{\boldsymbol{\beta}}) + \varepsilon_i.$$

This can be thought as a nonlinear AR model with homoscedastic errors and hence the above definitions.

A way to obtain a preliminary $n^{1/2}$ - consistent estimator of $\boldsymbol{\beta}$ is to proceed as follows. First, estimate $\boldsymbol{\alpha}$ in (8.3.1) by a preliminary consistent estimator $\hat{\boldsymbol{\alpha}}_p$ which only considers the nonlinear autoregressive structure of (8.3.1) but does not take into account the heteroscedasticity of the model. Next, use $\hat{\boldsymbol{\alpha}}_p$ to construct an estimator $\hat{\boldsymbol{\beta}}$ of the parameter $\boldsymbol{\beta}$. More precisely, let κ be a nondecreasing real valued function on $\mathbb{R}$. Define, for $\mathbf{t} \in \mathbb{R}^m$,

$$\begin{aligned} \boldsymbol{\mathcal{H}}(\mathbf{t}_1) &:= n^{-1/2} \sum_{i=1}^{n} \dot{\boldsymbol{\mu}}(\mathbf{Y}_{i-1}, \mathbf{t}_1)\Big(X_i - \mu(\mathbf{Y}_{i-1}, \mathbf{t}_1)\Big), \\ \boldsymbol{M}_s(\mathbf{t}) &:= n^{-1/2} \sum_{i=1}^{n} \frac{\dot{\boldsymbol{\sigma}}(\mathbf{Y}_{i-1}, \mathbf{t}_2)}{\sigma(\mathbf{Y}_{i-1}, \mathbf{t}_2)} \Big[\varepsilon_i(\mathbf{t})\kappa(\varepsilon_i(\mathbf{t})) - 1\Big]. \end{aligned}$$

A preliminary least squares estimator of $\boldsymbol{\alpha}$ is defined by the relation

$$\hat{\boldsymbol{\alpha}}_p := \text{argmin}_{\mathbf{t}_1 \in \Omega_1} \|\boldsymbol{\mathcal{H}}(\mathbf{t}_1)\|. \tag{8.3.6}$$

Its consistency is assured because under (8.3.1), $E[\boldsymbol{\mathcal{H}}(\boldsymbol{\alpha})] = \mathbf{0}$.

Next, let κ be such that $E\{\varepsilon_1 \kappa(\varepsilon_1)\} = 1$. This condition is satisfied, for example, when either κ is the identity function because $E\varepsilon^2 = 1$, or when κ is the score function for location of the maximum likelihood estimator at the error distribution F. Since $E[\boldsymbol{M}_s(\boldsymbol{\alpha}, \boldsymbol{\beta})] = \mathbf{0}$, an M-estimator of $\boldsymbol{\beta}$ is defined by the relation

$$\hat{\boldsymbol{\beta}} := \text{argmin}_{\mathbf{t}_2 \in \Omega_2} \|\boldsymbol{M}_s(\hat{\boldsymbol{\alpha}}_p, \mathbf{t}_2)\|. \tag{8.3.7}$$

In the next section we establish the asymptotic distributions of all of the above estimators.

8.3.2 Main Results in ARCH models

To begin with we shall state some additional assumptions needed to obtain the limiting distribution of these estimators. For the sake of brevity, write for $\mathbf{t}_1 \in \mathbb{R}^q$, $\mathbf{t}_2 \in \mathbb{R}^r$, $1 \le i \le n$,

$$\begin{aligned}
\mu_{ni}(\mathbf{t}_1) &:= \frac{\mu(\mathbf{Y}_{i-1}, \boldsymbol{\alpha} + n^{-1/2}\mathbf{t}_1)}{\sigma(\mathbf{Y}_{i-1}, \boldsymbol{\beta})}, \quad \dot{\boldsymbol{\mu}}_{ni}(\mathbf{t}_1) := \frac{\dot{\boldsymbol{\mu}}(\mathbf{Y}_{i-1}, \boldsymbol{\alpha} + n^{-1/2}\mathbf{t}_1)}{\sigma(\mathbf{Y}_{i-1}, \boldsymbol{\beta})}, \\
\sigma_{ni}(\mathbf{t}_2) &:= \frac{\sigma(\mathbf{Y}_{i-1}, \boldsymbol{\beta} + n^{-1/2}\mathbf{t}_2)}{\sigma(\mathbf{Y}_{i-1}, \boldsymbol{\beta})}, \quad \dot{\boldsymbol{\sigma}}_{ni}(\mathbf{t}_2) := \frac{\dot{\boldsymbol{\sigma}}(\mathbf{Y}_{i-1}, \boldsymbol{\beta} + n^{-1/2}\mathbf{t}_2)}{\sigma(\mathbf{Y}_{i-1}, \boldsymbol{\beta})}, \\
\mathbf{r}_{ni}(\mathbf{t}_2) &:= \frac{\dot{\boldsymbol{\sigma}}_{ni}(\mathbf{t}_2)}{\sigma_{ni}(\mathbf{t}_2)}.
\end{aligned}$$

Note that $\dot{\boldsymbol{\sigma}}_{ni}(0) = \dot{\boldsymbol{\sigma}}(\mathbf{Y}_{i-1}, \boldsymbol{\beta})/\sigma(\mathbf{Y}_{i-1}, \boldsymbol{\beta})$, $\sigma_{ni}(0) := 1$. In the sequel, $\dot{\boldsymbol{\mu}}_i$, μ_i, $\dot{\boldsymbol{\sigma}}_i$, $\mathbf{r}_i$ will stand for $\dot{\boldsymbol{\mu}}_{ni}(\mathbf{0})$, $\mu_i(\mathbf{0})$, $\dot{\boldsymbol{\sigma}}_{ni}(\mathbf{0}), \mathbf{r}_{ni}(\mathbf{0})$, respectively, as they also do not depend on n. Also, let $\dot{\mu}_{ni,j}$ and $\dot{\mu}_{i,j}$, respectively, denote the jth co-ordinate of $\dot{\boldsymbol{\mu}}_{ni}$ and $\dot{\boldsymbol{\mu}}_i$, $1 \le j \le p$. All expectations and probabilities below depend on $\boldsymbol{\theta} := (\boldsymbol{\alpha}', \boldsymbol{\beta}')'$, but this dependence is not exhibited for the sake of convenience. We now state additional assumptions.

(8.3.8) There exist positive definite matrices $\boldsymbol{\Lambda}$, $\dot{\boldsymbol{\Sigma}}$, $\dot{\mathbf{M}}$, and a matrix $\boldsymbol{\Gamma}$, all possibly depending on $\boldsymbol{\theta}$, such that

$$n^{-1} \sum_{i=1}^n \dot{\boldsymbol{\mu}}_i \dot{\boldsymbol{\mu}}_i' = \boldsymbol{\Lambda} + o_p(1), \quad n^{-1} \sum_{i=1}^n \dot{\boldsymbol{\sigma}}_i \dot{\boldsymbol{\sigma}}_i' = \dot{\boldsymbol{\Sigma}} + o_p(1),$$

$$n^{-1}\sum_{i=1}^{n} \dot{\boldsymbol{\mu}}(\mathbf{Y}_{i-1}, \boldsymbol{\alpha})\dot{\boldsymbol{\mu}}(\mathbf{Y}_{i-1}, \boldsymbol{\alpha})' = \dot{\mathbf{M}} + o_p(1),$$

$$n^{-1}\sum_{i=1}^{n} \dot{\boldsymbol{\mu}}_i \dot{\boldsymbol{\sigma}}_i' = \boldsymbol{\Gamma} + o_p(1).$$

(8.3.9) $$n^{-1}\sum_{i=1}^{n} E\Big(\frac{\dot{\mu}_{ni,j}(\mathbf{t}_1)}{\sigma_{ni}(\mathbf{t}_2)}\Big)^4 = O(1), \quad \forall 1 \le j \le p, \mathbf{t} \in \mathbb{R}^m.$$

(8.3.10) $$\max_{1\le i \le n} n^{-1/2}(\|\dot{\boldsymbol{\mu}}_i\| + \|\dot{\boldsymbol{\sigma}}_i\|) = o_p(1).$$

(8.3.11) $$n^{-1}\sum_{i=1}^{n} E\Big\{\|\dot{\boldsymbol{\mu}}_{ni}(\mathbf{t}_1) - \dot{\boldsymbol{\mu}}_i\|^2 + \|\dot{\boldsymbol{\sigma}}_{ni}(\mathbf{t}_2) - \dot{\boldsymbol{\sigma}}_i\|^2\Big\} = o(1), \quad \mathbf{t} \in \mathbb{R}^{\mathbf{m}}.$$

(8.3.12) $$n^{-1/2}\sum_{i=1}^{n} \Big\{\|\dot{\boldsymbol{\mu}}_{ni}(\mathbf{t}_1) - \dot{\boldsymbol{\mu}}_i\| + \|\dot{\boldsymbol{\sigma}}_{ni}(\mathbf{t}_2) - \dot{\boldsymbol{\sigma}}_i\|\Big\} = O_p(1), \quad \mathbf{t} \in \mathbb{R}^{\mathbf{m}}.$$

(8.3.13) For every $\mathbf{t} \in \mathbb{R}^m$, $1 \le j \le p$,

$$n^{1/2} E\Big[n^{-1}\sum_{i=1}^{n} \Big\{\frac{\dot{\mu}_{ni,j}(\mathbf{t}_1)}{\sigma_{ni}(\mathbf{t}_2)}\Big\}^2 \times \{|\mu_{ni,j}(\mathbf{t}_1) - \mu_{i,j}| + |\sigma_{ni}(\mathbf{t}_2) - 1|\}\Big]^2 = o(1).$$

(8.3.14) $\forall \epsilon > 0$, $\exists$ a $\delta > 0$, and an $N < \infty$, $\ni \forall 0 < b < \infty$, $\forall \|s\| \le b$, $\forall n > N$,

$$P\Big(\sup_{\|\mathbf{t}-\mathbf{s}\|<\delta} n^{-1/2}\sum_{i=1}^{n} \Big\|\frac{\dot{\boldsymbol{\mu}}_{ni}(\mathbf{t}_1)}{\sigma_{ni}(\mathbf{t}_2)} - \frac{\dot{\boldsymbol{\mu}}_{ni}(\mathbf{s}_1)}{\sigma_{ni}(\mathbf{s}_2)}\Big\| \le \epsilon\Big) \ge 1 - \epsilon.$$

Many of the above assumptions are the analogues of the assumptions (8.2.6) - (8.2.10) needed for the AR models. Attention should be paid to the difference between the $\dot{\boldsymbol{\mu}}_{ni}$ here and the one appearing in the previous section due to the presence of the conditional standard heteroscedasticity in the ARCH model.

A relatively easily verifiable sufficient condition for (8.3.13) is the following: For every $1 \le j \le p$, $\mathbf{t} \in \mathbb{R}^m$,

(8.3.15) $$n^{-1/2}\sum_{i=1}^{n} E\Big[\Big\{\frac{\dot{\mu}_{ni,j}(\mathbf{t}_1)}{\sigma_{ni}(\mathbf{t}_2)}\Big\}^4 \Big\{|\mu_{ni,j}(\mathbf{t}_1) - \mu_{i,j}|^2 + |\sigma_{ni}(\mathbf{t}_2) - 1|^2\Big\}\Big] = o(1).$$

Note also that if the underlying process is stationary and ergodic, then under appropriate moment conditions, (8.3.8) - (8.3.10), are *a priori* satisfied.

The first two theorems below give the asymptotic behavior of the preliminary estimators $\hat{\boldsymbol{\alpha}}_p$ and $\hat{\boldsymbol{\beta}}$ of (8.3.6) and (8.3.7), respectively. To state the first result we need to introduce

$$\mathcal{H}_n := n^{-1/2} \sum_{i=1}^{n} \dot{\mu}(\mathbf{Y}_{i-1}, \boldsymbol{\alpha}) \Big\{ \sigma(\mathbf{Y}_{i-1}, \boldsymbol{\beta}) - 1 \Big\} \varepsilon_i.$$

Theorem 8.3.1 *Suppose that the model assumptions (8.3.1), (8.3.2), and (8.3.3) hold and (8.3.8) holds. In addition, suppose the following holds: There exist a real matrix-valued function* $\ddot{\boldsymbol{M}}$ *on* $\mathbb{R}^p \times \Omega_1$ *such that* $\forall\, k < \infty$, $\mathbf{s}_1 \in \Omega_1$,

$$\text{(8.3.16)} \qquad \sup n^{1/2} \|\dot{\boldsymbol{\mu}}_{ni}(\mathbf{t}_1) - \dot{\boldsymbol{\mu}}_{ni}(\mathbf{s}_1) - \ddot{\boldsymbol{M}}(\mathbf{Y}_{i-1}, \mathbf{s}_1)(\mathbf{t}_1 - \mathbf{s}_1)\| = o_p(1),$$

$$\text{(8.3.17)} \quad \max_{i \geq 1} E\|\ddot{\boldsymbol{M}}(\mathbf{Y}_{i-1}, \boldsymbol{\alpha})\| = O(1), \ \ \|n^{-1} \sum_{i=1}^{n} \ddot{\boldsymbol{M}}(\mathbf{Y}_{i-1}, \boldsymbol{\alpha})\, \varepsilon_i\| = o_p(1),$$

where the supremum in (8.3.16) is over $1 \leq i \leq n, n^{1/2}\|\mathbf{t}_1 - \mathbf{s}_1\| \leq k$.

Then, for every $0 < b < \infty$,

$$\sup_{\|\mathbf{t}_1\| \leq b} \left\| \mathcal{H}(\boldsymbol{\alpha} + n^{-1/2}\mathbf{t}_1) - \{\mathcal{H}(\boldsymbol{\alpha}) + \mathcal{H}_n\} + \dot{\mathbf{M}}\,\mathbf{t}_1 \right\| = o_p(1).$$

The proof of this theorem is routine. As an immediate consequence we have the following corollary.

Corollary 8.3.1 *In addition to the assumptions of Theorem 8.3.1, assume that*

$$\text{(8.3.18)} \qquad (a) \quad \|n^{1/2}(\hat{\boldsymbol{\alpha}}_p - \boldsymbol{\alpha})\| = O_p(1). \qquad (b) \quad \|\mathcal{H}_n\| = O_p(1).$$

Then,

$$n^{1/2}(\hat{\boldsymbol{\alpha}}_p - \boldsymbol{\alpha}) = (\dot{\mathbf{M}})^{-1}\{\mathcal{H}(\boldsymbol{\alpha}) + \mathcal{H}_n\} + o_p(1).$$

The additional random vector $\mathcal{H}_n$ coming into picture is identically zero when $\sigma \equiv 1$.

Under additional smoothness assumptions on the function μ, we can use any other preliminary $n^{1/2}$-consistent estimator of $\boldsymbol{\alpha}$. For example, let ϕ be a nondecreasing score function on $\mathbb{R}$ such that $E\{\phi(c\varepsilon)\} = 0$, for every $c > 0$. This is satisfied for example when ϕ is skew-symmetric and

ε is symmetrically distributed. In this case, a preliminary estimator of $\boldsymbol{\alpha}$ can be defined as

$$\tilde{\boldsymbol{\alpha}} := \operatorname{argmin}_{\mathbf{t}\in\Omega_1} \|n^{-1/2}\sum_{i=1}^{n} \dot{\mu}(\mathbf{Y}_{i-1},\mathbf{t})\phi(X_i - \mu(\mathbf{Y}_{i-1},\mathbf{t}))\|.$$

The next theorem gives a similar linearity result about the scores $\boldsymbol{M}_s$. Its proof uses usual Taylor expansion and hence is not given here.

Theorem 8.3.2 *Suppose that the assumptions (8.3.1), (8.3.2), and (8.3.3) hold. In addition, suppose the following hold. The function κ is nondecreasing, twice differentiable and satisfies:*

$$(i)\ \int x\kappa(x)F(dx) = 1, \qquad (ii)\ \int x^2|\dot{\kappa}(x)|F(dx) < \infty,$$

(iii) the second derivative of κ is bounded.

There exist a matrix-valued functions $\ddot{\boldsymbol{R}}$ on $\mathbb{R}^p \times \Omega_2$, such that for every $k < \infty$, $\mathbf{s}_2 \in \Omega_2$,

$$\text{(8.3.19)} \quad \sup n^{1/2}\|\mathbf{r}_{ni}(\mathbf{t}_2) - \mathbf{r}_{ni}(\mathbf{s}_2) - \ddot{\boldsymbol{R}}(\mathbf{Y}_{i-1},\mathbf{s}_2)(\mathbf{t}_2 - \mathbf{s}_2)\| = o_p(1),$$

$$\text{(8.3.20)} \quad \max_{i\ge 1} E\|\ddot{\boldsymbol{R}}(\mathbf{Y}_{i-1},\boldsymbol{\beta})\| = O(1), \ \|n^{-1}\sum_{i=1}^{n} \ddot{\boldsymbol{R}}(\mathbf{Y}_{i-1},\boldsymbol{\alpha})\,\varepsilon_i\| = o_p(1).$$

where the supremum in (8.3.19) is over $1 \le i \le n, n^{1/2}\|\mathbf{t}_2 - \mathbf{s}_2\| \le k$.

Then, for every $0 < b < \infty$,

$$\sup_{\|\mathbf{t}\|\le b} \Big\| \boldsymbol{M}_s(\boldsymbol{\theta} + n^{-1/2}\mathbf{t}) - \boldsymbol{M}_s(\boldsymbol{\theta}) + \Big[\int \kappa(x)F(dx) + \int x\dot{\kappa}(x)\,F(dx)\Big]\,\boldsymbol{\Gamma}'\,\mathbf{t}_1 + \Big[\int x\kappa(x)F(dx) + \int x^2\dot{\kappa}(x)\,F(dx)\Big]\,\dot{\boldsymbol{\Sigma}}\,\mathbf{t}_2 \Big\| = o_p(1).$$

Consequently, we have the following corollary.

Corollary 8.3.2 *In addition to the assumptions of Theorem 8.3.2, assume that $\int \kappa(x)F(dx) = 0 = \int x\dot{\kappa}(x)F(dx)$ and that*

$$\text{(8.3.21)} \qquad \|n^{1/2}(\hat{\boldsymbol{\beta}} - \boldsymbol{\beta})\| = O_p(1).$$

Then,

$$\Big[\int x\kappa(x)F(dx) + \int x^2\dot{\kappa}(x)\,F(dx)\Big]\, n^{1/2}(\hat{\boldsymbol{\beta}} - \boldsymbol{\beta}) = \dot{\boldsymbol{\Sigma}}^{-1}\mathbf{M}_s(\boldsymbol{\theta}) + o_p(1).$$

Note that the asymptotic distribution of $\hat{\beta}$ does not depend on the preliminary estimator $\hat{\alpha}_p$ used in defining $\hat{\beta}$. Also, the conditions (*i*)-(*iii*) and those of the above corollary involving κ are satisfied by $\kappa(x) \equiv x$, because $E\varepsilon^2 = 1$. Again, if the underlying process is stationary and ergodic then (8.3.17) and (8.3.20) will be typically satisfied under appropriate moment conditions.

Now, we address the problem of obtaining the limiting distributions of the estimators defined at (8.3.5). The first ingredient needed is the AUL property of the $\boldsymbol{M}$ score and the ULAQ of the score $\mathcal{K}$. The following lemma is basic to proving these results when the underlying functions ψ and L are not smooth. Its role here is similar to that of its analogue given by the Lemma 8.2.1 in the AR models. Let, for $\mathbf{t} = (\mathbf{t}_1', \mathbf{t}_2')'$, $\mathbf{t}_1 \in \mathbb{R}^q$, $\mathbf{t}_2 \in \mathbb{R}^r$, $m = q + r$, and $x \in \mathbb{R}$,

$$\mathbf{W}(x,\mathbf{t}) := n^{-1/2} \sum_{i=1}^{n} \frac{\dot{\boldsymbol{\mu}}_{ni}(\mathbf{t}_1)}{\sigma_{ni}(\mathbf{t}_2)} I\Big(\varepsilon_i \le x + x(\sigma_{ni}(\mathbf{t}_2) - 1) + (\mu_{ni}(\mathbf{t}_1) - \mu_i)\Big),$$

$$\boldsymbol{\nu}(x,\mathbf{t}) := n^{-1/2} \sum_{i=1}^{n} \frac{\dot{\boldsymbol{\mu}}_{ni}(\mathbf{t}_1)}{\sigma_{ni}(\mathbf{t}_2)} F\Big(x + x(\sigma_{ni}(\mathbf{t}_2) - 1) + (\mu_{ni}(\mathbf{t}_1) - \mu_i)\Big),$$

$$\mathcal{W}(x,\mathbf{t}) := \mathbf{W}(x,\mathbf{t}) - \boldsymbol{\nu}(x,\mathbf{t}),$$

$$\mathcal{W}^*(x,\mathbf{t}) := n^{-1/2} \sum_{i=1}^{n} \frac{\dot{\boldsymbol{\mu}}_{ni}(\mathbf{t}_1)}{\sigma_{ni}(\mathbf{t}_2)} \Big[I(\varepsilon_i \le x) - F(x)\Big].$$

The basic result needed is given in the following lemma whose proof appears later.

Lemma 8.3.1 *Suppose the assumptions (8.3.1)- (8.3.3), (8.3.8)-(8.3.14) hold and that the error d.f. F satisfies (2.2.49), (2.2.50) and (2.2.51). Then, for every* $0 < b < \infty$,

(8.3.22) $$\|\mathcal{W}(x,\mathbf{t}) - \mathcal{W}^*(x,\mathbf{t})\| = u_p(1).$$

(8.3.23) $$\|\mathcal{W}(x,\mathbf{t}) - \mathcal{W}(x,\mathbf{0})\| = u_p(1).$$

(8.3.24) $$\Big\|\mathbf{W}(x,\mathbf{t}) - \mathbf{W}(x,\mathbf{0}) - \Big\{f(x)\boldsymbol{\Lambda}\,\mathbf{t}_1 + xf(x)\boldsymbol{\Gamma}'\,\mathbf{t}_2\Big\} - n^{-1/2}\sum_{i=1}^{n}\Big[\frac{\dot{\boldsymbol{\mu}}_{ni}(\mathbf{t}_1)}{\sigma_{ni}(\mathbf{t}_2)} - \dot{\boldsymbol{\mu}}_i\Big] F(x)\Big\| = u_p(1),$$

where $u_p(1)$ is a sequence of stochastic processes in x, $\mathbf{t}$, converging to zero, uniformly over the set $\{x \in \mathbb{R}, \|\mathbf{t}\| \le b\}$, in probability.

The claim (8.3.24) follows from (8.3.23) and the assumptions (8.3.8) - (8.3.14), and the assumption that F satisfies (2.2.49), (2.2.50) and (2.2.51). Note that the assumptions (8.3.8)-(8.3.14) ensure that for every $0 < b < \infty$,

$$\sup_{\|\mathbf{t}\| \le b} \Big\| n^{-1/2} \sum_{i=1}^{n} \frac{\dot{\boldsymbol{\mu}}_{ni}(\mathbf{t}_1)}{\sigma_{ni}(\mathbf{t}_2)} \{\mu_{ni}(\mathbf{t}_1) - \mu_i\} - \boldsymbol{\Lambda} \Big\| = o_p(1),$$

$$\sup_{\|\mathbf{t}\| \le b} \Big\| n^{-1/2} \sum_{i=1}^{n} \{\dot{\mu}_{ni}(\mathbf{t}_1)/\sigma_{ni}(\mathbf{t}_2)\}\{\sigma_{ni}(\mathbf{t}_2) - 1\} - \boldsymbol{\Gamma} \Big\| = o_p(1).$$

The proofs of these two claims are routine and left out for an interested reader. The proofs of (8.3.22) and (8.3.23) appear in the last section as a consequence of Theorem 2.2.5. The next result gives the AUL result for $\boldsymbol{M}$-scores.

Theorem 8.3.3 *Under the assumption (8.3.1)- (8.3.3), (2.2.49), (2.2.50), (2.2.51) with G replaced by F, and (8.3.8)-(8.3.14), for every $0 < b < \infty$, and for every bounded nondecreasing ψ with $\int \psi dF = 0$,*

$$\sup_{\|\mathbf{t}\| \le b} \Big\| \boldsymbol{M}(\boldsymbol{\theta} + n^{-1/2}\mathbf{t}) - \boldsymbol{M}(\boldsymbol{\theta}) - \Big(\int f d\psi \, \boldsymbol{\Lambda} \, \mathbf{t}_1 + \int x f(x) d\psi(x) \, \boldsymbol{\Gamma} \, \mathbf{t}_2 \Big) \Big\| = o_p(1).$$

This theorem follows from the Lemma 8.3.1 in the same way as Theorem 8.2.1 from Lemma 8.2.1, using the relation

$$\int [\mathbf{W}(x,\mathbf{t}) - \mathbf{W}(x,\mathbf{0})] \, d\psi(x) \equiv n^{-1/2} \sum_{i=1}^{n} \Big[\frac{\dot{\boldsymbol{\mu}}_{ni}(\mathbf{t}_1)}{\sigma_{ni}(\mathbf{t}_2)} - \dot{\mu}_i \Big] \psi(\infty) - [\boldsymbol{M}(\boldsymbol{\theta} + n^{-1/2}t) - \boldsymbol{M}(\boldsymbol{\theta})].$$

Next, we have the following immediate corollary.

Corollary 8.3.3 *In addition to the assumptions of Theorem 8.3.3, assume that $\int f d\psi > 0$, (8.3.21) holds, and that*

$$\|n^{1/2}(\hat{\boldsymbol{\alpha}} - \boldsymbol{\alpha})\| = O_p(1). \qquad (8.3.25)$$

Then,

$$\text{(8.3.26)} \qquad \int f d\psi\, n^{1/2}(\hat{\boldsymbol{\alpha}} - \boldsymbol{\alpha}) = -\boldsymbol{\Lambda}^{-1}\Big[\boldsymbol{M}(\boldsymbol{\theta}) + \boldsymbol{\Gamma} n^{1/2}(\hat{\boldsymbol{\beta}} - \boldsymbol{\beta}) \int x f(x) d\psi(x)\Big] + o_p(1).$$

From this corollary it is apparent that the asymptotic distribution of $\hat{\boldsymbol{\alpha}}$ depends on the preliminary estimator of the scale parameter in general. However, if either $\int x f(x) d\psi(x) = 0$ or if $\boldsymbol{\Gamma} = \boldsymbol{0}$, then the second term in the right hand side of (8.3.26) disappears and the preliminary estimation of the scale parameter has no effect on the asymptotic distribution of the estimation of $\boldsymbol{\alpha}$. Also, in this case, the asymptotic distribution of $\hat{\boldsymbol{\alpha}}$ is the same as that of an M-estimator of $\boldsymbol{\alpha}$ for the model $X_i/\sigma(\mathbf{Y}_{i-1}, \beta) = \mu(\mathbf{Y}_{i-1}, \alpha)/\sigma(\mathbf{Y}_{i-1}, \beta) + \varepsilon_i$ with $\boldsymbol{\beta}$ known. We summarize this in the following

Corollary 8.3.4 *In addition to the assumptions of Corollary 8.3.3, suppose either $\int x f(x) d\psi(x) = 0$, or $\boldsymbol{\Gamma} = \boldsymbol{0}$. Then,*

$$n^{1/2}(\hat{\boldsymbol{\alpha}} - \boldsymbol{\alpha}) \longrightarrow_d \mathcal{N}(\boldsymbol{0}, \boldsymbol{\Lambda}^{-1} v(\psi, F)),$$

where $v(\psi, F)$ is as in (7.3.4).

A sufficient condition for $\int x f(x) d\psi(x) = 0$ is that f is symmetric and ψ skew symmetric, i.e., $f(-x) = f(x)$, $\psi(-x) = -\psi(x)$, for every $x \in \mathbb{R}$.

To give the analogous results for the process $\mathcal{K}$ and the corresponding minimum distance estimator $\hat{\boldsymbol{\alpha}}_{md}$ based on ranks, we need to introduce

$$\begin{aligned} \hat{\mathbb{Z}}(u) &:= n^{-1/2} \sum_{i=1}^{n} \big[\dot{\boldsymbol{\mu}}_i - \bar{\dot{\boldsymbol{\mu}}}\big]\Big\{I(G(\varepsilon_i) \le u) - u\Big\}, \ \bar{\dot{\boldsymbol{\mu}}} := n^{-1} \sum_{i=1}^{n} \dot{\boldsymbol{\mu}}_i, \\ q(u) &:= f(F^{-1}(u)), \quad s(u) := F^{-1}(u) f(F^{-1}(u)), \quad u \in [0, 1]. \end{aligned}$$

Theorem 8.3.4 *In addition to the assumptions (8.3.1) - (8.3.3), (2.2.49), (2.2.50), (2.2.51), and (8.3.8) - (8.3.14), suppose that for some positive definite matrix $\boldsymbol{D}(\boldsymbol{\theta})$,*

$$n^{-1} \sum_{i=1}^{n} \big[\dot{\boldsymbol{\mu}}_i - \bar{\dot{\boldsymbol{\mu}}}\big]\big[\dot{\boldsymbol{\mu}}_i - \bar{\dot{\boldsymbol{\mu}}}\big]' = \boldsymbol{D}(\boldsymbol{\theta}) + o_p(1).$$

Then, for every $0 < b < \infty$,

$$\sup_{\|\mathbf{t}\| \le b} \left| \mathcal{K}(\mathbf{t}) - \int \left\| \hat{\mathbb{Z}}(u) + \{q(u)\mathbf{\Lambda}(\boldsymbol{\theta})\,\mathbf{t}_1 + s(u)\mathbf{\Gamma}(\boldsymbol{\theta})\,\mathbf{t}_2\} \right\|^2 L(du) \right| = o_p(1).$$

Moreover, if (8.3.21) holds and if $\|n^{1/2}(\hat{\boldsymbol{\alpha}}_{md} - \boldsymbol{\alpha})\| = O_p(1)$, *then*

$$\begin{aligned} n^{1/2}&(\hat{\boldsymbol{\alpha}}_{md} - \boldsymbol{\alpha}) \\ &= -\left(\int q dL\, \mathbf{\Lambda}\right)^{-1} \Big[\int \hat{\mathbb{Z}}(u) q(u) L(du) \\ &\qquad + \mathbf{\Gamma}\, n^{1/2}(\hat{\boldsymbol{\beta}} - \boldsymbol{\beta}) \int s dL \Big] + o_p(1). \end{aligned}$$

Additionally, if either $\int_0^1 s(u) dL(u) = 0$, *or if* $\mathbf{\Gamma} = \mathbf{0}$, *then*

$$n^{1/2}(\hat{\boldsymbol{\alpha}}_{md} - \boldsymbol{\alpha}) \longrightarrow_d \mathcal{N}\left(\mathbf{0}, \sigma_0^2\, \mathbf{\Lambda}^{-1}\mathbf{D}\mathbf{\Lambda}^{-1}\right),$$

where σ_0^2 *is as in (5.6.21) and (5.6.15).*

This theorem follows from Lemma 8.3.1 in a similar fashion as do Theorem 8.2.2 and Corollary 8.2.3 from Lemma 8.2.1. Note that f symmetric around zero and $L(u) \equiv -L(1-u)$ implies that $\int_0^1 s(u) dL(u) = 0$.

Next, we shall state an analogue of the Theorem 8.2.2 for sequential weighted empirical processes suitable here. Accordingly, let h_{ni} be as before and independent of ε_i, $1 \le i \le n$. Define, for a $\mathbf{t}' = (\mathbf{t}_1', \mathbf{t}_2')$, $\mathbf{t}_1 \in \mathbb{R}^q$, $\mathbf{t}_2 \in \mathbb{R}^r$,

$$\begin{aligned} \mathcal{S}(x, \mathbf{t}, u) &:= n^{-1/2} \sum_{i=1}^{[nu]} h_{ni} I\Big(\varepsilon_i \le x + x(\sigma_{ni}(\mathbf{t}_2) - 1) \\ &\qquad\qquad + (\mu_{ni}(\mathbf{t}_1) - \mu_i)\Big), \\ \mu(x, \mathbf{t}, u) &:= n^{-1/2} \sum_{i=1}^{[nu]} h_{ni} F\Big(x + x(\sigma_{ni}(\mathbf{t}_2) - 1) \\ &\qquad\qquad + (\mu_{ni}(\mathbf{t}_1) - \mu_i)\Big), \\ S(x, \mathbf{t}, u) &:= \mathcal{S}(x, \mathbf{t}, u) - \mu(x, \mathbf{t}, u), \quad x \in \mathbb{R},\ u \in [0, 1]. \end{aligned}$$

The next result is the analogue of Theorem 8.2.3 suitable for the current ARCH models.

Theorem 8.3.5 *Suppose the assumptions (8.3.1) - (8.3.3), (2.2.49), (2.2.50), (2.2.51) with G replaced by F, (2.2.52), and (8.3.10) hold. In addition, suppose the following hold.*

$$(8.3.27) \quad n^{-1}\sum_{i=1}^{n} Eh_{ni}^4\left(\|\dot{\boldsymbol{\mu}}_i\|^2 + \|\dot{\boldsymbol{\sigma}}_i\|^2\right) = O(1).$$

$$(8.3.28) \quad \sum_{i=1}^{n} Eh_{ni}^4\left[\left\{\mu_{ni}(\mathbf{t}_1) - \mu_i\right\}^2 + \left\{\sigma_{ni}(\mathbf{t}_2) - 1\right\}^2\right] = O(1), \ \mathbf{t} \in \mathbb{R}^m.$$

Then, for every $0 < b < \infty$,

$$(8.3.29) \qquad \sup_{x,\mathbf{t},u} |\mathcal{S}(x,\mathbf{t},u) - \mathcal{S}(x,\mathbf{0},u)| = o_p(1),$$

$$(8.3.30) \qquad \sup_{x,\mathbf{t},u} \Big| S(x,\mathbf{t},u) - S(x,\mathbf{0},u) - n^{-1}\sum_{i=1}^{[nu]} h_{ni}\left\{\dot{\boldsymbol{\sigma}}_i'\mathbf{t}_2 x f(x) + \dot{\boldsymbol{\mu}}_i'\mathbf{t}_1 f(x)\right\}\Big| = o_p(1),$$

where the supremum is taken over $x \in \mathbb{R}$, $\|\mathbf{t}\| \le b, 0 \le u \le 1$.

The proof of this theorem is similar to that of Theorem 8.2.3. No details will be given.

Because of the importance of the residual empirical processes, we give an AUL result for it obtainable from the above theorem. Accordingly, let $\hat{\boldsymbol{\alpha}}$, $\hat{\boldsymbol{\beta}}$ be any $n^{1/2}$-consistent estimators of $\boldsymbol{\alpha}$, $\boldsymbol{\beta}$, and let $\hat{F}_n(x,u)$, $F_n(x,u)$ denote, respectively, the sequential empiricals of the residuals $\hat{\varepsilon}_i := \big(X_i - \mu(\mathbf{Y}_{i-1},\hat{\boldsymbol{\alpha}})\big)/\sigma(\mathbf{Y}_{i-1},\hat{\boldsymbol{\beta}})$, and the errors ε_i, $1 \le i \le n$, i.e., for $x \in \mathbb{R}, 0 \le u \le 1$,

$$\hat{F}_n(x,u) := n^{-1}\sum_{i=1}^{[nu]} I\big(X_i \le x\sigma(\mathbf{Y}_{i-1},\hat{\boldsymbol{\beta}}) + \mu(\mathbf{Y}_{i-1},\hat{\boldsymbol{\alpha}})\big),$$

$$F_n(x,u) := n^{-1}\sum_{i=1}^{[nu]} I\big(X_i \le x\sigma(\mathbf{Y}_{i-1},\boldsymbol{\beta}) + \mu(\mathbf{Y}_{i-1},\boldsymbol{\alpha})\big).$$

Then upon specializing the above theorem to the case when $h_{ni} \equiv 1$, we obtain the following corollary. In its statement the assumption about time series being stationary and ergodic is made for the sake of transparency of the statement.

Corollary 8.3.5 *Suppose the assumptions (8.3.1), (8.3.2) and (8.3.3) hold and that the underlying time series is stationary and ergodic. In addition,* *suppose the error d.f. F has a positive bounded density f such that $f(F^{-1})$ is uniformly continuous on $[0,1]$ and satisfies* (**F3**)*; $\|\dot{\mu}_1\|$, $\|\dot{\sigma}_1\|$ are square integrable; and*

$$nE\Big[\Big\{\mu_{n1}(\mathbf{t}_1)-\mu_1\Big\}^2+\Big\{\sigma_{n1}(\mathbf{t}_2)-1\Big\}^2\Big]=O(1),\ \mathbf{t}\in\mathbb{R}^m.$$

Then,

$$\sup_{x\in\mathbb{R},0\le u\le 1}\Big|n^{1/2}\big[\hat{F}_n(x,u)-F_n(x,u)\big]-u\Big\{n^{1/2}(\hat{\boldsymbol{\alpha}}-\boldsymbol{\alpha})'E(\dot{\boldsymbol{\mu}}_1)f(x)$$
$$+n^{1/2}(\hat{\boldsymbol{\beta}}-\boldsymbol{\beta})'E(\dot{\boldsymbol{\sigma}}_1)xf(x)\Big\}\Big|=o_p(1),$$

This corollary may be used to obtain the limiting distributions of some tests of fit or for some tests of a change point in the errors of an ARCH model in a fashion similar to AR models.

We now begin to give proofs of some of the above results. Recall Theorem 2.2.5.This theorem is not enough to cover the cases where the weights h_{ni} and the disturbances τ_{ni}, δ_{ni} are functions of certain parameters and where one desires to obtain various approximations uniformly in these parameters, as is the case in Lemma 8.3.1. The next result thus gives the needed extension of this theorem to cover these cases. Accordingly, let ζ_{ni} be as in (2.2.48), with F denoting its d.f., and let l_{ni}, v_{ni}, u_{ni} be measurable functions from $\mathbb{R}^m$ to $\mathbb{R}$ such that for every $\mathbf{t}\in\mathbb{R}^m$, $(l_{nj}(\mathbf{t}), v_{nj}(\mathbf{t}), u_{nj}(\mathbf{t}))$, $1\le j\le i$, are independent of ζ_{ni}, for each $1\le i\le n$. Let P_n, E_n stand for the underlying probability and expectation. Let, for $x\in\mathbb{R}$, $\mathbf{t}\in\mathbb{R}^m$,

$$\text{(8.3.31)}\qquad \mathcal{V}(x,\mathbf{t}):=n^{-1/2}\sum_{i=1}^n l_{ni}(\mathbf{t})I\Big(\zeta_{ni}\le x+xv_{ni}(\mathbf{t})+u_{ni}(\mathbf{t})\Big),$$
$$\mathcal{J}(x,\mathbf{t}):=n^{-1/2}\sum_{i=1}^n l_{ni}(\mathbf{t})\,F\Big(x+xv_{ni}(\mathbf{t})+u_{ni}(\mathbf{t})\Big),$$
$$\tilde{\mathcal{U}}(x,\mathbf{t}):=\mathcal{V}(x,\mathbf{t})-\mathcal{J}(x,\mathbf{t}),$$
$$\mathcal{U}^*(x,\mathbf{t}):=n^{-1/2}\sum_{i=1}^n l_{ni}(\mathbf{t})\Big[I(\zeta_{ni}\le x)-F(x)\Big].$$

To state the needed result we need the following assumptions. For some positive random process $\ell(\mathbf{t})$ and for each $\mathbf{t}\in\mathbb{R}^m$,

$$\text{(8.3.32)}\qquad E_n\left(n^{-1}\sum_{i=1}^n l_{ni}^4(\mathbf{t})\right)=O(1),$$

(8.3.33) $$\max_{1 \le i \le n} n^{-1/2} |l_{ni}(\mathbf{t})| = o_p(1),$$

(8.3.34) $$\max_{1 \le i \le n} \{|v_{ni}(\mathbf{t})| + |u_{ni}(\mathbf{t})|\} = o_p(1),$$

(8.3.35) $$n^{1/2} E_n \left[n^{-1} \sum_{i=1}^{n} l_{ni}^2(\mathbf{t}) \{|u_{ni}(\mathbf{t})| + |v_{ni}(\mathbf{t})\} \right]^2 = o(1),$$

(8.3.36) $$n^{-1/2} \sum_{i=1}^{n} |l_{ni}(\mathbf{t})| \left[|v_{ni}(\mathbf{t})| + |u_{ni}(\mathbf{t})| \right] = O_p(1).$$

(8.3.37) $\forall \epsilon > 0, \exists \delta > 0$, and an $n_1 \ni \forall 0 < b < \infty$,

$\forall \|\mathbf{s}\| \le b, \forall n > n_1$,

$$P_n \Big(n^{-1/2} \sum_{i=1}^{n} |l_{ni}(\mathbf{s})| \Big\{ \sup_{\|\mathbf{t}-\mathbf{s}\| < \delta} |v_{ni}(\mathbf{t}) - v_{ni}(\mathbf{s})| + \sup_{\|\mathbf{t}-\mathbf{s}\| < \delta} |u_{ni}(\mathbf{t}) - u_{ni}(\mathbf{s})| \Big\} \le \epsilon \Big) > 1 - \epsilon.$$

(8.3.38) $\forall \epsilon > 0, \exists$ a $\delta > 0$, and an $n_2 < \infty, \ni, \forall 0 < b < \infty$,

$\forall \|\mathbf{s}\| \le b, \forall n > n_2$,

$$P_n \left(\sup_{\|\mathbf{t}-\mathbf{s}\| < \delta} n^{-1/2} \sum_{i=1}^{n} |l_{ni}(\mathbf{t}) - l_{ni}(\mathbf{s})| \le \epsilon \right) > 1 - \epsilon.$$

The following lemma gives the needed result.

Lemma 8.3.2 *Under the above setup and under the assumptions (8.3.2), (8.3.3), (2.2.49), (2.2.50), (2.2.51) with G replaced by F, and (8.3.32) - (8.3.38), for every $0 < b < \infty$,*

(8.3.39) $$\sup_{x \in \mathbb{R}, \|\mathbf{t}\| \le b} |\tilde{\mathcal{U}}(x, \mathbf{t}) - \mathcal{U}^*(x, \mathbf{t})| = o_p(1).$$

Proof. The proof uses Theorem 2.2.5. Fix a $0 < b < \infty$. Observe that if in (2.2.48), we take

(8.3.40) $$h_{ni} = l_{ni}(\mathbf{t}), \quad \tau_{ni} = v_{ni}(\mathbf{t}), \quad \delta_{ni} = u_{ni}(\mathbf{t}), \quad 1 \le i \le n,$$

then, $\tilde{U}_n(x)$ and $U_n^*(x)$ are equal to $\tilde{\mathcal{U}}(x, \mathbf{t})$ and $\mathcal{U}^*(x, \mathbf{t})$, respectively, for all $x \in \mathbb{R}$, $\mathbf{t} \in \mathbb{R}^m$. Clearly the assumption (8.3.32)-(8.3.36) imply (2.2.52)-(2.2.55) for each fixed $\mathbf{t}$. Note that in this application one takes

$$\mathcal{A}_{n1} := \sigma - \text{field}\{(l_{n1}(\mathbf{t}), v_{n1}(\mathbf{t}), u_{n1}(\mathbf{t}))\},$$
$$\mathcal{A}_{ni} := \sigma - \text{field}\{\zeta_{n1}, \cdots, \zeta_{ni-1}; (l_{nj}(\mathbf{t}), v_{nj}(\mathbf{t}), u_{nj}(\mathbf{t})), 1 \le j \le i\},$$

for $2 \le i \le n$. Hence, (2.2.57) implies that

$$\sup_{x \in \mathbb{R}} |\tilde{\mathcal{U}}(x, \mathbf{t}) - \mathcal{U}^*(x, \mathbf{t})| = o_p(1), \quad \mathbf{t} \in \mathbb{R}^m. \tag{8.3.41}$$

To complete the proof of (8.3.39), in view of the compactness of the ball $\|\mathbf{t}\| \le b$, it suffices to show that $\forall \epsilon > 0$, $\exists$ a $\delta > 0$, and an $n_0 < \infty$, $\ni$ $\forall \|\mathbf{s}\| \le b$,

$$P\left(\sup_{x \in \mathbb{R}, \|\mathbf{t}-\mathbf{s}\| \le \delta} |\mathcal{D}(x, \mathbf{t}) - \mathcal{D}(x, \mathbf{s})| \ge \epsilon \right) \le \epsilon, \quad n > n_0, \tag{8.3.42}$$

where $\mathcal{D} := \tilde{\mathcal{U}} - \mathcal{U}^*$. Details that follow are similar to the proof of Lemma 8.2.1.

For convenience, write for $1 \le i \le n$, $\mathbf{t} \in \mathbb{R}^m$, $x \in \mathbb{R}$,

$$\begin{aligned} \beta_i(x, \mathbf{t}) &:= I\Big(\varepsilon_i \le x + x v_{ni}(\mathbf{t}) + u_{ni}(\mathbf{t})\Big) \\ &\qquad\qquad - F\Big(x + x v_{ni}(\mathbf{t}) + u_{ni}(\mathbf{t})\Big), \\ \beta_i(x) &:= I(\varepsilon_i \le x) - F(x). \end{aligned}$$

Then

$$\tilde{\mathcal{U}}(x, \mathbf{t}) = n^{-1/2} \sum_{i=1}^n l_{ni}(\mathbf{t}) \beta_i(x, \mathbf{t}), \quad \mathcal{U}^*(x, \mathbf{t}) = n^{-1/2} \sum_{i=1}^n l_{ni}(\mathbf{t}) \beta_i(x),$$

and

$$\begin{aligned} &\mathcal{D}(x, \mathbf{t}) - \mathcal{D}(x, \mathbf{s}) \\ &= n^{-1/2} \sum_{i=1}^n [l_{ni}(\mathbf{t}) - l_{ni}(\mathbf{s})] [\beta_i(x, \mathbf{t}) - \beta_i(x)] \\ &\qquad + n^{-1/2} \sum_{i=1}^n l_{ni}(\mathbf{s}) [\beta_i(x, \mathbf{t}) - \beta_i(x, \mathbf{s})] \\ &= \mathcal{D}_1(x, \mathbf{s}, \mathbf{t}) + \mathcal{D}_2(x, \mathbf{s}, \mathbf{t}), \quad \text{say.} \end{aligned}$$

It thus suffices to prove the analog of (8.3.42) for $\mathcal{D}_1$, $\mathcal{D}_2$.

Consider $\mathcal{D}_1$ first. Note that because $\beta_i(x, \mathbf{t}) - \beta_i(x)$ are uniformly bounded by 1, we obtain

$$|\mathcal{D}_1(x, \mathbf{s}, \mathbf{t})| \le n^{-1/2} \sum_{i=1}^n |l_{ni}(\mathbf{t}) - l_{ni}(\mathbf{s})|.$$

This and the assumption (8.3.38) then readily verifies (8.3.42) for $\mathcal{D}_1$.

Now consider $\mathcal{D}_2$. For a $\delta > 0$, and $\mathbf{s}$ fixed, let

$$\begin{aligned} d_{ni}(\mathbf{s}) &\equiv \sup_{\|\mathbf{t}-\mathbf{s}\|<\delta} |v_{ni}(\mathbf{t}) - v_{ni}(\mathbf{s})|, \\ c_{ni}(\mathbf{s}) &\equiv \sup_{\|\mathbf{t}-\mathbf{s}\|<\delta} |u_{ni}(\mathbf{t}) - u_{ni}(\mathbf{s})|, \quad 1 \le i \le n, \\ \mathcal{B}_n &:= \left\{ n^{-1/2} \sum_{i=1}^{n} |l_{ni}(\mathbf{s})| \Big[d_{ni}(\mathbf{s}) + c_{ni}(\mathbf{s}) \Big] \le \epsilon \right\}. \end{aligned}$$

By (8.3.37), for an $\epsilon > 0$, there exists a $\delta > 0$ and an $n_1 < \infty$, such that

$$P_n\Big(\mathcal{B}_n\Big) \ge 1 - \epsilon, \quad n > n_1. \tag{8.3.43}$$

Next, write $l_{ni} = l_{ni}^{+} - l_{ni}^{-}$ and $\mathcal{D}_2 = \mathcal{D}_2^{+} - \mathcal{D}_2^{-}$, where $\mathcal{D}_2^{\pm}$ correspond to $\mathcal{D}_2$ with l_{ni} replaced by $l_{ni}^{\pm}$. Let

$$\begin{aligned} &D_2^{\pm}(x, \mathbf{s}, a) \\ &:= n^{-1/2} \sum_{i=1}^{n} l_{ni}^{\pm}(\mathbf{s}) \Big[I\Big(\varepsilon_i \le x + x\{v_{ni}(\mathbf{s}) + a d_{ni}(\mathbf{s})\} \\ &\qquad\qquad + u_{ni}(\mathbf{s}) + a c_{ni}(\mathbf{s})\Big) \\ &\qquad\qquad - F\Big(x + x\{v_{ni}(\mathbf{s}) + a d_{ni}(\mathbf{s})\} \\ &\qquad\qquad + u_{ni}(\mathbf{s}) + a c_{ni}(\mathbf{s})\Big)\Big]. \end{aligned}$$

Arguing as for (8.3.41), verify that $h_{ni} \equiv l_{ni}^{\pm}(\mathbf{s})$, $\tau_{ni} \equiv v_{ni}(\mathbf{s}) + a d_{ni}(\mathbf{s})$, $\delta_{ni} \equiv u_{ni}(\mathbf{s}) + a c_{ni}(\mathbf{s})$ satisfy the conditions of Theorem 2.2.5. Hence, one more application of (2.2.57) yields that for each $\mathbf{s} \in \mathbb{R}^m$, $a \in \mathbb{R}$,

$$\sup_{x \in \mathbb{R}} |D_2^{\pm}(x, \mathbf{s}, a) - D_2^{\pm}(x, \mathbf{s}, 0)| = o_p(1). \tag{8.3.44}$$

Now, suppose $x > 0$. Then, again using monotonicity of the indicator function and G, we obtain that on $\mathcal{B}_n$, for all $\|\mathbf{t} - \mathbf{s}\| < \delta$, $\mathbf{t}, \mathbf{s} \in \mathbb{R}^m$,

$$\begin{aligned} &|\mathcal{D}_2^{\pm}(x, \mathbf{s}, \mathbf{t})| \\ &\le |D_2^{\pm}(x, \mathbf{s}, 1) - D_2^{\pm}(x, \mathbf{s}, 0)| + |D_2^{\pm}(x, \mathbf{s}, -1) - D_2^{\pm}(x, \mathbf{s}, 0)| \\ &\quad + n^{-1/2} \sum_{i=1}^{n} l_{ni}^{\pm}(\mathbf{s}) \Big[F\Big(x + x(v_{ni}(\mathbf{s}) + d_{ni}(\mathbf{s})) \\ &\qquad\qquad + u_{ni}(\mathbf{s}) + c_{ni}(\mathbf{s})\Big) \\ &\qquad\qquad - F\Big(x + x(v_{ni}(\mathbf{s}) - d_{ni}(\mathbf{s})) \\ &\qquad\qquad + u_{ni}(\mathbf{s}) - c_{ni}(\mathbf{s})\Big)\Big]. \end{aligned}$$

Again, under the conditions (2.2.49)-(2.2.51) and (8.3.37), there exists a $\delta > 0$ such that the last term in this upper bound is $O_p(\epsilon)$, while the first two terms are $o_p(1)$, by (8.3.44). This completes the proof of (8.3.42) for D_2 in the case $x > 0$. The proof is similar in the case $x \le 0$. Hence the proof of (8.3.39) is complete. □

Proof of Lemma 8.3.1. First, consider (8.3.22). Let $m = q + r$ and write $\mathbf{t} = (\mathbf{t}_1', \mathbf{t}_2')'$, $\mathbf{t}_1 \in \mathbb{R}^q$, $\mathbf{t}_2 \in \mathbb{R}^r$ and let $\mathcal{W}_j$, $\mathcal{W}_j^*$ etc. denote the jth coordinate of $\mathcal{W}$, $\mathcal{W}^*$, etc. Take

$$l_{ni}(\mathbf{t}) := \frac{\dot{\mu}_{ni,j}(\mathbf{t}_1)}{\sigma_{ni}(\mathbf{t}_2)}, \quad u_{ni}(\mathbf{t}) := \mu_{ni}(\mathbf{t}_1) - \mu_i, \tag{8.3.45}$$

$$v_{ni}(\mathbf{t}) := \sigma_{ni}(\mathbf{t}_2) - 1, \quad 1 \le i \le n, \tag{8.3.46}$$

in $\tilde{\mathcal{U}}$ to see that now $\tilde{\mathcal{U}}$ equals $\mathcal{W}_j$. Thus (8.3.22) will follow from (8.3.39) once we verify the conditions of Lemma 8.3.2 for the quantities in (8.3.45) for each $1 \le j \le q$.

To that effect, note that by (8.3.2) and (8.3.3), $\forall\, \epsilon > 0$, $\exists\, N$, such that $\forall n > N$,

$$P\Bigg(\max\Big\{ \sup_{i,\mathbf{t}_1} \Big|\mu_{ni}(\mathbf{t}_1) - \mu_i - n^{-1/2}\mathbf{t}_1'\dot{\boldsymbol{\mu}}_i\Big|, \sup_{i,t_2} \Big|\sigma_{ni}(\mathbf{t}_2) - 1 - n^{-1/2}\mathbf{t}_2'\dot{\boldsymbol{\sigma}}_i\Big| \Big\} \le b\epsilon n^{-1/2} \Bigg) \ge 1 - \epsilon, \tag{8.3.47}$$

where, here and in the sequel, $i, \mathbf{t}_1, \mathbf{t}_2$ in the supremum vary over the range $1 \le i \le n, \|\mathbf{t}\| \le b$, $\mathbf{t}_1 \in \mathbb{R}^q, \mathbf{t}_2 \in \mathbb{R}^r$, unless specified otherwise. From (8.3.47) and the assumption (8.3.10) we obtain that

$$\max_{i,\mathbf{t}_1} |\mu_{ni}(\mathbf{t}_1) - \mu_i| = o_p(1) = \max_{i,\mathbf{t}_2} |\sigma_{ni}(\mathbf{t}_2) - 1|. \tag{8.3.48}$$

This verifies (8.3.34) for the v_{ni}, u_{ni} of (8.3.45). The condition (8.3.32) follows from (8.3.9).

Next, let

$$\begin{aligned}
b_{ni} &\equiv n^{-1/2}(\delta\, \|\dot{\boldsymbol{\mu}}_i\| + 2b\epsilon), & b_n &:= \max_{1 \le i \le n} b_{ni}\,;\\
c_{ni} &\equiv n^{-1/2}(\delta\, \|\dot{\boldsymbol{\sigma}}_i\| + 2b\epsilon), & c_n &:= \max_{1 \le i \le n} c_{ni}\,;\\
z_{ni} &\equiv bn^{-1/2}(\|\dot{\boldsymbol{\sigma}}_i\| + \epsilon), & z_n &:= \max_{1 \le i \le n} z_{ni}\,;\\
w_{ni} &\equiv bn^{-1/2}(\|\dot{\boldsymbol{\mu}}_i\| + \epsilon), & w_n &:= \max_{1 \le i \le n} w_{ni}\,.
\end{aligned}$$

Note that by (8.3.10), $b_n = o_p(1) = z_n = w_n$. Now let, for an $\|\mathbf{s}\| \le b$,

$$\begin{aligned} \mathcal{C}_n \;:=\; \Big\{ & \sup_{\|\mathbf{t}_1 - s_1\| < \delta} |\mu_{ni}(\mathbf{t}_1) - \mu_{ni}(s_1)| \le b_{ni}, \\ & \sup_{\|\mathbf{t}_2 - \mathbf{s}_2\| < \delta} |\sigma_{ni}(\mathbf{t}_2) - \sigma_{ni}(s_2)| \le c_{ni}, \\ & \sup_{\|\mathbf{t}_1\| \le b} |\mu_{ni}(\mathbf{t}_1) - \boldsymbol{\mu}_i\| \le w_{ni}, \\ & \sup_{\|\mathbf{t}_2\| \le b} |\sigma_{ni}(\mathbf{t}_2) - 1| \le z_{ni},\ 1 \le i \le n;\ z_n \le 1/2 \Big\}. \end{aligned}$$

From (8.3.47) and (8.3.10), there exists an N_1 such that

$$P\Big(\mathcal{C}_n\Big) \ge 1 - \epsilon, \qquad n > N_1. \tag{8.3.49}$$

Because on $\mathcal{C}_n$, $1 - z_n \le \sup_{i,\mathbf{t}_2} \sigma_{ni}(\mathbf{t}_2) \le 1 + z_n$, we obtain that on $\mathcal{C}_n$,

$$\max_i \frac{|\dot\mu_{ni,j}(\mathbf{t}_1)|}{\sigma_{ni}(\mathbf{t}_2)} \le (1 - z_n)^{-1} \max_i |\dot\mu_{ni,j}(\mathbf{t}_1)|, \qquad 1 \le j \le q.$$

Moreover, for all $1 \le j \le q$, $\mathbf{t}_1 \in \mathbb{R}^q$,

$$\begin{aligned} n^{-1/2}|\dot\mu_{ni,j}(\mathbf{t}_1)| &\le n^{-1/2} \max_i |\dot\mu_{ni,j}(\mathbf{t}_1) - \dot\mu_{i,j}| + n^{-1/2} \max_i |\dot{\boldsymbol{\mu}}_{i,j}|, \\ n^{-1/2} \max_i |\dot\mu_{ni,j}(\mathbf{t}_1) - \dot\mu_{i,j}| &\le \Big(n^{-1} \sum_{i=1}^n \{\dot\mu_{ni,j}(\mathbf{t}_1) - \dot\mu_{i,j}\}^2 \Big)^{1/2}. \end{aligned}$$

These facts together with (8.3.10) and (8.3.11) imply that (8.3.33) is satisfied by the l_{ni} of (8.3.45).

Using the definitions of (8.3.45), one sees that on $\mathcal{C}_n$,

$$\begin{aligned} & n^{-1/2} \sum_{i=1}^n |l_{ni}(\mathbf{s})| \Big\{ |v_{ni}(\mathbf{s})| + |u_{ni}(\mathbf{s})| \Big\} \\ \le\; & 2n^{-1/2} \sum_{i=1}^n \frac{|\dot\mu_{ni,j}(\mathbf{s}_1)|}{\sigma_{ni}(\mathbf{s}_2)} [z_{ni} + w_{ni}] \\ \le\; & 2b(1 - z_n)^{-1} n^{-1} \sum_{i=1}^n |\dot\mu_{ni,j}(\mathbf{s}_1)| \Big[2\epsilon + \|\dot{\boldsymbol{\mu}}_i\| + \|\dot{\boldsymbol{\sigma}}_i\| \Big] \\ =\; & O_p(1), \end{aligned}$$

for each $1 \le j \le q$, $\|s\| \le b$, by (8.3.8) and (8.3.11). This verifies (8.3.36) for the entities given at (8.3.45). Also, (8.3.35) follows directly from the assumption (8.3.13). Finally, again on $\mathcal{C}_n$,

$$\begin{aligned}
n^{-1/2}\sum_{i=1}^{n} |l_{ni}(\mathbf{s})| \Big\{ \sup_{\|\mathbf{t}-\mathbf{s}\|<\delta} |v_{ni}(\mathbf{t}) - v_{ni}(\mathbf{s})| \\
+ \sup_{\|\mathbf{t}-\mathbf{s}\|<\delta} |u_{ni}(\mathbf{t}) - u_{ni}(\mathbf{s})| \Big\} \\
\le\ 2n^{-1/2}\sum_{i=1}^{n} \frac{|\dot{\mu}_{ni,j}(\mathbf{s}_1)|}{\sigma_{ni}(\mathbf{s}_2)} \Big\{ \sup_{\|\mathbf{t}_1-\mathbf{s}_1\|<\delta} |\mu_{ni}(\mathbf{t}_1) - \mu_{ni}(\mathbf{s}_1)| \\
+ \sup_{\|\mathbf{t}_2-\mathbf{s}_2\|<\delta} |\sigma_{ni}(\mathbf{t}_2) - \sigma_{ni}(\mathbf{s}_2)| \Big\} \\
\le\ 2(1-z_n)^{-1} n^{-1/2} \sum_{i=1}^{n} |\dot{\mu}_{ni,j}(\mathbf{s}_1)| [b_{ni} + c_{ni}] \\
\le\ 2(1-z_n)^{-1} n^{-1} \sum_{i=1}^{n} |\dot{\mu}_{ni,j}(\mathbf{s}_1)| \{\delta\, [\|\dot{\boldsymbol{\mu}}_i\| + \|\dot{\boldsymbol{\sigma}}_i\|] + 2b\epsilon\}.
\end{aligned}$$

In view of (8.3.8), this bound can be seen to be $O_p(\delta) + O_p(\epsilon)$, thereby verifying the condition (8.3.37) for the entities at (8.3.45), where as (8.3.38) is implied by (8.3.14). This completes the proof of (8.3.22).

To prove (8.3.23), write

$$\begin{aligned}
\mathcal{W}(x,\mathbf{t}) - \mathcal{W}(x,\mathbf{0}) &= \mathcal{W}(x,\mathbf{t}) - \mathcal{W}^*(x,\mathbf{t}) + \mathbf{U}(x,\mathbf{t}), \\
\mathbf{U}(x,\mathbf{t}) \equiv \mathcal{W}^*(x,\mathbf{t}) - \mathcal{W}(x,\mathbf{0}) &= n^{-1/2}\sum_{i=1}^{n} \left[\frac{\dot{\boldsymbol{\mu}}_{ni}(\mathbf{t}_1)}{\sigma_{ni}(\mathbf{t}_2)} - \frac{\dot{\boldsymbol{\mu}}_i}{\sigma_i} \right] \beta_i(x).
\end{aligned}$$

Thus it suffices to prove

(8.3.50) $$\sup_{x\in\mathbb{R},\|\mathbf{t}\|\le b} \|\mathbf{U}(x,\mathbf{t})\| = o_p(1).$$

Fix a $1 \le j \le q$, $\mathbf{t} \in \mathbb{R}^m$ and now let

$$h_{ni} = \frac{\dot{\mu}_{ni,j}(\mathbf{t}_1)}{\sigma_{ni}(\mathbf{t}_2)} - \frac{\dot{\mu}_{i,j}}{\sigma_i}.$$

Again write $h_{ni} = h_{ni}^+ - h_{ni}^-$, so that the jth component of $\mathbf{U}$ can be rewritten as $U_j^+ - U_j^-$, where

$$U_j^{\pm}(x,\mathbf{t}) := n^{-1/2}\sum_{i=1}^{n} h_{ni}^{\pm}\beta_i(x).$$

Because h_{ni} is independent of ε_i, $1 \le i \le n$ and past measurable, using a conditioning argument and (8.3.11), one obtains that

(8.3.51) $$Var\,\{U_j^{\pm}(x,\mathbf{t})\} \ \le\ n^{-1}\sum_{i=1}^{n} E\{h_{ni}^{\pm}\}^2 = o(1).$$

Next, fix an $\epsilon > 0$ and let $-\infty = x_0 < x_1, \cdots < x_r = \infty$ be a partition of $\mathbb{R}$ such that $G(x_k) - G(x_{k-1}) \le \epsilon$, $k = 1, \cdots, r$. Then, again using the monotonicity of the indicator and G, we have

$$\sup_{x \in \mathbb{R}} \left|U_j^{\pm}(x, \mathbf{t})\right| \le 2 \max_{1 \le k \le r} \left|U_j^{\pm}(x_k, \mathbf{t})\right| + \epsilon\, n^{-1/2} \sum_{i=1}^{n} |h_{ni}|.$$

This, (8.3.51), (8.3.12), and the arbitrariness of ϵ implies

$$\sup_{x \in \mathbb{R}} \|\mathbf{U}(x, \mathbf{t})\| = o_p(1), \qquad \forall\, \mathbf{t} \in \mathbb{R}^m.$$

To finish the proof of (8.3.50), we need to prove that an analog of (8.3.42) holds for the $\mathbf{U}(x, \mathbf{t})$-processes. But this is implied by (8.3.14), because

$$\mathbf{U}(x, \mathbf{t}) - \mathbf{U}(x, \mathbf{s}) = n^{-1/2} \sum_{i=1}^{n} \left[\frac{\dot{\boldsymbol{\mu}}_{ni}(\mathbf{t}_1)}{\sigma_{ni}(\mathbf{t}_2)} - \frac{\dot{\boldsymbol{\mu}}_{ni}(\mathbf{s}_1)}{\gamma_{ni}(\mathbf{s}_2)} \right] \beta_i(x).$$

This completes the proof of (8.3.23) and hence of the Lemma 8.3.1. □

8.3.3 Examples of ARCH models

We shall now discuss some details for verifying the general conditions of the previous section in three examples.

Example 8.3.1 (ARCH MODEL).
In the ARCH model of Engle (1982), one observes $\{Z_i, 1 - p \le i \le n\}$ such that for some $\alpha_0 > 0$, $\alpha_j \ge 0$, $j = 1, \cdots, p$,

$$\eta_i := Z_i / (\alpha_0 + \alpha_1 Z_{i-1}^2 + \cdots + \alpha_p Z_{i-p}^2)^{1/2} \tag{8.3.52}$$

are i.i.d. with mean zero and variance 1. Squaring both sides and writing $\varepsilon_i := \eta_i^2 - 1$, $X_i = Z_i^2$, $\mathbf{Y}_{i-1} = [X_{i-1}, \cdots, X_{i-p}]' = [Z_{i-1}^2, \cdots, Z_{i-p}^2]'$, and $\mathbf{W}_{i-1}' = [1, \mathbf{Y}_{i-1}']$, $\alpha = [\alpha_0, \ldots, \alpha_p]'$, the above model can be rewritten as

$$X_i = \boldsymbol{\alpha}' \mathbf{W}_{i-1} + \boldsymbol{\alpha}' \mathbf{W}_{i-1}\, \varepsilon_i, \qquad i \ge 1. \tag{8.3.53}$$

This is an example of the model (8.3.2) with $\boldsymbol{\alpha} = \boldsymbol{\beta}$, $q = p + 1$, $r = p + 1$, $\mu(\mathbf{y}, \boldsymbol{\alpha}) = \alpha' \mathbf{w} = \sigma(\mathbf{y}, \boldsymbol{\alpha})$, $\mathbf{w}' = (1, \mathbf{y}')$, $\mathbf{y} \in [0, \infty)^p$. Here, F denotes the d.f. of the error $\varepsilon := \eta^2 - 1$, assumed to satisfy (2.2.49), (2.2.50) and (2.2.51), and to have $E\varepsilon^2 = 1$.

Assume that the process $\{Z_i; 1 - p \le i\}$ is *stationary and ergodic* and $EZ_0^4 < \infty$. In the case $p = 1$, a sufficient condition for this to hold is

$-E\{\log(\alpha_1\varepsilon_1^2)\} > 0$, c.f. Nelson (1990). For $p \geq 1$, Bougerol and Picard (1992, Theorem 1.3) give some sufficient conditions based on Lyapunov exponent.

We shall now show that under no additional assumptions, the model (8.3.53) satisfies all the assumptions of the previous section except (8.3.9) and (8.3.13). These assumptions also hold provided, additionally, either $\alpha_j > 0$, for all $j = 1, \cdots, p$, or $EZ_0^8 < \infty$.

The assumptions (8.3.2) and (8.3.3), are readily seen to hold here with

$$\dot{\boldsymbol{\mu}}(\mathbf{Y}_{i-1}, \mathbf{s}) \equiv \mathbf{W}_{i-1} \equiv \dot{\boldsymbol{\sigma}}(\mathbf{Y}_{i-1}, \mathbf{s}).$$

From this we obtain

$$\dot{\boldsymbol{\mu}}_i = \frac{\mathbf{W}_{i-1}}{\boldsymbol{\alpha}'\mathbf{W}_{i-1}} = \dot{\boldsymbol{\sigma}}_i, \qquad 1 \leq i \leq n.$$

Observe that $E(Z_0^4) < \infty$ and $\alpha_0 > 0$, $\alpha_k \geq 0$, $k = 1, 2, \cdots, p$, imply that

$$E\|\dot{\boldsymbol{\mu}}_1\|^2 \leq C(1 + EZ_0^4)/\alpha_0^2 < \infty,$$

which in turn, together with the stationarity of the process, implies that $n^{-1/2}\max_{1\leq i\leq n}\|\dot{\boldsymbol{\mu}}_i\| = o_p(1) = n^{-1/2}\max_{1\leq i\leq n}\|\dot{\boldsymbol{\sigma}}_i\|$, thereby ensuring the satisfaction of (8.3.10), and that of (8.3.8) with

$$\boldsymbol{\Lambda} = E\left[\frac{\mathbf{W}_0\mathbf{W}_0'}{(\boldsymbol{\alpha}'\mathbf{W}_0)^2}\right] = \dot{\boldsymbol{\Sigma}} = \boldsymbol{\Gamma}, \quad \dot{\mathbf{M}} = E\mathbf{W}_0\mathbf{W}_0',$$

Since, here $\boldsymbol{\theta} = (\boldsymbol{\alpha}', \boldsymbol{\alpha}')'$, the above expectations depend only on $\boldsymbol{\alpha}$.

The same considerations and the linearity of μ and σ in the parameters enables one to guarantee the satisfaction of the conditions (8.3.11), (8.3.12), (8.3.16), (8.3.17), (8.3.19) and (8.3.20) with $\ddot{\mathbf{M}} = \mathbf{0}$, and

$$\dot{\mathbf{R}}(\mathbf{Y}_{i-1}, \mathbf{s}) = -\mathbf{W}_{i-1}\mathbf{W}_{i-1}'/(\mathbf{W}_{i-1}'\mathbf{s})^2.$$

Next, observe that

$$\text{(8.3.54)} \qquad \begin{aligned} \frac{\mu_{ni,j}(\mathbf{t}_1)}{\sigma_{ni}(\mathbf{t}_2)} &= \frac{1}{(\boldsymbol{\alpha} + n^{-1/2}\mathbf{t}_2)'\mathbf{W}_{i-1}}, && j = 1, \\ &= \frac{Z_{i-j+1}^2}{(\boldsymbol{\alpha} + n^{-1/2}\mathbf{t}_2)'\mathbf{W}_{i-1}}, && 2 \leq j \leq q; \\ \frac{\dot{\boldsymbol{\mu}}_{ni}(\mathbf{t}_1)}{\sigma_{ni}(\mathbf{t}_2)} &= \frac{\mathbf{W}_{i-1}}{(\boldsymbol{\alpha} + n^{-1/2}\mathbf{t}_2)'\mathbf{W}_{i-1}}, && 1 \leq i \leq n. \end{aligned}$$

From this, (8.3.14) is readily seen to hold, again using the D.C.T. and $EZ_0^4 < \infty$.

The condition (8.3.13) for $j = 1$ in the current case is seen to hold trivially. To check it for $2 \le j \le q = p+1$, we consider the case when $\alpha_{j-1} > 0$ for all $j = 2, \cdots, q$ and the case when $\alpha_{j-1} = 0$ for some $j = 2, \cdots, q$, separately. In the first case we have the following fact: $\forall\, \boldsymbol{a}, \boldsymbol{b} \in \mathbb{R}^{p+1}$, $n \ge 1$,

$$\text{(8.3.55)} \qquad \mathbf{w} \mapsto \boldsymbol{a}'\mathbf{w}/(\boldsymbol{\alpha} + n^{-1/2}\boldsymbol{b})'\mathbf{w}, \ \ \mathbf{w} \in [0,\infty)^{p+1}, \text{ is bounded.}$$

Use this fact and (8.3.54) to obtain that for some $k \in (0,\infty)$, possibly depending on $\mathbf{t}$,

$$\begin{aligned}
& n^{-1}\sum_{i=1}^{n}\left\{\frac{\dot\mu_{ni,j}(\mathbf{t}_1)}{\sigma_{ni}(\mathbf{t}_2)}\right\}^2 \{|\mu_{ni}(\mathbf{t}_1) - \mu_{i,j}| + |\sigma_{ni}(\mathbf{t}_2) - 1|\} \\
&= 2n^{-3/2}\sum_{i=1}^{n}\frac{Z_{i-j+1}^2}{\boldsymbol{\alpha}'\mathbf{W}_{i-1}}\,\frac{Z_{i-j+1}^2}{(\boldsymbol{\alpha}' + n^{-1/2}\mathbf{t}_2)'\mathbf{W}_{i-1}}\,\frac{\mathbf{t}_1'\mathbf{W}_{i-1}}{(\boldsymbol{\alpha}' + n^{-1/2}\mathbf{t}_2)'\mathbf{W}_{i-1}} \\
&\le k n^{-1/2} n^{-1}\sum_{i=1}^{n}\frac{Z_{i-j+1}^2}{\boldsymbol{\alpha}'\mathbf{W}_{i-1}} \\
&\le k n^{-1/2}\,(1/\alpha_0)\, n^{-1}\sum_{i=1}^{n} Z_{i-j}^2, \qquad j = 2, \cdots, p+1.
\end{aligned}$$

This bound together with the stationarity of $\{Z_i\}$'s and $EZ_0^4 < \infty$ readily imply that (8.3.13) holds in the first case. In the second case a similar argument and $EZ_0^8 < \infty$ yields the satisfaction of (8.3.13). Finally, the condition (8.3.9) is verified similarly, using (8.3.54) and (8.3.55). Note that the condition $EZ_0^8 < \infty$ is needed in verifying (8.3.13) and (8.3.9) only in the case when some $\alpha_j = 0$, $j = 1, \cdots, p$.

Since here $\boldsymbol{\alpha} = \boldsymbol{\beta}$, for estimation in this model, we use just a two-step procedure, i.e., use $\hat{\boldsymbol{\alpha}}_p$ instead of $\hat{\boldsymbol{\beta}}$ to define final $\hat{\boldsymbol{\alpha}}$. Since in this case $\hat{\boldsymbol{\alpha}}_p$ has the explicit expression (least squares estimator), it is easy to see that condition (8.3.18)(a) is guaranteed. Because $\mathcal{H}_n$ is a sum of square integrable martingales differences of stationary and ergodic r.v.'s, $\mathcal{H}_n$ converges weakly to a normal r.v., and hence here (8.3.18)(b) is *a priori* satisfied. Moreover, because of the linearity of μ in $\boldsymbol{\alpha}$, the condition (8.3.25) is seen to be satisfied as in Section 5.5. Therefore, from Corollary 8.3.4, if $\int x f(x) d\psi(x) = 0$, then $n^{1/2}(\hat{\boldsymbol{\alpha}} - \boldsymbol{\alpha}) \longrightarrow_d \mathcal{N}(\mathbf{0}, \boldsymbol{\Sigma}(\boldsymbol{\alpha}))$, where

$$\boldsymbol{\Sigma}(\boldsymbol{\alpha}) := \left(E\left[\mathbf{W}_0\mathbf{W}_0'/(\boldsymbol{\alpha}'\mathbf{W}_0)^2\right]\right)^{-1} v(\psi, F).$$

Now, recall from Weiss (1986) that under the stationarity of $\{X_i\}$'s and the finite fourth moment assumption on the i.i.d. errors ε_i, the asymptotic distribution of the widely used quasi maximum likelihood estimator $\hat{\boldsymbol{\alpha}}_{qmle}$ is as follows:

$$n^{1/2}(\hat{\boldsymbol{\alpha}}_{qmle} - \boldsymbol{\alpha}) \longrightarrow_d \mathcal{N}(\mathbf{0}, \boldsymbol{\Sigma}_{qmle}),$$
$$\boldsymbol{\Sigma}_{qmle} := \left(E\left[\mathbf{W}_0\mathbf{W}_0'/(\boldsymbol{\alpha}'\mathbf{W}_0)^2\right]\right)^{-1} Var(\varepsilon).$$

Thus it follows that the asymptotic relative efficiency of an M- estimator $\hat{\boldsymbol{\alpha}}$, relative to the widely used quasi maximum likelihood estimator in Engle's ARCH model is exactly the same as that of the M- estimator relative to the widely used least squared estimator in the one sample location model or in the linear regression model.

Fitting an error d.f. Consider the model (8.3.52). This model is a special case of the general model (8.3.1) with

$$\mu(\mathbf{y}, \boldsymbol{\alpha}) \equiv 0, \qquad \sigma(\mathbf{y}, \boldsymbol{\alpha}) \equiv (\boldsymbol{\alpha}'\mathbf{w})^{1/2}, \qquad \mathbf{w}' = (1, \mathbf{y}').$$

Let now F denote the common d.f. of η_i having mean zero and unit variance and F_0 be a known d.f. of a standard r.v. Consider the problem of testing $H_0 : F = F_0$ against the alternative that H_0 is not true, where F_0 is a known distribution function. Let $\hat{\boldsymbol{\alpha}}$ be $n^{1/2}$- consistent estimator of $\boldsymbol{\alpha}$ and $\hat{F}_n$ denote the empirical d.f. of the residuals $\hat{\eta}_i := Z_i/(\hat{\boldsymbol{\alpha}}'\mathbf{W}_{i-1})^{1/2}$, $1 \le i \le n$. A natural test of H_0 is to reject it if $D_n := \sup_x n^{1/2}|\hat{F}_n(x) - F_0(x)|$ is large.

The following corollary gives the asymptotic behavior of $\hat{F}_n$. It is obtained from Theorem 8.3.5 upon taking $h_{ni} \equiv 1$, $\mu_i \equiv 0$, $\dot{\boldsymbol{\mu}}_i \equiv \mathbf{0}$ in there. See also Remark 2.2.5.

Corollary 8.3.6 *Suppose the process given by the model (8.3.52) is stationary, $E\eta^2 < \infty$, and the d.f. F satisfies* (**F1**), (**F2**) *and* (**F3**). *Moreover, suppose $\hat{\boldsymbol{\alpha}}$ is any estimator of $\boldsymbol{\alpha}$ with $n^{1/2}\|\hat{\boldsymbol{\alpha}} - \boldsymbol{\alpha}\| = O_p(1)$. Then*

$$\sup_{x\in\mathbb{R}} \Big| n^{1/2}[\hat{F}_n(x) - F(x)] - n^{-1}\sum_{i=1}^{n} \frac{\mathbf{W}_{i-1}'}{2(\boldsymbol{\alpha}'\mathbf{W}_{i-1})^{1/2}} n^{1/2}(\hat{\boldsymbol{\alpha}} - \boldsymbol{\alpha})\, x f(x) \Big| = o_p(1).$$

This result is then useful in assessing the limiting behavior of D_n under H_0 and any alternatives satisfying the assumed conditions. The conclusions here are thus similar to those in Section 6.4. For example, tests based in $\hat{F}_n$ will not be ADF in general.

Example 8.3.2 (AR MODEL WITH ARCH ERRORS).
Consider the first order autoregressive model with heteroscedastic errors where the conditional variance of the ith observation depends linearly on the past as follows:

$$X_i = \alpha X_{i-1} + \{\boldsymbol{\beta}'\mathbf{Z}_{i-1}\}^{1/2}\varepsilon_i, \qquad i \geq 1, \tag{8.3.56}$$

where, $\alpha \in \mathbb{R}$, $\boldsymbol{\beta} = (\beta_0, \beta_1)'$ with $\beta_0 > 0$ and $\beta_1 \geq 0$, $\mathbf{Z}_{i-1} = (1, X_{i-1}^2)'$. Here, now F denotes the d.f. of the error ε, assumed to satisfy **(F1)**, **(F2)** and **(F3)**.

This is an example of the model (8.3.1) with $p = 1$, $q = p$, $r = p+1$, $\mathbf{Y}_{i-1} \equiv X_{i-1}$, $\mu(y, \alpha) = \alpha y$, and $\sigma(y, \boldsymbol{\beta}) = (\boldsymbol{\beta}'\mathbf{z})^{1/2}$, $\mathbf{z}' = (1, y^2)$, $y \in \mathbb{R}$. Throughout the discussion of this example, we assume $EX_0^4 < \infty$, which in turn guarantees that $E\varepsilon^4 < \infty$.

An additional assumption needed on the parameters under which this model is stationary and ergodic is as follows:

$$|\alpha| + E|\varepsilon_1|\{\beta_0\beta_1/(\beta_0 + \beta_1)\}^{1/2} < 1. \tag{8.3.57}$$

This follows with the help of Lemma 3.1 of Härdle and Tsybakov (1997, p 227) upon taking $C_1 = |\alpha|$ and $C_2 = \{\beta_0\beta_1/(\beta_0 + \beta_1)\}^{1/2} = \sup\{(\beta_0 + \beta_1 x^2)^{1/2}/(1 + |x|); x \in \mathbb{R}\}$ in there, to conclude that under (8.3.57), the process $\{X_i; i \geq 0\}$ of (8.3.56) is stationary and ergodic.

The assumptions (8.3.2) and (8.3.3) are readily seen to hold with

$$\dot{\mu}(\mathbf{Y}_{i-1}, \mathbf{s}) = X_{i-1}, \qquad \dot{\sigma}(\mathbf{Y}_{i-1}, \mathbf{s}) = \mathbf{Z}_{i-1}/\{2(\mathbf{Z}_{i-1}'\mathbf{s})^{1/2}\}.$$

Note that here $\dot{\mu}_{i1} \equiv X_{i-1}/(\beta_0 + \beta_1 X_{i-1}^2)^{1/2}$. Use the boundedness of the function $x \to x/(\beta_0 + \beta_1 x^2)^{1/2}$ on $[0, \infty)$ when $\beta_1 > 0$, $EX_0^4 < \infty$, and the stationarity and the ergodicity of $\{X_i\}$ to verify (8.3.8), (8.3.9), and (8.3.10) here with $\dot{\mathbf{M}} = E[X_0^2]$, and

$$\Lambda = E\left[\frac{X_0^2}{\boldsymbol{\beta}'\mathbf{Z}_0}\right], \quad \dot{\boldsymbol{\Sigma}} = E\left[\frac{\mathbf{Z}_0\mathbf{Z}_0'}{4(\boldsymbol{\beta}'\mathbf{Z}_0)^2}\right], \quad \boldsymbol{\Gamma} = E\left[\frac{X_0\mathbf{Z}_0'}{2(\boldsymbol{\beta}'\mathbf{Z}_0)^{3/2}}\right].$$

To verify (8.3.11), note that $\dot{\mu}_{ni}(t_1) - \dot{\mu}_i \equiv 0$. Hence, by the stationarity of the process the left hand side of (8.3.11) here equals

$$\mathbb{E}\left\| \left(\frac{\mathbf{Z}_0}{2\boldsymbol{\beta}'\mathbf{Z}_0}\right)\left[\left\{\frac{\boldsymbol{\beta}'\mathbf{Z}_0}{(\boldsymbol{\beta} + n^{-1/2}\mathbf{t}_2)'\mathbf{Z}_0}\right\}^{1/2} - 1\right]\right\|^2.$$

But, clearly the sequence of r.v.'s $[\{\boldsymbol{\beta}'\mathbf{Z}_0/(\boldsymbol{\beta}+n^{-1/2}\mathbf{t}_2)'\mathbf{Z}_0\}^{1/2}-1]$ is bounded and tends to 0, a.s. These facts together with the D.C.T. imply (8.3.11) in this example.

Next, to verify (8.3.12), note that the derivative of the function $s \mapsto [x/(x+s)]^{1/2}$ at $s=0$ is $-1/(2x)$. Now, rewrite the left hand side of (8.3.12) as

$$\frac{1}{2\sqrt{n}}\sum_{i=1}^{n}\frac{\|\mathbf{Z}_{i-1}\|}{\boldsymbol{\beta}'\mathbf{Z}_{i-1}}\left|\left\{\frac{\boldsymbol{\beta}'\mathbf{Z}_{i-1}}{(\boldsymbol{\beta}+n^{-1/2}\mathbf{t}_2)'\mathbf{Z}_{i-1}}\right\}^{1/2}-1\right|$$

$$\leq \frac{1}{2\sqrt{n}}\sum_{i=1}^{n}\frac{\|\mathbf{Z}_{i-1}\|}{\boldsymbol{\beta}'\mathbf{Z}_{i-1}}\Bigg|\left\{\frac{\boldsymbol{\beta}'\mathbf{Z}_{i-1}}{(\boldsymbol{\beta}+n^{-1/2}\mathbf{t}_2)'\mathbf{Z}_{i-1}}\right\}^{1/2} -1+\frac{n^{-1/2}\mathbf{Z}'_{i-1}\mathbf{t}_2}{2\,\boldsymbol{\beta}'\mathbf{Z}_{i-1}}\Bigg|$$

$$+\frac{1}{4n}\sum_{i=1}^{n}\frac{\|\mathbf{Z}_{i-1}\|^2}{(\boldsymbol{\beta}'\mathbf{Z}_{i-1})^2}\|\mathbf{t}_2\|.$$

Because $E\|\mathbf{Z}_0\|^2<\infty$, we have $\max_{1\le i\le n}|n^{-1/2}\mathbf{Z}'_{i-1}t_2| = o_p(1)$. This and the stationarity implies that the first term tends to zero in probability. The Ergodic Theorem implies that the r.v.'s in the second term converges in probability to $E[\|\mathbf{Z}_0\|^2/4(\boldsymbol{\beta}'\mathbf{Z}_0)^2]$, thereby verifying (8.3.12) here.

To verify (8.3.13) we shall use (8.3.15). The stationarity implies that in this example the left hand side of (8.3.15) is equal to

$$\text{(8.3.58)}\qquad n^{-1/2}E\left[\frac{X_0^4}{\{(\boldsymbol{\beta}+n^{-1/2}\mathbf{t}_2)'\mathbf{Z}_0\}^2}\times\frac{(t_1X_0)^2}{(\boldsymbol{\beta}'\mathbf{Z}_0)}\right]$$

$$+n^{1/2}E\left[\frac{X_0^4}{\{(\boldsymbol{\beta}+n^{-1/2}\mathbf{t}_2)'\mathbf{Z}_0\}^2}\left|\left\{\frac{(\boldsymbol{\beta}+n^{-1/2}\mathbf{t}_2)'\mathbf{Z}_0}{\boldsymbol{\beta}'\mathbf{Z}_0}\right\}^{1/2}-1\right|^2\right].$$

If $\beta_1>0$, then the expectation in the first term of (8.3.58) stays bounded, as in this case the integrands are bounded uniformly in n. The same remains true under the additional assumption $EX_0^6<\infty$, when $\beta_1=0$. In either case this shows that the first term in (8.3.58) is $O(n^{-1/2})$.

To handle the second term, apply the mean value theorem to the function $s\mapsto\{(x+s)/x\}^{1/2}$ around $s=0$, to obtain that for some $0<\xi<1$, the second term in (8.3.58) equals

$$n^{-1/2}E\left[\frac{X_0^4}{4\{(\boldsymbol{\beta}+n^{-1/2}\mathbf{t}_2)'\mathbf{Z}_0\}^2\boldsymbol{\beta}'\mathbf{Z}_0}\frac{(\mathbf{t}_2'\mathbf{Z}_0)^2}{(\boldsymbol{\beta}'\mathbf{Z}_0+\xi n^{-1/2}\mathbf{t}_2'\mathbf{Z}_0)}\right]$$

$$= O(n^{-1/2}),$$

by arguing as for the first term of (8.3.58). Therefore, (8.3.58), and hence the left hand side of (8.3.15), is $O(n^{-1/2})$, thereby verifying this condition. Because $\dot{\mu}(y, t_1) \equiv y$, a constant in t_1, we see that the condition (8.3.14) readily holds. Conditions (8.3.16) and (8.3.19) are satisfied with $\ddot{\mathbf{M}} \equiv 0$ and $\dot{\boldsymbol{R}}(\mathbf{Y}_{i-1}, \mathbf{s}) \equiv -\mathbf{Z}_{i-1}\mathbf{Z}'_{i-1}(2\mathbf{Z}'_{i-1}\mathbf{s})^{-2}$.

Observe that here $\hat{\alpha}_p = \sum_{i=1}^n X_i X_{i-1} / \sum_{i=1}^n X_{i-1}^2$. It is easy to see that in this example $n^{1/2}(\hat{\alpha}_p - \alpha) \Longrightarrow \mathcal{N}_1(0, a^2\gamma^2)$, where $a^2 = E\{X_0^2(\beta_0 + \beta_1 X_0^2)\}/\{E(X_0^2)\}^2$, $\gamma^2 := Var(\varepsilon)$, thereby guaranteeing the satisfaction of (8.3.18). The condition (8.3.25) is implied here by the monotonicity of score function $M(t_1, \mathbf{t}_2)$ in t_1, for every $\mathbf{t}_2$ fixed.

Therefore, to summarize, we obtain that if either $\beta_1 > 0$ and $EX_0^4 < \infty$, or $\beta_1 = 0$ and $EX_0^6 < \infty$, and if either $\int x f(x) d\psi(x) = 0$ or $\boldsymbol{\Gamma}(\boldsymbol{\theta}) = \mathbf{0}$, then, $n^{1/2}(\hat{\alpha} - \alpha) \longrightarrow_d \mathcal{N}_1\big(0, \tau^2(\boldsymbol{\theta}) v(\psi, F)\big)$, where

$$\tau^2(\boldsymbol{\theta}) := \frac{1}{E[X_0^2/(\beta_0 + \beta_1 X_0^2)]}.$$

Again, it follows that the asymptotic relative efficiency of the M- estimator corresponding to the score function ψ, relative to the least square estimator, in the above ARCH model is the same as in the one sample location or in the linear regression and autoregressive models.

Example 8.3.3 (THRESHOLD ARCH MODEL). Consider the pth order autoregressive model with self exciting threshold heteroscedastic errors where the conditional standard deviation of the ith observation is piecewise linear on the past as follows: For $i \geq 1$,

$$\begin{aligned} X_i &= \boldsymbol{\alpha}'\mathbf{Y}_{i-1} + \Big\{\beta_1 X_{i-1} I(X_{i-1} > 0) - \beta_2 X_{i-1} I(X_{i-1} \leq 0) \\ &\qquad + \cdots + \beta_{2p-1} X_{i-p} I(X_{i-p} > 0) - \beta_{2p} X_{i-p} I(X_{i-p} \leq 0)\Big\}\varepsilon_i, \end{aligned}$$

where all β_j's are positive. For details on the applications and many probabilistic properties of this model and for the conditions on the stationarity and ergodicity, see Rabemananjara and Zakoian (1993). For a discussion on the difficulties associated with the asymptotics of the robust estimation in this model, see Rabemananjara and Zakoian (1993, p 38).

This model is again an example of the model (8.3.1) with $q = p$, $r = 2p$, $\mu(\mathbf{y}, \boldsymbol{\alpha}) = \boldsymbol{\alpha}'\mathbf{y}$, and

$$\sigma(\mathbf{y}, \boldsymbol{\beta}) = \sum_{j=1}^{p} \beta_{2j-1} y_j I(y_j \geq 0) + \sum_{j=1}^{p} \beta_{2j}(-y_j) I(y_j < 0),$$

for $\mathbf{y} \in \mathbb{R}^p$, $\boldsymbol{\beta} \in (0,\infty)^{2p}$. We shall now verify the assumptions (8.3.2), (8.3.3), (8.3.8) - (8.3.19) in this model. Define

$$\begin{aligned} \mathbf{W}'_{i-1} &= [X_{i-1}I(X_{i-1} > 0), -X_{i-1}I(X_{i-1} \le 0), \cdots, \\ &\qquad X_{i-p}I(X_{i-p} > 0), -X_{i-p}I(X_{i-p} \le 0)]'. \end{aligned}$$

The assumptions (8.3.2) and (8.3.3) are trivially satisfied with

$$\dot{\mu}(\mathbf{Y}_{i-1}, \mathbf{s}) = \mathbf{Y}_{i-1}, \qquad \dot{\sigma}(\mathbf{Y}_{i-1}, \mathbf{s}) = \mathbf{W}_{i-1}.$$

Assuming $\{X_i\}$'s are stationary and ergodic, (8.3.8) is satisfied with

$$\begin{aligned} \mathbf{\Lambda}(\boldsymbol{\theta}) = E\left[\frac{\mathbf{Y}_0\mathbf{Y}'_0}{(\boldsymbol{\beta}'\mathbf{W}_0)^2}\right], \quad \dot{\mathbf{\Sigma}}(\boldsymbol{\theta}) = E\left[\frac{\mathbf{W}_0\mathbf{W}'_0}{(\boldsymbol{\beta}'\mathbf{W}_0)^2}\right], \\ \dot{\mathbf{M}}(\theta) = E[\mathbf{Y}_0\mathbf{Y}'_0], \quad \mathbf{\Gamma}(\boldsymbol{\theta}) = E\left[\frac{\mathbf{Y}_0\mathbf{W}'_0}{(\boldsymbol{\beta}'\mathbf{W}_0)^2}\right]. \end{aligned}$$

Since the functions $x \to x/(\beta_{2j-1}xI(x \ge 0) - \beta_{2j}xI(x < 0))$ are bounded, $\dot{\mu}_{ij}$ are bounded in this case, uniformly in i, j. Moreover, (8.3.10) is seen to hold by the stationarity and the finite fourth moment assumption. Conditions (8.3.11), (8.3.12) (8.3.14)-(8.3.16) are satisfied since the functions μ and σ are linear in parameters and (8.3.13) is seen to hold as in Example 8.3.2.

Finally, (8.3.20) and (8.3.19) are seen to be satisfied with

$$\dot{\mathbf{R}}(\mathbf{Y}_{i-1}, \mathbf{s}) = -\mathbf{W}_{i-1}\mathbf{W}'_{i-1}/(\mathbf{W}'_{i-1}\mathbf{s})^2.$$

Therefore, from Corollary 8.3.4, if $f(-x) = f(x)$, $\psi(-x) = -\psi(x)$, for every $x \in \mathbb{R}$, then, $n^{1/2}(\hat{\boldsymbol{\alpha}} - \boldsymbol{\alpha}) \longrightarrow_d \mathcal{N}(\mathbf{0}, \mathbf{\Sigma}(\boldsymbol{\theta})v(\psi, F))$, where

$$\mathbf{\Sigma}(\boldsymbol{\theta}) := \left(E\left[\mathbf{Y}_0\mathbf{Y}'_0/(\boldsymbol{\beta}'\mathbf{W}_0)^2\right]\right)^{-1}.$$

Again, a relative efficiency statement similar to the one in the previous two examples holds here also.

We end this section by mentioning that similar asymptotic normality and efficiency comparison statements can be deduced in all these examples from Theorem 8.3.4 pertaining to the minimum distance estimators $\hat{\boldsymbol{\alpha}}_{md}$.

Note: The results in this chapter are based on the works of Koul (1996) and Koul and Mukherjee (2001). A special case of Corollary 8.3.6 is also obtained by Boldin (1998).

Bollerslev, Chou and Kroner (1992), Bera and Higgins (1993), Shephard (1996) and the books by Taylor (1986) and Gouriéroux (1997), among others, discuss numerous aspects of ARCH models. In particular, when $\mu \equiv 0$, the asymptotic distribution of the quasi-maximum likelihood estimator of $\boldsymbol{\beta}$ appears in Weiss (1986) and many probabilistic properties of the model are investigated in Nelson (1990). Adaptive estimation for linear regression models with Engle's ARCH errors was discussed by Linton (1993). Robust L-estimation of the heteroscedastic parameter $\boldsymbol{\beta}$ based on a preliminary estimator of $\boldsymbol{\alpha}$ in a special case of the above model is discussed in Koenker and Zhou (1996).

9

Appendix

9.1 Appendix

We include here some results relevant to the weak convergence of processes in $\mathcal{D}[0,1]$ and $C[0,1]$ for the sake of easy reference and without proofs. Our source is the book by Billingsley (1968) (B) on *Convergence of Probability Measures*.

To begin with, let $\xi_1, \cdots, \xi_m$ be r.v.'s, not necessarily independent and define

$$S_k := \sum_{j=1}^{k} \xi_j, \quad 1 \le k \le m; \qquad M_m := \max_{1 \le k \le m} |S_k|.$$

The following lemma is obtained by combining (12.5), (12.10) and Theorem 12.1 from pp. 87-89 of (B).

Lemma 9.1.1 *Suppose there exist nonnegative numbers $u_1, u_2, \cdots, u_m$, a $\gamma \ge 0$ and an $\alpha > 0$ such that*

$$E\{|S_k - S_j|^\gamma |S_j - S_i|^\gamma\} \le (\sum_{r=i+1}^{k} u_r)^{2\alpha}), \quad 0 \le i \le j \le k \le m.$$

Then, $\forall\ \lambda > 0,$

$$P(M_m \ge \lambda) \le K_{\gamma,\alpha} \cdot \lambda^{-2\gamma}\Big(\sum_{r=1}^{m} u_r\Big)^{2\alpha} + P(|S_m| \ge \frac{\lambda}{2}),$$

where $K_{\gamma,\alpha}$ is a constant depending only on γ and α.

The following inequality is given as Corollary 8.3 in (B).

Lemma 9.1.2 *Let $\{\zeta(t), 0 \le t \le 1\}$ be a stochastic process on some probability space. Let $\delta > 0$, $0 = t_0 < t_1 < \cdots < t_r = 1$ with $t_i - t_{i-1} \ge \delta$, $2 \le i \le r-1$, be a partition of $[0,1]$. Then, $\forall\ \epsilon > 0, \forall\ 0 < \delta \le 1$,*

$$P(\sup_{|t-s|<\delta} |\zeta(t) - \zeta(s)| \ge 3\epsilon)$$
$$\le \sum_{i=1}^{r} P(\sup_{t_{i-1} \le t \le t_i} |\zeta(t) - \zeta(t_{i-1})| \ge \epsilon).$$

Definition: A sequence of stochastic processes $\{\zeta_n\}$ in $\mathcal{D}[0,1]$ is said to converge weakly to a stochastic process $\zeta \in \mathcal{C}[0,1]$ if every finite dimensional distribution of $\{\zeta_n\}$ converges weakly to that of ζ and if $\{\zeta_n\}$ is tight with respect to the uniform metric.

The following theorem gives sufficient conditions for the weak convergence of a sequence of stochastic processes in $\mathcal{D}[0,1]$ to a limit in $\mathcal{C}[0,1]$. It is essentially Theorem 15.5, p. 127 of (B).

Theorem 9.1.1 *Let $\{\zeta_n(t), 0 \le t \le 1\}$ be a sequence of stochastic processes in $\mathcal{D}[0,1]$. Suppose that $|\zeta_n(0)| = O_p(1)$ and that $\forall\ \epsilon > 0$,*

$$\lim_{\eta \to 0} \limsup_{n} P\left(\sup_{|s-t|<\eta} |\zeta_n(s) - \zeta_n(t)| \ge \epsilon\right) = 0.$$

Then the sequence $\{\zeta_n(t), 0 \le t \le 1\}$ is tight, and of ζ is the weak limit of a subsequence $\{\zeta_{n'}(t), 0 \le t \le 1\}$, then $P(\zeta \in \mathcal{C}[0,1]) = 1$.

The following theorem gives sufficient conditions for the weak convergence of a sequence of stochastic processes in $\mathcal{C}[0,1]$ to a limit in $\mathcal{C}[0,1]$. It is essentially Theorem 12.3, p. 95 of (B).

Theorem 9.1.2 *Let $\{\zeta_n(t), 0 \le t \le 1\}$ be a sequence of stochastic processes in $\mathcal{C}[0,1]$. Suppose that $|\zeta_n(0)| = O_p(1)$ and that there exist a $\gamma \ge 0, \alpha > 1$ and a nondecreasing continuous function F on $[0,1]$ such that,*

$$P(|\zeta_n(t) - \zeta_n(s)| \ge \lambda) \le \lambda^{-\gamma} |F(t) - F(s)|^{\alpha}$$

holds for all s, t in $[0,1]$ and for all $\lambda > 0$.

Then the sequence $\{\zeta_n(t), 0 \le t \le 1\}$ is tight, and if ζ is the weak limit of a subsequence $\{\zeta_{m_n}(t), 0 \le t \le 1\}$, then $P(\zeta \in \mathcal{C}[0,1]) = 1$.

Next, we state a central limit theorem for martingale arrays. Let $(\Omega, \mathcal{F}, P)$ be a probability space; $\{\mathcal{F}_{n,i}, 1 \le i \le n\}$, be an array of sub σ-fields such that $\mathcal{F}_{n,i} \subset \mathcal{F}_{n,i+1}$, $1 \le i \le n$; X_{ni} be $\mathcal{F}_{n,i}$ measurable r.v. with $EX_{ni}^2 < \infty, E(X_{ni}|\mathcal{F}_{n,i-1}) = 0$, $1 \le i \le n$; and let $S_{nj} = \sum_{i \le j} X_{ni}, 1 \le j \le n$. Then $\{S_{ni}, \mathcal{F}_{n,i}; 1 \le i \le n, n \ge 1\}$ is called a *zero-mean square-integrable martingale array with differences* $\{X_{ni}; 1 \le i \le n, n \ge 1\}$.

The central limit theorem we find useful is Corollary 3.1 of Hall and Heyde (1980) which we state here for an easy reference.

Lemma 9.1.3 *Let $\{S_{ni}, \mathcal{F}_{n,i}; 1 \le i \le n, n \ge 1\}$ be a zero - mean square - integrable martingale array with differences $\{X_{ni}\}$ satisfying the following conditions.*

$$\forall \ \epsilon > 0, \quad \sum_{i=1}^{n} E[X_{ni}^2 I(|X_{ni}| > \epsilon)|\mathcal{F}_{n,i-1}] = o_p(1). \tag{9.1.1}$$

$$\sum_{i=1}^{n} E[X_{ni}^2|\mathcal{F}_{n,i-1}] \to \textit{a r.v. } \eta^2, \textit{ in probability.} \tag{9.1.2}$$

$$\mathcal{F}_{n,i} \subset \mathcal{F}_{n+1,i}, 1 \le i \le n, \; n \ge 1. \tag{9.1.3}$$

Then S_{nn} converges in distribution to a r.v. whose characteristic function at t is $E\exp(-\eta^2 t^2)$, $t \in \mathbb{R}$. □□

The following inequality on the tail probability of a sum of martingale differences is obtained by combining the Doob and Rossenthal inequalities: cf. Hall and Heyde (1980, Corollary 2.1 and Theorem 2.12).

Suppose $M_j = \sum_{i=1}^{j} D_i$ is a sum of martingale differences with respect to the underlying increasing filtration $\{\mathcal{D}_i\}$ and $E|D_i|^p < \infty, 1 \le i \le n$, for some $p \ge 2$. Then, there exists a constant $C = C(p)$ such that for any $\epsilon > 0$,

$$\begin{aligned} P\left[\max_{1 \le j \le n} |M_j| > \epsilon\right] \\ \le \; C\epsilon^{-p}\left[\sum_{i=1}^{n} E|D_i|^p + E\left\{\sum_{i=1}^{n} E(D_i^2|\mathcal{D}_{i-1})\right\}^{p/2}\right]. \end{aligned} \tag{9.1.4}$$

Next, we state and prove three lemmas of general interest. The first lemma is due to Scheffé (1947) while the second has its origin in Theorem II.4.2.1 of of Hájek - Sǐdák (op. cit.). The third lemma is the same as Theorem V.1.3.1 of Hájek - Sǐdák (op. cit.). All these results are reproduced here for the sake of completeness.

Lemma 9.1.4 *Let $(\Omega, \mathcal{A}, \nu)$ be a σ-finite measure space. Let ℓ, ℓ_n, $n \geq 1$ be sequence of probability densities w.r.t. ν such that $\ell_n \to \ell$, a.e. ν. Then,*

$$\int |\ell_n - \ell| \, d\nu \longrightarrow 0.$$

Proof. Let $\delta_n := \ell_n - \ell$, $\delta_n^+ := \max(\delta_n, 0)$, $\delta_n^- := \max(-\delta_n, 0)$. By assumption, $\delta_n^- \to 0$, a.e., ν. Moreover, $\delta_n^- \leq \ell$. Thus, by the DCT, $\int \delta_n^- \, d\nu \to 0$. This in turn along with the fact that $\int \delta_n \, d\nu = 0$, implies that $\int \delta_n^+ \, d\nu \to 0$. The claim now follows from these facts and the relation $\int |\ell_n - \ell| \, d\nu = \int \delta_n^- \, d\nu + \int \delta_n^+ \, d\nu$. □

Lemma 9.1.5 *Let $(\Omega, \mathcal{A}, \nu)$ be a σ-finite measure space. Let $\{g_n\}$, g be a sequence of measurable functions such that*

$$g_n \to g, \text{ a.e. } \nu, \tag{9.1.5}$$

$$\limsup_n \int |g_n| \, d\nu \leq \int |g| \, d\nu < \infty, \tag{9.1.6}$$

then, for any measurable function φ from $\mathbb{R}$ to $[0, 1]$,

$$\int \varphi g_n \, d\nu \to \int \varphi g \, d\nu. \tag{9.1.7}$$

Proof. For any function h, let $h^+ = \max(0, h)$, $h^- = \max(-h, 0)$. By (9.1.5), $g_n^\pm \to g^\pm$, a.e. Hence, by Fatou, because $0 \leq g_n^\pm$,

$$\liminf_n \int g_n^\pm \, d\nu \geq \int g^\pm \, d\nu. \tag{9.1.8}$$

But (9.1.8) is compatible with (9.1.6) only if

$$\limsup_n \int g_n^\pm \, d\nu \leq \int g^\pm \, d\nu. \tag{9.1.9}$$

For, otherwise, suppose that $\limsup_n \int g_n^+ \, d\nu > \int g^+ \, d\nu$. Then,

$$\begin{aligned}
\limsup_n \int |g_n| \, d\nu &= \limsup_n \Big(\int g_n^+ \, d\nu + \int g_n^- \, d\nu \Big) \\
&\geq \limsup_n \int g_n^+ \, d\nu + \liminf_n \int g_n^- \, d\nu \\
&\geq \limsup_n \int g_n^+ \, d\nu + \int g^- \, d\nu, \qquad \text{by (9.1.8}^-) \\
&> \int g^+ \, d\nu + \int g^- \, d\nu = \int |g| \, d\nu,
\end{aligned}$$

which violates (9.1.6). The other case can be handled similarly. This proves (9.1.9).

But (9.1.9) together with (9.1.8) imply that $\int g_n^\pm \, d\nu \to \int g^\pm \, d\nu$, which in turn imply that $\int g_n \, d\nu \to \int g \, d\nu$. Thus we have proved that (9.1.5) and (9.1.6) imply

$$\int g_n^\pm \, d\nu \;\to\; \int g^\pm \, d\nu, \tag{9.1.10}$$

$$\int g_n \, d\nu \;\to\; \int g \, d\nu. \tag{9.1.11}$$

Again, by Fatou Lemma applied to $g_n^\pm \varphi$ and $g_n^\pm(1-\varphi)$, we obtain that

$$\liminf_n \int \varphi g_n^\pm \, d\nu \;\geq\; \int \varphi g^\pm \, d\nu, \tag{9.1.12}$$

$$\liminf_n \int (1-\varphi) g_n^\pm \, d\nu \;\geq\; \int (1-\varphi) g^\pm \, d\nu. \tag{9.1.13}$$

But, in view of (9.1.10), (9.1.13) is equivalent to

$$\limsup_n \int \varphi g_n^\pm \, d\nu \;\leq\; \int \varphi g^\pm \, d\nu,$$

which together with (9.1.12) completes the proof of (9.1.7). □

Lemma 9.1.6 *Let $(\Omega, \mathcal{A}, \nu)$ be a σ-finite measure space. Let $\{g_n\}$, g be a sequence of square integrable functions such that (9.1.5) holds and that*

$$\limsup_n \int g_n^2 \, d\nu \;\leq\; \int g^2 \, d\nu. \tag{9.1.14}$$

Then,

$$\int (g_n - g)^2 d\nu \to 0.$$

Proof. The Fatou Lemma and (9.1.14) imply

$$\lim_n \int g_n^2 \, d\nu \;=\; \int g^2 \, d\nu. \tag{9.1.15}$$

By the Cauchy-Schwarz inequality,

$$\int |g_n g| d\nu \leq \left(\int g_n^2 d\nu \int g^2 \, d\nu \right)^{1/2}, \tag{9.1.16}$$

so that

$$\limsup_n \int |g_n g| d\nu \leq \int g^2 \, d\nu.$$

Hence by Lemma 9.1.5,

$$\lim \int g_n g d\nu = \int g^2 \, d\nu.$$

The lemma now follows from this, (9.1.15), and some algebra. □□□

10

Bibliography

Adichie, J. (1967). Estimation of regression parameters based on rank tests. *Ann. Math. Statist.*, **38**, 894-904.

An, H.-s. and Bing, C. (1991). A Kolmogorov-Smirnov type statistic with application to test for nonlinearity in time series. *Int. Statist. Rev.*, **59**, 287-307.

Anderson, T. W. (1971). *The statistical analysis of time series.* J. Wiley & Sons, Inc. New York.

Anderson, T. W. and Darling, D. A. (1952). Asymptotic theory of certain 'goodness of fit' criteria based on stochastic processes. *Ann. Math. Statist.*, **23**, 193-212.

Babu, J. and Singh, K. (1983). Inference on means using bootstrap. *Ann. Statist.*, **11**, 999-1003.

——— (1984). On one term Edgeworth correction by Efron's bootstrap. *Sankhya, Series A, Pt. 2*, **46**, 219-232.

Basawa, I. V. and Koul, H. L. (1988). Large sample statistics based on quadratic dispersions. *Int. Statist. Review*, **56**, 199-219.

Bera, A. K. and Higgins, M. L. (1993). ARCH models: properties, estimation and testing. *J. Economic Surveys*, **8**(4), 305-366.

Beran, R. J. (1977). Minimum Helinger distance estimates for parametric models. *Ann. Statist.*, **5**, 445-463.

——— (1978). An efficient and robust adaptive estimator of location. *Ann. Statist.*, **6**, 292-313.

——— (1982). Robust estimation in models for independent non - identically distributed data. *Ann. Statist.*, **2**, 415-428.

Bickel, P. J. (1982). On adaptive estimation. *Ann. Statist.*, **10**, 647-671.

Bickel, P. J. and Freedman, D. (1981). Some asymptotic theory for the bootstrap. *Ann. Statist.*, **2**, 415-428.

Bickel, P. J. and Lehmann, E. L. (1976). Descriptive statistics for nonparametric models. III Dispersion. *Ann. Statist.*, **4**, 1139-1158.

Bickel, P. J. and Wichura, M. (1971). Convergence criteria for multiparameter stochastic processes and some applications. *Ann. Math. Statist.*, **42**, 1656-1670.

Billingsley, P. (1968). *Convergence of Probability Measures.* John Wiley and Sons, New York.

Boldin, M. V. (1982). Estimation of the distribution of noise in autoregression scheme. *Theor. Probab. Appl.*, **27**, 866-871.

———- (1989). On testing hypotheses in the sliding average scheme by the Kolmogorov - Smirnov and ω^2 tests. *Theor. Probab. Appl.*, **34**, 699-704.

———- (1998). On residual empirical distributions in ARCH models with applications to testing and estimation. *Mitteilungen aus dem Mathem. Seminar*, Giessen. **235**.

Bollerslev, T., Chou, R. Y. and Kroner, K. F. (1992). ARCH modeling in finance; a review of the theory and empirical evidence. *J. Econometrics*, **52**, 115-127.

Boos, D. D. (1981). Minimum distance estimators for location and goodness of fit. *J. Amer. Statist. Assoc.*, **76**, 663-670.

——— (1982a). Minimum Anderson - Darling estimation. *Comm. Stat. Theor. Meth.*, **11** (24), 2747-2774.

——— (1982b). A test of symmetry associated with the Hodges - Lehmann estimator. *J. Amer. Statist. Assoc.*, **77**, 647-651.

Bougerol, P. and Picard, N. (1992). Stationarity of GARCH processes and of some nonnegative time series. *J. Econometrics*, **52**, 115-127.

Brockwell, P. J. and Davis, R. A. (1987). *Time Series : Theory and Methods.* Springer-Verlag, New York.

Brown, L. D. and Purves, R. (1973). Measurable selections of extrema. *Ann. Statist.*, **1**, 902-912.

Bustos, O. H. (1982). General M-estimates for contaminated p-th order autoregressive processes: consistency and asymptotic normality. *Zeit fur Wahrscheinlichkeitstheorie*, **59**, 491-504.

Chang, N. M. (1990). Weak convergence of a self-consistent estimator of a survival function with doubly censored data. *Ann. Statist.*, **18**, 391-404.

Cheng, K. F. and Sefling, R. J. (1981). On estimation of a class of efficiency - related parameters. *Scand. Act. J.*, 83-92.

Chow, Y.S., Robbins, H. and Teicher, H. (1965). Moments of randomly stopped sum. *Ann. Math. Statist.*, **36**, 789-799.

Chow, Y.S. and Teicher, H. (1978). *Probability Theory: Independence, Interchangeability, Martingales.* Springer-Verlag, New York.

Dehling, H. and Taqqu, M. S. (1989). The empirical process of some long range dependent sequences with an application to U-statistics. *Ann. Statist.*, **17**, 1767 - 1783.

Delgado, M. A. (1993). Testing the equality of nonparametric curves. *Statist. and Probab. Letters*, **17**, 199-204.

Delong, D. (1983). Personal Communication. S.A.S. Institute, Inc. Cary. N. C., 27511-8000.

Denby, L. and Martin, D. (1979). Robust estimation of the first order autoregressive parameter. *J. Amer. Statist. Assoc.*, **74**, 140-146.

Dhar, S. K. (1991a). Minimum distance estimation in an additive effects outliers model. *Ann. Statist.*, **19**, 205-228.

——— (1991b). Computation of minimum distance estimators in multiple lienar regression model. *Comm. Statist. Theor. and Meth. Ser. B*, **21(1)**, (1992), 18 pps.

——— (1991c). Compulation of certain minimum distance estimators in AR(k) model. (1993). *J. Amer. Statist. Assoc.*, **88**, 278-283.

Diebolt, J. (1995). A nonparametric test for the regression function: Asymptotic theory. *J. Statist. Plann. Infer.*, **44**, 1-17.

Donoho, D. L. and Liu, R. C. (1988). The "automatic" robustness of minimum distance functionals. *Ann. Statist.*, **16**, 552-586.

——— (1988). Pathologies of some minimum distance estimators. *Ann. Statist.*, **16**, 552-586.

Dupač, V. and Hájek, J. (1969). Asymptotic normality of linear rank statistics II. *Ann. Math. Statist.*, **40**, 1992-2017.

Durbin, J. (1973). *Distribution theory for tests based on the sample d.f.*. Philadelphia : SIAM.

——— (1976). Kolomogorov - Smirnov tests when parameters are estimated. *Empirical distributions and Processes; Springer - Verlag Lecture Notes in Math*, # 566.

Durbin, J., Knott, M. and Taylor, C.C. (1975). Components of Cramér-von Mises statistics. II. *J. Roy. Statist. Soc. B,* **37**, 216 - 237.

Dvoretzky, A., Kiefer, J., and Wolfowitz, J. (1956). Asymptotic minimax character of the sample distribution function and of the classical multinomial estimator. *Ann. Math. Statist.*, **27**, 642-669.

Efron, B. (1979). Bootstrap methods; another look at the jackknife. *Ann. Statist.*, **7**, 1 - 26.

——— (1982). *The Jackknife, the Bootstrap and Other resampling plans.* CBMS - NSF Reg. Conf. Ser. **38**, S.I.A.M.

Einmahl, U. and Mason, D. (1992). Approximations to permutation and exchangeable processes. *J. Theor. Probab.*, **5**, 101-126.

Engle, R.F. (1982). Autoregressive conditional heteroskedasticity and estimates of the variance of UK inflation. *Econometrica,* **50**, 987-1008.

Eyster, J. (1977). Asymptotic normality of simple linear random rank statistics under the alternativs. Ph.D. thesis, Michigan State University, East Lansing.

Fabian, V. and Hannan, J. (1982). On estimation and adaptive estimation for locally asymptotically normal families. *Z. Wahrsch. Verw. Gebiete*, **59**, 459-478.

Feller, W. (1966). *An Introduction to Probability Theory and its Applications* **II**. John Wiley & Sons, New York.

Ferretti, N.E, Klemansky, D.M., and Yohai, V.J. (1991). Estimators based on ranks for ARMA models. *Communications in Statistics A*, **20**, 3879-3907.

Fine, T. (1966). On Hodges - Lehmann shift estimator in the two sample problem. *Ann. Math. Statist.*, **37**, 1814-1818.

Finkelstein, H. (1971). The law of the iterated logarithm for empirical distributions. *Ann. Math. Statist.*, **42**, 607-615.

Freedman, D. (1981). Bootstrapping regression models. *Ann. Statist.*, **9**, 1218-1228.

Ghosh, M. and Sen, P. K. (1972). On bounded confidence intervals for regression coefficients basedon a class of rank statistics. *Sankhya* A, **34**, 33-52.

Gouriéroux C. (1997). *ARCH models and Financial Applications.* Springer - Verlag. New York.

Grutenbrunner, C. and Jurečková, J. (1992). Regression rank scores and regression quantiles. *Ann. Statist.*, **20**, 305-329.

Hájek, J. and Šidák, Z. (1967). *Theory of rank tests.* Academic Press, New York.

Hájek, J. (1969). *Nonparametric Statistics.* Holden Day, San Francisco.

——— (1972). Local asymptotic minimax and admissibility in estimation. *Proc. Sixth. Berkeley Symp. Math. Statist. Probability*, **I**, 597-616. University of California Press.

Hall, P. and Heyde, C. C. (1980). *Martingale limit theory and its applications*, Academic Press, New York.

Hampel, F. (1971). A general qualitative definition of robustness. *Ann. Math. Statist.*, **42**, 1887-1896.

Härdle, W. and Tsybakov, A. (1997). Local polynomial estimators of the volatility function in nonparametric autoregression. *J. of Econometrics.*, **81**, 223-242.

Hardy, G. H., Littlewood, J. E. and Polya, G. (1952). *Inequalities* (2nd Ed.). Cambridge University Press.

Heiler, S. and Willers, R. (1988). Asymptotic normality of R-estimates in the linear model. *Statistics*, **19**, 173-184.

Hettmansperger, T. P. (1984). *Statistical inference based on ranks.* J. Wiley & Sons. Inc. New York.

Hodges, J. L., Jr. and Lehmann, E. L. (1963). Estimates of location based on rank tests. *Ann. Math. Statist.*, **34**, 598-611.

Hoeffding, W. (1963). Probability for sums of bounded random variables. *J. Amer. Statist. Assoc.*, **58**, 13-30.

Huber, P. (1973). Robust regression: Asymptotics, conjectures and Monte Carlo. *Ann. Statist.*, **1**, 799-821.

Huber, P. (1981). *Robust Statistics.* Wiley, New York.

Jaeckel, L. A. (1972). Estimating regression coefficient by minimizing the dispersion of residuals. *Ann. Math. Statist.*, **43**, 1449-1458.

Johnson, W. B., Schechtman, G. and Zinn, J. (1985). Best constants in moment inequalities for linear combinations of independent and exchangeable random variables. *Ann. Probab.*, **13**, 234-253.

Jurečková, J. (1969). Asymptotic linearity of rank test statistics in regression parameters. *Ann. Math. Statist.*, **40**, 1889-1900.

——— (1971). Nonparametric estimates of regression coefficients. *Ann. Math. Statist.*, **42**, 1328 - 1338.

Kac, M., Kiefer, J. and Wolfowitz, J. (1955). On tests of normality and other tests of goodness-of-fit based on distance methods. *Ann. Math. Statist.*, **26**, 189-211.

Khmaladze, E. V. (1981). Martingale approach in the theory of goodness-of-fit tests. *Theory Probab. Appl.*, **26**, 240-257.

——— (1988). An innovation approach to goodness-of-fit tests in $\mathbb{R}^m$. *Ann. Statist.*, **16**, 1503-1516.

——— (1993). Goodness of fit problem and scanning innovation martingales. *Ann. Statist.*, **21**, 798-829.

Kiefer, J. (1959). K-sample analogues of the Kolmogorov - Smirnov and Cramér - von Mises tests. *Ann. Math. Statist.*, **30**, 420-447.

Koenker, R. and Bassett, G. (1978). Regression quantiles. *Econometrica*, **46**, 33-50.

Koenker, R. and d'Orey, V. (1987). Algorithm AS 229: Computing regression quantiles. *J. Royal Statist. Soc., Ser. C*, **36**, 383-393.

Koenker, R. and d'Orey, V. (1993). Remark AS R92. A remark on Algorithm AS 229: Computing dual regression quantiles and regression rank scores. *J. Royal Statist. Soc., Ser. C*, **43**, 410-414.

Koenker, R. and Zhao, Q. (1996). Conditional quantile estimation and inference for ARCH models. *Econometric Theory*, **12**, 793-813.

Koul, H. L. (1969). Asymptotic behavior of Wilcoxon type confidence regions in multile linear regression. *Ann. Math. Statist.*, **40**, 1950-1979.

Koul, H. L. (1970). Some convergence theorems for ranks and weighted empirical cumulatives. *Ann. Math. Statist.*, **41**, 1768-1773.

——— (1971). Asymptotic behavior of a class of confidence regions based on rank in regression. *Ann. Math. Statist.*, **42**, 466-476.

——— (1977). Behavior of robust estimators in the regression model with dependent errors. *Ann. Statist.*, **5**, 681-699.

——— (1979). Weighted empirical processes and the regression model. *J. Indian Statist. Assoc.*, **17**, 83-91.

——— (1980). Some weighted empirical inferential procedures for a simple regression model. *Colloq. Math. Soc. Janos Bolyai*, **32**, *Nonpar. Statist. Inf.* (537-565).

——— (1984). Tests of goodness-of-fit in linear regression. *Colloq. Math. Soc. Janos Bolyai*, **45**, *Goodness-of-Fit.* (279-315).

——— (1985a). Minimum distance estimation in multiple linear regression. *Sankhya, Ser. A.*, **47, Part 1**, 57-74.

——— (1985b). Minimum distance estimation in linear regression with unknown errors. *Statist. & Prob. Letters*, **3**, 1-8.

——— (1986). Minimum distance estimation and goodness of fit in first-order autoregression. *Ann. Statist.*, **14**, 1194-1213.

——— (1989). A quadraticity limit theorem useful in linear models. *Probab. Th. Rel. Fields*, **82**, 371-386.

——— (1991). A weak convergence result useful in robust autoregression. *J. Statist. Plannig and Inference*, **29**, 291-308.

——— (1996). Asymptotics of some estimators and sequential residual empiricals in non-linear time series. *Ann. Statist.*, **24**, 380-404.

Koul, H. L. and DeWet, T. (1983). Minimum distance estimation in a linear regression model. *Ann. Statist.*, **11**, 921-932.

Koul, H. L. and Levental, S. (1989). Weak convergence of residual empirical processes in explosive autoregression. *Ann. Statist.*, **17**, 1784-1794.

Koul, H. L. and Mukherjee, K. (1993). Asymptotics of R-, MD- and LAD-estimators in linear regression models with long range dependent errors. *Probab. Theory Related Fields*, **95**, 535–553.

Koul, H. L. and Mukherjee, K. (2001). On weighted and sequential residual empiricals in ARCH models with some applications. MSU Stat. & Prob. RM 601. April, 2001.

Koul, H. L. and Ossiander, M. (1994). Weak convergence of randomly weighted dependent residual empiricals with application to autoregression. *Ann. Statist.*, **22**, 540–562.

Koul, H. L. and Saleh, A.K.Md.E. (1995). Autoregression quantiles and related rank-scores processes. *Ann. Statist.*, **23**, 670-689.

Koul, H. L., Sievers, G. L., and McKean, J. W. (1987). An estimator of the scale parameter for the rank analysis of linear models under general score functions. *Scand. J. Statist.*, **14**, 131-141.

Koul, H. L. and Staudte, R. G., Jr. (1972). Weak convergence of weighted empirical cumulatives based on ranks. *Ann. Math. Statist.*, **43**, 832-841.

Koul, Hira L. and Stute, Winfried. (1999). Nonparametric model checks in time series. (1999). *Ann. Statist.*, **27**, 204-237.

Koul, H. L. and Susarla, V. (1983). Adaptive estimation in linear regression. *Statist. and Decis.*, **1**, 379-400.

Koul, H. L. and Zhu, Z. (1993). Bahadur representations for some minimum distance estimators in linear models. **Statist. and Probab: A Raghu Raj Bahadur Festschrift**, 349-364. **Eds.** J.K. Ghosh, S.K. Mitra, K.R. Parthasarathy, and B.L.S. Prakas Rao. Wiley Eastern Lmtd.

Kreiss, P. (1991). Estimation of the distribution function of noise in stationary processes. *Metrika*, **38**, 285-297.

Kuelbs, J. (1976). A strong convergence theorem for Banach spaces valued r.v.'s *Ann. Prob.*, **4**, 744-771.

Lahiri, S. (1989). Bootstrap approximations to the distributions of M-estimators. Ph.D. thesis. Michigan State Univ., East Lansing.

——— (1992). Bootstrapping M-estimators of a multiple regression parameter. *Ann. Statist.*, **20**, 1548–1570.

Le Cam, L. (1972). Limits of experiments. *Proc. Sixth. Berkeley Symp. Math. Statist. Probability*, **I**, 245-261. University of California Press.

——— (1986). *Asymptotical methods in statistical theory.* Springer, New York.

Lehmann, E. L. (1963). Nonparametric confidence intervals for a shift parameter. *Ann. Math. Statist.*, **34**, 1507-1512.

——— (1975). *Nonparametrics.* Holden Day, San Francisco.

Levental, S. (1989). A uniform CLT for uniformly bounded families of martingale - differences. *J. Theoret. Probab.*, **2**, 271-287.

Linton, O. (1993). Adaptive estimation in ARCH models. *Econometric Theory*, **9**, 539-569.

Loéve, M. (1963). *Probability Theory.* D. Van Nostrand Co., Inc. Princeton, New Jersey.

Loynes, R. M. (1980). The empirical d.f. of residuals from generalized regression. *Ann. Statist.*, **8**, 285-298.

Marcus, M. B. and Zinn, J. (1984). The bounded law of iterated logarithm for the weighted empirical distribution process in the non i.i.d. case. *Ann. Prob.*, **12**, 335-360.

MacKinnon, J. G. (1992). Model specification tests and artificial regression. *J. Econometric Literature*, **XXX** (March 1992), 102-146.

Martynov, G. V. (1975). Computation of distribution functions of quadratic forms of normally distributed r.v.'s. *Theor. Probab. Appl.*, **20**, 782-793.

——— (1976). Computation of limit distributions of statistics for normality tests of type ω^2. *Theor. Probab. Appl.*, **21**, 1-13.

Massart, P. (1990). The tight constant in the Dvoretzky-Kiefer-Wolfowitz inequality. *Ann. of Probab.*, **18**, 1269-1283.

Mehra, K. L. and Rao, M. S. (1975). Weak convergence of generalized empirical processes relative to d_q under strong mixing. *Ann. Prob.*, **3**, 979-991.

Millar, P. W. (1981). Robust estimation via minimum distance methods. *Zeit fur Wahrscheinlichkeitstheorie*, **55**, 73-89.

——— (1982). Optimal estimation of a general regression function. *Ann. Statist.*, **10**, 717-740.

——— (1984). A general approach to the optimality of minimum distance estimators. *Trans. Amer. Math. Soc.*, **286**, 377-418.

Mukherjee, K. and Bai, Z.D. (2001). R-estimation in autoregression under square-integrable score function. To appear in *J. Mult. Analysis*, 2002.

Nelson, D. B. (1990). Stationarity and persistence in the GARCH (1, 1) model. *Econometric Theory*, **6**, 318-334.

Noether, G. E. (1949). On a theorem by Wald and Wolfowitz. *Ann. Math. Statist.*, **20**, 455-458.

Ozaki, T. and Oda, H. (1978). Non-linear time series model identification by Akaike's information criterion. *Proc. IFAC Workshop on Information and Systems,* Compiegn, France. October 1977.

Parr, W. (1981). Minimum distance estimation: a bibliography. *Comm. Statist. Theor. Meth.*, **A10** (12), 1205-1224.

Parr, W. C. and Schucany, W. R. (1979). Minimum distance and robust estimation. *J. Amer. Statist. Assoc.*, **75**, 616-624.

Parthasarathy, K. R. (1967). *Probability measures on metric spaces.* Academic Press, New York.

Pollard, D., (1984). *Convergence of Stochastic Processes.* Springer, N. Y.

——— (1991). Asymptotics for least absolute deviation regression estimators. *Econometric Theor.*, **7**, 186-199.

Puri, M. L. and Sen, P. K. (1969). *Nonparametric methods in Multivariate Analysis.* John Wiley and Sons, New York.

Rabemananjara, R. and Zakoian, J.M. (1993). Threshold ARCH models and asymmetry in volatility. *J. Applied Econometrics*, **8**, 31-49.

Rao, K. C. (1972). The Kolmogoroff, Cramér-von Mises chi squares statistics for goodness-of-fit tests in the parametric case. (abstract). *Bull. Inst. Math. Statist.*, **1**, 87.

Rao, P. V., Schuster, E., and Littel, R. (1975). Estimation of shift and center of symmetry based on Kolmogorov - Smirnov statistics. *Ann. Statist.*, **3**, 862-873.

Robinson, P. M. (1984). Robust nonparametric autoregression. *Robust and nonlinear time series analysis.* Springer - Verlag Lecture Notes in Statistics, **26**, 247-255. Eds: J. Franke, W. Härdle and D. Martin.

Rockafeller, R. T. (1970). *Convex Analysis.* Princeton University Press, Princeton, N. J.

Schweder, T. (1975). Window estimation of the asymptotic variance of rank estimators of location. *Scand. J. Statist.*, **2**, 113-126.

Scheffé, H. (1947). A useful convergence theorem for probability distributions. *Ann. Math. Statist.*, **18**, 434-438.

Shephard, N. (1996). Statistical aspects of ARCH and stochastic volatility. In *Time Series Models in Econometric, Finance and other fields*, Ed by Cox, Hinkley, and Barndorff-Nielsen, 1-67.

Shorack, G. (1973). Convergence of reduced empiricals and quantile processes with applications to functions of order statistics in non-i.i.d. case. *Ann. Statist.*, **1**, 146-152.

——— (1979). Weighted empirical process of row independent r.v.'s with arbitrary df's. *Statistica Neerlandica*, **35**, 169-189.

——— (1982). Bootstrapping robust regression. *Comm. Statist. Theor. Meth.*, **A11** (9), 1205-1224.

——— (1991). Embedding the finite sampling process at a rate. *Ann. Probab.*, **19**, 826-842.

Sen, P.K. (1966). On a distribution - free method of estimating asymptotic efficiency of a class of nonparametric tests. *Ann. Math. Statist.*, **37**, 1759 - 1770.

Serfling, R. J. (1980). *Approximation Theorems of Mathematical Statistics.* J. Wiley & Sons, Inc., N. Y.

Singh, K. (1981). On asymptotic accuracy of Efron's bootstrap. *Ann. Statist.*, **9**, 1187-1195.

Sinha, A. & Sen, P. K. (1979). Progressively censored tests for clinical experiments and life testing problems based on weighted empirical distributions. *Comm. Statist. Theor. and Meth.*, **A8**, 871-898.

Smirnov, N. V. (1947). Sur un critēre de symétrie de la loi de distribution d'une variable aléatoire. *Compte Rendu (Doklady) de l Academic des Sciences de l' URSS*, **LVI**, No. **1**.

Stephens, M. A. (1976). Asymptotic results for goodness-of-fit statistics with unknown parameters. *Ann. Statist.*, **4**, 357-369.

——— (1979). Tests of fit for the logistic distribution based on the empirical df. *Biometrica*, **66**, 3, 591-595.

Stute, W. (1997). Nonparametric model checks for regression. *Ann. Statist.*, **25**, 613-641.

Stute, W., Thies, S. and Zhu, L.X. (1998). Model checks for regression: An innovation process approach. Tentatively accepted by *Ann. Statist.*

Su, J. Q. and Wei, L. J. (1991). A lack-of-fit test for the mean function in a generalized linear model. *J. Amer. Statist. Assoc.*, **86**, 420-426.

Taylor, S. J. (1986). *Modelling Financial Time Series.* John Wiley and Sons, Chickester, U. K.

Tjøstheim, D. (1986). Estimation in nonlinear time series models I: stationary series. *Stochastic Process. Appl.*, **21**, 251-273.

Tong, H. (1990). *Non-linear Time Series Analysis: A Dynamical Approach.* Oxford Univ. Press.

Vanderzanden, A. J. (1980). Some results for the weighted empirical processes concerning the laws of iterated logarithm and weak convergence. Ph.D. Thesis, MSU.

——— (1984). A functional law of the iterated logarithm for the weighted empirical processes. *J. Ind. Statist. Assoc.*, **22**, 97-110.

van Eeden, C. (1972). An analogue, for signed rank statistics, of Jurečková's asymptotic linearity theorem for rank statistics. *Ann. Math. Statist.*, **43**, 791-802.

Weiss, A. A. (1986). Asymptotic Theory for ARCH models: estimation and testing. *Econometric Theory* **2**, 107-131.

Wheeden, R. L. and Zygmund, A. (1977). *Measure and Integral: An Introduction to Real Analysis.* Marcel Dekker Inc. New York.

Williamson, M. A. (1979). Weighted empirical - type estimators of regression parameter. Ph.D. Thesis, MSU.

——— (1982). Cramér - von Mises estimations of regression parameter; The rank analogue. *J. Mult. Anal.*, **12**, 248-255.

Withers, C. S. (1975). Convergence of empirical processes of mixing r.v.'s on $[0, 1]$. *Ann. Statist.*, **3**, 1101-1108.

Wolfowitz, J. (1953). Estimation by minimum distance method. *Ann. Inst. Stat. Math.*, **5**, 9-23.

——— (1954). Estimation of the components of stochastic structures. *Proc. Nat. Acad. Sci.*, **40**, 602-606.

——— (1957). Minimum distance estimation method. *Ann. Math. Statist.*, **28**, 75-88.

Lecture Notes in Statistics

For information about Volumes 1 to 111, please contact Springer-Verlag

112: Doug Fischer, Hans-J. Lenz (Editors), Learning from Data. xii, 450 pp., 1996.

113: Rainer Schwabe, Optimum Designs for Multi-Factor Models. viii, 124 pp., 1996.

114: C.C. Heyde, Yu. V. Prohorov, R. Pyke, and S. T. Rachev (Editors), Athens Conference on Applied Probability and Time Series Analysis, Volume I: Applied Probability In Honor of J.M. Gani. viii, 424 pp., 1996.

115: P.M. Robinson and M. Rosenblatt (Editors), Athens Conference on Applied Probability and Time Series Analysis, Volume II: Time Series Analysis In Memory of E.J. Hannan. viii, 448 pp., 1996.

116: Genshiro Kitagawa and Will Gersch, Smoothness Priors Analysis of Time Series. x, 261 pp., 1996.

117: Paul Glasserman, Karl Sigman, and David D. Yao (Editors), Stochastic Networks. xii, 298 pp., 1996.

118: Radford M. Neal, Bayesian Learning for Neural Networks. xv, 183 pp., 1996.

119: Masanao Aoki and Arthur M. Havenner, Applications of Computer Aided Time Series Modeling. ix, 329 pp., 1997.

120: Maia Berkane, Latent Variable Modeling and Applications to Causality. vi, 288 pp., 1997.

121: Constantine Gatsonis, James S. Hodges, Robert E. Kass, Robert McCulloch, Peter Rossi, and Nozer D. Singpurwalla (Editors), Case Studies in Bayesian Statistics, Volume III. xvi, 487 pp., 1997.

122: Timothy G. Gregoire, David R. Brillinger, Peter J. Diggle, Estelle Russek-Cohen, William G. Warren, and Russell D. Wolfinger (Editors), Modeling Longitudinal and Spatially Correlated Data. x, 402 pp., 1997.

123: D. Y. Lin and T. R. Fleming (Editors), Proceedings of the First Seattle Symposium in Biostatistics: Survival Analysis. xiii, 308 pp., 1997.

124: Christine H. Müller, Robust Planning and Analysis of Experiments. x, 234 pp., 1997.

125: Valerii V. Fedorov and Peter Hackl, Model-Oriented Design of Experiments. viii, 117 pp., 1997.

126: Geert Verbeke and Geert Molenberghs, Linear Mixed Models in Practice: A SAS-Oriented Approach. xiii, 306 pp., 1997.

127: Harald Niederreiter, Peter Hellekalek, Gerhard Larcher, and Peter Zinterhof (Editors), Monte Carlo and Quasi-Monte Carlo Methods 1996. xii, 448 pp., 1997.

128: L. Accardi and C.C. Heyde (Editors), Probability Towards 2000. x, 356 pp., 1998.

129: Wolfgang Härdle, Gerard Kerkyacharian, Dominique Picard, and Alexander Tsybakov, Wavelets, Approximation, and Statistical Applications. xvi, 265 pp., 1998.

130: Bo-Cheng Wei, Exponential Family Nonlinear Models. ix, 240 pp., 1998.

131: Joel L. Horowitz, Semiparametric Methods in Econometrics. ix, 204 pp., 1998.

132: Douglas Nychka, Walter W. Piegorsch, and Lawrence H. Cox (Editors), Case Studies in Environmental Statistics. viii, 200 pp., 1998.

133: Dipak Dey, Peter Müller, and Debajyoti Sinha (Editors), Practical Nonparametric and Semiparametric Bayesian Statistics. xv, 408 pp., 1998.

134: Yu. A. Kutoyants, Statistical Inference For Spatial Poisson Processes. vii, 284 pp., 1998.

135: Christian P. Robert, Discretization and MCMC Convergence Assessment. x, 192 pp., 1998.

136: Gregory C. Reinsel, Raja P. Velu, Multivariate Reduced-Rank Regression. xiii, 272 pp., 1998.

137: V. Seshadri, The Inverse Gaussian Distribution: Statistical Theory and Applications. xii, 360 pp., 1998.

138: Peter Hellekalek and Gerhard Larcher (Editors), Random and Quasi-Random Point Sets. xi, 352 pp., 1998.

139: Roger B. Nelsen, An Introduction to Copulas. xi, 232 pp., 1999.

140: Constantine Gatsonis, Robert E. Kass, Bradley Carlin, Alicia Carriquiry, Andrew Gelman, Isabella Verdinelli, and Mike West (Editors), Case Studies in Bayesian Statistics, Volume IV. xvi, 456 pp., 1999.

141: Peter Müller and Brani Vidakovic (Editors), Bayesian Inference in Wavelet Based Models. xiii, 394 pp., 1999.

142: György Terdik, Bilinear Stochastic Models and Related Problems of Nonlinear Time Series Analysis: A Frequency Domain Approach. xi, 258 pp., 1999.

143: Russell Barton, Graphical Methods for the Design of Experiments. x, 208 pp., 1999.

144: L. Mark Berliner, Douglas Nychka, and Timothy Hoar (Editors), Case Studies in Statistics and the Atmospheric Sciences. x, 208 pp., 2000.

145: James H. Matis and Thomas R. Kiffe, Stochastic Population Models. viii, 220 pp., 2000.

146: Wim Schoutens, Stochastic Processes and Orthogonal Polynomials. xiv, 163 pp., 2000.

147: Jürgen Franke, Wolfgang Härdle, and Gerhard Stahl, Measuring Risk in Complex Stochastic Systems. xvi, 272 pp., 2000.

148: S.E. Ahmed and Nancy Reid, Empirical Bayes and Likelihood Inference. x, 200 pp., 2000.

149: D. Bosq, Linear Processes in Function Spaces: Theory and Applications. xv, 296 pp., 2000.

150: Tadeusz Caliński and Sanpei Kageyama, Block Designs: A Randomization Approach, Volume I: Analysis. ix, 313 pp., 2000.

151: Håkan Andersson and Tom Britton, Stochastic Epidemic Models and Their Statistical Analysis. ix, 152 pp., 2000.

152: David Ríos Insua and Fabrizio Ruggeri, Robust Bayesian Analysis. xiii, 435 pp., 2000.

153: Parimal Mukhopadhyay, Topics in Survey Sampling. x, 303 pp., 2000.

154: Regina Kaiser and Agustín Maravall, Measuring Business Cycles in Economic Time Series. vi, 190 pp., 2000.

155: Leon Willenborg and Ton de Waal, Elements of Statistical Disclosure Control. xvii, 289 pp., 2000.

156: Gordon Willmot and X. Sheldon Lin, Lundberg Approximations for Compound Distributions with Insurance Applications. xi, 272 pp., 2000.

157: Anne Boomsma, Marijtje A.J. van Duijn, and Tom A.B. Snijders (Editors), Essays on Item Response Theory. xv, 448 pp., 2000.

158: Dominique Ladiray and Benoît Quenneville, Seasonal Adjustment with the X-11 Method. xxii, 220 pp., 2001.

159: Marc Moore (Editor), Spatial Statistics: Methodological Aspects and Some Applications. xvi, 282 pp., 2001.

160: Tomasz Rychlik, Projecting Statistical Functionals. viii, 184 pp., 2001.

161: Maarten Jansen, Noise Reduction by Wavelet Thresholding. xxii, 224 pp., 2001.

162: Constantine Gatsonis, Bradley Carlin, Alicia Carriquiry, Andrew Gelman, Robert E. Kass Isabella Verdinelli, and Mike West (Editors), Case Studies in Bayesian Statistics, Volume V. xiv, 448 pp., 2001.

163: Erkki P. Liski, Nripes K. Mandal, Kirti R. Shah, and Bikas K. Sinha, Topics in Optimal Design. xi, 164 pp., 2002.

164: Peter Goos, The Optimal Design of Blocked and Split-Plot Experiments. xiii, 256 pp., 2002.

165: Karl Mosler, Multivariate Dispersion, Central Regions and Depth: The Lift Zonoid Approach. x, 312 pp., 2002.

166: Hira L. Koul, Weighted Empirical Processes in Dynamic Nonlinear Models, Second Edition. xviii, 425 pp., 2002.